Integrating Mycorrhizal Fungi and Technology to Mitigate Environmental Stresses

Editors

Kamel A. Abd-Elsalam
Agricultural Research Center
Plant Pathology Research Institute
Giza, Egypt

Sozan E. El-Abeid
Agricultural Research Center
Plant Pathology Research Institute
Giza, Egypt

Mohamed A. Mosa
Agricultural Research Center
Plant Pathology Research Institute
Giza, Egypt

Sajeewa Maharachchikumbura
School of Life Science and Technology
University of Electronic Science and Technology of China
Chengdu, China

CRC Press
Taylor & Francis Group
Boca Raton London New York

CRC Press is an imprint of the
Taylor & Francis Group, an **informa** business
A SCIENCE PUBLISHERS BOOK

Cover credit:
Species: *Entrophospora infrequens*. Credit: Photo courtesy of Dr. Sozan E. El-Abeid.

First edition published 2027
by CRC Press
2385 NW Executive Center Drive, Suite 320, Boca Raton FL 33431

and by CRC Press
4 Park Square, Milton Park, Abingdon, Oxon, OX14 4RN

Library of Congress Cataloging-in-Publication Data (applied for)

ISBN: 978-1-041-09736-5 (hbk)
ISBN: 978-1-041-13428-2 (pbk)
ISBN: 978-1-003-66977-7 (ebk)

DOI: 10.1201/9781003669777

Typeset in Times New Roman
by Prime Publishing Services

Preface

Climate change and environmental stressors present a significant threat to global food security, natural ecosystems, and the sustainability of agriculture. Rising temperatures, soil degradation, and pollutant accumulation are disrupting plant growth and productivity worldwide, necessitating urgent, sustainable, and innovative solutions. A promising strategy involves the synergistic interaction between arbuscular mycorrhizal fungi (AMF) and modern technologies to enhance plant resilience and ecosystem stability.

"**Integrating Mycorrhizal Fungi and Technology to Mitigate Environmental Stresses**" is a book with 12 chapters that looks at AMF as natural allies of plants when they are under stress from other organisms or the environment. These contributions collectively demonstrate the mechanistic knowledge and practical application of AMF to address significant agricultural and environmental challenges.

The initial chapters focus on the crucial functions that AMF perform in agro-ecosystems, including influencing secondary metabolism and nutrient cycling, controlling carbon emissions, and protecting against oxidative damage. These insights highlight a fundamental ecological importance of AMF in facilitating plant survival under heat stress and various environmental challenges. Following these foundations, the following chapters explore the role of AMF in regulating plant defense molecules, flavonoids, and antioxidant systems, illustrating how these fungi make the plant more resistant to stress.

A notable strength of this volume is its integration of AMF research with complementary technologies. Chapters on brassinosteroids, signaling molecules, and the synergism between vermicompost and AMF show how combining natural and innovative technologies can make processes resilient and productive. Applied perspectives on removing pollutants and managing systems in an approach that is helpful for the environment also show how AMF can help with more than just farming, offering solutions for soil remediation and keeping ecosystems healthy.

This book is a great resource for plant scientists, agronomists, ecologists, and policymakers who want to find all-around ways to promote sustainable development. It brings together molecular, physiological, and applied research. The interdisciplinary insights presented here confirm that AMF has the potential to be a key part of reducing environmental stresses while also helping to make agriculture more climate-resilient and sustainable around the world.

We hope that this book will encourage more research, partnerships, and practical uses of mycorrhizal fungi and new technologies to make agriculture more resilient in the future. I sincerely thank all the authors who contributed to sharing their expertise and valuable insights, which have been instrumental in shaping this compilation. We are deeply grateful to the entire team, including Raju Primlani, the Editorial Manager, Jyotsna Jangra, and Soumya Jain, the Editorial Assistants from CRC Press, Taylor and Francis Books, for their generous cooperation and tireless efforts in bringing this book to fruition. We would also like to extend our sincere appreciation to all the contributing authors for their excellent work and unwavering support in crafting these thought-provoking and scholarly chapters. Furthermore, we want to express our heartfelt gratitude to each reviewer who generously dedicated their valuable time to complete the review process. Lastly, we would like to thank our friends, family, and colleagues for their continuous support and encouragement throughout this endeavor. We extend our sincere gratitude to all the contributing authors, reviewers, and researchers whose work forms the foundation of this volume. Their dedication and vision have made this book a valuable contribution to the field.

Kamel A. Abd-Elsalam
Sozan E. El-Abeid
Mohamed A. Mosa
Sajeewa Maharachchikumbura

Contents

CHAPTER
1

Incorporating Technology with Mycorrhizal Fungi to Resist Heat Stress

P. Saranraj[1*], Riyaz Sayyed[2], B. Lokeshwari[1], K. Gayathri[1], M. Manigandan[1] and P. Sivasakthivelan[3]

[1] PG and Research Department of Microbiology, Sacred Heart College (Autonomous), Tirupattur – 635601, Tamil Nadu, India

[2] Department of Biological Science and Chemistry, College of Arts and Science, University of Nizwa, 616, Oman

[3] Tamil Nadu Agricultural University, Agricultural Research Station, Bhavanisagar, Tamil Nadu, India

1. Introduction

Climate change and global warming pose a significant threat to food security, biodiversity, and ecosystem resilience in agriculture and natural environments. In tropical and subtropical locations, heat stress can permanently harm plant development, physiology, and productivity (Chandrasekaran et al., 2019). High temperatures impair nutrient and water uptake, chlorophyll stability, photosynthetic efficiency, membrane integrity, and hormone regulation, leading to oxidative stress from ROS overproduction and ultimately reducing growth potential. AMF colonization activates improved root hydraulic conductivity, phosphorus and micronutrient uptake, antioxidant defense systems, the accumulation of osmolytes such as proline and glycine betaine, the regulation of stress-responsive genes, and the modulation of plant hormones (Bender et al., 2023). AMF species vary across soils and climates, large-scale production methods are unstandardized, farmers are unaware, and socio-economic barriers make field-scale adoption (Zhu et al., 2022). These challenges necessitate the use of microbial bioinoculants in technology-driven farming systems.

Incorporating AMF into an innovation-driven agricultural framework where Biotechnology, Nanoscience, AI, and Sustainable development goals form resilient

*Corresponding author: microsaranraj@gmail.com

DOI: 10.1201/9781003669777-1

agroecosystems that improve plant survival under rising global temperatures and support the UN Sustainable Development Goals (SDGs) for zero climate action, living on land, and sustainable communities (Kaur and Suseela, 2023). Arbuscular Mycorrhizal Fungi (AMF) play a crucial role in regulating plant stress responses by modulating hormone synthesis, signalling, and catabolism. They play a vital role in maintaining a balance between growth, defense, and resource allocation, enabling plants to navigate the complex trade-offs imposed by thermal extremes. AMF colonization often increases root ABA levels, reducing water loss through transpiration and triggering the expression of ABA-responsive genes involved in osmoregulation. It also enhances JA biosynthesis and signalling, upregulating the production of protective secondary metabolites and proteins (Basa et al., 2021). AMF also manipulate auxin distribution and signalling, maintaining roots and exploratory growth even under stress. AMF symbiosis increases shoot cytokinin production or signalling for the effects of ABA and ethylene, thereby maintaining chlorophyll content and the integrity of photosynthetic machinery, and promoting cell division and nutrient mobilisation (Quiroga et al., 2020).

Advanced phytohormone profiling technologies, such as ultra-high-performance liquid chromatography coupled with mass spectrometry (UHPLC-MS/MS), can quantify multiple hormone classes and their conjugates in specific tissues over time. Nanosensors and genetically encoded biosensors can monitor real-time hormone fluxes in living root and leaf tissues as the AMF colonization alters hormones and signalling in response to rising temperatures (Liu et al., 2020). Transcriptomic and proteomic analyses identify symbiosis-differentially regulated genes and enzymes involved in hormone biosynthesis, conjugation, and signalling pathways. AI and ML platforms integrate and analyze massive, multidimensional datasets from hormone concentrations, gene expression, protein abundance, and sensor outputs to model complex, non-linear interactions between hormonal pathways for the specific AMF taxa that affect a crop cultivar's hormonal profile under heat conditions (Van der Heijden and Hartmann, 2023). This rational design of AMF inoculants to optimise hormonal balance for stress resilience, such as selecting fungal strains that modulate ABA and cytokinin ratios to maintain stomatal conductance and photosynthetic rates under moderate stress.

2. Mechanisms of AM Fungi in Heat Stress Mitigation

Arbuscular mycorrhizal (AM) fungi play a crucial role in enhancing heat stress tolerance in host plants. Their vast, extraradical mycelial networks extend the root system, increasing the absorptive surface area and accessing water from soil micropores that are beyond the reach of even the finest root hairs (Ryan and Graham, 2018). This enhances hydraulic conductivity and maintains a more favorable plant water status, which is monitored through soil moisture sensors and tensiometers linked to IoT-based irrigation platforms. The improved nutritional status, particularly of phosphorus (P) and nitrogen (N), orchestrated by AM fungal partners, is crucial for metabolic resilience under high temperatures. The fungal mycelium efficiently mines immobile soil phosphorus (P) and accesses organic nitrogen (N) sources, transporting these vital nutrients directly to the plant cortical cells via arbuscules

(Aliferis and Jabaji, 2021). AM fungi also instigate a robust biochemical defense arsenal within the host plant, including the upregulation of key osmolytes such as proline, glycine betaine, and soluble sugars, which act as compatible solutes to lower cellular osmotic potential, stabilize proteins and membranes, and prevent oxidative damage (Fig. 1) (Jansson and Hofmockel, 2020).

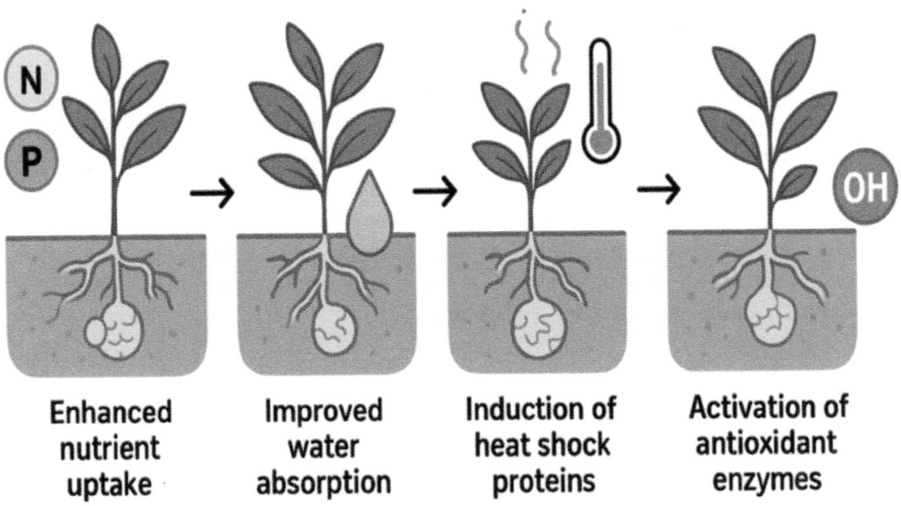

Fig. 1. Microbial symbiosis for enhanced plant abiotic stress tolerance.

AM Fungi stimulate the antioxidant enzyme system, which can be monitored in real-time using hyperspectral imaging and drone technology to assess plant health and stress levels. AM fungi directly protect the photosynthetic machinery, one of the most heat-labile systems in plants (He et al., 2023). Colonization has been shown to help maintain chlorophyll content, stabilize the efficiency of Photosystem II (PSII), and sustain higher rates of net photosynthesis under stress. This can be quantified at scale using chlorophyll fluorescence sensors and canopy gas exchange analyzers integrated with the rovers or drones. AM fungi modulate the host plant's gene expression, priming it for a more effective stress response by manipulating complex phytohormone signalling pathways (Kopittke et al., 2019). Digital farming platforms facilitate the management of the entire agroecosystem to support the persistent and beneficial activity of these introduced or indigenous AM fungal networks.

2.1. Enhanced Physiological Resilience

Arbuscular Mycorrhizal Fungi (AMF) and modern technology can boost plant physiological resilience, especially during global heat stress. AMF, beneficial soil fungi that live symbiotically with over 80 percent of terrestrial plants, improve water uptake and hydraulic conductivity, reducing physiological dryness caused by heatwave evaporative demand. It also enhances nutrient absorption, especially phosphorus, zinc, and other immobile micronutrients needed for energy metabolism and biomolecule production, including antioxidants and heat shock proteins (Wu and

Zou, 2023). Abscisic Acid is upregulated for stomatal optimization, while Jasmonic and Salicylic acids are upregulated to prime the Systemic Acquired Resistance (SAR) mechanisms (Chareesri et al., 2020). Large-scale, consistent agricultural applications have been complex due to the variability of AMF-plant symbiosis, which is influenced by fungal species/strain specificity, soil conditions, host genotype, and environmental variables.

Advanced technology enables a data-driven resilience strategy rather than blanket immunisation. Next-generation sequencing and metagenomics can characterize and fingerprint soil microbiomes, detecting thermally stressed super-performers. As the central nervous system, AI and machine learning algorithms process massive datasets from molecular analyses, sensor networks, soil probes, and weather patterns to their predictive models that accurately predict heat stress events and recommend several interventions (Bhardwaj et al., 2022). Nano-encapsulated biofertilizers protect sensitive AMF spores during storage and application, ensure controlled release in the rhizosphere, and deliver essential nutrients or "priming" agents like silicon nanoparticles, enhancing AMF efficacy (Saiz Rubio and Rovira Mas, 2020).

2.1.1. Improved Water Uptake

Root membranes and soil water potential in plants can be damaged by heat stress. Nevertheless, Arbuscular Mycorrhizal Fungi (AMF) increase water uptake. Fungal hyphae such as *Rhizophagus irregularis* and *Funneliformis mosseae,* the volume of soil compared to even the most delicate root hairs. Higher hydraulic conductivity allows these fungal hyphae to transfer water effectively along a potential gradient from the root cortex to the soil (Latef et al., 2021). The intraradical hyphae of the root cortex develop the arbuscules, which serve as sites for nutrient exchange, altering the osmotic dynamics of the root cells and inducing the aquatic uptake of the plant through aquaporins, thereby allowing water uptake from the soil-plant-fungi continuum. Its efficiency relies on an intricate combination of interactions between fungal strains and species, soil structure and texture, intensity and duration of heat stress, and host plant genotypes (Banerjee et al., 2023). High-tech soil sensor networks, such as IoT-based tensiometers, capacitance sensors, and TDR sensors, yield real-time information on soil water potential and volumetric water content at various depths. Nanotechnology offers new delivery technologies to maximize hydraulic benefits for biological interactions, making it an active, predictable, and controllable component of a comprehensive strategy for protecting plants against heat-induced drought.

2.1.2. Nutrient Acquisition

Heat stress reduces the uptake and metabolism of nutrients in plants, leading to nutritional deficiency and metabolic degradation (Rillig et al., 2021). The ancient symbiotic association with Arbuscular Mycorrhizal Fungus (AMF) has made it an effective designed thermo-tolerance device. Nutrition augmentation AMF is an advanced type of exchange economy that occurs in the root cortex, where the fungus forages for immobile soil nutrients, such as phosphorus, zinc, copper, and ammonium, which are bound to soil particles and inaccessible to plant roots. The fungus leaches micronutrients from the soil that the plant cannot obtain through acidification, organic acid secretion, enzymes, and chelating compounds (Begna, 2023). They are pumped quickly through the arbuscules by fungal hyphae, which in turn exchange carbon photosynthates with plant-based carbon in a strictly regulated interfacial apoplast. This role increases exponentially as the energy requirement of the plant to generate heat shock proteins, antioxidants, and suitable osmolytes increases due to heat stress. The novel technology is transcending complexity and variability in this symbiosis for its analysis and control. This improved nutrient uptake for the resilience-promoting biochemical apparatus, allowing the plant to maintain photosynthetic rates, membrane integrity, and growth during extreme heat stress (Auge et al., 2023).

2.1.3. Photosynthetic Protection

Plant survival relies on photosynthetic equipment, which is threatened by heat stress. LHCII denaturation, PSII reaction centre disintegration, Rubisco activase inhibition, and ROS overproduction can occur due to high temperatures. In this vulnerable photobiological setting, Arbuscular Mycorrhizal Fungi (AMF) create a vital protective layer. The chloroplast's antioxidant defense mechanism is significantly enhanced by AMF colonisation, which increases the production and activity of key enzymes such as SOD, CAT, APX, and GR (Abdallah et al., 2018). This scavenges superoxide, hydrogen peroxide, and hydroxyl radicals that result from the over-reduction of the photosynthetic electron transport chain at high temperatures, thereby preventing the oxidation of the thylakoid membrane and the PSII D1 protein. AMF symbiosis affects the xanthophyll cycle, enhancing the conversion of violaxanthin to zeaxanthin and facilitating non-photochemical quenching (NPQ), a safe mechanism for dissipating excess light energy as heat. This photosynthetic protection depends on the fungal isolate, host plant genotype, colonization timing and success, and environmental circumstances. Integration platforms are crucial for measuring, modelling, and regulating symbiosis to achieve maximum photoprotection (Zhang et al., 2022).

2.2. Oxidative Stress Reduction

Heat stress causes oxidative stress, which in turn reduces crop yields. AMF symbiosis is a potent antioxidant system, but precision requires modern technology. AMF colonization enhances the host plant's nutrition, particularly in terms of phosphorus, zinc, and copper uptake. Upregulation activates the plant's antioxidant defense network, thereby enhancing its oxidative protection (Gianinazzi et al., 2022). The protective impact varies depending on the AMF genotype, the host plant's genetic background, the soil environment, and the severity and duration of heat stress. Continuous and maximum efficacy need technological intervention. Integrated technologies assess, analyze, and regulate plant oxidative state and fungal partner functionality. Biosensor technology, particularly implanted or nanoprobe-based electrochemical sensors. CRISPR-Cas9 gene editing and RNA interference can modulate fungal genes that trigger the plant's systemic antioxidant responses

(Kumar et al., 2020). Advanced AI and machine learning platforms combine and process multidimensional datasets from biosensor ROS readings, spectral indices, soil microbiome sequencing, and plant transcriptomic data on antioxidant gene expression (Li et al., 2023). This holistic approach develops a closed-loop, adaptive shield against oxidative stress, protecting membrane integrity, enzyme function, genetic material, cellular homeostasis, and metabolic function under extreme heat, as well as agricultural productivity from climate-induced oxidative damage.

2.2.1. Antioxidant Defense Activation

The antioxidant defense mechanism of the plant is an essential component of the protective association with Arbuscular Mycorrhizal Fungi (AMF). This biological priming pathway alleviates heat stress damage by neutralizing Reactive Oxygen Species (ROS). Through this biological priming mechanism, the entire antioxidant mechanism (both enzymatic and non-enzymatic) is triggered due to the wide communication of signal molecules (Jiang et al., 2022). The symbiosis also increases the uptake of micronutrients (zinc, copper, and manganese), which act as cofactors in the activity of large antioxidant enzymes e.g., Superoxide dismutase (SOD). The symbiosis promotes increased activity and gene expression in most H_2O_2-detoxifying enzymatic systems, including Catalase (CAT), as well as in the Ascorbate-Glutathione (AsA-GSH) cycle (Seleiman et al., 2021).

AMF stimulates the biosynthesis and accumulation of low-molecular-weight non-enzymatic antioxidants, such as Ascorbate (AsA), Glutathione (GSH), tocopherols, carotenoids, and phenolic compounds, which act as versatile scavengers, directly neutralizing ROS, regenerating oxidised ascorbate, and stabilizing membranes against lipid peroxidation. Integrated technology platforms are indispensable in harnessing this capacity reliably in agriculture (Nadeem et al., 2019). CRISPR-Cas9-based gene editing and RNA interference can be used to create plant lines with fluorescent reporter genes fused to the promoters of key antioxidant genes, allowing for real-time visualization of the priming phenomenon in living root and leaf tissues upon AMF colonization and during stress. Nanotechnology offers revolutionary biosensing capabilities, allowing implantable, nanoparticle-based electrochemical sensors to provide *in vivo* monitoring of ROS dynamics and key antioxidants like glutathione (Begum et al., 2022).

2.2.2. Osmolyte Accumulation

Plants utilize osmotic adjustment to prevent heat stress-induced dehydration and imbalance by strategically accumulating osmolytes, which are low-molecular-weight organic molecules. Arbuscular Mycorrhizal Fungi (AMF) symbiosis enhances the biosynthesis and compartmentalisation of key osmolytes, including proline, glycine betaine, soluble sugars, and polyamines, while stabilising cellular structures, maintaining turgor pressure, protecting enzymatic activity, and scavenging reactive oxygen species (Cabral et al., 2021). Genotypic differences, environmental factors, and management practices influence the induction and magnitude of osmolyte accumulation, necessitating the integration of sophisticated technologies to decode, monitor, and optimise this process for consistent and maximal thermo-tolerance.

High-throughput mass spectrometry and nuclear magnetic resonance spectroscopy can quantify the complete osmolyte profile in heat-stressed plant tissues with and without AMF inoculation. Real-time detection of proline using implantable nano-biosensors enables *in vivo* monitoring of osmotic adjustment dynamics during heat stress (Hashem et al., 2019). This is enhanced by the sensors on drones or satellites that detect small changes related to water content and osmotic potential. Metagenomic and meta-transcriptomic sequencing of the root and soil microbiome identifies AMF taxa and their functional capacity by analyzing fungal genes involved in nutrient transport and osmolytes (Pozo et al., 2021).

2.3. Molecular and Genetic Regulation

Plants and Arbuscular Mycorrhizal Fungi (AMF) interact under heat stress, which is crucial for thermo-tolerance. Transcriptional reprogramming, epigenetic modification, protein-level interaction, and hormone interaction define the thermal assault response. Presymbiotic signaling by strigolactones from the plant induces fungal metabolism and branching ofhyphae, invoking this complex process (Schlapfer et al., 2019). Plant receptor-like kinases perceive fungal chitooligosaccharides and lipochitooligosaccharides, triggering the Common Symbiosis Signalling Pathway (CSSP) and inducing transcription factors that condition the plant for colonization.

Heat stress becomes more severe and alters the symbiosis functions as a potent priming agent that systemically influences the genetic expression of the host. The AMF symbiosis tightly controls plant hormonal circuitry, increasing Abscisic Acid (ABA) to maximize stomatal conductance and altering Jasmonic Acid (JA) and Salicylic Acid (SA) signalling to balance defense and development (Karaki, 2023). Under stress conditions, it binds with cytokinin and auxin pathways to maintain roots and photosynthetic strength. Deep learning models, such as CNNs and RNNs, can combine these levels of biological information to make predictions about gene regulatory networks, identify master transcriptional regulators, and simulate the impact of different AMF taxa on the plant's molecular response to heat stress. AMF strains with CRISPR-based overexpression of fungal effectors or plant heat shock response, or antioxidant genes, for the synthetic biology solutions (Riaz et al., 2021). Nanotechnology offers the platform for gene-informed interventions, e.g., nanocarriers such as LDH clay nanosheets or carbon nanoparticles for transient editing of the gene to the molecular interactions without developing GMOs for the technologically boosted resilience.

2.3.1. Stress-Responsive Gene Expression

Arbuscular Mycorrhizal Fungi (AMF) play a crucial role in enhancing thermo-tolerance by modulating stress-responsive gene expression. This is activated through symbiotic signalling by plant receptor-like kinases, which in turn activate calcium and the symbiosis signaling pathways (Bowles et al., 2022). This is marked by the control of genes that encode Heat Shock Proteins (HSPs), which are molecular chaperones that protect against protein denaturation and aggregation, and facilitate refolding to ensure proteostasis during conditions that would otherwise be proteotoxic. AMF colonization significantly upregulates the genes encoding the synthesis of osmotic key osmolytes, such as P5CS for proline accumulation, BADH for the synthesis of glycine betaine, and genes involved in sugar biosynthesis, including trehalose and raffinose family oligosaccharides (Trivedi et al., 2021).

The symbiosis systemically upregulates the entire antioxidant defense machinery at the transcript level by increasing the expression of genes encoding Superoxide Dismutase (SOD), Catalase (CAT), Ascorbate Peroxidase (APX), Glutathione Reductase (GR), and enzymes involved in the ascorbate-glutathione cycle (Zandalinas et al., 2021). AMF also affects the expression of genes related to hormone signaling pathways, specifically by enhancing genes involved in Abscisic Acid (ABA) biosynthesis and response, thereby maximizing stomatal control. In addition to this, AMF also modulates the expression of genes for hormone signalling pathways, including upregulation of genes for Abscisic Acid (ABA) biosynthesis and response to improve stomatal regulation, regulating Jasmonic Acid (JA) and Salicylic Acid (SA) pathways for improving systemic acquired resistance, and changing auxin and cytokinin transporters for sustaining root growth under stress. Sophisticated technology is essential for deciphering, tracking, and unlocking this intricate transcriptional regulation (Li et al., 2021). The latest RNA sequencing (RNA-Seq) and single-cell RNA sequencing (scRNA-Seq) technologies enable the comprehensive characterisation of transcriptional alterations in specific cell types, such as arbuscule-containing cells, under controlled heat stress conditions (Oehl et al., 2021).

2.3.2. Hormonal Modulation

Arbuscular Mycorrhizal Fungi (AMF) regulate plant hormone networks, helping plants manage high temperatures and their complex interactions. AMF colonization increases root ABA, enabling stomatal guard cells to respond more sensitively and effectively to heat-induced hydraulic stress. It enhances JA biosynthesis and signaling, leading to increased production of protective secondary metabolites and proteins. AMF regulate auxin distribution and signalling to preserve roots and their growth (Kumar et al., 2021). They influence photosynthetic efficiency and source-sink connections with cytokinin dynamics. The AMF symbiosis enhances shoot cytokinin synthesis and signalling, counteracting the senescence-inducing actions of ABA and ethylene, thereby sustaining chlorophyll and photosynthetic machinery, and promoting cell division and nutrient mobilisation. Ethylene bursts from heat stress increase senescence and abscission. AMF increases ACC deaminase activity or alters ethylene biosynthesis gene expression to delay stress-induced senescence and sustain reproductive potential (Dhiman et al., 2022). Ultra-high-performance liquid chromatography combined with tandem mass spectrometry (UHPLC-MS/MS) can quantify numerous hormone classes and their conjugates in specific tissues throughout time.

Nanosensors and genetically encoded biosensors can measure real-time hormone fluxes in living root and leaf tissues to determine the AMF colonization changes hormone gradients and signaling hotspots in response to rising temperatures. Symbiosis regulates the production ofhormones, conjugation, and signalling pathway genes and enzymes, as identified through transcriptomic and proteomic analysis

(Frew, 2023). AI and ML platforms model the complex, non-linear interactions of hormonal pathways and predict that AMF taxa will affect a crop cultivar's hormonal profile under forecasted heat conditions, utilizing massive, multidimensional datasets from hormone concentrations, gene expression, protein abundance, and sensors. In precision agriculture, IoT-enabled soil sensors monitor moisture and temperature, triggering irrigation or shading based on AI analysis of plant hormonal conditions (Begum et al., 2022). This comprehensive, technology-driven approach leverages its variable hormonal influence to effectively manage plant stress responses, ensuring crops maintain optimal hormonal balance for growth, defense, and production even under heat stress.

2.4. Technological Integration Strategies

The new technology, combined with Arbuscular Mycorrhizal Fungi (AMF), is transforming the agricultural management process by incorporating the management of microbiomes, nano-enabled delivery systems, in situ soil sensors, artificial intelligence, and predictive analytics. These technologies ensure the viability and efficacy of fungal inoculants, as well as their protection against drying and thermal degradation, in rhizosphere conditions. Plant-based biosensors are sensors that monitor hormonal flux and reactive oxygen species, or in situ soil sensors used to record continuous data on rhizosphere conditions (Liu et al., 2020). Massive datasets are analysed through AI and predictive analytics to identify complex patterns and correlations and predict heat stress and AMF-crop performance across conditions. Biotechnology and synthetic biology, based on CRISPR, can augment the innate potential of each symbiosis partner, including gene editing to tune plant symbiosis signaling genes or stress response. It is possible to create genomic selection-based breeding that produces plant varieties better suited to AMF colonization during stress (Nouri et al., 2022). Digital twins and transparency are ensured through blockchain-enabled traceability. The use of digital twins to simulate interventions in silico before real-world use, combined with blockchains, facilitates supply chain decisions during inoculant production and field application. With nanotechnology, sensing, AI, and biotechnology coming together as an integrated component of a resilient, precision agriculture system, AMF offers an unprecedented level of heat stress resistance, yields protection, water and nutrient conservation, and a reduction in chemical input during climate crises (Jansa and Gryndler, 2023).

Table 1 outlines seven synergistic strategies for enhancing Arbuscular Mycorrhizal (AM) symbiosis under thermal stress, integrating advanced technologies to improve plant resilience. These strategies range from selecting heat-resistant AM fungal strains using AI and genomics, to engineering the soil microbiome with metagenomics and nano-biomaterials, and employing nano-carriers for targeted delivery of biofertilizers. Additional approaches include real-time monitoring via IoT and UAV-based sensors, CRISPR-Cas9 gene editing of plant and fungal genomes for improved heat tolerance, transcriptomic and metabolomic profiling to understand molecular responses, and the application of precision agriculture tools like AI and autonomous drones to dynamically optimize symbiotic outcomes in stressful environments.

Table 1. Synergistic strategies for augmenting arbuscular mycorrhizal symbiosis under thermal stress

S.No	Strategy	Technologies	Functions	References
1	Strain Selection and Bioprospecting	Remote sensing (drones/satellites), Robotics, AI/ Machine Learning, Genomics	• Identifying native strains from resilient ecosystems • Identifying and Characterising Heat-Resistant AMF Strains	Nadeem et al. (2019)
2	Soil Microbiome Engineering	Metagenomics, Computational Biology, Nano-biomaterials, Digital Twins	• Buffers plants against heat • Maintaining Synergistic Microbial Community	Begum et al. (2022)
3	Nano-Carriers and Biofertilizers	Nanotechnology (e.g., chitosan, silica NPs), Materials Science, Variable-Rate Tech (GPS)	• AMF Inoculants & Synergistic Compounds • Protect, deliver, enhance effectiveness • Directly reach the root zone	Cabral et al. (2021)
4	Sensor-Based Monitoring	IoT Soil Sensors, Plant Biosensors, UAVs with hyperspectral/ thermal cameras, AI	• Real-Time Soil and Plant Health • Predictive and precise management	Hashem et al. (2019)
5	CRISPR/Cas9 Gene Editing	CRISPR-Cas9, Multi-omics, High-Throughput Phenotyping	• Plant and AMF Genomes for Symbiotic and Heat-Stress Tolerance	Pozo et al. (2021)
6	Transcriptomic and Metabolomic Profiling	RNA Sequencing, Mass Spectrometry, AI/ML Integration	• Systems-Level AMF-Plant under Stress • Molecular and metabolic changes • Resilience under stress	Schlapfer et al. (2019)
7	Precision Agriculture	IoT, AI, Autonomous Vehicles (Drones), Nanotechnology, Yield Monitors	• Integrated Technology Management System • Dynamic and self-optimising • Maximizes thermo-tolerance	Karaki (2023)

2.4.1. Strain Selection and Bioprospecting

Heat-tolerant Arbuscular Mycorrhizal Fungi (AMF) and their genes are complex to identify and screen using molecular biology, bioinformatics, remote sensing, and artificial intelligence technology (Thirkell et al. 2020). In both natural and agricultural ecosystems where plant communities exhibit abnormal resistance, satellite and drone-based remote sensing using thermal and multispectral sensors is used to detect hotspots within native plant communities. They are also examined on a large scale, such as through whole-genome sequencing, transcriptomic and proteomic experiments, and metabolomic profiling. In this method, synthetic microbial communities and microcosm experiments are employed to determine the degree of interaction between potential AMF strains and the soil microbiome during heat stress (Raya Gonzalez et al., 2021). These multi-omic studies generate large datasets that are analyzed using machine learning and phylogenetic software to map genetic markers, transcriptional signatures, and metabolic pathways associated with thermotolerance. Automated phenotyping systems utilise sensors to detect growth, water use efficiency, photosynthetic activity, and spectral indices related to plant growth and stress (Zhang et al., 2022). Global use of AMF inoculants in food production against heat stress through symbiosis.

2.4.2. Soil Microbiome Engineering

Engineering soil microbiomes is a technology-based approach that utilizes advanced technologies to design a self-regulating, heat-tolerant, and synergistic microbial community that protects plants against heat stress (Santander et al., 2021). It is done using high-resolution soil sensors, drones, metagenomic sequencing, and metatranscriptomic analysis to generate high-resolution maps of physicochemical heterogeneity. Their interactions with heat-stable AMF strains and symbiosis are modelled using the computationally bioengineered precision microbial consortia in the presence of symbiotic Plant Growth Promoting Rhizobacteria (PGPR), including *Bacillus subtilis*. The microbial cells are encased, stored, and utilized with nano-biomaterials, which also have outer shells that can detect the soil temperature of microorganisms. Selective reinforcement options are employed to maintain an engineered microbiome (e.g., the activity of the sensors and the well-being of the plants). The next-generation technology is synthetic biology and gene editing (CRISPR-Cas systems) (Evelin et al., 2019). They can also induce small mutations in the significant genes of the PGPR component to produce protective compounds in excess, thereby inducing heat stress without generating transgenic organisms. It is a form of combined technology that transforms the rhizosphere into a biologically optimized interface, allowing for active manipulation to significantly improve the health and productivity of plants in severely stressful environments (Bitterlich et al., 2018).

2.4.2.1. Nano-carriers and Biofertilizers

Arbuscular Mycorrhizal Fungi (AMF) are microorganisms that have been utilized in controlling heat stress using nano-carriers and biofertilizers. Biodegradable and biocompatible polymers, such as chitosan, alginate, PLGA, or silica-based mesoporous nanoparticles, are used to prepare such carriers. They are produced and designed through materials science and nanotechnology and have a high surface-to-volume ratio, which enables them to interact effectively with plant roots and avoid desiccation, UV radiation, and antagonism by microbes. The nano-carriers are active and functionalized to generate an all-in-one biofertilizer system (Chen et al., 2020). Coat ligands to bind to root receptors may be added to the surface of the nanoparticles, and a synergistic mixture of nutrients, osmolytes, and signaling molecules can be loaded onto the core matrix. This food provides the hyphae with all the energy they need to grow actively and colonize, and the host plant receives compounds that help prevent stress. Nano-enabled biofertilizers of variable-rate-based technology and the data from soil sensors are sprayed onto the root zone. These nano-carriers increase colonization and functioning of the soil, which suppresses pathogens and activates the plant immune system, a temporary effect. A biofertilizer formula can contain quantum dot-based sensors or carbon nanotube arrays to monitor the soil and microbial activity in real-time (Gebbers and Adamchuk 2020). This combined solution makes AMF a climate-resilience technology that accepts biological input and transforms it into a precision instrument to generate benefits such as improved water and nutrient uptake, hormonal control, and antioxidant defense.

2.4.2.2. Sensor-Based Monitoring

The heat stress response of Arbuscular Mycorrhizal Fungi can be enhanced through real-time sensor-based measurements of both the soil environment, fungi, and the physiological activity of the host plant. It enables accurate and predictive management interventions that will maximize the symbiotic relationship between AMF and plants. *In situ* soil sensor networks integrate the rhizosphere with IoT-capable instruments to measure pH, soil moisture content, electrical conductivity, and drought stress (Sarma et al., 2022). Sophisticated biosensors, which enable the monitoring of water consumption and tissue water potential, include stem psychrometers and minimally invasive sap flow sensors. Biosensors based on MFC directly sense biological activity, while nanomaterial-based biosensors also directly sense biological activity. These have been combined with remote sensors, such as UAVs equipped with thermal and hyperspectral sensors, to achieve field-level monitoring. Deep learning algorithms enable cloud-based AI and machine learning systems to predict the response of plants to short-term stressors by analyzing data on soil chemistry, microbiology, and plant physiology collected by sensors (Ahanger et al., 2020). Digital twins in natural environments model water, nutrient, and AMF-plant symbiosis. This sensor-based ambient monitoring environment renders AMF heat stress control proactive, accurate, and predictive, ensuring practical symbiotic engagement to shield crops against climate-caused extremes of temperatures.

2.4.3. Genetic and Molecular Tools

A revolution in the genetic and molecular technology for studying and utilizing Arbuscular Mycorrhizal Fungi (AMF) to confer heat stress resistance. The technique involves analyzing and refining the fundamental properties of the symbiotic dialogue, including nucleotide sequences, gene transcripts, protein interactions, and epigenetics (Liakos et al., 2018). The difficulties of Arbuscular Mycorrhiza Fungi (AMF) are overcome by using advanced genomics and functional genetics to generate high-quality genomes of species such as *Rhizophagus irregularis* and *Funneliformis mosseae*. CRISPR-Cas9 editing methods produce targeted mutations in the host plant. snRNA-Seq (single-nucleus RNA sequencing) of colonized root tissues indicates the specific gene expression programs triggered in cells containing the arbuscule during heat stress (Mathur et al., 2019). Epigenomic analyses indicate that AMF colonization alters the host plant's chromatin in response to heat, potentially activating stress-responsive genes. Engineered plant roots where the protective gene expressions (Navarro et al., 2022) are redesigned are a part of synthetic biology. Predictive models of machine learning predict AMF genotypes to regulate stress transcriptomes. Their genetic modification is not necessary to achieve colonization using nanoparticle-mediated delivery techniques in stressful conditions.

2.4.3.1. CRISPR/Cas9

In CRISPR/Cas9 gene-editing technology, agricultural biotechnology is being transformed by allowing the introduction of accurate, targeted, and heritable changes to the genome, enhancing symbiotic interactions between crops and Arbuscular Mycorrhizal Fungi (AMF) to withstand heat stress (Rizvi et al., 2022). This is a multi-omics approach to identify key genetic targets, including Heat Shock Proteins (HSPs), chaperonins, and enzymes involved in trehalose biosynthesis. It is then engineered to be used as CRISPR/Cas9 to precisely edit the genome in plants, such as knocking out negative regulators of symbiosis or knocking in more potent promoters of stress-responsive genes. Multiplexed CRISPR approaches aim to target entire pathways, such as editing multiple HSF transcription factors or reconfiguring plant and fungal genomes to facilitate nutrient exchange during stress (Mohanty et al., 2020). High-throughput phenotyping and functional confirmation of genetic alterations are performed to ensure that genetic alterations do not have pleiotropic effects and can produce the desired phenotype. Automated phenotyping systems are used to measure root architecture, rate of colonization, photosynthetic efficiency, and biomass accretion at increasing heat levels in plants and fungal strains. CRISPR ribonucleoproteins can be delivered directly to roots or seeds in nanoparticle-based carriers, eliminating the need for GMO regulatory barriers (Hart et al., 2018). This universal solution can overcome the inherent limitations of AMF-plant symbiosis in crops, enabling them to survive heat stresses previously unexperienced, thereby guaranteeing food production on a global scale with genetic precision.

2.4.3.2. Transcriptomic and Metabolomic Profiling

Transcriptomic and metabolomic profiling are employed to investigate and utilize Arbuscular Mycorrhizal Fungi (AMF) for enhancing resistance to heat stress. The approach offers a molecular conversation and symbiotic resilience as characterised by AMF. Heat stress transcriptomic analyses also indicate that heat shock factors are upregulated and energy metabolism is downregulated overall. Heat-stressed mycorrhizal plants have been found to exhibit metabolomic signatures characterized by a higher proportion of protective osmolytes, antioxidants, and stress-inducible phytohormones (Vangelisti et al., 2020). Multi-omics integration enhances these methods by adding bioinformatic tools, such as network analysis and metabolic pathway mapping, to identify regulators of the stress response. Machine Learning algorithms trained on transcriptomic and metabolomic data can predict the most efficient molecular profiles from various fungi strains (Miller et al., 2020). Nanotechnology offers precise biofertilizers that utilize substances identified by metabolomics and modified through stress or RNA interference mechanisms, which inhibit plant symbiosis.

2.4.4. Precision Agriculture

Precision agriculture utilises IoT sensors, autonomous vehicles, AI analytics, and nanotechnology to optimise soil-plant-atmosphere interactions for maximum thermotolerance. Hyperspectral and thermal drones measure soil organic matter, moisture-holding capacity, and thermal patterns, while EMI sensors map soil texture and electrical conductivity (Kah et al., 2019). A network of *in-situ* IoT sensors monitors the rhizosphere in real-time, notifying plant-based sensors when root-zone temperatures exceed 35 °C or soil moisture levels drop below a specific threshold. AI-driven analytics solutions integrate weather forecasts with real-time predictions, analyzing connections between environmental variables, AMF activity, and plant health using machine learning algorithms. AI-directed irrigation systems and drone spraying of smart nano-fertilizers enhance precision in interventions (Hameed et al., 2021). Post-stress, yield monitors, and quality sensors on harvest equipment collect crop performance data for AI models, ensuring crop productivity and sustainability in variable climates.

3. Applications of AM Fungi in Combating Heat Stress

Arbuscular Mycorrhizal Fungi (AMF) are being utilized in modern crop production to enhance physiological resilience, optimize resource utilization, and secure yields in the face of increasing thermal stress. Precision soil mapping, utilising electromagnetic induction and gamma radiometry, identifies high-vulnerability zones, followed by the variable-rate application of custom-formulated AMF consortia (Kalia and Kaur 2023). Automated interventions, such as drip irrigation with nano-encapsulated AMF spores, are used. Soil sensors with IoT capabilities monitor root zone temperature and moisture, feeding data into AI models that predict stress events and initiate irrigation cycles. Agriculture utilizes CRISPR-edited crop varieties in conjunction with AMF strains to develop a climate control system.

The Figure 2 illustrates the multifaceted contributions of arbuscular mycorrhizal (AM) fungi in enhancing plant tolerance to heat stress. By forming a symbiotic association with plant roots, AM fungi improve nutrient and water uptake, which is critical under stressful conditions. They also enhance the plant's antioxidant defenses, reduce oxidative damage, and support photosynthetic efficiency. Additionally, AM fungi stimulate the activation of plant defense hormones and other protective mechanisms, thereby strengthening overall plant resilience and promoting survival during periods of thermal stress.

3.1. Nutrient Uptake

Arbuscular Mycorrhizal Fungi (AMF) play a crucial role in maintaining the functionality of soil microsites during heat stress. Their extensive extraradical hyphal network acts as a dynamic pipeline, exploring a vastly greater soil volume to access immobilized nutrients and maintain functionality. AMF hyphae use Acid phosphatases and organic acids to solubilize fixed Phosphorous forms and transport them rapidly to the plant via specific phosphate transporters. This directly boosts ATP synthesis, enabling AMF to release chelating agents, lowers soil pH, and enhances

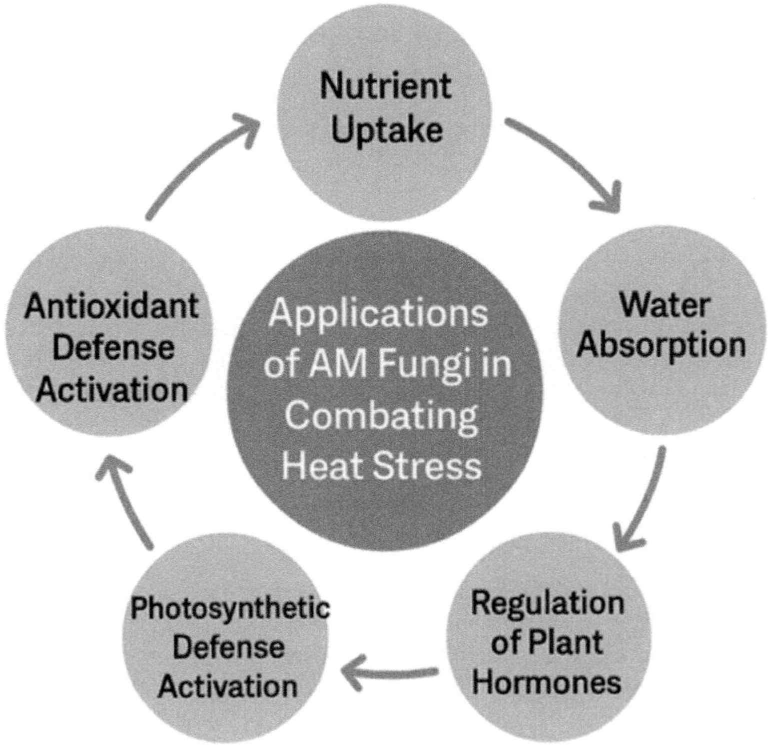

Fig. 2. Multifunctional roles of AM fungi in plant heat stress tolerance.

the bioavailability and translocation of micronutrients (Bowles et al., 2022). The symbiosis of fungal ammonium transporters improves nitrogen uptake, ensuring a steady supply for protective compounds. Technology integration transforms biological nutrient acquisition into a precision-managed system, using soil sensors, drone-based hyperspectral imaging, machine learning algorithms, and weather forecasts to predict heat stress-induced nutrient lockups. Automated, variable-rate applications of nano-enabled biofertilizers are applied through irrigation systems or autonomous drones (Oehl et al., 2021). CRISPR-based gene editing can enhance nutrient uptake efficiency in plants and fungal strains by increasing sugar transporters and modifying AMF strains to overexpress acid phosphatase and nutrient transporter genes under heat stress (Trivedi et al., 2021). This holistic method sustains the AMF-mediated nutrient absorption mechanism, allowing plants to maintain metabolic stability, develop defense chemicals, and develop and reproduce under nutrient-deprived and crop-loss conditions.

3.2. Water Absorption

Arbuscular Mycorrhizal Fungi (AMF) improve water absorption and heat stress mitigation in agriculture. A dynamic extension of the root system, the extraradical hyphal network maintains hydraulic activity by accessing larger soil volumes to identify water in micropores that roots are unable to penetrate, even as soil moisture becomes insufficient and temperatures rise (Kumar et al., 2021). The symbiosis increases the host plant's aquaporin genes, enhancing the root's intrinsic water intake and smooth water transfer from the fungal interface to the xylem. Advanced technology controls and adapts this biological water management system. IoT-enabled soil moisture sensors and plant-based sensors monitor crop water status and transpiration rates, demonstrating the benefits of the AMF network to plant health. AI technologies, particularly machine learning models, link soil moisture changes to weather projections to predict water shortages. This enables the AMF hyphal networks to remain hydrated and functional without over-irrigation, using adjustable-rate irrigation systems that only provide water to areas where sensors indicate soil wilting (Dhiman et al., 2022). Smart carriers with osmoprotectants sustain hyphal integrity under osmotic stress in this hydraulic system using nanotechnology (Begum et al., 2022). Drone-based misting devices are used to cool and reduce nano-fluid sprays, including water, in a fungal-mediated water delivery system, stimulating silica nanoparticles through AMF.

3.3. Regulation of Plant Hormones

The fungal symbiont (AFM) plays a crucial role in regulating the synthesis, signaling, and catabolism of key phytohormones in plants, allowing them to navigate the delicate trade-offs imposed by thermal extroversion. AMF colonization often primes the plant for a more efficient ABA response, leading to earlier stomatal closure during initial heat exposure, which reduces water loss (Quiroga et al., 2020). The fungi also enhance the shoot's cytokinin (CK) synthesis, particularly trans-zeatin, which promotes cell division, maintains chlorophyll content, and delays senescence, preserving photosynthetic capacity. The symbiosis also alters the Jasmonic Acid (JA) and Salicylic Acid (SA) pathways, increasing JA to enhance heat shock protein (HSP) and antioxidant enzyme production, and modulating SA to prevent excessive resource allocation to defense, thereby preserving energy for root growth and nutrient assimilation. AMF also regulates ethylene biosynthesis by downregulating the ACC synthase (ACS) and ACC oxidase (ACO) genes, or through associated bacteria that express ACC deaminase, thereby reducing the ethylene-induced acceleration of senescence and abscission (van der Heijden and Hartmann, 2023). UHPLC-MS/MS technology quantifies multiple hormone classes in plant tissues, enabling the detection of hormonal shifts induced by AMF strains under various heat regimes. AI algorithms analyze complex datasets to predict hormonal imbalances based on weather and soil conditions, transforming AMF-mediated hormone regulation into a proactive system for crop resilience, productivity, and heat stress survival (Ryan and Graham, 2018).

3.4. Photosynthetic Protection

Arbuscular Mycorrhizal Fungi (AMF) play a crucial role in mitigating heat stress on the photosynthetic apparatus. They employ a complex preventive approach to maintain the structural integrity of chloroplasts, the stability of their electron transport chain, and photo-oxidative damage to ensure sustainable production (Jansson and Hofmockel, 2020). The extensive hyphe network enhances the uptake and translocation of water and primary nutrients, including phosphorus, magnesium, and zinc, which are vital in chlorophyll biosynthesis, ATP generation, and the activation of carbon fixation enzymes. Systemic AMF colonization also increases the expression of Heat shock protein (HSPs) genes in chloroplasts, which are molecular chaperones that ensure photosynthetic proteins do not aggregate or denature in high temperatures. It also enhances the creation and functioning of antioxidant enzymes (He et al., 2023).

The symbiosis also adjusts the xanthophyll cycle to enhance the ratio of violaxanthin to zeaxanthin, thereby stimulating Non-photochemical quenching (NPQ). High-tech biological protection becomes a controlled system, utilizing drones or satellites equipped with hyperspectral and chlorophyll fluorescence sensors to measure indices such as the Photochemical Reflectance Index (PRI) and the Normalized Difference Vegetation Index (NDVI). Data on photosynthetic efficiency is analyzed using the IoT-enabled soil sensors and artificial intelligence algorithms, especially machine learning models. It is applied to forecast the susceptibility of heatwaves to activate artificial measures such as variable-rate water sprinkler systems and drones (Wu and Zou, 2023). This combination of biological understanding and technology enhances AMF-based photosynthesis protection, allowing crops to retain high rates of carbon assimilation.

3.5. Antioxidant Defense Activation

AMF induces the antioxidant defense system in the presence of heat stress, decreasing the accumulation of ROS and lipid peroxidation. AMF complements nutrient intake by increasing the uptake of zinc, copper and manganese, antioxidant enzyme cofactors. These enzymes neutralize ROS and increase the level of selenium. The direct advantage of this is that AMF symbiosis directly prepares the genetic expression of the plant, reinforcing the ascorbate-glutathione cycle and redox homeostasis of the cells (Chareesri et al., 2020). AMF also encourages the synthesis of antioxidants such as ascorbate, glutathione, tocopherols, carotenoids and phenolic compounds, which are ROS scavengers and prevent lipid peroxidation in membranes. Plant tissue nanosensors continuously measure the levels of ROS and redox state in real-time, which can be used to provide feedback on the effectiveness of AMF intervention and oxidative stress. The analytics solutions are driven by Artificial Intelligence (AI) to detect oxidative increases based on the physiological conditions of plants and environmental predictions, as well as by drone-based hyperspectral imaging to assess short-term changes in vegetation indices. CRISPR-Cas9 gene editing enables innovative irrigation systems to enhance the antioxidant defense of plants, thereby maintaining cellular integrity and function in the face of heat stress (Saiz Rubio and Rovira Mas, 2020). Such a system can be manipulated to achieve optimal soil moisture and overproduce ROS-scavenging compounds, thereby converting the AMF-mediated antioxidant defense into an active, predictable, and optimized system.

4. Challenges to Resist the Heat Stress

Arbuscular Mycorrhizal Fungi (AMF) and technology integration offer a potential solution to alleviate heat stress; however, several essential biological, technological, environmental, and socioeconomic obstacles must be addressed. The specific effectiveness of AMF symbiosis, influenced by fungal genotype, host plant species, soil type, pH, nutritional condition, and microbial population, presents a biological obstacle (Begna, 2023). It's costly and challenging to mass-produce clean, contamination-free inoculants, and storage and shelf life limit their viability and colonization. Sensor networks and monitoring systems in harsh agricultural situations have durability, calibration, and interoperability challenges. Rural farming areas often lack reliable high-speed connectivity and adequate computational resources for the real-time processing of spectral imaging, soil sensors, and DNA sequencers' massive and complex records for microbiome analysis (Rillig et al., 2021).

The application of nanotechnology is used to develop nano-amendments or innovative delivery systems that enhance heat tolerance; however, concerns regarding long-term environmental impact currently constrain its use, including potential for bioaccumulation in food chains and a lack of clear regulatory frameworks. Climate change poses a critical environmental hurdle that may surpass the adaptive threshold of even the most resilient AMF strains and their host plants. Advanced technological tools for smallholder farmers remain expensive, exacerbating global inequalities (Chen et al., 2020). A multidisciplinary approach involving plant microbiologists, soil scientists, nanotechnologists, policymakers, and economists is needed to develop scalable, affordable, and environmentally sustainable solutions.

Table 2 provides a detailed overview of the key challenges and their associated impacts in the application of mycorrhizal technology, categorizing them into environmental, technological, agricultural, and economic-regulatory domains. Environmentally, issues such as fungal and plant specificity, heat sensitivity of fungi (which reduces spore germination and colonization), and complex microbiome interactions (where native microbes may outcompete introduced fungi) limit consistent efficacy. Technologically, hurdles include the difficulty and cost of mass production and formulation of viable inoculants, the need for precision application tools, and the development of real-time monitoring systems to assess fungal colonization and plant response. Agriculturally, variable field performance due to diverse soil types, climates, and farming practices such as tillage disrupting hyphal networks and phosphate fertilization reducing plant dependency on fungi complicates integration, while low farmer awareness hinders adoption. Economically and regulatory-wise, challenges encompass proving cost-effectiveness and return on investment under stress, the absence of global standards for product quality and efficacy, and difficulties in scaling innovations from lab to market due to intellectual property and production bottlenecks. These multifaceted obstacles highlight the need for targeted research, technological innovation, and supportive policies to realize the full potential of mycorrhizal technology in sustainable agriculture.

Table 2. Challenges and impacts in the application of mycorrhizal technology

Categories	Challenges	Impacts	References
Environmental Impact & Life Cycle Assessment (LCA)	Fungal & Plant Specificity	• Specificity of mycorrhizal strains • Benefits may vary across plants • Symbiotic relationship in plants	Zhang et al. (2022)
	Heat Sensitivity of Fungi	• Reduces spore germination, hyphal growth, and colonization rates • High soil temperatures impact on arbuscular mycorrhizal fungi	Santander et al. (2021)
	Microbiome Complexity	• Native microbial communities can outcompete or suppress fungi	Evelin et al. (2019)
Technology Production	Mass Production & Formulation	• Difficulty and cost • Need for technological advances in fermentation and formulation • Ensures viability and shelf-life • Commercial mycorrhizal inoculants production	Bitterlich et al. (2018)
	Precision Application Technology	• Precision delivery of fragile fungal inoculants	Chen et al. (2020)
	Real-Time Monitoring & Sensing	• Biosensors for real-time measurement • Assessing fungal colonization's physiological impact on plants • Fungal colonization	Gebbers and Adamchuk (2020)
Agriculture	Variable Field Performance	• Variable soil types • Climate conditions • Existing microbial life • Agricultural practices	Sarma et al. (2022)
	Integration with Farming Practices	• Tillage disrupts hyphal networks • High-phosphate fertilization reduces a plant's fungi dependency • Technology adaptation needed for no-till, low-input systems	Ahanger et al. (2020)
	Farmer Awareness & Adoption	• Yield benefits under heat stress	Liakos et al. (2018)
Economic & Regulatory	Cost-Effectiveness & ROI	• Reliable, significant increase in crop yield and quality • Guaranteed under stress conditions	Mathur et al. (2019)
	Lack of Standardisation & Regulation	• Lack of global standards for mycorrhizal concentration, viability, and efficacy • Market variability and low-quality products	Navarro et al. (2022)
	Intellectual Property & Scaling	• Challenges in scaling from lab discovery to mass production	Oehl et al. (2021)

5. Future Prospects

The integration of technology with Arbuscular Mycorrhizal Fungi (AMF) is poised to fundamentally revolutionize agricultural resilience, developing a new paradigm of adaptive, self-optimizing biosystems capable of preventing and mitigating the devastating impacts of escalating thermal extremes, thereby securing global food systems against climate volatility (Kaur and Suseela, 2023). The transformation of agriculture will be done by advanced synthetic biology and genomics, enabling the precise engineering of tailored AMF strains with hyper-enhanced heat tolerance. This will involve the design of sophisticated multi-species consortia that integrate complementary Plant Growth Promoting Bacteria (PGPB) to fix atmospheric nitrogen, produce ACC deaminase to cleave stress ethylene precursors, and secrete robust exopolysaccharides as biohydrants to improve soil aggregation and moisture retention (Quiroga et al., 2020). Next-generation nano-carriers will revolutionize the delivery of these advanced bio-inoculants, constructed from innovative, biodegradable materials like chitosan, cellulose nanocrystals, or lignin-derived polymers.

Nanosensors on these carriers will monitor microbial viability, colonization rates, and soil characteristics in real-time (Liu et al., 2020). Large-scale IoT networks will provide macro-scale data for agricultural landscapes, including autonomous drones, hyperspectral Earth observation satellites, and in-field phyto monitoring devices. AMF, plant, and environment interactions will be simulated and predicted using AI and quantum computing in diverse climate conditions (He et al., 2023). Augmented reality interfaces could assist agronomists and farmers in direct variable-rate application of inoculation, irrigation, and micronutrients in areas with nutrient deficiencies (Aliferis and Jabaji, 2021). Blockchain solutions ensure traceability and verification, enhancing customer trust in sustainable food. Agroecosystems that thrive under heat stress will be more perceptive, responsive, and self-regulating due to this technological symbiosis, ensuring ecological sustainability and food security for future generations.

6. Conclusion

The integration of Arbuscular Mycorrhizal Fungi (AMF) with advanced technological innovations presents a transformative approach in enhancing plant resilience against heat stress, thereby addressing one of the most critical challenges posed by climate change to global food security. AMF symbiosis fundamentally enhances plant physiological and biochemical capacities through multiple mechanisms, including improved water and nutrient uptake, increased photosynthetic efficiency, activation of antioxidant defense systems, accumulation of protective osmolytes, and sophisticated hormonal and genetic regulation. These symbiotic benefits are significantly amplified when coupled with modern technologies such as nano-carrier-based biofertilizers, real-time sensor networks, AI-driven predictive analytics, CRISPR-Cas9 gene editing, and precision agriculture systems. These tools enable the precise selection, delivery, and monitoring of heat-tolerant AMF strains, optimize soil microbiome engineering, and facilitate dynamic, data-informed management practices that ensure maximal symbiotic efficacy under fluctuating environmental conditions. Despite existing challenges, including biological specificity, scalability of inoculant production, technological accessibility, socioeconomic barriers, and environmental concerns, the synergistic potential of AMF and technology offers a scalable and sustainable pathway toward climate-resilient agriculture. Future advancements in synthetic biology, multi-omics profiling, innovative materials science, and digital farming platforms promise to refine this synergy further, enabling the development of self-optimizing agricultural systems capable of adapting to thermal extremes. The integration of biological symbiosis with advanced technology not only enhances crop productivity and stability under heat stress but also promotes ecological sustainability, reduces chemical inputs, and improves global food system resilience in the face of climatic volatility.

7. References

Abdallah, E.F., Alqarawi, A.A., Hashem, A., Radhakrishnan, R., Al-Huqail, A.A., Al-Otibi, F.O.N. et al. (2018). Endophytic bacterium *Bacillus subtilis* (BERA 71) improves salt tolerance in chickpea plants by regulating the plant defence mechanisms. J. Plant Interact. 13, 37–44.

Ahanger, M.A., Tittal, M., Mir, R.A. & Agarwal, R.M. (2020). Arbuscular mycorrhiza in crop improvement under environmental stress. pp. 159–176. *In*: M.R. Hakeem and R.A. Bhat [eds.]. Emerging Technologies and Management of Crop Stress Tolerance, Vol. 2. Academic Press, London, UK.

Aliferis, K.A. & S. Jabaji. (2021). Metabolomics – A robust bioanalytical approach for the discovery of the modes of action of pesticides and biostimulants. J. Agric. Food Chem. 69, 2883–2894.

Augé, R.M., Toler, H.D. & Saxton, A.M. (2023). Mycorrhizal stimulation of leaf gas exchange in well-watered and droughted plants: A meta-analysis. Plant Soil 484, 1–24.

Banerjee, S., Walder, F., Schlaeppi, K. & van der Heijden, M.G.A. (2023). Bioprospecting of native Arbuscular Mycorrhizal Fungi for sustainable agriculture in arid and semi-arid regions. Appl. Soil Ecol. 181, 104647.

Basu, A., Prasad, P., Das, S.N., Kalam, S., Sayyed, R.Z., Reddy, M.S. et al. (2021). Nanofertilizers and nanopesticides: The new edge in agriculture. ACS Agric. Sci. Technol. 1, 493–495.

Begna, T. (2023). Role of CRISPR/Cas9 in improving plant abiotic stress tolerance. J. Biotechnol. 367, 12–24.

Begum, N., Ahanger, M.A., Su, Y., Lei, Y., Mustafa, N.S.A., Ahmad, P. et al. (2022). Role of Arbuscular Mycorrhizal Fungi in plant growth regulation: Implications in abiotic stress tolerance. Front. Plant Sci. 13, 793610.

Begum, N., Qin, C., Ahanger, M.A., Raza, S., Khan, M.I., Ashraf, M. et al. (2022). AMF-induced regulation of phytohormones for enhanced drought tolerance in plants. Plant Stress 5, 100091.

Bender, S.F., Schlaeppi, K. & van der Heijden, M.G.A. (2023). Symbiotic soil fungi as a cornerstone of climate-smart agriculture. Nat. Rev. Earth Environ. 4, 84–99.

Bhardwaj, S., Sharma, D., Singh, S., Ramamurthy, P.C., Verma, T., Pujari, M. et al. (2022). Engineered mycorrhizal biofertilizers: A potential tool for sustainable agriculture. J. Clean. Prod. 370, 133561.

Bitterlich, M., Rouphael, Y., Graefe, J. & Franken, P. (2018). The rhizosphere: A playground for mycorrhizal hyphae and their helpers. Plant Soil 424, 1–17.

Bowles, T.M., Jackson, L.E., Locher, M. & Cavagnaro, T.R. (2022). Addressing the context-dependency of microbial inoculant success in agriculture. Nat. Rev. Microbiol. 20, 321–334.

Cabral, C., Ravnskov, S., Tringovska, I. & Wollenweber, B. (2021). Arbuscular Mycorrhizal Fungi in plant nutrition and health: A potential biotechnological tool. J. Soil Sci. Plant Nutr. 21, 2459–2480.

Chandrasekaran, M., Chanratana, M., Kim, K., Seshadri, S. & Sa, T. (2019). A meta-analysis of arbuscular mycorrhizal effects on plants grown under salt stress. Mycorrhiza 29, 197–217.

Chareesri, A., De Deyn, G.B., Sergeeva, L., Polthanee, A. & Kuyper, T.W. (2020). The context-dependency of AMF-mediated plant growth and stress tolerance: Implications for agricultural applications. Mycorrhiza 30, 425–439.

Chen, S., Zhao, X., Ruan, Y., Li, Y. & Zou, Y. (2020). Arbuscular Mycorrhizal Fungi enhance drought tolerance in citrus by altering root hydraulic conductivity and hormone levels. Front. Plant Sci. 11, 548.

Dhiman, S., Sharma, N., Singh, V., Singh, A. & Pal, R. (2022). Chitosan nanoparticles as a carrier for mycorrhizal spores: A novel bio-nanofertilizer for sustainable agriculture. Int. J. Biol. Macromol. 220, 1497–1507.

Evelin, H., Devi, T.S., Gupta, S. & Kapoor, R. (2019). Mitigation of salinity stress in plants by arbuscular mycorrhizal symbiosis: Current understanding and new challenges. Front. Plant Sci. 10, 470.

Frew, A. (2023). Arbuscular mycorrhizal fungal diversity and its role in ecosystem resilience to climate change. Funct. Ecol. 37, 234–247.

Gebbers, R. &. Adamchuk, V.I (2020). Precision agriculture and food security. Science 327, 828–831.

Gianinazzi, S., Gianinazzi-Pearson, V. & Franken, P. (2022). The challenges of applying mycorrhizal inoculants in the field: From research to commercialization. Agronomy 12, 865.

Hameed, A., Dilfuza, E., Abd_Allah, E.F., Hashem, A., Kumar, A. & Ahmad, P. (2021). Arbuscular Mycorrhizal Fungi and plant symbiosis under stress conditions: Mechanisms and adaptations. Symbiosis 84, 167–189.

Hart, M.M., Antunes, P.M. & Abbott, L.K. (2018). Fungal inoculants in the field: Is the reward greater than the risk? Funct. Ecol. 32, 126–135.

Hashem, A., Alqarawi, A.A., Radhakrishnan, R., Al-Arjani, A.-B.F., Aldehaish, H.A., Egamberdieva, D. et al. (2019). Arbuscular Mycorrhizal Fungi and plant stress tolerance: Current knowledge and future perspectives. pp. 267–289. *In*: A. Varma, D. Tripathi & R. Prasad [eds.]. Plant Microbe Interface. Springer, Cham, Switzerland.

He, J.D., Dong, T., Wu, H.H., Jing, Y.X. & Zhang, J.H. (2023). Mycorrhizal-induced changes in root transcriptome and metabolome of maize under drought stress. J. Plant Physiol. 280, 153890.

Jansa, J. & Gryndler, M. (2023). Towards a more predictable use of Arbuscular Mycorrhizal Fungi in sustainable agriculture: Challenges and perspectives. Plant Soil 484, 1–22.

Jansson, J.K. & Hofmockel, K.S. (2020). Soil microbiomes and climate change. Nat. Rev. Microbiol. 18, 35–46.

Jiang, Y., Wang, W., Xie, Q., Liu, S. & Zhang, Z. (2022). Arbuscular Mycorrhizal Fungi alter gene expression and metabolite profile of tomatoes under heat stress. Hortic. Res. 9, uhac076.

Kah, M., Kookana, R.S., Gogos, A. & Bucheli, T.D. (2019). A critical evaluation of nanopesticides and nanofertilizers against their conventional analogues. Nat. Nanotechnol. 14, 517–522.

Kalia, A. & Kaur, H. (2023). Nano-biofertilizers: Harnessing the potential of nanomaterials for enhanced nutrient use efficiency. Plant Physiol. Biochem. 194, 1–12.

Karaki, G.N. (2023). The role of mycorrhizae in improving water and mineral uptake by plants under abiotic stress conditions. J. Agron. Crop Sci. 209, 1–12.

Kaur, S. & Suseela, V. (2023). Unraveling the role of Arbuscular Mycorrhizal Fungi in mitigating oxidative stress in plants. Plant Soil 482, 1–20.

Kopittke, P.M., Menzies, N.W., Wang, P., McKenna, B.A. & Lombi, E. (2019). The role of soil in defining planetary health. Sci. Total Environ. 658, 1434–1440.

Kumar, A., Singh, S., Mukherjee, A., Rastogi, R.P. & Verma, J.P. (2021). Nano-based bioformulations for plant growth promotion and stress tolerance. pp. 353–370. *In*: J.P. Verma & A.K. Macdonald [eds.]. Microbial Bioprospecting for Sustainable Development. Springer, Singapore.

Kumar, K., Gambhir, G., Dass, A., Tripathi, A.K., Singh, A., Jha, A.K. et al. (2020). CRISPR-based genome editing for nutrient enrichment in crops: A promising approach toward global food security. Food Energy Secur. 9, e242.

Latef, A.A.H.A., Alhmad, M.F.A. & Abdelfattah, K.E. (2021). The role of Arbuscular Mycorrhizal Fungi in plant growth, photosynthesis, and antioxidant defense under abiotic stress. Agronomy 11, 1218.

Li, J., Zhang, C., Zhang, J., Li, X. & Zhang, F. (2021). The *Medicago truncatula* transcription factor MtERF1 enhances drought tolerance by regulating auxin biosynthesis and mycorrhizal symbiosis. Plant J. 108, 781–797.

Li, Z., Liu, Y. & Walker, R. (2023). Real-time monitoring of soil health parameters using IoT-based sensors for precision agriculture. Comput. Electron. Agric. 204, 107540.

Liakos, K.G., Busato, P., Moshou, D., Pearson, S. & Bochtis, D. (2018). Machine learning in agriculture: A review. Sensors 18, 2674.

Liu, C.Y., Zhang, Y., Zhang, K., An, H., Hu, K. & Wen, G. (2020). Arbuscular Mycorrhizal Fungi improve the growth and drought tolerance of *Zenia insignis* seedlings under drought stress. New For. 51, 637–657.

Liu, J., Guo, C., Chen, Z.L., He, J.D. & Zou, Y.N. (2020). Mycorrhizal symbiosis modulates the expression of stress-responsive genes and proteins in maize under heat stress. J. Plant Growth Regul. 39, 1415–1427.

Mathur, S., Sharma, M.P. & Jajoo, A. (2019). Arbuscular Mycorrhizal Fungi (AMF) protect the photosynthetic apparatus of wheat under drought stress. Photosynth. Res. 139, 227–238.

Miller, G., Schlauch, K., Tam, R., Cortes, D., Torres, M.A., Shulaev, V. et al. (2020). Reactive oxygen species homeostasis and signalling in plants. Plant Cell Environ. 43, 283–289.

Mohanty, S.P., Hughes, D.P. & Salathé, M. (2020). AI and IoT for sustainable agriculture: A survey. IEEE Internet Things J. 7, 6808–6827.

Nadeem, S.M., Ahmad, M., Naveed, M., Imran, M., Zahir, Z.A. & Crowley, D.E. (2019). Role of mycorrhizae in alleviation of drought stress in plants. pp. 227–248. *In*: D.P. Singh & V.S. Gupta [eds.]. Microbial Interventions in Agriculture and Environment. Springer, Singapore.

Navarro, J.M., Pérez-Tornero, O. & Morte, A. (2022). Contribution of Arbuscular Mycorrhizal Fungi to the antioxidant response of lettuce plants under water stress. Antioxidants 11, 558.

Nouri, E., Breuillin-Sessoms, F., Feller, U. & Reinhardt, D. (2022). The road to market: Regulatory and commercial challenges for mycorrhizal biofertilizers. Front. Plant Sci. 13, 867890.

Oehl, F., van der Heijden, M.G.A. & Jansa, J. (2021). Diversity and ecology of Arbuscular Mycorrhizal Fungi in agricultural soils. Biol. Fertil. Soils 57, 453–467.

Pozo, M.J., Zabalgogeazcoa, I., Vázquez de Aldana, B.R. & Martínez-Medina, A. (2021). Phytohormones as integrators of environmental signals in the regulation of mycorrhizal symbioses. New Phytol. 229, 2451–2465.

Quiroga, G., Erice, G., Aroca, R., Chaumont, F. & Ruiz-Lozano, J.M. (2020). Arbuscular Mycorrhizal Fungi regulate the oxidative stress system and photosynthesis of host plants under heat stress. Microorganisms 8, 204.

Raya-González, J., López-Bucio, J.S. & López-Bucio, J. (2021). The jasmonic acid pathway and mycorrhizal symbiosis: A dialogue in the face of stress. Plant Cell Environ. 44, 1391–1404.

Riaz, M., Kamran, M., Rizwan, M., Ali, S., Parveen, A. & Malik, Z. (2021). Arbuscular mycorrhizal fungi-induced mitigation of heat stress in plants: A comprehensive review. Plants 10, 2004.

Rillig, M.C., Ryo, M., Lehmann, A., Aguilar-Trigueros, C.A., Buchert, S., Wulf, A. et al. (2019). The role of multiple global change factors in driving soil functions and microbial biodiversity. Science 366, 886–890.

Rillig, M.C., Sosa-Hernández, M.A., J. Roy, C.A. Aguilar-Trigueros, K. Vályi & A. Lehmann. (2021). Mycorrhizal fungi as mediators of soil organic matter dynamics. Annu. Rev. Ecol. Evol. Syst. 52, 237–259.

Rizvi, A., Ahmed, B., Khan, M.S., Umar, S. & Lee, J. (2022). Mycorrhizal symbiosis: An effective tool for drought mitigation in crops. Physiol. Plant. 174, e13639.

Ryan, M.H. & Graham, J.H. (2018). Little evidence that farmers should consider the abundance or diversity of Arbuscular Mycorrhizal Fungi when managing crops. New Phytol. 220, 1092–1107.

Saiz-Rubio, V. & Rovira-Más, F. (2020). From smart farming towards agriculture 5.0: A review on crop data management. Agronomy 10, 207.

Santander, C., Aroca, R., Ruiz-Lozano, J.M., Olave, J., Cartes, P., Borie, F. et al. (2021). Arbuscular mycorrhizal colonization and its role in antioxidant metabolism of plants under abiotic stress. J. Soil Sci. Plant Nutr. 21, 2481–2493.

Sarma, B.K., Yadav, S.K., Patel, J.S. & Singh, H.B. (2022). Next-generation bioformulations for sustainable agriculture: Engineering microbial synergism. Curr. Microbiol. 79, 332.

Schläpfer, P., Zhang, P., Wang, C., Kim, T., Banf, M., Chae, L. et al. (2019). Multi-omics data integration reveals the molecular networks regulating mycorrhizal symbiosis. Cell Rep. 29, 3977–3990.

Seleiman, M.F., Al-Suhaibani, N., Ali, N., Akmal, M., Alotaibi, M., Refay, Y. et al. (2021). Drought stress impacts on plants and different approaches to alleviate its adverse effects. Plants 10, 259.

Thirkell, T.J., Pastok, D. & Field, K.J. (2020). The future of AM fungal inoculants in agriculture: Challenges and opportunities. Funct. Ecol. 34, 1031–1039.

Trivedi, P., Mattupalli, C., Eversole, K. & Leach, J.E. (2021). Plant–microbiome interactions under a changing world: Responses, consequences, and perspectives. New Phytol. 234, 1951–1959.

van der Heijden, M.G.A. & Hartmann, M. (2023). Networking in the plant microbiome: A key to sustainable agriculture. PLOS Biol. 21, e3002062.

Vangelisti, A., Natali, L., Bernardi, R., Sbrana, C., Turrini, A., Hassani-Pak, K. et al. (2020). Transcriptome changes induced by Arbuscular Mycorrhizal Fungi in sunflower (*Helianthus annuus* L.) roots. Sci. Rep. 10, 1–14.

Wu, Q.S. & Zou, Y.N. (2023). Arbuscular Mycorrhizal Fungi and tolerance of drought stress in plants. pp. 25–41. *In*: Q.S. Wu [ed.]. Arbuscular Mycorrhizas and Stress Tolerance of Plants. Springer, Singapore.

Zandalinas, S.I., Fritschi, F.B. & Mittler, R. (2021). Global warming, climate change, and environmental pollution: Recipe for a multifactorial stress combination disaster. Trends Plant Sci. 26, 588–599.

Zhang, F., Huo, Y., Cobb, A.B., Luo, G., Zhou, J., Yang, G. et al. (2022). The role of Arbuscular Mycorrhizal Fungi in plant invasion and ecosystem processes. J. Ecol. 110, 3–15.

Zhang, Y., Li, T., Xu, M., Guo, J., Zhang, C. & Feng, Z. (2022). Integrated transcriptomic and metabolomic analysis reveals the molecular mechanisms of mycorrhizal-enhanced heat tolerance in cucumber. Horticulturae 8, 172.

Zhu, X., Song, F., Liu, S. & Liu, F. (2022). Arbuscular Mycorrhizal Fungi enhance heat stress tolerance of wheat plants by modulating photosynthetic efficiency and antioxidant defense. Plant Growth Regul. 96, 91–102.

CHAPTER
2

Arbuscular Mycorrhizal Fungi: Impact on Agroecosystems and Plant Secondary Metabolism

Zhaleh Soheilikhah[1*], Zakira Naureen[2] and Riyaz Sayyed[2]

[1] Department of Biology, Faculty of Science, Razi University, Kermanshah, Iran
[2] Department of Biological Sciences and Chemistry, College of Arts and Science, University of Nizwa, Nizwa, 616, Oman

1. Introduction

Plant are not only the source of food to feed the enormously increasing population of the world but also provide raw material for necessary commodities such as, paper, clothes andmedicine. In addition, crop plants play an important role in providing income for farmers and farm workers. Therefore, the role of plants in meeting the needs of society is far more than providing food products (Fazeli-Nasab et al., 2023; Hosseini et al., 2023; Sharma et al., 2023).

The global usage of plants as herbal medicines has drawn the attention of many researchers to the identification and exploration of therapeutic potential of medicinal plants that will not only play a significant role in economic growth and encourage entrepreneurship but will also add to the repertoire of knowledge in medicinal plants (Goyal et al., 2014; Hashemi, 2014; Hashemi et al., 2014; Piri et al., 2020; Piri et al., 2019). Today, due to the identification of the side effects of chemical drugs, the general approach to the use of herbal drugs is increasing day by day. In general, medicinal plants are plants whose organs contain effective substances that have medicinal properties that are of particular importance in terms of both the treatment and prevention of disease and in providing health and safety to communities. According to the statistics of the World Health Organization (WHO), more than 80% of the world's people, especially in developing countries and poor areas, meet their main medical needs from medicinal plants (Alizadeh Salteh et al., 2023; Fazeli-Nasab et al., 2023; Hosseini et al., 2023; Sharma et al., 2023; Zadali et al., 2022; Zhao et al., 2022).

*Corresponding author: zhalehsoheili@gmail.com

DOI: 10.1201/9781003669777-2

Medicinal plants are not only huge reservoirs of valuable secondary metabolites, but are also provide effective replacement of chemical drugs that have many adverse side effects on human health. For instance, chemical drugs with all their effectiveness have many adverse effects, and there are only a few pure substances that do not have adverse effects, in contrast to the effective substances present in medicinal plants, which, are in a state of biological balance (Ahmadi, 2022; Alavi et al., 2023; Hoseini et al., 2022).

2. Arbuscular Mycorrhiza (AMF)

AMF are obligate root biotrophs that are members of the fungus phylum Glomeromycota that create symbiotic relationships with 70 - 80% of plants. The most ancient plant-mutualistic interaction that has existed for more than 400 million years is the symbiosis between plants and AMF (Chiu and Paszkowski, 2019; Kaur and Suseela, 2020; Torres-Cortes et al., 2015; Wipf et al., 2019).

The advantages of this mutualistic connection for plants are numerous and wide-ranging, with the transfer of nutrients such as phosphorous, nitrogen, copper and zinc—due to their large hyphal network (Saboor et al., 2021; Vafa et al., 2021). Due to their thin fungal hyphae, AMF is said to obtain more nutrients since they can access portions of the soil that the roots cannot because they can investigate a wider volume of soil. AMF may get P from less accessible P-minerals by chelation, acidification, and enzyme activity in addition to expanding the rooting volume (Andrino et al., 2021; Kaur and Suseela, 2020; Keyes et al., 2022; Rezacova et al., 2018; Zhang et al., 2016).

AMF help their host mostly by boosting absorption of phosphate ions that are relatively stationary because of the fungal ERM's ability to grow outside of the phosphate depletion zone that quickly forms around the root. The host plant gives the fungus carbon (C) in exchange. Additional advantages for the host include greater drought resilience, higher resistance to soil diseases, increased tolerance to salt, and increased tolerance to heavy metals. In addition to higher absorption of certain micronutrients, greater uptake of other macronutrients has also been seen, including nitrogen (N), potassium (K), and magnesium (Mg). Mycorrhizas have also been proven to be crucial in preserving the integrity of the soil aggregate (Gosling et al., 2006).

AMF symbiosis additionally improves resilience to pathogens, illnesses, salt stress, and drought. Through enhanced soil aggregation and hyphal input, AMF not only provides mineral nutrition but also enhances the quality of plant products and significantly contributes to the buildup of soil organic matter (Jabborova et al., 2022). As the most prevalent symbiotic relationship in nature, AMF symbiosis offers enormous potential for utilization in agriculture, particularly in low input farming and sustainable agriculture. This technique would cut agricultural expenses and further minimize environmental degradation viz., export of surplus P to water bodies by reducing the demand for P fertilizer (Bahadur et al., 2019; Kaur and Suseela, 2020; Pavinato et al., 2020; Plouznikoff et al., 2019; Verzeaux et al., 2017).

In addition to promoting plant growth and nutrient acquisition, the AMF influence the host plants' accumulation of secondary metabolites (Bastami et al.,

2021; Hodge and Storer, 2015; Kapoor et al., 2017; Thirkell et al., 2016). Numerous studies have demonstrated a relationship between plant secondary metabolism and the growth of AM fungus. AM fungus can influence secondary metabolic processes in plants either directly or indirectly. Terpenoids, phenolics, and alkaloids are the three classes into which secondary metabolites are divided (ZHAO and YAN, 2006). In this chapter, we are going to examine the types of secondary metabolites and the effect of mycorrhizal fungi on the expression of these secondary metabolites.

3. AMF's Impact on Agroecosystems

3.1. Crop Nutrition

Improved P absorption is the main benefit that AMF provide to their host plant, and plant P status is often the main controlling factor in the plant-fungal relationship. Because it improves overall absorption and, in some cases, P utilisation efficiency, AMF has a significant effect on crop P nutrition. This could be connected to increased growth and yield. The disturbance of AMF colonization can significantly affect P uptake, growth, and yield in certain cases. Often, this is brought on by high levels of soil P that are available to plants. In these situations, AMF often suppresses its root colonization. If there is consistently significant AMF colonization in the presence of high soil P concentrations, it could impede crop development. Plants may not respond to native AMF or AMF inoculation, even on soils with low levels of accessible P, even though colonization may be significantly increased for unclear reasons (Gosling et al., 2006; Keyes et al., 2022; Wang et al., 2023).

3.2. Crop Pests and Disease

Although much of the study on the AM connection has been on nutrient intake, there is evidence that AMF also contributes to the control of crop pests and diseases, notably soil-borne fungal diseases (Dave et al., 2024). Even while pathogen infection often lowers AM colonization, disease severity reductions rather than complete inhibition can nonetheless lead to a considerable gain in yield over plants not inoculated with AMF (Carlucci et al., 2022; Chandrasekaran, 2022; Chen et al., 2022; Gosling et al., 2006).

3.3. Crop Water Relations

There is compelling evidence that AMF can strengthen the host plant's resistance to water stress, particularly those caused by high salinity. Several mechanisms have been proposed to explain the effect, including increased root hydraulic conductivity, improved stomatal control, osmotic host adjustment, and improved interaction with soil particles via the binding effect of hyphae, which enables water to be extracted from smaller holes. Frequently, drought-stressed mycorrhizal plants absorb more water and nutrients than non-mycorrhizal plants (Vafa et al., 2024; Begum et al., 2019; Chun et al., 2018; Gosling et al., 2006; Ma et al., 2022; Silva et al., 2023).

3.4. Soil Structure

In addition to enhance the plant development AMF has a direct impact on soil structure, which is crucial in agricultural settings since cultivations, trafficking, and low levels of soil organic matter all frequently lead to weakened soil structures. The fungal partner receives up to 20% of all the fixed carbon from the host plant, and in agricultural soils, AMF may generate a sizable amount of biomass. It was discovered that on a sandy soil with linseed (*Linum usitatissimus*) , AMF made up as much as 50% of the total soil microbial biomass (Gosling et al., 2006; Kohl et al., 2016; Rola et al., 2023; Wen et al., 2023; Zhang et al., 2021b; Zubek et al., 2019).

The influences on soil structure are many. AMF uses the hyphal enmeshing properties of other soil fungi to link soil microaggregates into bigger macroaggregates. In a pot experiment infected with AMF, it has been shown that the growth of extraradical hyphae and soil aggregation are directly related. AMF also create glomalin, an extracellular glycoprotein that adheres hyphae to the ground. Glomalin builds up in soils, research suggests that it has a significant impact on the stability of the soil aggregate. Fast hyphal turnover and widespread hyphal exudation provide C to other soil microorganisms.The overall impact of hyphal enmeshment and C inputs result in significant improvement in soil structural stability (Gosling et al., 2006; Li et al., 2022; Miao et al., 2022; Wang et al., 2022b; Yang et al., 2022).

4. Plant Essential Oils

Majority of plant essential oils derived from medicinal plants exhibit anti-fungal, anti-parasitic, anti-bacterial, and anti-viral properties. Therefore, much research has been done on the use of plant essential oils in pharmacology, herbal pharmacology, medical and clinical microbiology, phytopathology, and the preservation of food, fruits, and vegetables (Doosti-Moghaddam et al., 2022; Fazeli-Nasab et al., 2023; Mirzaei and Shakoory-Moghadam, 2022). Plant Essences are oily and aromatic compounds that are obtained from different organs of medicinal and aromatic plants and are widely used as food flavorings. The industrial and technological application of essential oils in food requires the use of higher concentrations of plant essential oils compared to laboratory culture (Arsana et al., 2022; Salehi-Sardoei and Khalili, 2022; Tabin et al., 2022). Essential oils exist in the form of various compounds and are also known as volatile oils. Essential oils are irregularly distributed throughout plants. These substances exist in individual secretory cells and hairs or in the surface and internal parts of various plant organs such as leaves, fruits, buds, and branches of plants (Bagherniya et al., 2018; Bakal et al., 2017).

They are often liquids at room temperature and can be dissolved in alcohol and fat, and if they are fresh, they mostly look like colorless substances, but they do not resemble oily substances. These materials are known as antiseptics, medicinal, and aromatic compounds. They do not have the same chemical structure and have a lower specific gravity than water, and rarely, some of them have a higher specific gravity than water (Chiej, 1984; Chukwuka et al., 2011). According to the theory put forth by the International Organization for Standardization, essential oils are products that are obtained by steam or water distillation from plant sources or fruits, and their chemical compositions vary greatly depending on the type of plant, the environment, and the method of extraction (Ahmed et al., 2021; Aleebrahim-Dehkordy et al., 2017). Owing to their their antibacterial, anti-oxidant, anti-inflammatory, and anti-cancer capabilities these essential oils can replace their chemical counterparts in food and pharamaceutical industries. Thymol, alpha-pinene, beta-pinene, carvacrol, linalool, germacrene, and cumin aldehyde are the principal chemical constituents of plant essential oils (Aali et al., 2017). Resin and gum also belong to this category, which are liquid at high temperatures and solid at low temperatures (Foroughi et al., 2017; Ghasemi, 2009).

5. The Importance and Position of Natural Essential Oils

Around 700 different kinds of essential oils have so far been recognized as being present in diverse plants. Terpenoids, hydrocarbons, phenols, aldehydes, esters, and ketones are only a few of the many distinct substances that make up an essential oil's composition. Frequently, one or more of these components affect the essential oil's scent (Shojaiee et al., 2018; Tabin et al., 2022). Any kind of essential oil in aromatic plants and in different parts of them, such as ginger (root), pine needles (wood), cinnamon (bark), rosemary (leaf), jasmine (flower), fruit (fruit), fennel (seed), etc., is made and stored in the space between the cells of the epidermis and the mesophyll in multiple secretory structures and have a low boiling point and can be extracted in different ways. Although the exact mechanism of the synthesis of essential oils in plants is unclear, its existence is known because of the resistance of plants against adverse environmental conditions, the attraction of pollinating insects, and the development and spread of these plants. In the exploitation of plants essential oil, the evidence available over the past 50 years shows that the demand for essential oils and their products has increased. This evolution in consumption has been influenced by several factors, including the following (Ganjali et al., 2022; Yeşil et al., 2020; Zingue et al., 2015; Zomorodian et al., 2015):

1. There are several uses for these plants and their essential oils. They can be used to make medications since some of their component chemicals have therapeutic qualities. Their smell is utilized in the air freshener and perfume industries as well as in aromatherapy, a form of therapeutic technique. Essential oils' flavoring ability is employed in a variety of meals, toothpaste, and drinks. Its saturated fumes, which are utilized in some industrial poisons, have the power to drive away pests and dangerous insects (Chaghakaboodi et al., 2022; Kaul et al., 2012; Kheiri and Salehi, 2014; Kheiry et al., 2017).
2. The enrichment of global societies has changed their consumption culture so that they demand more variety in their diet and show more desire to consume delicious products with natural flavors (Kazemi et al., 2012; Manandhar et al., 2019).
3. Increasing attention to environmental issues and food hygiene to reduce the consumption of artificial chemicals has led to a trend towards natural foods and additives. Along with flavoring products, essential oils act as antioxidants

against harmful compounds and play an important role in people's health (Mata et al., 2019; Mickymaray et al., 2016).

6. Secondary Metabolites

Secondary metabolites are compounds in plants that have no direct role in processes such as photosynthesis, respiration, and transport of substances, differentiation, or the formation of carbohydrates, proteins, or lipids and are side products in the process of producing primary metabolites inside plant cells. Unlike primary metabolites, these compounds have a limited distribution in the plant kingdom; in other words, specific secondary metabolites are often found only in one species or plants related to that species (Ghamarnia et al., 2022; Hamidian et al., 2023; Yousefi and Safari, 2022).

In plants, organic compounds are produced that are structurally and chemically different from each other and have no direct relationship to plant metabolism or growth. Although they are not essential for plant growth and development, they are economically important for human well-being in various fields, including pharmaceuticals, nutrients, food additives, and agrochemicals. Lack of secondary metabolites does not cause the immediate death of the organism. Rather, in the long term, it can cause disruptions in survival, fertility, or appearance (Toussaint et al., 2007).

Phytochemicals, also known as active substances, are known to be found in medicinal herbs and are highly sought after for their beneficial effects on human health. Since ancient times, members of the Lamiaceae have been employed as sources of spices and flavorings as well as for their medicinal benefits. It is widely acknowledged that secondary metabolites like anthocyanins and phenolic chemicals (including flavonoids and phenylpropanoids) are what give this family its therapeutic benefits. Rosmarinic acid (RA), an ester of caffeic acid (CA), is a significant phenolic molecule found in the Lamiaceae. It is well known to have a multitude of biological actions, mostly as an antioxidant but also as an antibacterial, antiviral, and anti-inflammatory. Over 50 years ago, the food processing industry began using antioxidants. Recently, however, natural antioxidants like RA have gained popularity due to serious concerns on the toxic side effects of synthetic antioxidants like BHT (butylated hydroxytoluene) and BHA (butylated hydroxyanisole) in food (Toussaint et al., 2007). P addition clearly affects the synthesis of RA and CA in NM plants and that there is a negative correlation between shoot P concentrations and shoot RA concentrations in AM plants. Overall, our results demonstrate that sweet basil plants can be grown with AMF or additional P supplies just as successfully, producing more biomass with higher concentrations of active components (Toussaint et al., 2007).

Considering the economic potential of secondary metabolites in food additives, pharmaceuticals, nutrients, and chemicals, various strategies have been investigated to increase their production from plant sources. Based on the structure and chemical nature of secondary metabolites, they are classified into three groups: 1) Terpenes, 2) Phenolics and 3) Compounds containing nitrogen. The compounds in these groups are defensive compounds against herbivores and include alkaloids and cyanogenic glycosides. These compounds have been highly considered due to their toxicity in humans and their medicinal properties (Ganjali et al., 2022; Wink,

2012). The essential properties of plant secondary metabolites are the shapes of their unique carbon structures (Almasi, 2021; Pagare et al., 2015; Shaafi et al., 2021). Their molecular weight is low, and they have a complex chemical structure that is made from the biosynthesis of primary metabolites (Tabin et al., 2022; Valizadeh et al., 2021).

Flowers, fruits, and seeds, especially in annual plants, are usually rich in secondary compounds. In perennial plants, large amounts of secondary metabolites are found in bulbs, roots, rhizomes, and stems. The place of accumulation and storage of secondary metabolites is different depending on the nature of these compounds; Hydrophilic compounds accumulate inside the vacuole, and cell walls, and lipophilic substances accumulate inside the canals, glandular hairs, trichomes, and cuticle (Kheyrodin et al., 2022; Yousefi and Safari, 2022; Zadali et al., 2022; Zhao et al., 2022).

Secondary metabolites are often responsible for the protection of plants against biotic or abiotic stresses. These compounds often play an important role in the defense systems of plants against herbivory, plant competition, and other defense systems between species (Pagare et al., 2015). Among the adaptive responses of plants to environmental stresses is the ability to synthesize many types of these chemical compounds (Karuppusamy, 2009). In other words, secondary metabolites "prevent damage to plants by viruses, bacteria, fungi, or herbivores and minimize their competition with other plants (Hasanloo et al., 2019; Ramachandra Rao and Ravishankar, 2002). It can be challenging to discern between primary and secondary metabolites sometimes. Terpenoids, for instance, include both primary and secondary metabolites, and this substance may play both main and secondary roles in the body (Figure 1) (Winkel-Shirley, 2001).

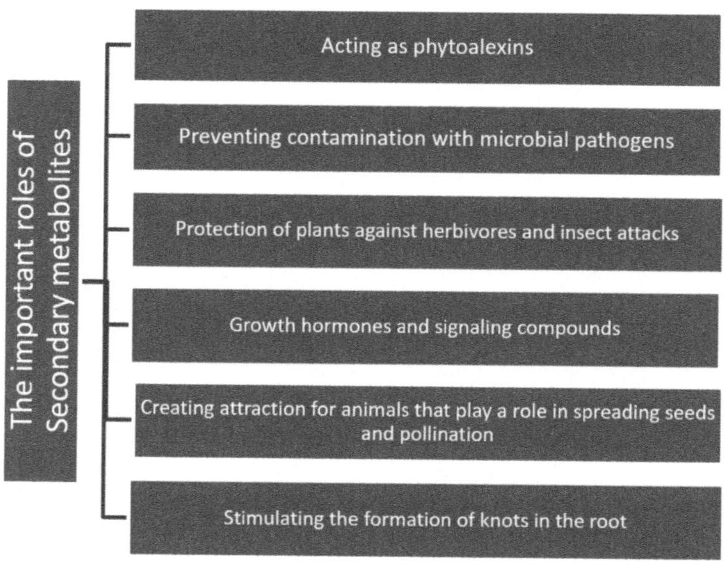

Fig. 1. The important roles of these compounds (Wink, 2015).

A unique source for making drugs are the secondary metabolites of plants, and various stresses such as elicitors or messenger molecules in plants cause the production of these substances (Zhao et al., 2005). For instance, inoculation with mycorrhizal fungi has been shown to increase the antioxidant properties of the *Ipomoea purpurea*, and this increase can probably be attributed to the positive role of mycorrhizal fungi in increasing the synthesis of secondary metabolites such as phenol, flavonoids, and terpenoids (Rashidi et al., 2020b).

7. Types of Secondary Metabolites

More than 2 million secondary metabolites are known, and they are usually classified based on their wide diversity in structure, function, and biosynthesis (McMurry, 2014). Most secondary metabolites show very different structural characteristics and different functions to protect plants. Based on the chemical structure and synthetic pathway, they include the following divisions: 1:Compounds containing nitrogen alkaloids and glycosides 2:Trines and steroids 3:phenols, flavonoids, tannin, and lignin, 4: Other secondary metabolites that do not fit into the previous three groups due to their variety and different chemical structures include compounds such as bitter substances such as mucilages, vitamins, and salicylic acid (Bartwal et al., 2013).

Although secondary metabolites are very diverse, the number of their main biosynthetic pathways is limited and specific. The structural precursors of secondary metabolites are imported as sources of the production of these metabolites in nine main pathways, including the acetate-malonate pathways and shikimic acid pathway (Dewick, 2002).

7.1. Alkaloids

Alkaloids are nitrogenous substances that have alkaline properties and produce salt in an acidic environment. Alkaloids in humans produce strong physiological reactions with special effects that affect the nervous system. The role of alkaloids in plants is not known yet, but some scientists believe that their bitter taste repels insects from plants. Some types of alkaloids have obvious effects and medicinal indicators, and in this sense, they are both very useful and expensive. In this case, we can mention morphine, quinine, strychnine, caffeine, and the like (Yun et al., 2021).

There are 20,000 or more known alkaloids, the majority of which have been isolated from plants. Alkaloids, however, are also present in microbes, terrestrial animals like insects, and aquatic species like algae. Numerous alkaloids have potent biological effects on people. The structural similarities between alkaloids and neurotransmitters like dopamine, acetylcholine, noradrenaline, and serotonin can help to explain some of this. Alkaloids have special features for medical application since they can be carried in the protonated form and can be transferred across membranes in the neutral form. This is because they are soluble in water under acidic conditions and soluble in lipids under neutral conditions (Bhambhani et al., 2021; Lichman, 2021; Othman et al., 2019; Rai et al., 2022).

In a glasshouse trial, the effectiveness of seven AMF, including *Acaulospora deligata*, *Glomus aggregatum*, *G feugianum*, *G fasciculatum*, *G rubiforme*, *Gigaspora margarita*, and *Scutellospora heterogama*, was evaluated. On the same host, they were examined for elevated amounts of total phenols, ortho dihydroxy phenols, flavonoids, alkaloids, tannins, and saponins as well as acid and alkaline phosphatase, nitrate reductase, peroxidase, and polyphenol oxidase activities. Depending on the kind of AM fungus that resided on the *W. chinensis* seedlings, the quantities of phytochemical components and enzymes varied. According to an analysis done 90 days after inoculation, *Glomus fasciculatum* had the most favorable impact in terms of improved phytochemical production (Nisha and Rajeshkumar, 2010)

7.2. Phenol, Flavonoids, and Anthocyanins

The most common natural antioxidants include flavonoids, cinnamic acid derivatives, tocopherols, amino acids, and peptides (Shaghaghi et al., 2019). The greatest class of secondary metabolites in plants are polyphenols, often known as phenolic chemicals. These substances have a wide range of chemical properties, from small, simple phenolic acids to extremely big, complicated polymers like tannins and lignins. The production of many phenolic compounds starts with the aromatic amino acids' phenylalanine, tyrosine, and tryptophan. The family of flavonoids includes flavones, flavonols, isoflavonoids, and anthocyanidins. These compounds have antiviral, antimicrobial, and high antioxidant properties (Reyes-Carmona et al., 2005). Flavonoids are secondary metabolites in plants that are produced through phenylpropanoid (Gürtler et al., 1994).

Flavones and flavonoids are found in many plants in free form or in combination with other glycosides and are present in the sap of plant cells in a soluble form (Fazeli-Nasab et al., 2017). Flavonoids prevent oxidation by mechanisms such as trapping free radicals, forming complexes with proteins, and chelating metals. In addition to their role in plant growth and adaptation to the environment, they act as antimicrobial agents, light-protective pigments, and protection against pathogens and herbivores. They play an important role in seed pollination, fertility, and the control of auxin transfer. Many of them are also necessary for flower color to attract insects for pollination and seed dispersal (Fazeli-Nasab and Mirzaei, 2018; Petrussa et al., 2013).

The rapid effects of these metabolites on health may be related to interactions with key enzymes in messenger chains, including cytokines, transcription factors, or antioxidant systems (Buer et al., 2010).

Among the most important medicinal properties of flavonoids in plants, the following can be mentioned (Figure 2)(Forouzandeh et al., 2019; Hasanloo et al., 2019; Liu et al., 2019; Liu et al., 2020; Rashidi et al., 2020a; Xie et al., 2018):

Flavonoids, phenolic chemicals, and polyphenolic derivatives are examples of the numerous secondary metabolites that are generated by plants. Pentose phosphate, shikimate, and phenylpropanoid pathways are responsible for their production in plants. In addition to aiding in plant development and reproduction, these substances also shield plants against infections and predators. They also contribute to the color qualities of fruits and vegetables (Chikezie et al., 2015). In a study, the simultaneous

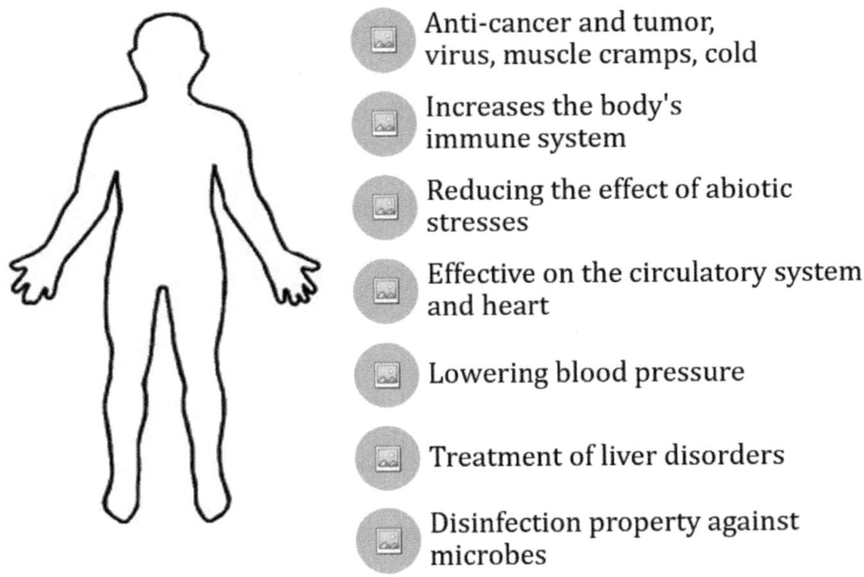

Fig. 2. The most important medicinal properties.

inoculation of licorice plant seeds with two species of AMF, *Glomus mosseae* and *G. intraradices*, the important secondary compounds of the plant, including phenolic compounds, total flavonoids, and glycyrrhizin, significantly increased in the roots of the treated plants after six months of growth compared to the control plants (Orujei et al., 2013).

More than 8,000 technical compound structures are known (Del Rio et al., 2013). The most important group of technical compounds in the human diet are phenolic acids, flavonoids, and tannins. Chemically, technical acids have at least one aromatic ring in which at least one hydrogen is replaced by a hydroxyl group (Heleno et al., 2015). It has revealed (Zhao et al., 2015) that maize's roots and leaves had more flavonoids after being inoculated with AMF. Also, inoculation with AMF improved phenolic composition and antioxidant activities in grains from wheat plants growing under hydric stress conditions (Nahuelcura et al., 2022).

7.2.1. Phenolic

Numerous different phenolic-containing secondary metabolites are produced by plants. The hydroxyl functional group that makes up the phenolic group is attached to an aromatic ring. About 10,000 different chemicals make up the chemically diverse class known as plant phenols, some of which are only soluble in organic solvents, while others are insoluble macropolymers, glycosides, and carboxylic acids (Fazeli-Nasab et al., 2022). Some phenols have anti-herbivore and anti-pathogen properties. Other phenols aid in mechanical support, fruit dissemination, attracting pollinators, absorbing harmful UV radiation, and increasing plant competitiveness by stunting neighboring plants' development. The shikimic acid pathway and the mevalonic acid pathway are the two fundamental processes involved in the synthesis of plant phenols. Phenylalanine is the source of the most prevalent class of secondary phenolic compounds in plants (Feizi et al., 2013; Fiori et al., 2000).

Arecent study has investigated the effects of two AMF, *Glomus mosseae* and *G. claroideum*, and a pathogenic fungus, *Pythium ultimum*, on the production of eight flavonoids in the roots of two cultivars of white clover (Trifolium repens L.). According to measurements of AM and pathogenic fungus in the roots, the AM symbiosis significantly reduced the biomass of *P. ultimum* and, in some cases, avoided infection. Depending on the presence of helpful and/or harmful fungus, a particular fungal strain, or the plant cultivar, different amounts of flavonoids were produced in clover roots (Carlsen et al., 2008; Lu et al., 2020; Shabani et al., 2016; Song et al., 2015).

Only plants infected with *G. claroideum* had measurable levels of coumestrol or kaempferol (depending on the cultivar). Additionally, coumestrol and medicarpin concentrations in cv. Sonja and cv. Milo were dramatically increased after receiving an injection with *G. claroideum*. It's possible that *G. mosseae* alone is responsible for mycorrhizal plants' low coumestrol and kaempferol synthesis. Only formononetin and daidzein concentrations rose in response to *P. ultimum* infection in clover roots. In response to pathogen infection, these flavonoids are allegedly stress metabolites that are synthesized or generated from glycosides. However, the formononetin and daidzein concentrations were dramatically reduced when one or both AMF were present, overriding *P. ultimum*'s inductive action. Therefore, another mechanism must be used for AM's antagonistic activity against the pathogen (Carlsen et al., 2008; Tosar et al., 2021; Velasquez et al., 2020; Zhu et al., 2015).

7.2.2. Flavonoids

One of the biggest subgroups of plant phenols are flavonoids. A flavonoid's fundamental carbon structure consists of 71 carbons organized in two aromatic rings joined by a 3-carbon bridge. Flavones and their derivatives (flavonoids) are substances that exist freely in many plants or in combination with glycosides. Different types of flavonoids show various functions in plants, which include pigmentation and defense. Some compounds belonging to this group can be mentioned as coumarins and anthocyanins (Wang et al., 2022a; Zadali et al., 2022; Zamani et al., 2021).

Flavonoids have been proposed as signaling molecules in the AM symbiosis because, interestingly, variations in the flavonoid pattern appear to depend on the developmental stage of the AM symbiosis. Flavonoids are recognized as crucial signals for the formation of the rhizobial symbiosis (Akiyama et al., 2002; Scervino et al., 2005; Vierheilig et al., 2000; Vierheilig et al., 2002).

Numerous studies have been conducted on the effects of flavonoids on the germination of AM spores, the growth of fungi, and root colonization; however, there are relatively few studies on the effects of flavonoids on other critical aspects such as hyphal branching and the creation of auxiliary cells or secondary spores. It has been shown that the flavonoid pattern in white clover changes after it has been infected by the AM fungus *Glomus intraradices*. Some flavonoids, such as 3,5,6,7,4′-hydroxy flavone (RR4-2) and 3,5,7,8-hydroxy-4′-methoxy flavone (RR4-2), were only present in the roots of white clover plants that were not mycorrhizal, but others, such as acacetin, quercetin, and rhamnetin, were freshly synthesized in mycorrhizal roots. White clover roots contained the flavonoid 3,5,6,7,8-hydroxy-4′-methoxy flavone in both mycorrhizal fungus-containing and mycorrhizal-fungus-free plants (Bécard et al. 1992; Ponce et al., 2004; Scervino et al., 2005).

Flavonoids may have a role in the control of the AM symbiosis at different phases if they are synthesized or suppressed in white clover roots that have been infected by an AM fungus. When applied to plants, flavonoids have been shown to boost AM hyphal development and improve root colonization by AM fungi. These flavonoids were extracted from the roots of non-mycorrhizal white clover plants. Although studies on the effects of the other flavonoids taken from mycorrhizal and non-mycorrhizal white clover roots are still ongoing, quercetin has been proven to have some impacts on AM fungus (Bécard et al. 1992; Chabot et al. 1992; Scervino et al., 2005; Vierheilig et al., 2002).

7.2.3. Anthocyanins

Anthocyanins are the most extensive group of colored flavonoids that are responsible for producing red, pink, purple, and blue colors in different parts of the plant (Figure 3).

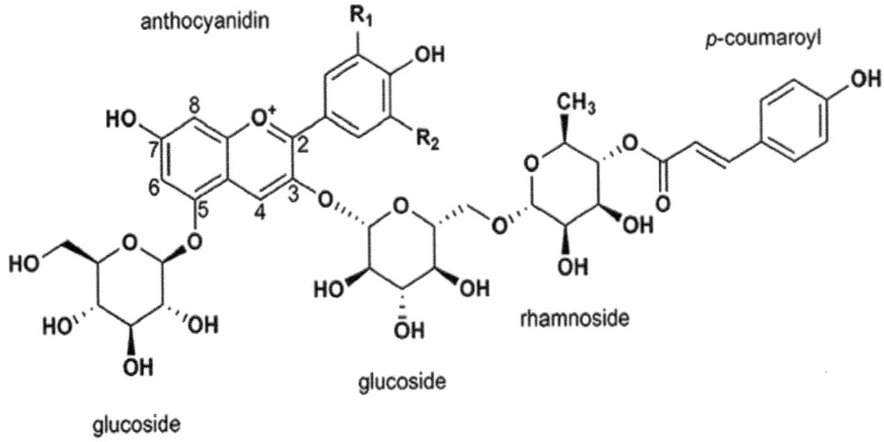

Fig. 3. Chemical structure of anthocyanin (Liu et al., 2018).

The biosynthetic pathway of anthocyanins is shown in Figure 4. They are vital in attracting animals for pollination and seed dispersal. The amount of anthocyanins compared to other metabolites is relatively high, so that one kilogram of raspberry contains about one gram and red and black beans about 20 mg of anthocyanin (Noroozi et al., 2017; Orejola et al., 2017; Talpur et al., 2017). In general, they have an important physiological role, for example, hormones to maintain membrane structure, and ecological roles, for example, defensive compounds to attract insects or animals (Zhou and Pichersky, 2020).

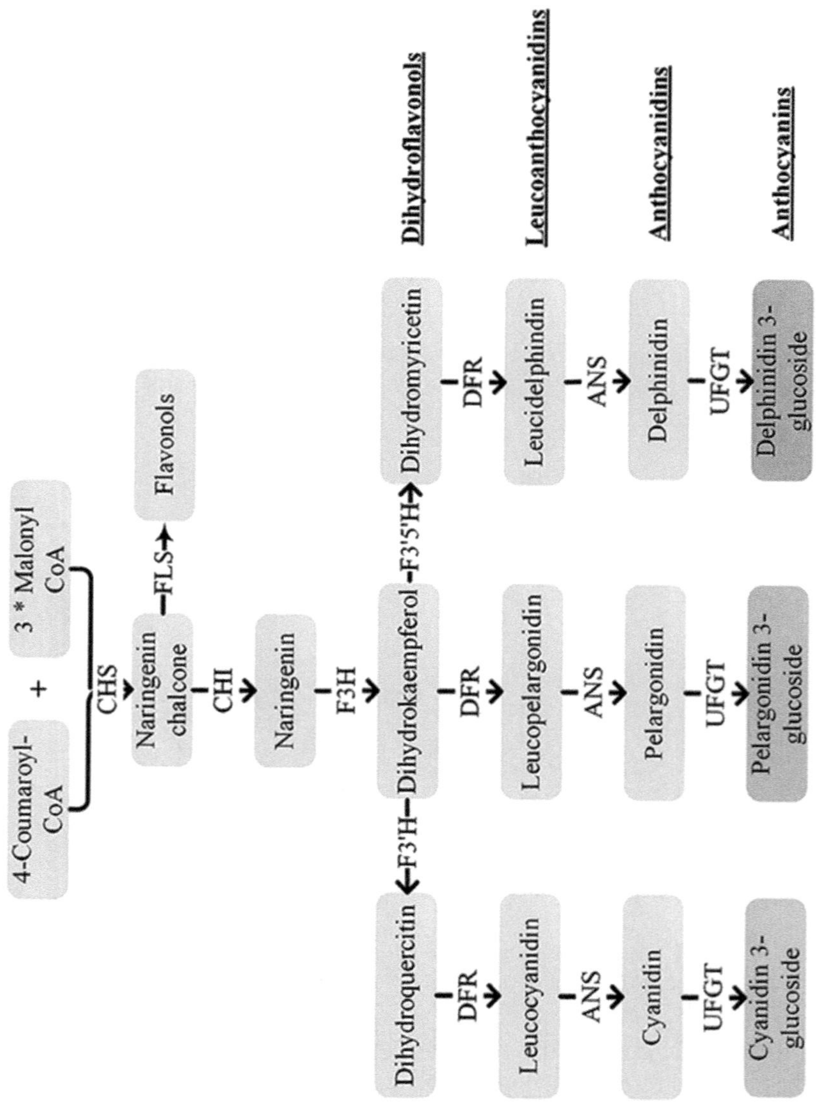

Fig. 4. Biosynthesis Pathway of anthocyanin (Liu et al., 2018).

7.3. Terpenes

Terpenes form a different group in terms of structure and function. So far, more than 22,000 of them have been identified. They are synthesized through two different pathways in plants, one in the cytoplasm and the other in the plastids. Their common biosynthetic origin is acetyl coenzyme A or glycolytic mediators. The majority are from the family of hydrocarbons and five-carbon compounds (Pagare et al., 2015).

Terpene compounds play an important role in plants' tolerance of temperature fluctuationsby producing isoprenes whichprotect photosystem II against reactive oxygen species (ROS). By directly connecting terpenes with ROS, they create a double bond and prevent their activity. One of the most important growth inhibitors in plants is abscisic acid, which is obtained from the decomposition of the 15-carbon containing carotenoid precursor. Gibberellin is also known as a plant growth regulator, one of the most important terpenoid compounds in plants that plays a significant role in controlling germination and flowering (De La Peña and Sattely, 2021; Ninkuu et al., 2021).

The link between terpenoids and AM fungus has been well investigated, and several studies have looked at molecularly interacting processes. Blumenin was originally isolated and identified from mycorrhizal cereals, and an isotopic labelling approach was used to demonstrate that it is synthesized via the Glyceraldehyde 3-phosphate/pyruvate route (MEP). Since then, it has been noted that AM fungi cause blumenin to accumulate, and that various AM fungi have blumenin levels (ZHAO and YAN, 2006).

The transcription of 1-deoxy-D-xylulose-5-phosphate reductoisomerase (DXR) and 1-deoxy-D-xylulose-5-phosphate synthase (DXS), two essential enzymes in the production of carotenoid metabolism via the MEP pathway, has been reported to increase in plants with AM fungus. Additionally, it had a brief and geographical correlation with the buildup of apocarotenoids. Two genes were subsequently discovered: TC78589, which encodes DXS2 and is highly expressed in roots infected with AM fungus, and TC77051, which encodes mevalonate disphosphate decarboxylase and catalyzes the production of terpenoids in the mevalonate pathway. Despite the fact that both genes individually code for enzymes in different pathways, an improvement in carotenoid production has been noted (ZHAO and YAN, 2006).

In-depth research has also been done on the relationship between phenolic substances (such as phytoalexin, wall-bound phenol, flavonoids, isoflavonoids, and their derivatives) and AM fungi. It has been demonstrated that certain flavonoids raised the levels of flavonoids prior to the infection of AM fungus, which encouraged the spore germination and hyphal development of AM fungi. Therefore, some researchers proposed that flavonoids served as a signal substance during the emergence of AM fungi. Additionally, according to certain tests, AM plants have considerably higher levels of peroxidase (POD), phenylalanine ammonia-lyase (PAL), and polyphenol oxidase (PPO) activity(ZHAO and YAN, 2006).

The buildup of secondary metabolites brought on by the mycorrhizal fungus involves two distinct signalling pathways in phenylpropamoid metabolism: one involves the stimulation of PAL and chalcone synthase (CHS), and the other involves the repression of isoflavone reductase (IFR). A recent study has demonstrated that the creation of AM is advantageous to the accumulation of alkaloids, even though there hasn't been much research on the interaction between alkaloids and AM fungus. This study also demonstrated how species uniqueness in AM affects alkaloids' production (ZHAO and YAN, 2006).

7.4. Glycosides

Glycosides are a group of natural compounds found in different plant families (Zhang et al., 2021a). In plants, glycosides are mostly obtained from secondary metabolite structure changes catalyzed by plant glycosyltransferase enzymes (Figure 5) (Beran et al., 2019). Glycoside means a molecule of sugar. They play an important role in living organisms. Most plants store their chemical resources in the form of inactive glycosides, which are broken down by hydrolyzing enzymes in the bodies of herbivores and converted into sugar.

Also, they are structurally diverse, and their biological activity and traditional use have been proven, and they are very important in pharmaceuticals. All glycosides are obtained from the secondary metabolism of plants, and their structure consists of two parts. The sugar part of the aglycone (genin), which is mainly lipophilic, and the non-sugar part of the glycone, which is hydrophobic (Kytidou et al., 2020).

Among the most important compounds of plant glycosides are coumarins, cardiac (heart medicine), anthraquinones (as anti-cancer medicine), saponosides, cyanogenic, and thioglycosides (Dong et al., 2019; Karre et al., 2019). In many cases, glycosides are produced as a response to environmental conditions such as moisture, soil composition, sunlight (temperature or abiotic factors) (Islam et al., 2020; Libik-Konieczny et al., 2021). Inoculation of the medicinal plant *Stevia rebaudiana* with mycorrhizal fungus improved the biomass of the branches and also increased the concentration of essential glycosides of this medicinal plant compared to the control treatment (no inoculation) (Mandal et al., 2013).

Although AMF have a significant impact on their host plant's metabolism, there is still much to learn about how they affect the plant's defense systems. It has been investigated how mechanical injury and insect herbivory impact the main direct defences (iridoid glycosides) and indirect defenses (volatile organic compounds) of *Plantago lanceolata* L. Volatile compounds were extracted and quantified from *P. lanceolata* plants that were subjected to one of three treatments: non-mycorrhizal, mycorrhizal, or both. The possible outcomes are insect herbivory, mechanical injury, or no harm (Fontana et al., 2009).

The same plants were used to extract and measure the iridoids catalpol and acubin. In comparison to the other treatments, insect herbivory dramatically increased the emission of terpenoid volatiles. However, compared to herbivore-damaged non-mycorrhizal plants, mycorrhizal plants produced less sesquiterpenes, but not monoterpenes. In contrast, mycorrhizal infection enhanced the emission of the green leaf volatile (Z)-3-hexenyl acetate in untreated control plants, bringing it on par with the emission from plants that had been mechanically injured or harmed by herbivores, whether those plants had mycorrhizal companions. Iridoid glycoside concentrations were unaffected by treatment or mycorrhization. Therefore, the amounts of direct

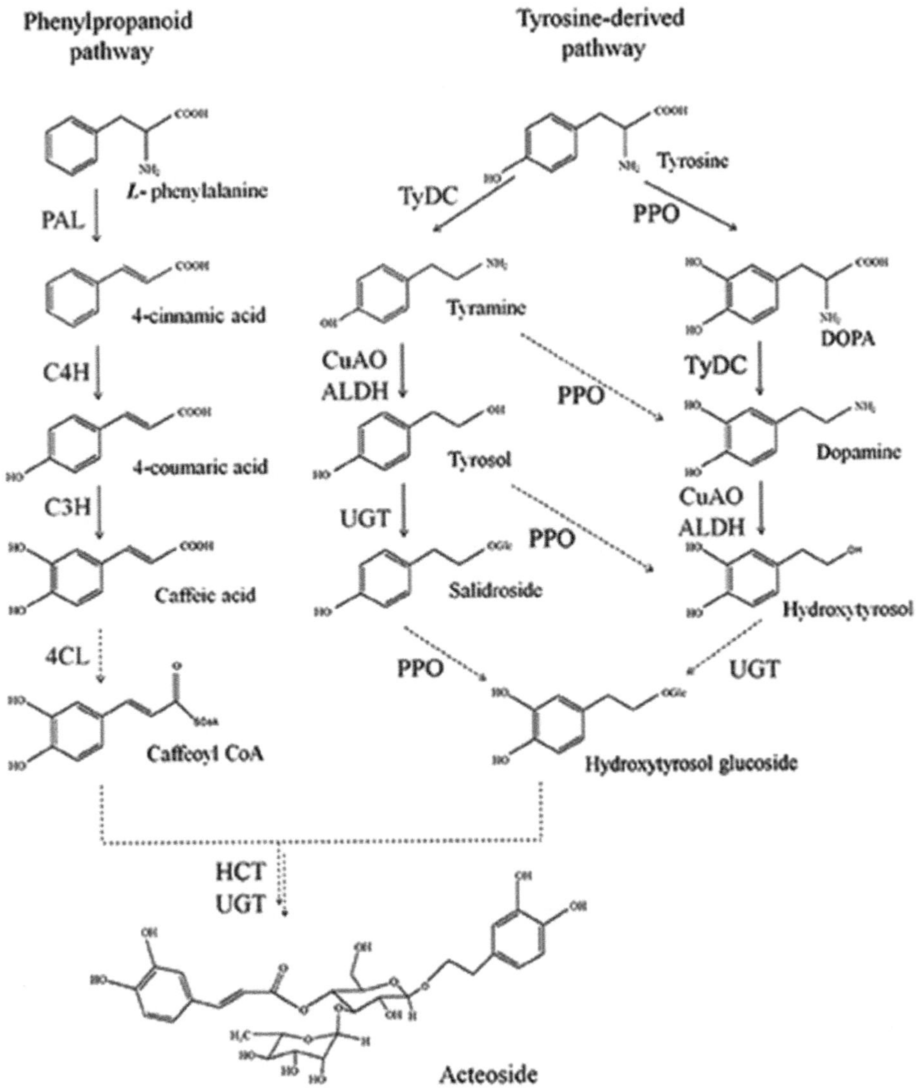

Fig. 5. Biosynthesis pathway of glycosides (Wang et al., 2017).

defense chemicals tested in *P. lanceolata* were unaffected by mycorrhizal infection (Fontana et al., 2009).

7.5. Saponins

Saponins are glycosides that are found in most medicinal plants and have a high molecular weight (Nasseri et al., 2020). Their structure consists of a sugar moiety attached to a steroid or triterpene. They are found in plants in the parts of roots, tuber,

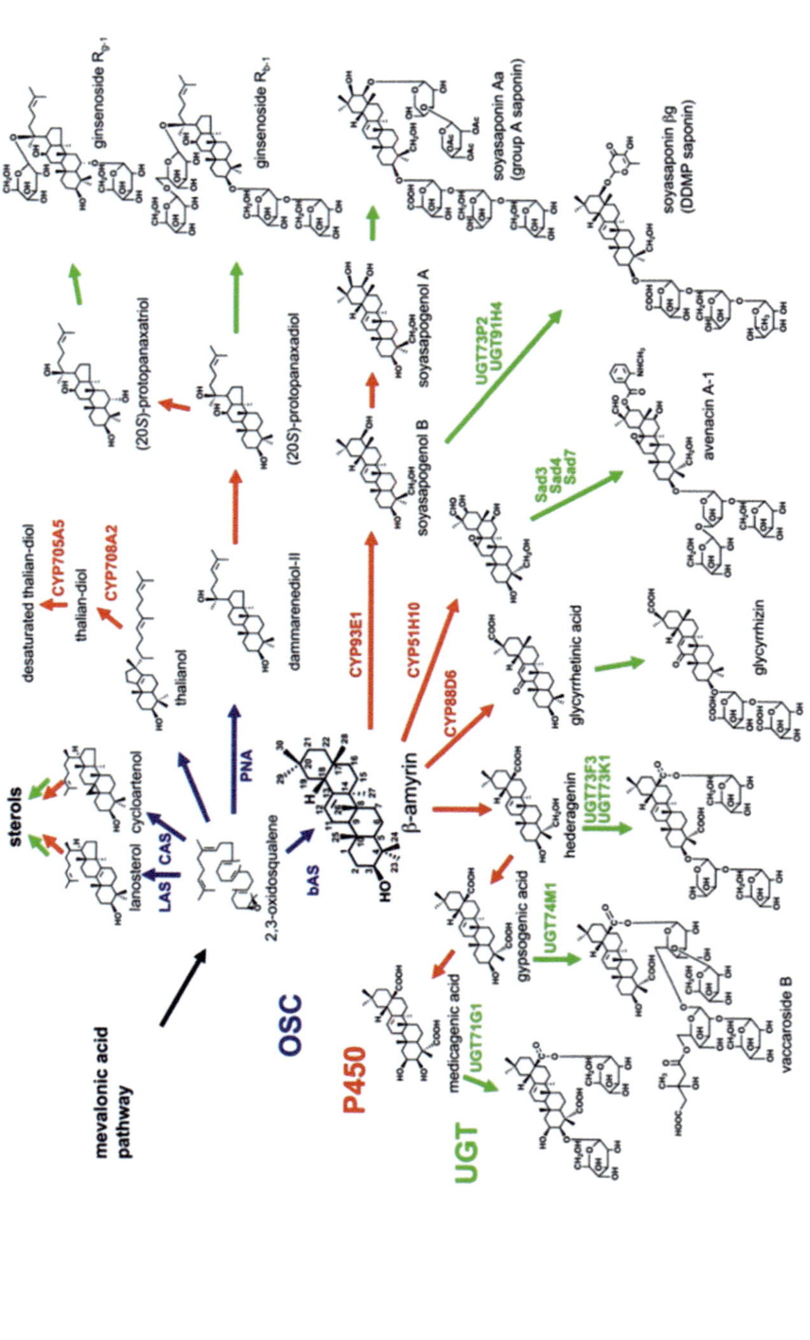

Fig. 6. The biosynthetic pathway of saponins production (Sawai and Saito, 2011). This figure is under the terms of the Creative Commons Attribution License (CC BY).

leaf bark, seed and fruit (Figure 6) (Liu et al., 2019; Siddiqui et al., 2021). Steroidal saponins are mainly found in monocots and triterpene saponins in dicots. Among the plants that contain saponin compounds and play a major role in human diet, we can mention soybean, garlic, spinach, and lentils.

Other plants that contain saponin compounds, but are not used as food, but are used as a medicinal plant or as a combination in medicinal materials, include fenugreek, alfalfa and licorice(Adjou et al., 2021; Radha and Supritha, 2021). Among the most important medicinal properties of saponins are anti-inflammatory, immunization and antifungal properties (Marrelli et al., 2016). In addition, they are also used in baking, making soft drinks, and preparing cosmetics (Chen et al., 2021; Yao et al., 2021). The *Passiflora alata* Phyto mass presented in one study has a high concentration of bioactive components, including saponins and phenols (Muniz et al., 2021).

Seven AMF, including *Acaulospora deligata*, *Glomus aggregatum*, *G feugianum*, *G fasciculatum*, *G rubiforme*, *Gigaspora margarita*, and *Scutellospora heterogama*, were collected from the rhizosphere soils of *Wedilia chinensis* (Osbeck) Merril, and their efficacy was tested in a glasshouse experiment. On the same host, they were assessed for increased levels of acid and alkaline phosphatase, nitrate reductase, peroxidase, and polyphenol oxidase activities as well as total phenols, ortho dihydroxy phenols, flavonoids, alkaloids, tannins, and saponins. The amount of phytochemical component and enzyme concentrations changed according on the kind of AM fungus associated with the *W. chinensis* seedlings. Analysis carried out 90 days after inoculation revealed that *Glomus fasciculatum* had the greatest favorable effect in terms of enhanced phytochemical components and defense-related enzyme levels (Nisha and Rajeshkumar, 2010).

8. Conclusion

Discovering and obtaining effective factors to increase medicinal properties and the concentration of effective substances in medicinal plants has always been considered by experts in different fields of agriculture, human and livestock. Although the production of secondary metabolites is under the control of genes, their production is significantly influenced by environmental conditions, among which the most important factors are the physical and chemical characteristics of the soil, beneficial biological agents and microbes in the soil include AMF and low and high consumption nutrients. Arbuscular mycorrhizal (AM) is a mutualistic relationship between plant roots and fungus that plays an important role in nutrient absorption, water retention, and plant growth promotion. AM has also been found in studies to boost the synthesis of secondary metabolites in plants. To respond to the increasing demand for plants used in the pharmaceutical industry, the emphasis in recent research has been on developing new methods to improve the performance and quality of effective plant substances, and the use of AMF during planting can be useful and effective in enhancing the production of medicinal compounds.

References

Aali, E., Mahmoudi, R., Kazeminia, M., Hazrati, R. & Azarpey, F. (2017). Essential oils as natural medicinal substances, Review article. [Review Article]. Tehran University Medical Journal 75(7), 480–489.

Adjou, E.S., Koudoro, A.Y., Nonviho, G., Ahoussi, E.D. & Sohounhloue, D.C. (2021). Phytochemical profile and potential pharmacological properties of leaves extract of Senna italica Mill. American Journal of Pharmacological Sciences 9(1), 36–39. https://doi.org/10.12691/ajps-9-1-3

Ahmadi, S. (2022). Antibacterial and antifungal activities of medicinal plant species and endophytes. Cellular, Molecular and Biomedical Reports 2(2), 109–115. https://doi.org/10.55705/cmbr.2022.340532.1042

Ahmed, S.R., Rabbee, M.F., Roy, A., Chowdhury, R., Banik, A. et al. (2021). Therapeutic promises of medicinal plants in Bangladesh and their bioactive compounds against ulcers and inflammatory diseases. Plants 10(7), 1348. https://doi.org/10.3390/plants10071348

Akiyama, K., Matsuoka, H. & Hayashi, H. (2002). Isolation and identification of a phosphate deficiency-induced C-glycosylflavonoid that stimulates arbuscular mycorrhiza formation in melon roots. Molecular Plant-Microbe Interactions 15(4), 334–340. https://doi.org/10.1094/MPMI.2002.15.4.334

Alavi, M., Hamblin, M.R., Aghaie, E., Mousavi, S.A.R. & Hajimolaali, M. (2023). Antibacterial and antioxidant activity of catechin, gallic acid, and epigallocatechin-3-gallate, focus on nanoformulations. Cellular, Molecular and Biomedical Reports 3(2), 62–72. https.//doi.org/10.55705/cmbr.2022.353962.1052

Aleebrahim-Dehkordy, E., Nasri, H., Baradaran, A., Nasri, P., Tamadon, M.R., Hedaiaty, M. et al. (2017). Medicinal plants, effective plant compounds (compositions) and their effects on stomach cancer. International Journal of Preventive Medicine 8. https://doi.org/10.4103/ijpvm.IJPVM_4_17

Alizadeh Salteh, S., Taghizadeh Baghchejooghi, H. & Matloobi, M. (2023). The effect of fulvic acid and triacantanol foliar application on some biochemical and physiological properties and active ingredients of *Calendula officinalis*. Journal of Horticultural Science, In Press. https://doi.org/10.22067/jhs.2022.74365.1119

Almasi, F. (2021). Organic fertilizer effects on morphological and biochemical traits and yield in coriander (*Coriandrum sativum* L.) as an industrial and medicinal plant. Agrotechniques in Industrial Crops 1(1), 19–23. https://doi.org/10.22126/etic.2021.6476.1011

Andrino, A., Guggenberger, G., Kernchen, S., Mikutta, R., Sauheitl, L. & Boy, J. (2021). Production of organic acids by Arbuscular Mycorrhizal Fungi and their contribution in the mobilization of phosphorus bound to iron oxides. Frontiers in Plant Science 12, 661842. https://doi.org/10.3389/fpls.2021.661842

Arsana, I.N., Juliasih, N.K.A., Ayu Sauca Sunia, Widyantari, A.A., Suriani, N.L. & Manto, A. (2022). GC-MS analysis of the active compound in ethanol extracts of white pepper (*Piper nigrum* L.) and pharmacological effects. Cellular, Molecular and Biomedical Reports 2(3), 151–161. https://doi.org/10.55705/cmbr.2022.351720.1051

Bagherniya, M., Nobili, V., Blesso, C.N. & Sahebkar, A. (2018). Medicinal plants and bioactive natural compounds in the treatment of non-alcoholic fatty liver disease: A clinical review. Pharmacological Research 130, 213–240. https://doi.org/10.1016/j.phrs.2017.12.020

Bahadur, A., Batool, A., Nasir, F., Jiang, S., Mingsen, Q., Zhang, Q. et al. (2019). Mechanistic insights into Arbuscular Mycorrhizal Fungi-mediated drought stress tolerance in plants. International Journal of Molecular Sciences 20(17). https://doi.org/10.3390/ijms20174199

Bakal, S.N., Bereswill, S. & Heimesaat, M.M. (2017). Finding novel antibiotic substances from medicinal plants—Antimicrobial properties of *Nigella sativa* directed against multidrug resistant bacteria. European Journal of Microbiology and Immunology 7(1), 92–98. https://doi.org/10.1556/1886.2017.00001

Bartwal, A., Mall, R., Lohani, P., Guru, S. & Arora, S. (2013). Role of secondary metabolites and brassinosteroids in plant defense against environmental stresses. Journal of Plant Growth Regulation 32, 216-232. https://doi.org/10.1007/s00344-012-9272-x

Bécard, G., Douds, D. & Pfeffer, P. (1992). Extensive in vitro hyphal growth of vesicular-arbuscular mycorrhizal fungi in the presence of CO_2 and flavonols. Applied and Environmental Microbiology 58(3), 821–825. https://doi.org/10.1128/aem.58.3.821-825.1992

Begum, N., Qin, C., Ahanger, M.A., Raza, S., Khan, M.I., Ashraf, M. et al. (2019). Role of *Arbuscular Mycorrhizal* fungi in plant growth regulation, implications in abiotic stress tolerance. Frontiers in Plant Science 10, 1068. https://doi.org/10.3389/fpls.2019.01068

Beran, F., Köllner, T.G., Gershenzon, J. & Tholl, D. (2019). Chemical convergence between plants and insects, biosynthetic origins and functions of common secondary metabolites. New Phytologist 223(1), 52–67. https://doi.org/10.1111/nph.15718

Bhambhani, S., Kondhare, K.R. & Giri, A.P. (2021). Diversity in chemical structures and biological properties of plant alkaloids. Molecules 26(11), 3374. https://doi.org/10.3390/molecules26113374

Buer, C.S., Imin, N. & Djordjevic, M.A. (2010). Flavonoids, new roles for old molecules. Journal of Integrative Plant Biology 52(1), 98–111. https://doi.org/10.1111/j.1744-7909.2010.00905.x

Carlsen, S.C.K., Understrup, A., Fomsgaard, I.S., Mortensen, A.G. & Ravnskov, S. (2008). Flavonoids in roots of white clover, interaction of arbuscular mycorrhizal fungi and a pathogenic fungus. Plant and Soil 302(1), 33–43. https://doi.org/10.1007/s11104-007-9452-9

Carlucci, A., Raimondo, M.L., Colucci, D. & Lops, F. (2022). Streptomyces albidoflavus strain CARA17 as a biocontrol agent against fungal soil-borne pathogens of fennel plants. Plants (Basel) 11(11). https://doi.org/10.3390/plants11111420

Chabot, S., Bel-Rhlid, R., Chenevert, R. & Piché, Y. (1992). Hyphal growth promotion *in vitro* of the VA mycorrhizal fungus, Gigaspora margarita Becker & Hall, by the activity of structurally specific flavonoid compounds under CO_2-enriched conditions. New Phytologist 122(3), 461–467. https://doi.org/10.1111/j.1469-8137.1992.tb00074.x

Chaghakaboodi, Z., Nasiri, J. & Farahani, S. (2022). Fumigation toxicity of the essential oils of Ferula persica against *Tribolium castaneum* and *Ephestia kuehniella*. Agrotechniques in Industrial Crops 2(3), 123–130. https://doi.org/10.22126/atic.2022.8344.1068

Chandrasekaran, M. (2022). Arbuscular Mycorrhizal Fungi mediated enhanced biomass, root morphological traits and nutrient uptake under drought stress, a meta-analysis. Journal of Fungi (Basel) 8(7). https://doi.org/10.3390/jof8070660

Chen, W., Ye, T., Sun, Q., Niu, T. & Zhang, J. (2022). *Arbuscular mycorrhizal* fungus alleviates anthracnose disease in tea seedlings. Frontiers in Plant Science 13, 1058092. https://doi.org/10.3389/fpls.2022.1058092

Chen, Y., Wu, J., Yu, D. & Du, X. (2021). Advances in steroidal saponins biosynthesis. Planta 254(5), 91. https://doi.org/10.1007/s00425-021-03732-y

Chiej, R. (1984). The Macdonald Encyclopedia of Medicinal Plants. Macdonald & Co (Publishers) Ltd. 448 Pages.

Chikezie, P., Ibegbulem, C. & Mbagwu, F. (2015). Medicinal potentials and toxicity concerns of bioactive principles. Medicinal and Aromatic Plants 4(3), 1–15. https://doi.org/10.4172/2167-0412.1000202

Chiu, C.H. & Paszkowski, U. (2019). Mechanisms and impact of symbiotic phosphate acquisition. Cold Spring Harb Perspect Biology 11(6). https://doi.org/10.1101/cshperspect.a034603

Chukwuka, K., Ikheloa, J., Okonko, I., Moody, J. & Mankinde, T. (2011). The antimicrobial activities of some medicinal plants on *Escherichia coli* as an agent of diarrhea in livestock. Advances in Applied Science Research 2(4), 37–48.

Chun, S.C., Paramasivan, M. & Chandrasekaran, M. (2018). Proline accumulation influenced by osmotic stress in *Arbuscular Mycorrhizal* symbiotic plants. Frontier Microbiology 9, 2525. https://doi.org/10.3389/fmicb.2018.02525

Dave, A., Ingle, S. Perveen, K., Bukhari, N.A., Sayyed, R. & Mastinu, A. (2024). Harnessing plant growth-promoting and wilt-controlling biopotential of a consortium of actinomycetes and mycorrhizae in pigeon pea. Journal of Phytopathology 172, e13399 https://doi.org/10.1111/jph.13399

De La Peña, R. & Sattely, E.S. (2021). Rerouting plant terpene biosynthesis enables momilactone pathway elucidation. Nature Chemical Biology 17(2), 205–212. https://doi.org/10.1038/s41589-020-00669-3

Del Rio, D., Rodriguez-Mateos, A., Spencer, J.P., Tognolini, M., Borges, G. & Crozier, A. (2013). Dietary (poly) phenolics in human health, structures, bioavailability, and evidence of protective effects against chronic diseases. Antioxidants & Redox Signaling 18(14), 1818–1892. https://doi.org/10.1089/ars.2012.4581

Dewick, P.M. (2002). Medicinal Natural Products: A Biosynthetic Approach. John Wiley & Sons. 507 Pages.

Dong, F., Hu, J., Shi, Y., Liu, M., Zhang, Q. & Ruan, J. (2019). Effects of nitrogen supply on flavonol glycoside biosynthesis and accumulation in tea leaves (*Camellia sinensis*). Plant Physiology and Biochemistry 138, 48–57. https://doi.org/10.1016/j.plaphy.2019.02.017

Doosti-Moghaddam, M., Miri, H.R., Ghahghaei, A., Hajinezhad, M.R. & Saboori, H. (2022). Effect of unripe fruit extract of Momordica charantia on total cholesterol, total triglyceride, and blood lipoproteins in the blood of rats with hyperlipidemia. Cellular, Molecular and Biomedical Reports 2(2), 74–86. https://doi.org/10.55705/cmbr.2022.338806.1038

Fazeli-Nasab, B., Sirousmehr, A., Mirzaei, N. & Solimani, M. (2017). Evaluation of total phenolic, flavonoid content, and antioxidant activity of Leaf and Fruit in 14 different genotypes of *Ziziphus mauritiana* L. in the South of Iran. Eco-Phytochemical Journal of Medicinal Plants 4(4), 1–14.

Fazeli-Nasab, B. & Mirzaei, N. (2018). Evaluation of total phenol and flavonoid content in a wide range of local and imported plants. Scientific Journal of Ilam University of Medical Sciences 26(2), 141–154. http, //doi.org/10.29252/sjimu.26.2.141

Fazeli-Nasab, B., Valizadeh, M. & Beigomi, M. (2022). Evaluation of antioxidant and antimicrobial activity of some medicinal plant extracts on *Escherichia coli* isolated from poultry feces. Journal of Medicinal Plants and Byproducts 11(2), 265–275. https://doi.org/10.22092/jmpb.2021.353243.1319

Fazeli-Nasab, B., Saeidi, S., Fazeli, F., Bidarnamani, F. & Beigomi, Z. (2023). Investigation of antimicrobial and antioxidant activity of 4 medicinal plants on 10 standard strains. New Findings in Veterinary Microbiology 5(2), 19–30. https://doi.org/10.22034/nfvm.2022.361192.1153

Feizi, H., Kamali, M. & Rezvani-Moghaddam, P. (2013). The effects of different concentrations of nanoparticles and nano titanium dioxide on seed germination and seedling growth of fennel. The first National Conference on the use of Medicinal Plants and Traditional Medicine, Lifestyle, University of Torbat-e Heidariyeh, Iran.

Fiori, A., Schwan-Estrada, K., Stangarlin, J., Vida, J., Scapim, C., Cruz, M. et al. (2000). Antifungal activity of leaf extracts and essential oils of some medicinal plants against Didymella bryoniae. Journal of Phytopathology 148(7–8), 483–487.

Fontana, A., Reichelt, M., Hempel, S., Gershenzon, J. & Unsicker, S.B. (2009). The effects of arbuscular mycorrhizal fungi on direct and indirect defense metabolites of *Plantago lanceolata* L. Journal of Chemical Ecology 35, 833–843. https://doi.org/10.1007/s10886-009-9654-0

Foroughi, A., Pournaghi, P., Najafi, F., Zangeneh, A., Zangeneh, M.M. & Moradi, R. (2017). Medicinal plants: Antibacterial effects and chemical composition of essential oil of *Foeniculum vulgare*. International Journal of Current Pharmaceutical Review and Research 8(1), 13–17.

Forouzandeh, M., Mohkami, Z. & Fazeli-Nasab, B. (2019). Evaluation of biotic elicitors foliar application on functional changes, physiological and biochemical parameters of fennel (*Foeniculum vulgare*). International Journal of Plant Production 25(4), 49–65. https://doi.org/10.22069/jopp.2018.14077.2262

Ganjali, S., Khajeh, H., Gholami, Z., Jomeh-ghasemabadi, Z. & Fazeli-Nasab, B. (2022). Evaluation of dormancy failure of Datura stramonium plant seeds under the influence of different treatments. Agrotechniques in Industrial Crops 2(1), 32–41. https://doi.org/10.22126/atic.2022.7656.1049

Ghamarnia, H., Palash, M. & Dousti, B. (2022). Camelina zoning for different climate conditions in Kurdistan Province. Agrotechniques in Industrial Crops 2(1), 49–56. https://doi.org/10.22126/atic.2022.7903.1056

Ghasemi, A. (2009). Medicinal and Aromatic Plants. Islamic Azad University Press of ShahreKord, pages 541.

Gosling, P., Hodge, A., Goodlass, G. & Bending, G.D. (2006). Arbuscular mycorrhizal fungi and organic farming. Agriculture, Ecosystems & Environment 113(1), 17–35. https://doi.org/10.1016/j.agee.2005.09.009

Goyal, S., Arora, J. & Ramawat, K.G. (2014). Biotechnological approaches to medicinal plants of Aravalli Hills. Conservation and Scientific Validation of Biological Activities. 4, 203–245. https://doi.org/10.1007/978-3-319-09381-9_11

Gürtler, H., Pedersen, R., Anthoni, U., Christophersen, C., Nielsen, P.H., Wellington, E.M. et al. (1994). Albaflavenone, a sesquiterpene ketone with a zizaene skeleton produced by a streptomycete with a new rope morphology. The Journal of Antibiotics 47(4), 434–439. https://doi.org/10.7164/antibiotics.47.434

Hamidian, M., Movahhedi-Dehnavi, M., Sayyed, R.Z., Almalki, W.H., Gafur, A. & Fazeli-Nasab, B. (2023). Co-inoculation of mycorrhiza and methyl jasmonate regulates morpho-physiological and antioxidant responses of *Crocus sativus* (Saffron) under salinity stress conditions. Scientific Reports 13(1), 7378. https://doi.org/10.1038/s41598-023-34359-6

Hasanloo, T., Jafarkhani Kermani, M., Dalvand, Y. & Rezazadeh, S. (2019). A complete review on the genus vaccinium and Iranian Ghareghat. [Review]. Journal of Medicinal Plants 18(72), 46–65. https://doi.org/10.29252/jmp.4.72.46

Hashemi, M. (2014). The effects of using medicinal plants on rumen fermentation parameters in ruminants. International Journal of Advanced Biological and Biomedical Research 2(4), 1318–1327.

Hashemi, S., Asgarpanah, J., Alaee, Z., Sadeghian, S., Hasani, H. & Azimi, A. (2014). In vitro antifungal activity of four medicinal plants used in Iranian Traditional Medicine. Research Journal of Pharmacognosy 1(1), 39–43.

Heleno, S.A., Martins, A., Queiroz, M.J.R. & Ferreira, I.C. (2015). Bioactivity of phenolic acids, metabolites versus parent compounds, A review. Food Chemistry 173, 501–513. https://doi.org/10.1016/j.foodchem.2014.10.057

Hodge, A. & Storer, K, (2015), Arbuscular mycorrhiza and nitrogen, implications for individual plants through to ecosystems. Plant and Soil 386, 1–19. https://doi.org/10.1007/s11104-014-2162-1

Hoseini, A., Salehi, A., Sayyed, R.Z., Balouchi, H., Moradi, A., Piri, R. et al. (2022). Efficacy of biological agents and filler seed coating in improving drought stress in anise

[Original Research]. Frontiers in Plant Science 13, 955512. https://doi.org/10.3389/fpls.2022.955512

Hosseini, E.S., Majidi, M.M., Ehtemam, M.H. & Hughes, N. (2023). Characterization of fennel germplasm for physiological persistence and drought recovery, Association with biochemical properties. Plant Physiology and Biochemistry 194, 499–512. https://doi.org/10.1016/j.plaphy.2022.11.037

Islam, A., Islam, M.S., Uddin, M.N., Hasan, M.M.I. & Akanda, M.R. (2020). The potential health benefits of the isoflavone glycoside genistin. Archives of Pharmacal Research 43, 395–408. https://doi.org/10.1007/s12272-020-01233-2

Jabborova, D., Annapurna, K., Azimov, A., Tyagi, S., Pengani, K.R., Sharma, S. et al. (2022). Co-inoculation of biochar and arbuscular mycorrhizae for growth promotion and nutrient fortification in soybean under drought conditions. Frontiers in Plant Sciences 13, 947547. https://doi.org/10.3389/fpls.2022.947547

Kanakdande, A.P. & Jadhav, P.B. (2023). Anti-urinary tract infection activity of selected herbal extract towards isolated Kosakonia cowanii (OQ 073698). Cellular, Molecular and Biomedical Reports 3(2), 114–121. https.//doi.org/10.55705/cmbr.2023.385428.1099

Kapoor, R., Anand, G., Gupta, P. & Mandal, S. (2017). Insight into the mechanisms of enhanced production of valuable terpenoids by arbuscular mycorrhiza. Phytochemistry Reviews 16, 677–692.

Karre, S., Kumar, A., Yogendra, K., Kage, U., Kushalappa, A. & Charron, J.-B. (2019). HvWRKY23 regulates flavonoid glycoside and hydroxycinnamic acid amide biosynthetic genes in barley to combat Fusarium head blight. Plant Molecular Biology 100, 591–605. https://doi.org/10.1007/s11103-019-00882-2

Karuppusamy, S. (2009). A review on trends in production of secondary metabolites from higher plants by in vitro tissue, organ and cell cultures. Journal of Medicinal Plants Research 3(13), 1222–1239.

Kaul, S., Gupta, S., Ahmed, M. & Dhar, M.K. (2012). Endophytic fungi from medicinal plants, a treasure hunt for bioactive metabolites. Phytochemistry Reviews 11(4), 487–505. https://doi.org/10.1007/s11101-012-9260-6

Kaur, S. & Suseela, V. (2020). Unraveling arbuscular mycorrhiza-induced changes in plant primary and secondary metabolome. Metabolites 10(8), 335. https://doi.org/10.3390/metabo10080335

Kazemi, M., Rostami, H. & Shafiei, S. (2012). Antibacterial and antifungal activity of some medicinal plants from Iran. Journal of Plant Science 7, 55–66.

Keyes, S., van Veelen, A., McKay Fletcher, D., Scotson, C., Koebernick, N., Petroselli, C. et al. (2022). Multimodal correlative imaging and modelling of phosphorus uptake from soil by hyphae of mycorrhizal fungi. New Phytologist 234(2), 688–703. https://doi.org/10.1111/nph.17980

Kheiri, A. & Salehi, F. (2014). Effective medicinal plants in the prevention and treatment of various types of cancer. Paper presented at the First National Conference on Medicinal Plants, Traditional Medicine, and Organic Agriculture, Hamedan, Iran.

Kheiry, A., Tori, H. & Mortazavi, N. (2017). Effects of drought stress and jasmonic acid elicitors on morphological and phytochemical characteristics of peppermint (*Mentha piperita* L.). Scientific Journal Management System 33(2), 268–280. https://doi.org/10.22092/ijmapr.2017.106481.1783

Kheyrodin, H., Jami, R. & Rehman, F.U. (2022). Cellular structure and molecular functions of plants, animals, bacteria, and viruses. Cellular molecular Biomedical Reports 2(1), 33–41. Https://doi.org/10.55705/cmbr.2022.330941.1021

Kohl, L., Lukasiewicz, C.E. & van der Heijden, M.G. (2016). Establishment and effectiveness of inoculated *Arbuscular mycorrhizal* fungi in agricultural soils. Plant Cell Environ 39(1), 136–146. https://doi.org/10.1111/pce.12600

Kytidou, K., Artola, M., Overkleeft, H.S. & Aerts, J.M. (2020). Plant glycosides and glycosidases, a treasure-trove for therapeutics. Frontiers in Plant Science 11, 357. https://doi.org/10.3389/fpls.2020.00357

Li, Y., Xu, J., Hu, J., Zhang, T., Wu, X. & Yang, Y. (2022). *Arbuscular mycorrhizal* fungi and glomalin play a crucial role in soil aggregate stability in Pb-contaminated soil. International Journal of Environmental Research and Public Health 19(9). https://doi.org/10.3390/ijerph19095029

Libik-Konieczny, M., Capecka, E., Tuleja, M. & Konieczny, R. (2021). Synthesis and production of steviol glycosides, recent research trends and perspectives. Applied Microbiology and Biotechnology 105(10), 3883–3900. https://doi.org/10.1007/s00253-021-11306-x

Lichman, B.R. (2021). The scaffold-forming steps of plant alkaloid biosynthesis. Natural Product Reports 38(1), 103–129. https://doi.org/10.1039/D0NP00031K

Liu, J., Tan, Y., Zhou, H., Mundo, J.L.M. & McClements, D.J. (2019). Protection of anthocyanin-rich extract from pH-induced color changes using water-in-oil-in-water emulsions. Journal of Food Engineering 254, 1–9. https://doi.org/10.1016/j.jfoodeng.2019.02.021

Liu, Y., Tikunov, Y., Schouten, R.E., Marcelis, L.F.M., Visser, R.G.F. & Bovy, A. (2018). Anthocyanin biosynthesis and degradation mechanisms in solanaceous vegetables: A review. [Review]. Frontiers in Chemistry 6. https://doi.org/10.3389/fchem.2018.00052

Liu, Y., Zhao, G., Li, X., Shen, Q., Wu, Q., Zhuang, J. et al. (2020). Comparative analysis of phenolic compound metabolism among tea plants in the section Thea of the genus Camellia. Food Research International 135, 109276. https://doi.org/10.1016/j.foodres.2020.109276

Lu, C.C., Guo, N., Yang, C., Sun, H.B. & Cai, B.Y. (2020). Transcriptome and metabolite profiling reveals the effects of Funneliformis mosseae on the roots of continuously cropped soybeans. BMC Plant Biology 20(1), 479. https://doi.org/10.1186/s12870-020-02647-2

Ma, W.Y., Qin, Q.Y., Zou, Y.N., Kuca, K., Giri, B., Wu, Q.S. et al. (2022. *Arbuscular mycorrhiza* induces low oxidative burst in drought-stressed walnut through activating antioxidant defense systems and heat shock transcription factor expression. Frontiers in Plant Science 13, 1089420. https://doi.org/10.3389/fpls.2022.1089420

Manandhar, S., Luitel, S. & Dahal, R.K. (2019). In vitro antimicrobial activity of some medicinal plants against human pathogenic bacteria. Journal of Tropical Medicine 2019, Article ID, 1895340. https://doi.org/10.1155/2019/1895340

Mandal, S., Evelin, H., Giri, B., Singh, V.P. & Kapoor, R. (2013). Arbuscular mycorrhiza enhances the production of stevioside and rebaudioside-A in *Stevia rebaudiana* via nutritional and non-nutritional mechanisms. Applied Soil Ecology 72, 187–194. https://doi.org/10.1016/j.apsoil.2013.07.003

Marrelli, M., Conforti, F., Araniti, F. & Statti, G.A. (2016). Effects of saponins on lipid metabolism: A review of potential health benefits in the treatment of obesity. Molecules 21(10), 1404. https://doi.org/10.3390/molecules21101404

Mata, R., Figueroa, M., Navarrete, A. & Rivero-Cruz, I. (2019). Chemistry and biology of selected Mexican medicinal plants. pp. 1–142. *In*: Kinghorn, A.D., Falk, H., Gibbons, S., Kobayashi, J.I., Asakawa, Y., Liu, J.-K. [eds.]. Progress in the Chemistry of Organic Natural Products 108. Cham, Springer International Publishing. https://doi.org/10.1007/978-3-030-01099-7_1

McMurry, J.E. (2014). Organic Chemistry with Biological Applications. Cengage Learning. 1224 Pages.

Miao, Y., Lin, Y., Chen, Z., Zheng, H., Niu, Y., Kuzyakov, Y. et al. (2022). Fungal key players of cellulose utilization, Microbial networks in aggregates of long-term fertilized soils disentangled using (13)C-DNA-stable isotope probing. Science Total Environment 832, 155051. https://doi.org/10.1016/j.scitotenv.2022.155051

Mickymaray, S., Al Aboody, M.S., Rath. P.K., Annamalai. P. & Nooruddin T. (2016). Screening and antibacterial efficacy of selected Indian medicinal plants. Asian Pacific Journal of Tropical Biomedicine 6(3), 185–191. https://doi.org/10.1016/j.apjtb.2015.12.005

Mirzaei, A.R. & Shakoory-Moghadam, V. (2022). Bioinformatics analysis and pharmacological effect of Stevia rebaudiana in the prevention of type-2 diabetes. Cellular, Molecular and Biomedical Reports 64–73. https://doi.org/10.55705/cmbr.2022.336232.1035

Muniz, B.C., Falcão, E.L., de Paula Monteiro, R., dos Santos, E.L., Bastos Filho, C.J.A. & da Silva, F.S.B. (2021). *Acaulospora longula* Spain & NC Schenck: A low-cost bioinsumption to optimize phenolics and saponins production in *Passiflora alata* Curtis. Industrial Crops and Products 167, 113498. https://doi.org/10.1016/j.indcrop.2021.113498

Nahuelcura, J., Ruiz, A., Gomez, F. & Cornejo, P. (2022). The effect of arbuscular mycorrhizal fungi on the phenolic compounds profile, antioxidant activity, and grain yields in wheat cultivars growing under hydric stress. Journal of the Science of Food and Agriculture 102(1), 407–416. https://doi.org/10.1002/jsfa.11370

Nasseri, M.A., Keshtkar, H., Kazemnejadi, M. & Allahresani, A. (2020). Phytochemical properties and antioxidant activity of Echinops persicus plant extract, green synthesis of carbon quantum dots from the plant extract. SN Applied Sciences 2, 1–12. https://doi.org/10.1007/s42452-020-2466-0

Ninkuu, V., Zhang, L,, Yan, J., Fu, Z., Yang, T. & Zeng, H. (2021). Biochemistry of terpenes and recent advances in plant protection. International Journal of Molecular Sciences 22(11), 5710. https://doi.org/10.3390/ijms22115710

Nisha, M.C. & Rajeshkumar, S. (2010). Influence of arbuscular mycorrhizal fungi on biochemical changes in *Wedilla chinensis* (Osbeck) Merril. Ancient Science of Life 29(3), 26. https://pubmed.ncbi.nlm.nih.gov/22557356

Noroozi, V., Yousefzadeh, S., Asilan, K. & Mansourifar, S. (2017). Investigating the variation of essential oil content, chlorophyll, carotenoid, anthocyanin, and flavonoid of (*Mentha longifolia* (L.) Hods. subsp. Longifolia) in several habitats of Marand. Eco-phytochemical Journal of Medicinal Plants 5(1), 52–66.

Orejola, J., Matsuo, Y., Saito, Y. & Tanaka, T. (2017). Characterization of Proanthocyanidin Oligomers of *Ephedra sinica*. Molecules 22(8), 1308. https://doi.org/10.3390/molecules22081308

Orujei, Y., Shabani, L., Sharifi Tehrani, M., Aghababaei, F. & Enteshari, S. (2013). Dual effects of two mycorrhizal fungi on production of glycyrrhizin, total phenolic and total flavonoids compounds in roots of *Glycyrrhiza glabra* L. Iranian Journal of Plant Biology 5(17), 75–88. https://dorl.net/dor/20.1001.1.20088264.1392.5.17.8.8

Othman, L., Sleiman, A. & Abdel-Massih, R.M. (2019). Antimicrobial activity of polyphenols and alkaloids in Middle Eastern plants. Frontiers in Microbiology 10, 911. https://doi.org/10.3389/fmicb.2019.00911

Pagare, S., Bhatia, M., Tripathi, N., Pagare, S. & Bansal, Y. (2015). Secondary metabolites of plants and their role: Overview. Current Trends in Biotechnology and Pharmacy 9(3), 293–304.

Pavinato, P.S., Cherubin, M.R., Soltangheisi, A., Rocha, G.C., Chadwick, D.R. & Jones, D.L. (2020). Revealing soil legacy phosphorus to promote sustainable agriculture in Brazil. Science Report 10(1), 15615. https://doi.org/10.1038/s41598-020-72302-1

Petrussa, E., Braidot, E., Zancani, M., Peresson, C., Bertolini, A., Patui, S. & Vianello, A. (2013). Plant flavonoids—Biosynthesis, transport and involvement in stress responses. International Journal of Molecular Sciences 14(7), 14950–14973. https://doi.org/10.3390/ijms140714950

Piri, R., Moradi, A., Balouchi, H. & Salehi, A. (2019). Improvement of cumin (*Cuminum cyminum*) seed performance under drought stress by seed coating and biopriming. Science Horticulture 257(November), 108667. https://doi.org/10.1016/j.scienta.2019.108667

Piri, R., Moradi, A. & Balouchi, H. (2020). Improvement of salinity stress in cumin (*Cuminum cyminum*) seedling by inoculation with Rhizobacteria. Indian Journal of Agricultural Sciences 90(2), 371–375.

Plouznikoff, K., Asins, M.J., de Boulois, H.D., Carbonell, E.A. & Declerck, S. (2019). Genetic analysis of tomato root colonization by arbuscular mycorrhizal fungi. Annals of Botany 124(6), 933–946. https://doi.org/10.1093/aob/mcy240

Ponce, M.A., Scervino, J.M., Erra-Balsells, R., Ocampo, J.A. & Godeas, A.M. (2004). Flavonoids from shoots and roots of *Trifolium repens* (white clover) grown in the presence or absence of the arbuscular mycorrhizal fungus *Glomus intraradices*. Phytochemistry 65(13), 1925–1930. https://doi.org/10.1016/j.phytochem.2004.06.005

Radha, K. & Supritha, P. (2021). An Investigation on the extraction of phytochemicals and antimicrobial properties of domestic plants found in southern India. Herbal Medicines Journal 6(1), 7–17. https://orcid.org/0000-0001-7517-1080

Rai, S., Omar, A.F., Rehan, M., Al-Turki, A., Sagar, A., Ilyas, N. et al. (2022). Crop microbiome, their role and advances in molecular and omic techniques for the sustenance of agriculture. Planta 257(2), 27. https://doi.org/10.1007/s00425-022-04052-5

Ramachandra Rao, S. & Ravishankar, G.A. (2002). Plant cell cultures: Chemical factories of secondary metabolites. Biotechnology Advances 20(2), 101–153. https://doi.org/10.1016/S0734-9750(02)00007-1

Rashidi, S., Yousefi, A.R., Pouryousef, M. & Goicoechea, N. (2020a). Total phenol, anthocyanin, and terpenoid content, photosynthetic rate, and nutrient uptake of *Solanum nigrum* L. and *Digitaria sanguinalis* L. as affected by arbuscular mycorrhizal fungi inoculation. Weed Biology and Management 20(3), 95–108. https://doi.org/10.1111/wbm.12209

Rashidi, S., Yousefi, A.R., Pouryousef, M. & Nieves Goicoechea, M. (2020b). Effect of three species of mycorrhizal fungus on photosynthesis, growth, and secondary metabolites content of (*Ipomoea purpurea* L.). Iranian Journal of Weed Science 16(2), 99–114. https://doi.org/10.22092/ijws.2020.127905.1352

Reyes-Carmona, J., Yousef, G.G., Martínez-Peniche, R.A. & Lila, M.A. (2005). Antioxidant capacity of fruit extracts of blackberry (*Rubus* sp.) produced in different climatic regions. Journal of Food Science 70(7), s497–s503. https://doi.org/10.1111/j.1365-2621.2005.tb11498.x

Rezacova, V., Zemkova, L., Beskid, O., Puschel, D., Konvalinkova, T., Hujslova, M. & et al. (2018). Little cross-feeding of the mycorrhizal networks shared between C(3)-Panicum bisulcatum and C(4)-Panicum maximum under different temperature regimes. Frontiers in Plant Science 9, 449. https://doi.org/10.3389/fpls.2018.00449

Rola, K., Rozek, K., Chowaniec, K., Blaszkowski, J., Gielas, I., Stanek, M. et al. (2023). Vascular plant and cryptogam abundance as well as soil chemical properties shape microbial communities in the successional gradient of glacier foreland soils. Science of the Total Environment 860, 160550. https://doi.org/10.1016/j.scitotenv.2022.160550

Saboor, A., Muhammad, A.A., Hussain, S., El Enshasy, H.E., Hussain, S., Ahmed, N. et al. (2021). Zinc nutrition and arbuscular mycorrhizal symbiosis effects on maize (*Zea mays* L.) growth and productivity. Saudi Journal of Biological Sciences 28(11), 6339–6351. https://doi.org/10.1016/j.sjbs.2021.06.096

Salehi-Sardoei, A. & Khalili, H. (2022). Nitric oxide signaling pathway in medicinal plants. Cellular, Molecular and Biomedical Reports 2(1), 1–10. https://doi.org/10.55705/cmbr.2022.330292.1019

Sawai, S. & Saito, K. (2011). Triterpenoid Biosynthesis and engineering in plants. [Perspective]. Frontiers in Plant Science 2. https://doi.org/10.3389/fpls.2011.00025

Scervino, J.M., Ponce, M.A., Erra-Bassells, R., Vierheilig, H., Ocampo, J.A., Godeas, A. (2005). Flavonoids exclusively present in mycorrhizal roots of white clover exhibit a different effect on arbuscular mycorrhizal fungi than flavonoids exclusively present in non-mycorrhizal roots of white clover. Journal of Plant Interactions 1(1), 15–22. https://doi.org/10.1080/17429140500192597

Shaafi, B., Mosavi, S.S., Abdollahi, M.R. & Sarikhani, H. (2021). The optimized protocols for production, adaptation, and keeping of the produced artificial seeds from encapsulated lateral buds in Stevia rebaudiana (Bertoni). Agrotechniques in Industrial Crops 1(1), 24–35. https://doi.org/10.22126/etic.2021.6246.1004

Shabani, L., Sabzalian, M.R. & Mostafavi Pour, S. (2016). Arbuscular mycorrhiza affects nickel translocation and expression of ABC transporter and metallothionein genes in Festuca arundinacea. Mycorrhiza 26(1), 67–76. https://doi.org/10.1007/s00572-015-0647-2

Shaghaghi, A., Nejad Ebrahimi, S. & Sonboli, A. (2019). Study of total phenolic, total flavonoid content, and antioxidant potential in various organs of the genus Papaver and Glaucium collected from Iran. Journal of Plant Production Research 26(2), 195–214. https://doi.org/10.22069/jopp.2019.15047.2348

Sharma, M., Bhushan, S., Sharma, D., Kaul, S. & Dhar, M.K. (2023). A brief review of plant cell transfection, gene transcript expression, and genotypic integration for enhancing compound production. Methods in Molecular Biology 2575, 153–179. https://doi.org/10.1007/978-1-0716-2716-7_8

Shojaiee, K., Ganji Moghadam, E. & Jajarmi, V. (2018). Investigation of the effects of culture media and growth regulators concentration on proliferation and rooting of *Rosa damascena* Mill cv Kashan 2. Journal of Medicinal Plants Biotechnology 4(First), 28–36.

Siddiqui, M., Shah, N., Dur-Re-Shahwar, M., Ali, S.Y., Muzammil, A. & Fatima, N. (2021). The phytochemical analysis of some medicinal plants. Liaquat Medical Research Journal 3(1), 1–12.

Silva, A.M.M., Jones, D.L., Chadwick, D.R., Qi, X., Cotta, S.R., Araujo, V. et al. (2023). Can *Arbuscular mycorrhizal* fungi and rhizobacteria facilitate (33)P uptake in maize plants under water stress? Microbiology Research 271, 127350. https://doi.org/10.1016/j.micres.2023.127350

Song, Y., Chen, D., Lu, K., Sun, Z. & Zeng, R. (2015). Enhanced tomato disease resistance primed by arbuscular mycorrhizal fungus. Frontiers in Plant Science 6, 786. https://doi.org/10.3389/fpls.2015.00786

Tabin, S., Gupta, R.C., Kamili, A.N. & Parray, J.A. (2022). Medical and medicinal importance of *Rheum* spp. collected from different altitudes of the Kashmir Himalayan range. Cellular, Molecular and Biomedical Reports 2(3), 187–201. https://doi.org/10.55705/cmbr.2022.349901.1050

Talpur, M., Talpur, F., Balouch, A., Nizamani, S. & Surhio, M. (2017). Analysis and characterization of anthocyanin from phalsa (*Grewia asiatica*). MOJ Food Processing & Technology 5(3), 00127, DOI.

Thirkell, T.J., Cameron, D.D. & Hodge, A. (2016). Resolving the 'nitrogen paradox' of *arbuscular mycorrhizas,* fertilization with organic matter brings considerable benefits for plant nutrition and growth. Plant, Cell & Environment 39(8), 1683–1690. https://doi.org/10.1111/pce.12667

Torres-Cortes, G., Ghignone, S., Bonfante, P. & Schussler, A. (2015). Mosaic genome of endobacteria in arbuscular mycorrhizal fungi, Transkingdom gene transfer in an ancient mycoplasma-fungus association. Proceedings of the National Academy of Sciences USA 112(25), 7785–7790. https://doi.org/10.1073/pnas.1501540112

Tosar, V., Rousseau, G., Decruyenaere, V., Beckers, Y., Larondelle, Y. & Froidmont, E. (2021). Red clover-rich grassland increases equol concentration in eggs from free-range laying hens. British Poultry Science 62(6), 840–845. https://doi.org/10.1080/00071668.2021.1929069

Toussaint, J.P., Smith, F.A. & Smith, S.E. (2007). *Arbuscular mycorrhizal* fungi can induce the production of phytochemicals in sweet basil irrespective of phosphorus nutrition. Mycorrhiza 17(4), 291–297. https://doi.org/10.1007/s00572-006-0104-3

Vafa, Z.N., Sohrabi, Y., Sayyed, R.Z., Suriani, N.L. & Datta, R. (2021). Effects of combinations of rhizobacteria, mycorrhizae, and seaweeds on growth and yields in wheat cultivars under the influence of supplementary irrigation. Plants. https://www.mdpi.com/2223-7747/10/4/811/htm

Vafa, Z.N., Sohrabi, Y., Heidari, G., Rizwan, M. & Sayyed, R.Z. (2024). Wheat growth and yield response are regulated by mycorrhizae application and supplemental irrigation. Chemosphere 364, 143068 https://doi.org/10.1016/j.chemosphere.2024.143068

Valizadeh, M., ur Rehman, F., Hassanzadeh, M.A., Beigomi, M., & Fazeli-Nasab, B. (2021). Investigating the antimicrobial effects of *Glycyrrhiza glabra* and *Salvia officinalis* ethanolic extract against *Helicobacter pylori*. [Research Article]. International Journal of Infection 8(4), e114563. https://doi.org/10.5812/iji.114563

Velasquez, A., Valenzuela, M., Carvajal, M., Fiaschi, G., Avio, L., Giovannetti, M. et al. (2020). The arbuscular mycorrhizal fungus Funneliformis mosseae induces changes and increases the concentration of volatile organic compounds in Vitis vinifera cv. Sangiovese leaf tissue. Plant Physiol Biochem 155, 437–443. https://doi.org/10.1016/j.plaphy.2020.06.048

Verzeaux, J., Hirel, B., Dubois, F., Lea, P.J. & Tetu, T. (2017). Agricultural practices to improve nitrogen use efficiency through the use of arbuscular mycorrhizae: Basic and agronomic aspects. Plant Science 264, 48–56. https://doi.org/10.1016/j.plantsci.2017.08.004

Vierheilig, H., Garcia-Garrido, J., Wyss, U. & Piché, Y. (2000). Systemic suppression of mycorrhizal colonization of barley roots already colonized by AM fungi. Soil Biology and Biochemistry 32(5), 589–595. https://doi.org/10.1016/S0038-0717(99)00155-8

Vierheilig, H., Piché, Y., Buslig, B. & Manthey, J. (2002). Flavonoids in Cell Function by B. Buslig and J. Manthey, Kluwer Academic/Plenum Publishers, 23–38.

Wang, F., Zhi, J., Zhang, Z., Wang, L., Suo, Y., Xie, C. et al. (2017). Transcriptome analysis of salicylic acid treatment in Rehmannia glutinosa hairy roots using RNA-seq technique for identification of genes involved in acteoside biosynthesis [Original Research]. Frontiers in Plant Science 8. https://doi.org/10.3389/fpls.2017.00787

Wang, M., Wang, Z., Guo, M., Qu, L. & Biere, A. (2023). Effects of arbuscular mycorrhizal fungi on plant growth and herbivore infestation depend on availability of soil water and nutrients. Frontiers in Plant Science 14, 1101932. https://doi.org/10.3389/fpls.2023.1101932

Wang, Y., Zhao, Y., Liu, X., Li, J., Zhang, J. & Liu, D. (2022a). Chemical constituents and pharmacological activities of medicinal plants from the Rosa genus. Chinese Herbal Medicines 14(2), 187–209. https://doi.org/10.1016/j.chmed.2022.01.005

Wang, Y.J., He, X.H., Meng, L.L., Zou, Y.N. & Wu, Q.S. (2022b). Extraradical mycorrhizal hyphae promote soil carbon sequestration through difficultly extractable glomalin-related soil protein in response to soil water stress. Microbial Ecology https://doi.org/10.1007/s00248-022-02153-y

Wen, Z., Yang, M., Han, H., Fazal, A., Liao, Y., Ren, R. et al. (2023). Mycorrhizae enhance soybean plant growth and aluminum stress tolerance by shaping the microbiome assembly in an acidic soil. Microbiology Spectrum 11(2), e0331022. https://doi.org/10.1128/spectrum.03310-22

Wink, M. (2012). Medicinal plants: A source of antiparasitic secondary metabolites. Molecules 17(11), 12771–12791. https://doi.org/10.3390/molecules171112771

Wink, M. (2015). Modes of action of herbal medicines and plant secondary metabolites. Medicines 2(3), 251–286. https://doi.org/10.3390/medicines2030251

Winkel-Shirley, B. (2001). Flavonoid biosynthesis: A colorful model for genetics, biochemistry, cell biology, and biotechnology. Plant Physiology 126(2), 485–493. https://doi.org/10.1104/pp.126.2.485

Wipf, D., Krajinski, F., van Tuinen, D., Recorbet, G. & Courty, P.E. (2019). Trading on the arbuscular mycorrhiza market, from arbuscules to common mycorrhizal networks. New Phytology 223(3), 1127–1142. https://doi.org/10.1111/nph.15775

Xie, Q., Yan, F., Hu, Z., Wei, S., Lai, J. & Chen, G. (2018). Accumulation of anthocyanin and its associated gene expression in purple tumorous stem mustard (*Brassica juncea* var. *tumida* Tsen et Lee) sprouts when exposed to light, dark, sugar, and methyl jasmonate. Journal of Agricultural and Food Chemistry 67(3), 856–866. https://doiorg/101021/acsjafc8b04706

Yang, Y., Luo, W., Xu, J., Guan, P., Chang, L., Wu, X. et al. (2022). Fallow land enhances carbon sequestration in glomalin and soil aggregates through regulating diversity and network complexity of *Arbuscular Mycorrhizal* fungi under climate change in relatively high-latitude regions. Frontiers Microbiology 13, 930622. https://doi.org/10.3389/fmicb.2022.930622

Yao, L., Wang, J., He, J., Huang, L. & Gao, W. (2021). Endophytes, biotransforming microorganisms, and engineering microbial factories for triterpenoid saponins production. Critical Reviews in Biotechnology 41(2), 249–272. https://doi.org/10.1080/07388551.2020.1869691

Yeşil, M., Öztürk, I., Duymuş, Z.Y. & Özcan, M.M. (2020). Evaluating the effect of some medicinal plants (*Mentha piperita*, *Ocimum basilicum*, *Rosmarinus officinalis*, *Salvia officinalis*) on whitening of the permanent teeth. Turkish Journal of Agriculture-Food Science and Technology 8(1), 1–6. https://doi.org/10.24925/turjaf.v8i1.1-6.2508

Yousefi, B. & Safari, H. (2022). Evaluation of essential oil, morphological traits, and crop yield in Satureja mutica Fisch. & C.A. Mey. in rainfed cultivation under different row spacing. Agrotechniques in Industrial Crops 2(4), 187–197. https://doi.org/10.22126/atic.2023.8424.1074

Yun, D., Yoon, S.Y., Park, S.J. & Park, Y.J. (2021). The anticancer effect of natural plant alkaloid isoquinolines. International Journal of Molecular Sciences 22(4), 1653. https://doi.org/10.3390/ijms22041653

Zadali, R., Baghery, M., Abbasi, M., Yavari, N., Miran, M. & Ebrahimi, S.N. (2022). Anticonvulsant activity of Iranian medicinal plants and molecular docking studies of isolated phytochemicals. South African Journal of Botany 149, 646–657. https://doi.org/10.1016/j.sajb.2022.06.044

Zamani, S., Ghasemnezhad, A., Alizadeh, M. & Alami, M. (2021). Enhancement of PAL enzyme activity and production of phenylpropanoids and antioxidant contents of *Cynara scolymus* L. callus under the influence of precursor and elicitor. Journal of Plant Bioinformatics and Biotechnology 1(2), 103–113. https://doi.org/10.22034/jpbb.2021.299818.1012

Zhang, L., Jiang, C., Zhou, J., Declerck, S., Tian, C. & Feng, G. (2016). Increasing phosphorus concentration in the extraradical hyphae of Rhizophagus irregularis DAOM 197198 leads to a concomitant increase in metal minerals. Mycorrhiza 26(8), 909–918. https://doi.org/10.1007/s00572-016-0722-3

Zhang, L., Yan, J., Fu, Z., Shi, W., Ninkuu, V., Li, G. et al. (2021a). FoEG1, a secreted glycoside hydrolase family 12 protein from Fusarium oxysporum, triggers cell death and modulates plant immunity. Molecular Plant Pathology 22(5), 522–538. https://doi.org/10.1111/mpp.13041

Zhang, Z.C., Yang, J.Y., Hao, B.H., Hao, L.J., Luo, J.Q., Li, X. et al. (2021b). Potential of *Arbuscular mycorrhizal* fungi, biochar, and combined amendment on sandy soil improvement driven by microbial community. Huan Jing Ke Xue 42(4), 2066–2079. https://doi.org/10.13227/j.hjkx.202008154

Zhao, J., Davis, L.C. & Verpoorte, R. (2005). Elicitor signal transduction leading to production of plant secondary metabolites. Biotechnology Advances 23(4), 283–333. https://doi.org/10.1016/j.biotechadv.2005.01.003

Zhao, R., Guo, W., Bi, N., Guo, J., Wang, L., Zhao, J. & Zhang, J. (2015). Arbuscular mycorrhizal fungi affect the growth, nutrient uptake, and water status of maize (*Zea mays* L.) grown in two types of coal mine spoils under drought stress. Applied Soil Ecology 88, 41–49. https://doi.org/10.1016/j.apsoil.2014.11.016

Zhao, X. & Yan, X.-F. (2006). Effects of arbuscular mycorrhizal fungi on plant secondary metabolism. Chinese Journal of Plant Ecology 30(3), 514. https://doi.org/10.17521/cjpe.2006.0068

Zhao, Y., Cartabia, A., Lalaymia, I. & Declerck, S. (2022). *Arbuscular mycorrhizal* fungi and production of secondary metabolites in medicinal plants. Mycorrhiza 32(3-4), 221–256. https://doi.org/10.1007/s00572-022-01079-0

Zhou, F. & Pichersky, E. (2020). More is better: The diversity of terpene metabolism in plants. Current Opinion in Plant Biology 55, 1–10. https://doi.org/10.1016/j.pbi.2020.01.005

Zhu, H., Zhang, R., Chen, W., Gu, Z., Xie, X., Zhao, H. & Yao, Q. (2015). The possible involvement of salicylic acid and hydrogen peroxide in the systemic promotion of phenolic biosynthesis in clover roots colonized by arbuscular mycorrhizal fungus. Journal of Plant Physiology 178, 27–34. https://doi.org/10.1016/j.jplph.2015.01.016

Zingue, S., Nde, C. B. & Njamen, D. (2015). The methanol-soluble fraction of Millettia macrophylla (Fabaceae) stem bark endowed with estrogenic properties has adverse effects on the male reproductive system of Wistar rats. [Research Support, Non-U.S. Gov't]. Journal of Basic and Clinical Physiology and Pharmacology 26(1), 53–64. https://doi.org/10.1515/jbcpp-2013-0136

Zomorodian, K., Ghadiri, P., Saharkhiz, M.J., Moein, M.R., Mehriar, P., Bahrani, F. et al. (2015). Antimicrobial activity of seven essential oils from Iranian aromatic plants against common causes of oral infections. Journal of Microbiology 8(2), e17766. https://doi.org/10.5812/jjm.17766

Zubek, S., Chmolowska, D., Jamrozek, D., Ciechanowska, A., Nobis, M., Blaszkowski, J. et al. (2019). Monitoring of fungal root colonisation, *arbuscular mycorrhizal* fungi diversity and soil microbial processes to assess the success of ecosystem translocation. Journal of Environmental Management 246, 538–546. https://doi.org/10.1016/j.jenvman.2019.04.044

CHAPTER
3

Symbiotic Fungi in Carbon Cycling and Antioxidant Defenses to Enhance Stress Resilience in Plants

Joy M Joel[1], Riya Johnson[1] and Jos T Puthur[1*]

[1] Plant Physiology and Biochemistry Division, Department of Botany, University of Calicut, C.U. Campus P.O. Kerala – 673635, India

1. Introduction

Earth's climate and its atmosphere is highly influenced by the organic carbon (C) existing on its terrestrial ecosystem as soil organic carbon (SOC). The C content found in the atmosphere and plant biomass both put together is still lesser than the SOC content (Even and Kotrofo, 2024). The steady utilization of land and its resources has significantly affected the global C cycling chain in terms of the amount of C stored in soil or released back to the atmosphere. Moreover, intensive fertilizer usage in the field of agriculture, has weakened the soil quality and has resulted in soil deterioration (Chen et al., 2024). Understanding how to effectively manage and stabilize carbon within soil for long term is essential for predicting future climate change scenarios. The SOC pool largely depends on plant productivity and how photosynthetically fixed carbon is distributed between above- and below-ground biomass (Pan et al., 2024). SOC forms a critical component of terrestrial carbon storage and ecosystem health. Its pool size is not static; rather, it is shaped by dynamic interactions between plant productivity and the fate of carbon fixed through photosynthesis. The carbon distribution is directly influenced by the structure and function of various soil microbial communities inhabiting in it (Brunn et al., 2025). In turn, these microbes play critical and multifaceted roles in regulating SOC fluxes through a variety of natural, interconnected soil processes such as microbial respiration, decomposition, mineralization, and immobilization (Thotakuri et al., 2024, Parasar and Agarwala, 2025). During respiration, microbes break down organic substrates for energy, releasing carbon in the form of carbon dioxide into the atmosphere as a metabolic byproduct. Decomposition involves the enzymatic

*Corresponding author: jtputhur@yahoo.com

DOI: 10.1201/9781003669777-3

degradation of plant and animal residues, transforming complex organic matter into simpler, more bioavailable compounds. Through mineralization, organic carbon is further converted into inorganic forms—such as carbon dioxide and other mineral nutrients—that can be taken up by plants and other soil organisms. In contrast, immobilization temporarily incorporates nutrients and carbon into microbial biomass, effectively sequestering them within the soil matrix and slowing their immediate availability, while contributing to longer-term carbon storage (Bhattacharyya et al., 2022, Solaiman, 2024).

Recent research has accentuated the essential role of plant-associated fungi in regulating soil carbon dynamics, with some studies indicating that these organisms may enhance carbon sequestration by as much as 70% (Basiru and Hijri, 2024, Gachara et al., 2024). Despite their apparent significance, fungal contributions—particularly those of symbiotic fungi—remain underrepresented in global climate models. Among these, arbuscular mycorrhizal fungi (AMF) are especially noteworthy due to their widespread occurrence and symbiotic associations with the majority of terrestrial plant species across diverse ecosystems (Rillig, 2004, Powell and Rilling, 2018). This symbiosis is fundamentally driven by a reciprocal exchange mechanism: plants allocate carbon to the fungi in return for nutrients that are otherwise poorly mobile in the soil, such as ammonium (NH_4^+), phosphorus (P), zinc (Zn), and copper (Cu) (Wang et al., 2017).

AMF are known to receive approximately 4–20% of the host plant's total carbon allocation (Wang et al., 2017). This carbon, together with their extensive extraradical hyphal networks, plays a substantial role in contributing to the active soil carbon pool (Luo et al., 2023). Notably, the intraradical structures of AMF (those within plant roots, excluding the hyphae extending into the soil) collectively store an estimated 0.5 petagrams (Pg) of carbon in the topsoil alone (Shukla et al., 2025). Given the right land management strategies, this carbon pathway mediated by AMF could be strategically leveraged for climate change mitigation (Solaiman, 2024).

However, the persistence of this carbon in soil systems is contingent upon its physical protection within soil aggregates and its mean residence time, both of which determine its stability and long-term sequestration potential. Additionally, rising atmospheric CO_2 levels—a result of continued global carbon emissions—may stimulate plant photosynthesis, increasing carbon inputs to the rhizosphere. This enhanced carbon flow could alter microbial community composition and activity, ultimately influencing the direction and magnitude of terrestrial carbon cycling (Wang et al., 2020).

In this context, the role of AMF in enhancing soil carbon storage is especially relevant, as soils act not only as a carbon sink but also as the foundation for sustainable crop production. However, modern agriculture is increasingly threatened by aggressions of abiotic stresses, which include drought, salinity, heavy metal toxicity, and temperature extremes—which disrupt critical plant physiological functions such as nutrient uptake, water balance, and metabolic activity (Surówka et al., 2020, Ghag et al., 2023, George and Ray, 2023). These stresses are responsible for an estimated 70% reduction in global crop yields, posing a grave risk to food security, particularly as the global population is expected to surpass 9.3 billion by 2050, necessitating a 60% increase in food production (George and Ray, 2023).

The challenge is further intensified by climate change, which has escalated the frequency and severity of extreme weather events, undermining the resilience of cropping systems (Abraham et al., 2014). At the same time, unsustainable agricultural practices—including excessive use of chemical fertilizers accelerate soil degradation and undermine soil fertility, water retention property, and structural stability (Hossain et al., 2020). Together, these interconnected factors emphasize the urgent need for nature-based solutions, such as fungal symbioses, to restore soil health and bolster plant stress tolerance under a changing climate.

In this chapter, we first discuss an overview of the carbon dynamics and then present various antioxidant defense mechanisms attributed to the fungal symbiosis, the synergy of C metabolism and antioxidant pathways. Future climate scenarios may significantly influence the interactions between AMF and other soil microorganisms, ultimately affecting soil carbon sequestration and turnover. However, a deeper understanding of these dynamics is essential. We conclude by offering recommendations for future research aimed at enhancing AMF-driven carbon storage in soil ecosystems, thereby increasing the antioxidant potential to alleviate various stress conditions in plants.

2. The Role of Arbuscular Mycorrhizal Fungi (AMF) in Plant Carbon Dynamics

AMF significantly influence plant carbon cycling processes due to their close, co-evolutionary relationships with terrestrial plants (Fig. 1). Contemporary studies have highlighted that AMF colonization makes plant roots a significant carbon sink, with a notable fraction of photosynthetically fixed carbon being allocated to the fungal symbiont (Wahab et al., 2023). To fully appreciate the carbon-sequestering potential of AMF, it is crucial to unravel the processes and pathways underlying carbon movement from host plant to fungus. Establishment of the mycorrhizal partnership is initiated by intricate molecular signaling. Plants release strigolactones—compounds recognized by AMF spores—prompting fungal germination and directing the hyphal growth toward the roots (Gutjahr, 2014, Lanfranco et al., 2018). Contact between the fungus and root is facilitated by specialized structures known as hyphopodia, which allow the formation of a physical connection. Subsequent cellular reprogramming in the root epidermis enables fungal penetration via a pre-penetration apparatus, guiding the hyphae toward the cortex. Within the cortical cells, highly branched arbuscules develop, enveloped by a periarbuscular membrane that establishes a compartmentalized interface (Genre and Bonfante, 2010). The bi-directional exchange of nutrients and carbon is mediated by these arbuscules. While once thought to be limited to sugars, recent studies have indicated that plant-derived lipids are also transferred to the AMF (Kameoka and Gutjahr, 2022, Bisht et al., 2024).

Early research had highlighted that starch reserves in host roots play a central role in supporting fungal metabolism during symbiosis (Subramanian 1998, Smith and Smith, 2012). However, more recent work with mutants with impaired starch metabolism pathways have shown that normal colonization is possible even without functional starch turnover (Gutjahr et al., 2011), suggesting that other carbon forms

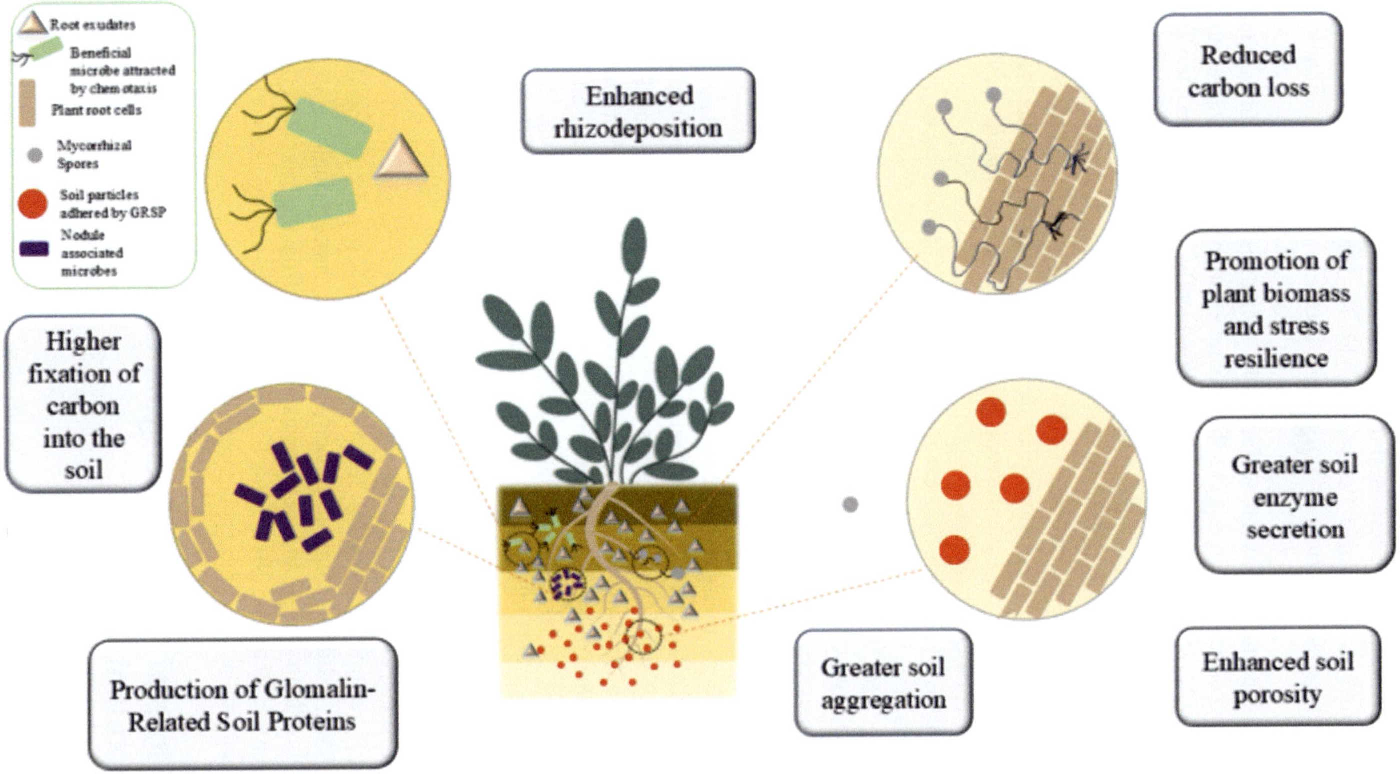

Fig. 1. The figure presents a schematic representation of the diverse interactions occurring among plants and arbuscular mycorrhiza fungi (AMF). These crucial relationships are pivotal for enhanced soil carbon sequestration mediated by the symbiotic partner, which enhances the tolerance capability of the host towards different abiotic stresses.

are predominantly involved. It is now well-established that AMF can assimilate hexoses derived from plant sucrose, though they cannot directly utilize sucrose molecules themselves (Roth and Paszkowski, 2017). This necessitates the prior cleavage of sucrose, derived from photosynthetic sources, into hexoses suitable for fungal uptake.

Cellular compartmentalization within AMF structures further refines carbon utilization: extraradical mycelium lacks the machinery to metabolize hexose sugars (Wang et al., 2022, Wang and Wu, 2023). Consequently, energy generation from hexoses is thought to occur mainly in the intraradical mycelium, which resides within root cortical cells—although definitive evidence for this "energy center" role is still required (van Aarle and Olsson, 2008, Hawkins et al., 2023). Even before colonization—the pre-symbiotic phase—AMF spores and hyphae are able to sustain their initial development via monosaccharide transporters (MSTs) which facilitate the direct absorption of simple sugars from the soil environment (Bhandari and Garg, 2017, Bhupenchandra et al., 2024).

Once symbiosis is established, the movement of sugars from plant to fungus can proceed through both passive and active pathways (Nehls et al., 2010, Genre and Bonfante, 2010, Wang and Wu, 2023). Current models propose that sucrose is delivered via the phloem to the root's apoplastic space, where invertase and sucrose synthase enzymes cleave it into glucose and fructose. These monosaccharides are subsequently allocated to the fungal partner within cortical cells (Helber et al., 2011, Wipf et al., 2019).

A suite of sugar transport proteins orchestrates this transfer. Plant sucrose transporters (SUTs), present in mycorrhizal roots, play prominent roles in allocating carbon to the root sink and in regulating the magnitude of carbon partitioned to the symbiosis (Doidy et al., 2012, Roth and Paszkowski, 2017). Additional sugar movement is facilitated by diverse transporter families from both plant and fungal origins (Gonçalves et al., 2016, Manck-Götzenberger and Requena, 2016). Of particular note are fungal MSTs localized to the plant–fungus interface, which enable the import of glucose and fructose split from sucrose by plant invertases (Liu et al., 2022). Genes such as *GpMST1* and *RiMST2* encode for glucose-specific transporters which mediate this carbon supply (Helber et al., 2011).

Another recently characterized transporter group are the Sugars Will Eventually be Exported Transporters (SWEETs) (Jeena et al., 2019). These proteins, found in both plants and AMF, exhibit the unique capacity to operate bidirectionally as exporters and importers of both sucrose and monosaccharides to or from the protoplast (Manck-Götzenberger and Requena, 2016). This highlights the complexity and versatility of sugar exchange at the plant–fungus boundary.

3. The Role of AMF in Strengthening Plant Antioxidant Defenses

3.1. AMF and Drought Stress

Drought is one among the most severe abiotic stresses, substantially reducing plant growth, posing a significant risk to global food security. As plants are immobile organisms, they are more susceptible to water deficits, which can drastically impair their development and biomass production. However, studies have shown that forming symbiotic relationships with AMF can improve plant resilience and adaptability under drought conditions (Jia et al., 2021). Prolonged dry spells make it difficult for plants to draw water from the soil, causing their cells to lose moisture and stunting growth, leaf expansion, and root development. However, plants colonized by AMF develop more extensive root systems, seeing gains in root length, lateral roots, and total root area (Abdalla et al., 2023). This extra root capacity lets plants reach deeper water and nutrients, giving them an edge during drought.

AMF not only play a protective role, but also modulate the plant's hormone levels under drought conditions, especially increasing the production of stress-related hormones like ABA and promoting others, such as auxins and cytokinins, to cope up with the stress levels (Samanta et al., 2025). Under colonization with AMF, accumulation of sugars and amino acids such as proline is often seen as "shock absorbers", which often safeguard the host cells from extensive damages caused by the drought stress. Alongside these protective compounds, AMF boosts plant's ability to synthesize and activate antioxidant enzymes. Recent studies reveal that AMF aid the plant in controlling water transport channels in the root (aquaporins), which play a vital role in regulating both water uptake inside plant cells. Such physiological and molecular changes extend beyond water uptake—AMF give plants better access to minerals like nitrogen, phosphorus, potassium, and magnesium, especially in dry environments (Kakouridis et al., 2022). Even the microscopic strands (hyphae) of the fungi penetrate soil areas where the accessibility of roots are limiting, extracting nutrients otherwise unavailable.

3.2. AMF and Salt Stress

Soils with excess salt, pose a significant hindrance to the plant development affecting the cells at physiological and molecular levels. Salt concentration above the threshold levels affects the plant's ionic and osmotic equilibrium, potentially leading to toxic cell damage (Alharbi et al., 2022). However, AMF which symbiotically partner with the host, impart tolerance under these tough soils conditions and frequently partner with plants in such habitats.

Under high salt levels in the soil, AMF colonized plants can accumulate higher levels of osmoregulatory molecules (such as proline, trehalose, and soluble sugars), which protect the cells against osmotic stress induced by the salt stress (Pooja et al., 2024). However, in some studies, AMF colonization was found to accumulate lower levels of proline in the leaves, especially in uninoculated plants—this could be likely because the reduced stress levels faced by the plants colonized with the AMF, reflecting the reduced severity of the osmotic stress levels, making excessive proline build-up unnecessary (Begum et al., 2022). Salinity often causes a buildup of ionic sodium inside the plant tissues, displacing the beneficial potassium inside plant tissues, which significantly affects the growth and metabolic functions. Plants colonized with AMF help to restore the ionic balance, enhancing the uptake of selective key minerals and reinstating the ionic homeostasis within the plant cells. They also upregulate key protein levels such as sodium exporters (like NHX and SOS1 proteins), which help in preventing the entry of sodium inside the cell cytoplasm and help in moving out the toxic sodium out of the cytoplasm or to accumulate inside the vacuole, preventing the toxic buildup (Zhang et al., 2023).

Besides ionic adjustments, AMF upregulate the expression and activity of various antioxidant enzymes, preventing the damage arising from reactive oxygen species (ROS) and reducing byproducts of stress-induced damage (like malondialdehyde). Research has revealed that wheat and rice have shown that seedlings treated with AMF under higher salt conditions maintain higher activity of superoxide dismutase (SOD), catalase (CAT), and peroxidase, while showing less evidence of membrane damage (Huang et al., 2023, Zhang et al., 2023).

3.3. AMF and Temperature Stress

Drastic fluctuations in soil temperatures disrupt the plant community dynamics, potentially modulating their reliance on AMF associations to sustain yield and productivity under changing environmental conditions. Heat stress adversely influences plant growth and development through multiple physiological and morphological impairments, including reduced seed germination and plant vigor, slowed growth rates, diminished biomass accumulation, leaf and reproductive tissue wilting or scorching, accelerated leaf senescence and abscission, fruit damage and discoloration, yield reduction, increased cell mortality, and heightened oxidative stress (Rai et al., 2021). However, colonized plants show noticeable improvements in response to the stress (Mathur and Jajoo, 2020). Maize plants inoculated with AMF and subjected to high-temperature stress showed significant improvements compared to non-mycorrhizal heat-stressed plants, which included better photosynthetic efficiency, increased levels of specific AMF fatty acid biomarkers, higher spore density and root colonization, and reduced cellular damage as indicated by lower lipid peroxidation. Overall, AMF symbiosis enhanced plant biomass, stabilized the photosynthetic system, and boosted stress tolerance through both above-ground and below-ground mechanisms under heat stress.

3.4. AMF and Heavy Metal Stress

AMF colonized plants, growing in soils contaminated with heavy metals, are widely believed to support plant growth and development because of their potential to support the defense system of the host where they are inhabiting. Heavy metals restrict the root activity and disrupt the essential nutrient intake, triggering ROS surges, causing extensive damage to plant tissues. AMF mitigate heavy metal toxicity through multiple mechanisms (Janeeshma et al., 2023). Their extensive hyphal networks extend beyond the root zone, enhancing nutrient uptake while minimizing the negative impact of heavy metals at damaging concentrations at the cellular level. The fungal cell walls, rich in components such as chitin, serve as physical adsorbents by binding toxic metals and limiting their translocation into plant tissues. Additionally, AMF produce and exudate glomalin, a sticky glycoprotein that chelates metal ions, thereby immobilizing them and reducing their bioavailability (Dhalaria et al., 2020). Furthermore, they facilitate the upregulation of plant antioxidant systems too and enhance the synthesis of metal-detoxifying proteins, including phytochelatins and metallothioneins, which chelate and neutralize toxic metals. Heavy metals can be accumulated within both the internal (intraradical) and external (extraradical) fungal hyphae (Alvarado-López et al., 2019), where they are immobilized by binding to the cell walls, being compartmentalized within vacuoles, or chelated by cytoplasmic molecules, thus reducing metal toxicity to the host plants.

3.5. AMF and Waterlogging Stress

Waterlogging represents a critical abiotic stress that adversely affects plant growth, productivity, and distribution by causing root-zone anoxia and hypoxia. These conditions impair hydraulic conductivity and induce stomatal closure, thereby disrupting water balance, photosynthesis, and nutrient uptake, ultimately reducing plant vigor. In response, plants develop morphological and physiological adaptations such as adventitious roots and lenticels to enhance root aeration and oxygen availability. AMF, which form symbiotic associations with terrestrial plants, are prevalent in wetland ecosystems and demonstrate resilience to waterlogged environments, including mangroves, floodplains, and peat swamps (Tuheteru and Wu, 2017). They enhance plant tolerance to waterlogging by improving nutrient acquisition, promoting growth and biomass accumulation, and facilitating aquatic plant community succession. For instance, inoculation with *Glomus clarum* and *Gigaspora decipiens* in *Dyera polyphylla* from tropical peat swamps increased waterlogging tolerance alongside elevated nitrogen and phosphorus uptake (Tuheteru and Wu, 2017). Moreover, AMF enhance phosphorus absorption in species such as *Panicum hemitomon* and *Oryza sativa*, and contribute to osmolyte accumulation (e.g., sugars and proline), which mitigates anaerobic stress by reducing toxic metabolites like ethanol (Das et al., 2022). The AMF also stimulates antioxidant enzyme activities (e.g., SOD and CAT), alleviating oxidative damage, as demonstrated in citrus species inoculated with *Diversispora spurca* (Wu et al., 2023). Increased chlorophyll content and photosynthetic efficiency have been observed in peach following AMF inoculation. Enhanced photosynthetic performance and stomatal conductance under waterlogged conditions are linked to improved water uptake, including upregulation of plant and fungal aquaporins (*SlPIP1;1* and *GintAQP1*) in tomato roots (Cheng et al., 2020).

4. AMF in Carbon Cycling and Antioxidant Defense: Synergistic Roles and Agricultural Implications

AMF significantly influence both carbon cycling and antioxidant defense response in plants. These fungi facilitate carbon movement in terrestrial ecosystems through three primary mechanisms: turnover of fungal structures (e.g., hyphae and spores) that deposit recalcitrant organic residues like chitin and GRSP; secretion of oxidative and hydrolytic enzymes that decompose plant litter and soil organic matter (SOM), thereby enhancing nutrient release and carbon transformation; and promotion of plant growth by imparting tolerance to various stressors, leading to increased biomass and rhizodeposition. AMF symbiosis amplifies root biomass and the exudation of carbon-rich compounds such as sugars, amino acids, and organic acids, which enrich the rhizosphere and stimulate microbial activity. Additionally, AMF hyphae exhibit functional differentiation – "runner" hyphae explore new soil regions while "absorptive" hyphae intensify nutrient uptake – both contributing significantly to carbon input and sequestration (Dodd et al., 2000). A central component of AMF-mediated carbon stabilization is the production of GRSP, a class of glycoproteins secreted by fungal hyphae. GRSP contributes a substantial portion of total soil carbon and nitrogen, particularly in upper soil layers. Chemically stable and highly resistant to degradation, GRSP enhances soil aggregation by binding the soil particles, organic matter, and minerals into stable micro- and macroaggregates (Yang et al., 2024). These aggregates physically protect organic carbon from microbial attack by occlusion and improve overall soil architecture. Furthermore, GRSP's ability to interact with polyvalent cations (especially the essential micronutrients Fe, Ca, Mg, and Al) strengthens aggregate stability, while its biofilm-like properties increase soil heterogeneity, promoting microbial colonization and enzymatic transformations that further stabilize carbon pools (Chen et al., 2023). Aggregate formation also plays a critical role in controlling the spatial distribution of carbon, allowing AMF to relocate exudates from zones of high microbial respiration to more stable soil regions, thereby reducing CO_2 losses.

The synergistic roles of AMF in carbon cycling and antioxidant defense bear immense implications for sustainable developmental goals in agriculture. Enhanced carbon sequestration, soil aggregation, and nutrient bioavailability contribute directly to improved soil structure, water retention, and fertility—factors critical for plant productivity under stress conditions. By strengthening the antioxidant systems of plants, AMF confer greater resilience against environmental stressors like drought, salinity, and temperature extremes (Wahab et al., 2023). This dual role not only reduces the need for chemical inputs through external fertilizers but also positions AMF as strategic agents for climate-resilient and low-input farming systems. Moreover, improved soil structure from AMF-mediated aggregation minimizes the soil erosion and nutrient leaching prevalent in stressed conditions, further safeguarding soil health (Wipf et al., 2019). The soil-enhancing property of AMF induces aggregation and heightened microbial activity, which helps to stabilize the SOC, while maintaining soil vitality, thereby supporting long-term agricultural productivity and ecological balance.

In addition to their direct contributions to soil carbon inputs and stabilization, AMF significantly influence plant physiological responses, particularly under abiotic stress conditions. One of the most critical responses is the modulation of antioxidant defense systems, which helps plants combat the damaging effects of ROS. Environmental stresses such as salinity, drought, and temperature extremes often trigger oxidative bursts in plant tissues, leading to membrane lipid peroxidation, protein denaturation, and nucleic acid damage. AMF colonization has been shown to enhance the enzymatic antioxidant capacity of plants by upregulating key enzymes such as SOD, CAT, ascorbate peroxidase (APX), and glutathione reductase (GR) (Garcia-Caparros et al., 2021). These enzymes scavenge harmful ROS, thus maintaining cellular redox homeostasis and preventing stress-induced cellular injury.

The antioxidant enhancement by AMF is tightly coupled to carbon allocation and energy metabolism. Improved photosynthetic efficiency in AMF-colonized plants ensures a steady supply of photosynthates to roots, where a significant portion is transferred to the fungal symbiont. In exchange, the fungus supports the plant with improved nutrient uptake, particularly phosphorus and micronutrients like zinc and copper, which are critical cofactors for antioxidant enzymes (Bhantana et al., 2021). This reciprocal exchange optimizes resource use, improves plant vigor, and maintains the energetic requirements for maintaining robust antioxidant and stress defense pathways. Simultaneously, this exchange increases rhizodeposition and contributes to belowground carbon inputs, reinforcing carbon cycling processes.

From an agroecological standpoint, integrating AMF into farming practices offers a sustainable route to build SOC while enhancing crop tolerance to environmental stresses. Long-term field studies have demonstrated that AMF inoculation, especially when combined with organic amendments such as biochar or compost, can lead to substantial increases in soil carbon content, nutrient retention, and crop yield. These benefits are further amplified when AMF symbiosis is supported by minimal tillage, reduced chemical inputs, and diverse crop rotations that sustain mycorrhizal networks. Moreover, AMF-mediated processes align with global climate goals by mitigating greenhouse gas emissions through enhanced carbon retention and reduced nitrous oxide losses. As a nature-based solution, AMF inoculation holds promise for scaling regenerative agriculture while contributing to soil restoration, carbon sequestration, and climate adaptation across diverse agroecosystems.

Recognizing the intricate networks governing carbon delivery from plant to fungus is vital for assessments of AMF in ecosystem carbon balance and soil fertility. Given that a considerable percentage of terrestrial plant photosynthates are funneled into the mycorrhizal pathway, advancing our understanding of the molecular machinery – particularly transporter proteins and metabolic regulation – will refine our knowledge of carbon sequestration potential in managed and natural environments. Furthermore, future investigation into the relative contributions of carbohydrates and lipids to AMF carbon nutrition will enhance our perspective on the dual roles of mycorrhizal fungi in both supporting plant growth and stabilizing soil carbon reservoirs.

5. Conclusion

The terrestrial carbon cycle is inextricably linked to the fate and dynamics of soil organic carbon (SOC), which not only surpasses the combined carbon stocks of the atmosphere and vegetation but also forms the backbone of terrestrial ecosystem health and climate mediation. Anthropogenic activities – especially those associated with intensive agriculture – have exerted profound impacts on the balance of SOC, accelerating rates of soil degradation, nutrient depletion, and carbon loss. These challenges, compounded by the increasing incidence of abiotic stresses and climate variability, underscore the critical need for sustainable and nature-based strategies to restore soil health and resilience. AMF have emerged as vital, yet often underappreciated, agents of soil carbon stabilization and plant stress tolerance. Through their extensive hyphal networks and close symbiotic associations with plant roots, AMF channel a significant fraction of photosynthetically fixed carbon belowground. This carbon transfer, coupled with the synthesis and accumulation of recalcitrant fungal compounds such as GRSP, greatly enhances soil aggregate formation and carbon retention. As this chapter has highlighted, the functional interplay between AMF and host plants not only drives soil carbon sequestration, but also strengthens the plant's physiological defenses—particularly antioxidant systems—thus reducing the deleterious impacts of drought, salinity, heavy metals, temperature extremes, and waterlogging. The integration of AMF into agricultural management provides a synergistic approach to address the dual goals of productivity and sustainability. By improving nutrient and water uptake, supporting root development, and fortifying plant antioxidant capacity, AMF offer a biological avenue to mitigate dependence on chemical inputs, bolster crop yields, and enhance climatic resilience. Furthermore, AMF-mediated improvements in soil structure, nutrient cycling, and microbial diversity serve as strong pillars for future regenerative agriculture and long-term ecosystem stability.

Acknowledgement

JMJ gratefully acknowledges the financial assistance provided in the form of Senior Research Fellowship from the University Grants Commission (UGC) with NTA Ref. No. 211610073457. RJ acknowledges Bayer's fellowship program MEDHA (PHD-2023-10454) and Department of Botany, University of Calicut. The authors also acknowledge the Department of Science & Technology (DST), Government of India, for granting fund under Fund for Improvement of S&T Infrastructure (FIST) programme (DST-FIST/15-16/28.05.2015).

References

Abdalla, M., Bitterlich, M., Jansa, J., Püschel, D. & Ahmed, M.A. (2023). The role of arbuscular mycorrhizal symbiosis in improving plant water status under drought. J. Exp. Bot. 74(16), 4808–4824. https://doi.org/10.1093/jxb/erad249

Abraham, B., Araya, H., Berhe, T., Edwards, S., Gujja, B., Khadka, R.B. et al. (2014). The system of crop intensification: Reports from the field on improving agricultural production, food security, and resilience to climate change for multiple crops. Agric. & Food Secur. 3(1), 4. https://doi.org/10.1186/2048-7010-3-4

Alharbi, K., Al-Osaimi, A.A. & Alghamdi, B.A. (2022). Sodium chloride (NaCl)-induced physiological alteration and oxidative stress generation in *Pisum sativum* (L.): A toxicity assessment. ACS Omega 7(24), 20819–20832.

Alvarado-López, C.J., Dasgupta-Schubert, N., Ambriz, J.E., Arteaga-Velazquez, J.C. & Villegas, J.A. (2019). Lead uptake by the symbiotic *Daucus carota* L.–*Glomus intraradices* system and its effect on the morphology of extra- and intraradical fungal microstructures. Environ. Sci. Pollut. Res. 26(1), 381–391. https://doi.org/10.1007/s11356-018-3569-7

Basiru, S. & Hijri, M. (2024). Trade-off between soil organic carbon sequestration and plant nutrient uptake in arbuscular mycorrhizal symbiosis. Fungal Biol. Rev. 49, 100381. https://doi.org/10.1016/j.fbr.2024.100381

Begum, N., Wang, L., Ahmad, H., Akhtar, K., Roy, R., Khan, M.I. & Zhao, T. (2022). Co-inoculation of arbuscular mycorrhizal fungi and the plant growth-promoting rhizobacteria improve growth and photosynthesis in tobacco under drought stress by up-regulating antioxidant and mineral nutrition metabolism. Micro. Ecol. 83(4), 971–988. https://doi.org/10.1007/s00248-021-01815-7

Bhandari, P. & Garg, N. (2017). Dynamics of arbuscular mycorrhizal symbiosis and its role in nutrient acquisition: An overview. *In*: Varma, A., Prasad, R., Tuteja, N. [eds.]. Mycorrhiza-Nutrient Uptake, Biocontrol, Ecorestoration. Springer, Cham. https://doi.org/10.1007/978-3-319-68867-1_2

Bhantana, P., Rana, M.S., Sun, X.C., Moussa, M.G., Saleem, M.H., Syaifudin, M. & Hu, C.X. (2021). Arbuscular mycorrhizal fungi and its major role in plant growth, zinc nutrition, phosphorous regulation and phytoremediation. Symbiosis 84(1), 19–37. https://doi.org/10.1007/s13199-021-00756-6

Bhattacharyya, S.S., Ros, G.H., Furtak, K., Iqbal, H.M. & Parra-Saldívar, R. (2022). Soil carbon sequestration – An interplay between soil microbial community and soil organic matter dynamics. Sci. Total. Environ. 815, 152928. https://doi.org/10.1016/j.scitotenv.2022.152928

Bhupenchandra, I., Chongtham, S.K., Devi, A.G., Dutta, P., Sahoo, M.R., Mohanty, S. et al. (2024). Unlocking the potential of arbuscular mycorrhizal fungi: Exploring role in plant growth promotion, nutrient uptake mechanisms, biotic stress alleviation, and sustaining agricultural production systems. J. Plant Growth Regul. 1–39. https://doi.org/10.1007/s00344-024-11467-9

Bisht, A., Lather, S., Goyal, P. & Garg, N. (2024). Molecular determinants and regulatory mechanisms of nutrient exchange between plant and AMF. In: Parihar, M., Rakshit, A., Adholeya, A., Chen, Y. [eds.]. Arbuscular Mycorrhizal Fungi in Sustainable Agriculture: Nutrient and Crop Management. Springer, Singapore. https://doi.org/10.1007/978-981-97-0300-5_3

Brunn, M., Mueller, C.W., Chari, N.R., Meier, I.C., Obersteiner, S., Phillips, R.P. et al. (2025). Tree carbon allocation to root exudates: implications for carbon budgets, soil sequestration and drought response. Tree Physiol. 45(4), tpaf026. https://doi.org/10.1093/treephys/tpaf026

Chen, S., Sheng, X., Qin, C., Waigi, M.G. & Gao, Y. (2019). Glomalin-related soil protein enhances the sorption of polycyclic aromatic hydrocarbons on cation-modified montmorillonite. Environ. Inter. 132, 105093. https://doi.org/10.1016/j.envint.2019.105093

Chen, X., Li, Z., Zhao, H., Li, Y., Wei, J., Ma, L. et al. (2024). Enhancing maize yield and nutrient utilization through improved soil quality under reduced fertilizer use: The efficacy of organic–inorganic compound fertilizer. Agriculture 14(9), 1482. https://doi.org/10.3390/agriculture14091482

Cheng, H.Q., Ding, Y.E., Shu, B., Zou, Y.N., Wu, Q.S. & Kuča, K. (2020). Plant aquaporin responses to mycorrhizal symbiosis under abiotic stress. Int. J. Agric. Biol. 23, 786–794.

Cheng, X.F., Wu, H.H., Zou, Y.N., Wu, Q.S. & Kuča, K. (2021). Mycorrhizal response strategies of trifoliate orange under well-watered, salt stress, and waterlogging stress by regulating leaf aquaporin expression. Plant Physiol. Biochem. 162, 27–35. https://doi.org/10.1016/j.plaphy.2021.02.026

Das, D., Ullah, H., Himanshu, S.K., Tisarum, R., Cha-Um, S. & Datta, A. (2022). Arbuscular mycorrhizal fungi inoculation and phosphorus application improve growth, physiological traits, and grain yield of rice under alternate wetting and drying irrigation. J. Plant Physiol. 278, 153829. https://doi.org/10.1016/j.jplph.2022.153829

Dhalaria, R., Kumar, D., Kumar, H., Nepovimova, E., Kuča, K., Torequl Islam, M. & Verma, R. (2020). Arbuscular mycorrhizal fungi as potential agents in ameliorating heavy metal stress in plants. Agronomy 10(6), 815. https://doi.org/10.3390/agronomy10060815

Dodd, J.C., Boddington, C.L., Rodriguez, A., Gonzalez-Chavez, C. & Mansur, I. (2000). Mycelium of arbuscular mycorrhizal fungi (AMF) from different genera: Form, function and detection. Plant and Soil 226(2), 131–151. https://doi.org/10.1023/A:1026574828169

Doidy, J., Van Tuinen, D., Lamotte, O., Corneillat, M., Alcaraz, G. & Wipf, D. (2012). The Medicago truncatula sucrose transporter family: characterization and implication of key members in carbon partitioning towards arbuscular mycorrhizal fungi. Mol. Plant. 5(6), 1346–1358. https://doi.org/10.1093/mp/sss079

Even, R.J. & Cotrufo, M.F. (2024. The ability of soils to aggregate, more than the state of aggregation, promotes protected soil organic matter formation. Geoderma 442, 116760.

Gachara, G., Kenfaoui, J., Suleiman, R., Kilima, B., Taoussi, M., Aberkani, K. et al. (2024). The role of soil microbiome in driving plant performance: An overview based on ecological and ecosystem advantages to the plant community. J. Crop Health 76(1), 3–17. https://doi.org/10.1007/s10343-023-00935-z

Garcia-Caparros, P., De Filippis, L., Gul, A., Hasanuzzaman, M., Ozturk, M., Altay, V. & Lao, M.T. (2021). Oxidative stress and antioxidant metabolism under adverse environmental conditions: A review. Bot. Rev. 87(4), 421–466. https://doi.org/10.1007/s12229-020-09231-1

Genre, A. & Bonfante, P. (2010). The making of symbiotic cells in arbuscular mycorrhizal roots. In: Koltai, H., Kapulnik, Y. [eds.]. Arbuscular Mycorrhizas: Physiology and Function. Springer, Dordrecht. https://doi.org/10.1007/978-90-481-9489-6_3

George, N.P. & Ray, J.G. (2023). The inevitability of arbuscular mycorrhiza for sustainability in organic agriculture—A critical review. Front. Sustain. Food Syst. 7, 1124688. https://doi.org/10.3389/fsufs.2023.1124688

Ghag, S.B., Alok, A., Rajam, M.V. & Penna, S. (2023). Designing climate-resilient crops for sustainable agriculture: a silent approach. J. Plant Growth Regul. 42(10), 6503–6522. https://doi.org/10.1007/s00344-022-10880-2

Gonçalves, C., Coelho, M.A., Salema-Oom, M. & Gonçalves, P. (2016). Stepwise functional evolution in a fungal sugar transporter family. Mol. Bio. Evol. 33(2), 352–366. https://doi.org/10.1093/molbev/msv220

Gutjahr, C. (2014). Phytohormone signaling in arbuscular mycorhiza development. Curr. Opin. Plant Biol. 20, 26–34. https://doi.org/10.1016/j.pbi.2014.04.003

Gutjahr, C., Novero, M., Welham, T., Wang, T. & Bonfante, P. (2011). Root starch accumulation in response to arbuscular mycorrhizal colonization differs among *Lotus japonicus* starch mutants. Planta, 234(3), 639–646. https://doi.org/10.1007/s00425-011-1472-2

Hawkins, H.J., Cargill, R.I., Van Nuland, M.E., Hagen, S.C., Field, K.J., Sheldrake, M. et al. (2023). Mycorrhizal mycelium as a global carbon pool. Curr. Biol. 33(11), R560–R573. https://doi.org/10.1016/j.cub.2023.02.027

Helber, N., Wippel, K., Sauer, N., Schaarschmidt, S., Hause, B. & Requena, N. (2011). A versatile monosaccharide transporter that operates in the arbuscular mycorrhizal fungus *Glomus* sp is crucial for the symbiotic relationship with plants. Plant Cell 23(10), 3812–3823. https://doi.org/10.1105/tpc.111.089813

Hossain, A., Krupnik, T.J., Timsina, J., Mahboob, M.G., Chaki, A K., Farooq, M. et al. (2020). Agricultural land degradation: Processes and problems undermining future food security. *In*: Environment, Climate, Plant and Vegetation Growth (pp. 17–61) Cham: Springer International Publishing. https://doi.org/10.1007/978-3-030-49732-3_2

Huang, S., Gill, S., Ramzan, M., Ahmad, M.Z., Danish, S., Huang, P. & Alharbi, S.A. (2023). Uncovering the impact of AM fungi on wheat nutrient uptake, ion homeostasis, oxidative stress, and antioxidant defense under salinity stress. Sci. Rep., 13(1), 8249. https://doi.org/10.1038/s41598-023-35148-x

Janeeshma, E., Joel, J.M., Shackira, A.M., Johnson, R. & Dhanya, T.T. (2023). Role of arbuscular mycorrhizal fungi in the alleviation of heavy metal stress in rice. *In*: Arbuscular Mycorrhizal Fungi (pp. 83–90). CRC Press.

Jeena, G.S., Kumar, S. & Shukla, R.K. (2019). Structure, evolution and diverse physiological roles of SWEET sugar transporters in plants. Plant Mol. Biol. 100, 351–365. https://doi.org/10.1007/s11103-019-00872-4

Jia, Y., van der Heijden, M.G., Wagg, C., Feng, G. & Walder, F. (2021). Symbiotic soil fungi enhance resistance and resilience of an experimental grassland to drought and nitrogen deposition. J. Ecol. 109(9), 3171–3181. https://doi.org/10.1111/1365-2745.13521

Kakouridis, A., Hagen, J.A., Kan, M.P., Mambelli, S., Feldman, L.J., Herman, D.J. & Firestone, M.K. (2022). Routes to roots: Direct evidence of water transport by arbuscular mycorrhizal fungi to host plants. New Phytol. 236(1), 210–221. https://doi.org/10.1111/nph.18281

Kameoka, H. & Gutjahr, C. (2022). Functions of lipids in development and reproduction of arbuscular mycorrhizal fungi. Plant. Cell. Physiol. 63(10), 1356–1365. https://doi.org/10.1093/pcp/pcad046

Lanfranco, L., Fiorilli, V. & Gutjahr, C. (2018). Partner communication and role of nutrients in the arbuscular mycorrhizal symbiosis. New Phytol. 220(4), 1031–1046. https://doi.org/10.1111/nph.15230

Liu, Y.H., Song, Y.H. & Ruan, Y.L. (2022). Sugar conundrum in plant–pathogen interactions: Roles of invertase and sugar transporters depend on pathosystems. J. Exp. Bot. 73(7), 1910–1925. https://doi.org/10.1093/jxb/erab562

Luo, X., Liu, Y., Li, S. & He, X. (2023). Interplant carbon and nitrogen transfers mediated by common arbuscular mycorrhizal networks: Beneficial pathways for system functionality. Front. Plant Sci. 14, 1169310. https://doi.org/10.3389/fpls.2023.1169310

Manck-Götzenberger, J. & Requena, N. (2016). Arbuscular mycorrhiza symbiosis induces a major transcriptional reprogramming of the potato SWEET sugar transporter family. Front. Plant Sci. 7, 487. https://doi.org/10.3389/fpls.2016.00487

Mathur, S. & Jajoo, A. (2020). Arbuscular mycorrhizal fungi protects maize plants from high temperature stress by regulating photosystem II heterogeneity. Ind. Crop. Prod., 143, 111934. https://doi.org/10.1016/j.indcrop.2019.111934

Nehls, U., Göhringer, F., Wittulsky, S. & Dietz, S. (2010). Fungal carbohydrate support in the ectomycorrhizal symbiosis: A review. Plant Biol. 12(2), 292–301. https://doi.org/10.1111/j.1438-8677.2009.00312.x

Pan, Y., Liu, J., Zhang, M., Huang, P., Hipesy, M., Dai, L. et al. (2024). The below-ground biomass contributes more to wetland soil carbon pools than the above-ground biomass—A survey based on global wetlands. J. Plant Ecol. 17(5), rtae017. https://doi.org/10.1093/jpe/rtae017

Parasar, B.J. & Agarwala, N. (2025). Unravelling the role of biochar-microbe-soil tripartite interaction in regulating soil carbon and nitrogen budget: A panacea to soil sustainability. Biochar, 7(1), 37. https://doi.org/10.1007/s42773-024-00411-5

Pooja, P., Tallapragada, S., Lamba, A. & Punia, S. (2024). Role played by arbuscular mycorrhizal fungi in amelioration of salinity stress: A review. Plant and Soil, 1–16. https://doi.org/10.1007/s11104-024-06976-1

Powell, J.R. & Rillig, M.C. (2018). Biodiversity of arbuscular mycorrhizal fungi and ecosystem function. New Phytol. 220(4), 1059–1075. https://doi.org/10.1111/nph.15119

Rai, A., Rai, G.K. & Dubey, R.S. (2021). Heat stress and its effects on plant growth and metabolism. *In*: Abiotic Stress Tolerance Mechanisms in Plants (pp. 203–265). CRC Press.

Rillig, M.C. (2004). Arbuscular mycorrhizae and terrestrial ecosystem processes. Ecol. Lett. 7(8), 740–754. https://doi.org/10.1111/j.1461-0248.2004.00620.x

Roth, R. & Paszkowski, U. (2017). Plant carbon nourishment of arbuscular mycorrhizal fungi. Curr. Opin. Plant Biol. 39, 50–56. https://doi.org/10.1016/j.pbi.2017.05.008

Samanta, I., Ghosh, K., Saikia, R., Savita, Maity, P.J. & Chowdhary, G. (2025). Arbuscular mycorrhizal fungi – A natural tool to impart abiotic stress tolerance in plants. Plant Signal. Behav. 20(1), 2525843. https://doi.org/10.1080/15592324.2025.2525843

Shukla, S., Didwania, N. & Choudhary, R. (2025). Arbuscular mycorrhizal fungi (AMF): A pathway to sustainable soil health, carbon sequestration, and greenhouse gas mitigation. J. Saudi Soc. Agric. Sci. 24(4), 22. https://doi.org/10.1007/s44447-025-00023-w

Smith, S.E. & Smith, F.A. (2012). Fresh perspectives on the roles of arbuscular mycorrhizal fungi in plant nutrition and growth. Mycologia, 104(1), 1–13. https://doi.org/10.3852/11-229

Solaiman, Z. (2024). Role of AMF in organic matter decomposition, carbon sequestration and climate change mitigation. *In*: Parihar, M., Rakshit, A., Adholeya, A., Chen, Y. [eds.]. Arbuscular Mycorrhizal Fungi in Sustainable Agriculture: Nutrient and Crop Management. Springer, Singapore. https://doi.org/10.1007/978-981-97-0300-5_6

Subramanian, K.S. (1998). Role of Arbuscular Mycorrhizal Fungus (*Glomus intraradices* Schenck and Smith) Colonization in Drought Tolerance of Maize (*Zea mays* L.). University of Ottawa, Canada. http://dx.doi.org/10.20381/ruor-10248

Surówka, E., Rapacz, M. & Janowiak, F. (2020). Climate change influences the interactive effects of simultaneous impact of abiotic and biotic stresses on plants. (pp. 1–50). *In*: Hasanuzzaman, M. [ed.]. Plant Ecophysiology and Adaptation under Climate Change: Mechanisms and Perspectives I: General Consequences and Plant Responses. Singapore: Springer Singapore. https://doi.org/10.1007/978-981-15-2156-0_1

Thotakuri, G., Angidi, S. & Athelly, A. (2024). Soil carbon pool as influenced by soil microbial activity—An overview. AJCC (American Journal of Climate Change), 13(2), 175–193. https://doi.org10.4236/ajcc.2024.132010

Tuheteru, F.D. & Wu, Q.S. (2017). Arbuscular mycorrhizal fungi and tolerance of waterlogging stress in plants. *In*: Arbuscular Mycorrhizas and Stress Tolerance of Plants (pp. 43–66). Singapore: Springer Singapore.

van Aarle, I.M. & Olsson, P.A. (2008). Resource partitioning between extraradical and intraradical AM fungal mycelium. *In*: Varma, A. [ed.]. Mycorrhiza. Springer, Berlin, Heidelberg. https://doi.org/10.1007/978-3-540-78826-3_16

Wahab, A., Muhammad, M., Munir, A., Abdi, G., Zaman, W., Ayaz, A. et al. (2023). Role of arbuscular mycorrhizal fungi in regulating growth, enhancing productivity, and potentially influencing ecosystems under abiotic and biotic stresses. Plants, 12(17), 3102. https://doi.org/10.3390/plants12173102

Wang, F., Zhang, L., Zhou, J., Rengel, Z., George, T.S. & Feng, G. (2022). Exploring the secrets of hyphosphere of arbuscular mycorrhizal fungi: Processes and ecological functions. Plant and Soil, 481(1), 1–22. https://doi.org/10.1007/s11104-022-05621-z

Wang, S., Zhang, Y., Ju, W., Chen, J.M., Ciais, P., Cescatti, A. et al. (2020). Recent global decline of CO_2 fertilization effects on vegetation photosynthesis. Sci. 370(6522), 1295–1300. https://doi.org/10.1126/science.abb77

Wang, W., Shi, J., Xie, Q., Jiang, Y., Yu, N. & Wang, E. (2017). Nutrient exchange and regulation in arbuscular mycorrhizal symbiosis. Mol. Plant. 10(9), 1147–1158. https://doi.org/10.1016/j.molp.2017.07.012

Wang, Y.J. & Wu, Q.S. (2023). Influence of sugar metabolism on the dialogue between arbuscular mycorrhizal fungi and plants. Hortic. Adv., 1(1), 2. https://doi.org/10.1007/s44281-023-00001-8

Wipf, D., Krajinski, F., van Tuinen, D., Recorbet, G. & Courty, P.E. (2019). Trading on the arbuscular mycorrhiza market: From arbuscules to common mycorrhizal networks. New Phytol. 223(3), 1127–1142. https://doi.org/10.1111/nph.15775

Wu, Q.S., Zou, Y.N. & Huang, Y.M. (2013). The arbuscular mycorrhizal fungus *Diversispora spurca* ameliorates effects of waterlogging on growth, root system architecture and antioxidant enzyme activities of citrus seedlings. Fungal Ecol. 6(1), 37–43. https://doi.org/10.1016/j.funeco.2012.09.002

Yang, H., Xiao, Q., Huang, Y., Cai, Z., Li, D., Wu, L. & Zhang, W. (2024). Long-term manuring facilitates glomalin-related soil proteins accumulation by chemical composition shifts and macro-aggregation formation. Soil. Till. Res., 235, 105904. https://doi.org/10.1016/j.still.2023.105904

Zhang, B., Shi, F., Zheng, X., Pan, H., Wen, Y. & Song, F. (2023). Effects of AMF compound inoculants on growth, ion homeostasis, and salt tolerance-related gene expression in *Oryza sativa* L. under salt treatments. Rice, 16(1), 18. https://doi.org/10.1186/s12284-023-00635-2

CHAPTER
4

The Role of AMF Networks in Nutrient Cycling: A Study of Genes and Mechanisms

Gökhan Boyno[1*], Younes Rezaee Danesh[1*], Necmettin Teniz[2], Rojbin Çevik[1], Oktay Calayır[1], Semra Demir[1], Ramila Fares[3,4,5], Beatrice Farda[3], Rihab Djebaili[3] and Marika Pellegrini[3]

[1] Department of Plant Protection, Faculty of Agriculture, Van Yuzuncu Yil University, Van-Türkiye
[2] Department of Agricultural Biotechnology, Faculty of Agriculture, Van Yuzuncu Yil University, Van-Türkiye
[3] Department of Life, Health and Environmental Sciences, University of L'Aquila, L'Aquila, Italy
[4] Laboratory Biotechnology, Water, Environment and Health, Abbes Laghrour University of Khenchela, Khenchela, Algeria
[5] Laboratory of Mycology, Biotechnology and Microbial Activity, Brothers Mentouri University of Constantine 1, Constantine, Algeria

1. Introduction

Arbuscular Mycorrhizal Fungi (AMF) are microorganisms that establish a symbiotic relationship with plant roots, enhancing water and nutrient uptake and playing a critical role in ecosystems (Bhupenchandra et al., 2024). This symbiotic interaction supports plant growth and development while improving soil health. AMF penetrate plant roots, expanding root surface area and increasing access to soil nutrients (Thangavel et al., 2022). This process enables plants to absorb essential nutrients, such as phosphorus, nitrogen, and micronutrients, more efficiently. Phosphorus, a vital nutrient, is often found in insoluble forms in the soil, and one of AMF's most important roles in plant-soil interactions is to solubilize and deliver this nutrient to plants.

AMF form microscopic tree-like structures called "arbuscules" within plant roots, facilitating nutrient exchange between the plant and the fungus (Boyno and Demir, 2022). These structures allow the plant to transfer carbon compounds

*Corresponding authors: gokhanboyno@yyu.edu.tr and y.rdanesh@yahoo.com

DOI: 10.1201/9781003669777-4

produced through photosynthesis to the fungus, while the fungus delivers water and nutrients from the soil to the plant. This mutualistic relationship ensures the survival and growth of both the plant and the fungus. Beyond optimizing nutrient exchange, AMF also contribute to improving soil structure (Zeng et al., 2023). By breaking down organic matter and promoting the formation of soil aggregates, they enhance soil stability, increase water retention capacity, and reduce soil erosion.

Ecologically speaking, AMF is necessary for plant adaptation and helplessness to face adverse environmental conditions which is of critical importance to ecosystem integrity and sustainable agriculture (Wahab et al., 2023). AMF have the ability to increase plant tolerance to environmental abiotic stresses such as drought, salinity and heavy metals in different ways (Demir et al., 2022). AMF allow plants to take up heavy metals less burdened by their toxic effects. Moreover, the AMF have great influences on soil microbiota. An extraordinary enzymatic activator of beneficial microbes in root zone Promotes soil microbiota balance functions as a biological barrier against pathogens (Boyno et al., 2022, ji et al., 2022).

Among others, AMF also contribute the most critical roles in nutrient cycles to ecosystems (Wahab et al., 2023). They also participate in nitrogen cycling, aside from phosphorous (Luo, et al., 2022). AMF also increase crop nutrition through greater uptake of micronutrients (e.g. zinc, copper and iron) (Bhantana et al., 2021). These attributes reduce the need for chemical fertilizers, and set the stage for sustainable agriculture. AMF also substantially affect the carbon cycle. AM fungi increase soil carbon content and support long-term soil fertility by sequestering a portion of atmospheric carbon in soil, stored as a glycoprotein called "glomalin" (Agnihotri et al., 2022).

AMF in soil ecosystems, the most important of which is their role in soil nutrient cycling. They provide significant contributions to the sustainability of ecosystems when they either preserve plant diversity and partition nutrients among plants (Ma et al., 2021). The extensive hyphal network built by AMF ensure stable soil aggregates, reduce soil erosion, and promote soil water retention (Fall et al., 2022).

AM fungi represent an important natural reservoir of these resources, with multifaceted benefits that could be valuable for environmentally friendly agricultural practices and ecosystem management. For example, AMF established their important symbiotic relationships with plants probably around 450 million years ago and are widely distributed in terrestrial ecosystems today (Liu et al., 2024). Knowledge of the genetic and mechanistic mechanisms underlying AMF could potentially provide valuable information for both biotechnological as well ecological research. In the following chapter I discuss the role of AMF in nutrient cycling within the context of genetic and mechanistic foundations. The ph:atn transport of AMF with their finely dished roots will be analyzed, particularly focused on the role of this fungus to the solubilisation of phosphates and to contribute towards the nitrogen cycle (Bhantana et al., 2021). It will detail the involved genetic factors in this process, the roles of genes that direct the symbiotic relationship and resource response regulation under environmental factors. The genetic potential for optimizing nutrient cycling and stabiliszng ecosystem functioning will be summarized across plant genomics and mechanistic range.

2. Overview of AMF Networks

2.1. Arbuscular Mycorrhizal Fungi (AMF) and Their Role in Nutrient Cycling

AMF play an important function in plant nutrient acquisition, soil health, and carbon cycle processes by virtue of their large-scale mutualisms (Wahab et al., 2023). These networks of fungi enable the redistribution of nutrients such as phosphorus, nitrogen, zinc and copper that are critical for plant growth and provide soil microbial communities and carbon storage proportionally to their important role (Thangavel et al., 2022). The cellular and molecular level is necessary for the interactive processes operating between AMF, its plant symbiont and soil microorganisms Hydroculture network models in web inferencing have been presented recently (Boyno et al., 2023).

AMF form intricate symbiotic networks between plant roots and soil, enhancing plant nutrient acquisition and contributing to the functioning of soil ecosystems (Khaliq et al., 2022). These networks consist of two main structures: intraradical hyphae (IRH) and extraradical hyphae (ERH) (Duan et al., 2024).

Intraradical Hyphae (IRH): AMF structures accessed by IRH into the plant root cells (Basyal et al., 2024). These hyphae grow inside the cortical cells of plant roots, which can result in shared direct nutrient transfer between the plant and fungus (Wang et al., 2022). As a few key structures of IRH is the arbuscule is one of the major ones it form within the cortical cells in plant root (El Malahi et al., 2025). Arbuscules are branched fungal networks that have high surface area allowing for fast exchange of nutrients including many essential elements to the plant such as carbon (Khan et al., 2022). Arbuscules supply carbon sources like hexose sugars and lipids to the plant and take phosphorous and other micronutrients (Salmeron-Santiago et al., 2023). This process is the backbone of the association between the plant and the fungus, with the formation and sustenance of arbuscules governed by genes such as RAM1 (Required for Arbuscule Maintenance), (Liu et al., 2022).

Extraradical Hyphae (ERH): ERH are structures that extend into the soil matrix beyond the root zone, searching for nutrients and enhancing plant nutrient uptake (He etr al., 2023). The ERH system significantly increases the absorptive surface area of plant roots, connecting nutrient-rich soil pockets to the host plant (Wendel et al., 2022). One of AMF's primary functions is the transport of phosphorus, a critical nutrient often limited in soils (Mitra et al., 2023). ERH take up inorganic phosphate (Pi) via phosphate transporter proteins such as GintPT and MtPT4 and actively transport it across fungal membranes (Duan et al., 2024). Phosphate is stored within ERH as polyphosphate granules and subsequently translocated to IRH, where it is released into plant cortical cells via the PT4 protein (Poirier et al., 2002). Symbiotic regulatory genes, including RAM1 and STR/STR2 (Symbiotic Transporter), ensure the stability of this nutrient exchange system (Vergara et al., 2024). In addition to phosphorus, AMF facilitate nitrogen transport, although they do not directly fix atmospheric nitrogen. Instead, they absorb ammonium (NH_4^+) and nitrate (NO_3^-) from the soil, using ammonium transporter proteins such as AMT2;3 and GintAMT

(Lilleskov et al., 2019). This nitrogen is assimilated before being delivered to the plant, thereby enhancing nitrogen use efficiency, particularly in nutrient-poor soils . ERH also extend into deeper soil layers, accessing nitrogen reserves and supporting plant nitrogen uptake (Munene et al., 2024).

In exchange for phosphorus and nitrogen, plants provide carbon to AMF in the form of hexose sugars and lipids (Wipf et al., 2019). This carbon transfer is vital for the sustainability of the symbiotic relationship. Genes such as RAM2, FatM and GPAT6 regulate lipid biosynthesis and transfer from plant cells to AMF, providing a continuous source of energy for fungal growth and function (Keymer et al., 2018). Thus, AMF networks both support growth and productivity by enhancing nutrient uptake by plants and contribute to the overall health of the soil ecosystem (Bano et al., 2021).

2.2. The Role of AMF in the Carbon Cycle and Soil Organic Matter Formation

AMF play a critical role in the carbon cycle and the formation of soil organic matter (Parihar et al., 2020). These organisms are deeply integrated into the soil carbon cycle, influencing carbon storage, microbial interactions, and soil aggregation (Schimel et al., 2012). Plants convert atmospheric CO_2 into organic compounds through photosynthesis and transfer these compounds to AMF via root exudates (Mohammed et al., 2022). These carbon compounds are utilized by AMF to sustain hyphal growth and metabolic activities (Mythili et al., 2022). Consequently, the carbon utilization by AMF contributes to the formation of extensive fungal networks in the soil, which play a key role in the functioning of soil ecosystems.

The carbon exchange facilitated by AMF not only supports fungal proliferation but also enhances the stability of soil organic matter (Solaiman et al., 2024). AMF hyphae decompose resistant organic matter in the soil, accelerating nutrient cycling and improving soil fertility (Wu et al., 2024). This process allows for the more efficient use of organic components in the soil, indirectly contributing to plant growth. Additionally, AMF produce glomalin-related soil proteins (GRSPs), which significantly contribute to soil carbon sequestration (Wang et al., 2023). Glomalin binds soil particles together, enhancing soil aggregation (Hossain et al., 2021). This not only improves soil structure but also increases water retention capacity, reduces erosion, and makes soils more resilient (Hartmann et al., 2023). Glomalin is considered a critical component for the long-term sustainability of soil ecosystems (Mishra et al., 2024).

As AMF hyphae age and decompose, they integrate their organic components into the soil organic matter pool (Qiu et al., 2021). This process enriches soil carbon storage and supports a diverse microbial community (Bhattacharyya et al., 2022). This function of AMF contributes to the long-term storage of carbon within soil ecosystems (Agnihotri et al., 2022). Furthermore, AMF networks promote the vertical transport of carbon into deeper soil layers (Xiao et al., 2023). This mechanism prevents carbon from remaining in surface soil layers and facilitates its long-term storage in deeper layers (Yang et al., 2023). This reduces the release of carbon back into the atmosphere, thereby playing a significant role in combating climate change (Holka et al., 2022).

Another critical impact of AMF on the soil carbon cycle lies in their interactions with microbial communities (Chourasiya et al., 2021). AMF enhance soil microbial diversity, fostering the establishment of symbiotic relationships (Zhang, et al., 2024). These symbiotic relationships optimize the functioning of soil ecosystems and enable more efficient utilization of carbon (Basiru et al., 2024). The carbon sources produced by AMF not only meet their own metabolic needs but also provide energy for other soil organisms, supporting soil biological activity (Ndeko et al., 2024).

In conclusion, AMF networks play a central role in the integration of carbon into soil ecosystems and the formation of organic matter (Tang et al., 2024). These processes, which facilitate the transfer of carbon from plants to AMF and subsequently to the soil organic matter pool, enhance soil fertility while contributing to climate change mitigation through carbon sequestration (Morya et al., 2023). Therefore, the role of AMF in the carbon cycle is of great importance for both the health of soil ecosystems and global carbon management (Mishra et al., 2024).

2.3. AMF Networks and Microbial Interactions

Not only do AMF have direct impacts on nutrient uptake by plants and the carbon cycle, but they also affect other soil microorganisms so encompass broader ecological roles. AMF promote soil ecosystem nutrient cycling and bioavailability (Ebbisa et al., 2022). This is established using interactions with phosphate solubilizing bacteria (PSB) and also nitrogen-fixing rhizobacteria (Pan et al., 2023). By collaborating with these microorganisms, AMF increase nutrient concentrations in the rhizosphere (the plant root zone) and optimize the bioavailability of essential nutrients (Bao et al., 2022). These interactions not only support plant growth and development but also improve soil fertility (Begum et al., 2019).

AMF indirectly impact ecosystem resilience and sustainability by promoting the formation of diverse microbial communities in the soil (Chourasiya et al., 2021). The symbiotic networks AMF establishes with plant roots provide habitat and energy sources for other soil microorganisms (Ullah et al., 2024). Carbon compounds secreted by AMF support the metabolic activities of soil microorganisms, increasing their diversity (Mythili, et al., 2022). This, in turn, strengthens the biological activity and functionality of the soil ecosystem. Moreover, AMF influence microbial communities in a way that suppresses pathogenic organisms while promoting beneficial microorganisms (Ji et al., 2022). This process positively affects soil health and plant productivity.

One of the significant effects of AMF on plants is their ability to enhance plant stress tolerance (Tahiri et al., 2022). Environmental stress factors such as drought, salinity, and pathogenic infections can negatively impact plant growth and productivity. However, AMF contribute to the development of resistance against these stress factors by forming symbiotic relationships with plants. The extensive hyphal network established by AMF with plant root systems increases water and nutrient uptake, supporting plants under abiotic stress conditions such as drought and salinity (Evelin et al., 2019). Additionally, glomalin and other biologically active compounds secreted by AMF modulate plant immune systems. Glomalin improves soil structure, enhances water retention capacity, and increases plant resilience to water stress (Mishra et al., 2024).

The mechanisms by which AMF strengthen plant immune systems involve complex molecular defense pathways. AMF trigger the production of defense components such as pathogenesis-related (PR) proteins and MYB transcription factors in plant roots (Song et al., 2011). These molecules elicit rapid and effective immune responses against pathogens in plants. Furthermore, AMF regulate microbial communities in the rhizosphere, limiting the spread of pathogens and reducing disease pressure (Ji et al., 2022). In this way, AMF enhance plant resistance to both biotic and abiotic stress factors.

In conclusion, AMF networks exert profound effects not only on plant nutrient uptake and the carbon cycle but also on ecosystem resilience, plant health, and soil fertility through their interactions with soil microorganisms (Thangavel et al., 2022). The regulatory role of AMF on microbial communities enhances the sustainability of soil ecosystems while strengthening plants' adaptive capacity to environmental stress factors. Therefore, the ecological functions of AMF hold strategic importance in the management of agricultural and natural ecosystems.

3. Nutrient Uptake and Transport Mechanisms

3.1. Phosphorus Uptake: The Role of AMF

Phosphorus is a macronutrient that plays a critical role in fundamental metabolic processes such as energy transfer, cell division, photosynthesis, and respiration in plants (Khan et al., 2023). However, phosphorus is often found in insoluble forms in the soil (e.g., calcium phosphate, iron phosphate, and aluminum phosphate), which limits its bioavailability to plants. AMF play a vital role in overcoming this limitation by enhancing phosphorus uptake in plants. The mechanism by which AMF facilitate phosphorus uptake is a complex biological system involving both extraradical and intraradical processes (Bhantana et al., 2021).

After establishing a symbiotic relationship with plant roots, AMF form an extensive hyphal network (hyphosphere) in the soil. This network extends into soil regions inaccessible to plant roots, thereby facilitating the mobilization and uptake of phosphorus. AMF secrete organic acids (e.g., citric acid, malic acid) and phosphatase enzymes to solubilize insoluble phosphate compounds in the soil (An et al., 2022). Organic acids form complexes with metal ions in the soil, releasing phosphate, while phosphatase enzymes hydrolyze organic phosphate compounds into inorganic phosphate forms (Rawat et al., 2021). The mobilized phosphate is actively absorbed by AMF hyphae through specific phosphate transporter proteins (e.g., Pi transporters) located in the AMF cell membrane (Etesami et al., 2021). Phosphate is then transported through cytoplasmic streaming within the hyphae toward the plant roots (Kakouridis et al., 2022).

Arbuscules, specialized structures formed by AMF within plant root cells, are the key sites where nutrient exchange between AMF and plants occurs. Phosphate absorbed from the soil is transferred from AMF hyphae to the arbuscule structure, where a symbiotic exchange takes place. AMF transfer phosphate to the plant root cells, while plants provide AMF with carbohydrates (e.g., hexoses and fatty acids) produced through photosynthesis (Salmeron-Santiago, et al., 2021). This exchange is regulated by phosphate transporter proteins (e.g., the PHT1 family) in the plant root cells (Chen et al., 2024).

This symbiotic process of phosphorus uptake is dynamically regulated based on both plant and soil conditions. When soil phosphorus levels are low, the phosphorus uptake mechanisms of AMF become more active. Similarly, plant roots secrete strigolactones that stimulate the formation of AMF hyphae and promotes their phosphorus uptake (Boyno et al., 2023). It is this mechanism using which the plant grows well in poor phosphorus soil and could reduce phosphorus fertilization requirement. It also manages to prevent phosphorus losses in the soil, what means that environmental pollution is also reduced.

As summarized, AMF phosphorus uptake is a challenging and efficient biological process involved in soil ecosystem functions to optimize nutrient cycle (Bhupenchandra et al., 2024). With the targeted support of plant growth, no two words can describe this better than to say that it also enhances agricultural sustainability and environmental safety. Phosphorus Uptake Strategies of AMF are of Strategeic importance To Increase Agriculture Productivity and Preserve Environment (Cao et al., 2025).

3.2. AMF-Mediated Nitrogen Cycle

P is another macronutrient which is foundational to nitrogen, a must have for plants and the backbone of proteins–nucleic acids–chlorophyll (Sandhu et al., 2021). Soil ecosystems, nitrogen may play a more important role in plant growth and productivity (Gou et al., 2023). AMF directly and indirectly contribute to this process in both direct and non-direct roles of the nitrogen cycle. Not only they will make the nitrogen in the soil available for plants but also operates on soil microbiota for efficient utilization of nitrogen cycling (He et al., 2023).

The role of AMF in the nitrogen cycle begins primarily with the transport of nitrogen to plants. AMF absorb inorganic nitrogen forms (e.g., ammonium and nitrate) from the soil through their extensive hyphal network and transport these nitrogen compounds to plant roots (Ebbisa et al., 2022). The large surface area of the hyphal network enables access to nitrogen sources in soil regions that plant roots cannot reach (Hammer et al., 2024). Nitrogen transporter proteins located in the AMF cell membrane actively mediate the transport of these compounds (Wang et al., 2024).

AMF also establish a synergistic relationship with nitrogen-fixing bacteria, supporting the conversion of atmospheric nitrogen into plant-available forms (Tharanath et al., 2024). Nitrogen-fixing bacteria transform atmospheric nitrogen into ammonia or ammonium, which are usable by plants, while AMF facilitate the transfer of these nitrogen compounds to plant roots (Matos et al., 2021). This symbiotic relationship optimizes nitrogen uptake by plants, particularly in soils with limited nitrogen availability, thereby promoting plant growth.

Another significant mechanism through which AMF contribute to the nitrogen cycle is their ability to decompose organic nitrogen compounds in the soil (Monika et al., 2022). AMF secrete enzymes such as proteases and amidases to solubilize organic nitrogen sources (Tirry et al., 2024). These enzymes break down organic nitrogen compounds into inorganic nitrogen forms, making them available for plant uptake (Zayed et al., 2023). This process not only reduces the accumulation of organic matter in the soil but also enhances the bioavailability of nitrogen (Bonassa et al., 2021).

An additional key role of AMF in the nitrogen cycle is their contribution to nitrogen retention in the soil. By improving soil structure, AMF reduce nitrogen losses through leaching. The hyphal network forms a physical matrix among soil particles, helping to stabilize nitrogen within the soil (Zhou et al., 2025). Furthermore, the regulatory effects of AMF on soil microbial communities balance nitrogen mineralization and immobilization processes, ensuring nitrogen remains in the soil for a longer duration (Yangyang et al., 2024).

These mechanisms provide both economic and environmental benefits by reducing the need for nitrogen fertilizers. The reduction of nitrogen losses lowers agricultural production costs while mitigating environmental issues such as nitrate pollution (Sitoe et al., 2022). The multifaceted role of AMF in the nitrogen cycle enhances the sustainability of soil ecosystems and holds strategic importance for improving plant growth and agricultural productivity.

3.3. Micronutrient Mobilization

Plants require micronutrients (elements required in minute concentrations) for healthy growth and metabolism and other vital functions including enzymatic reactions. Zinc, iron, copper and manganese are amongst the most common micronutrients in insoluble soil forms that leads to low bioavailable to the plants. AMF are essential for this mobilization and translocation of micronutrients to plants, preventing the occurrence of micronutrient deficiencies. AMF help in the uptake of micronutrients through their vast hyphal network and allows further transport from soil to plant root (Bhantana et al., 2021).

The mechanism of AMF-mediated micronutrient mobilization involves the solubilization of insoluble micronutrients in the soil (Thangavel et al., 2022). AMF secrete organic acids and chelating compounds to solubilize micronutrients. Organic acids, such as citric acid and malic acid, form complexes with metal ions in the soil, increasing the solubility of micronutrients (Justi et al., 2022). Chelating compounds bind micronutrients, ensuring they remain in soluble forms. Micronutrients mobilized in this way are transported to plant roots via the hyphal network of AMF.

The extensive hyphal network of AMF reaches soil regions inaccessible to plant roots, enabling the uptake of micronutrients from a broader area (Priyadarshani et al., 2025). Hyphae actively absorb micronutrients from the soil and transport them toward plant roots. Specific micronutrient transporter proteins located in the AMF cell membrane facilitate this process (Riaz et al., 2021). For instance, transporter proteins involved in the movement of micronutrients such as zinc and iron regulate the transfer of these elements to plant root cells. Arbuscules, specialized structures formed by AMF within plant root cells, play a crucial role in the transfer of micronutrients to plants (Xu et al., 2024).

Micronutrient mobilization not only benefits plant nutrition but also improves plant tolerances to stress. Micronutrient deficiency compromises the plant defense responses to abiotic (e.g., drought, salinity) and biotic (e.g., pathogens) stresses. AMF help the transport of micronutrients to plants which optimizes their metabolic processes and stress-resistance (Nie et al., 2024). Mobilization of micronutrients such as zinc and iron is of high importance to healthy functioning and growth of plant. Zinc is key for the activity and the gene expression of enzymes; iron, in the functioning of energy producing metabolic pathways such as photosynthesis and respiration (Sarma et al., 2023).

In addition to the above impacts of AMF on soil ecosystems, the AMFs influence also helps with remediation in soil nutrient cycling areas. AMF influence the soil architecture and holding capacity of micronutrients in soil, thus minimizes loss of nutrient. Further, the use of in situ solubilization by AMF in the micronutrient transport between the soil and host plant has reduced fertilizer use promoting agricultural sustainability (Bhantana et al., 2021). Micronutrient mobilization plays a significant role for multiple reasons; considering its strategic importance to the plant growth and development, stress tolerance and agricultural productivity, AMF in this process is the multifaceted player.

4. Molecular Basis of AMF Functionality

4.1. Key Genes Involved in AMF-Plant Symbiosis

The foundation of AMF-plant symbiosis lies in the precise regulation of specific genes and genetic mechanisms. This symbiotic relationship enhances nutrient uptake capacity, particularly through the activation of genes responsible for phosphate and nitrogen transport. During the colonization of plant roots by AMF, genes responsible for the remodeling of the plant root cell wall become active. For instance, early symbiotic signaling genes such as DMI1, DMI2, and DMI3 facilitate the entry of AMF into root cells and initiate the symbiotic interaction (Vergara et al., 2024). These genes regulate calcium signaling, which supports the growth of symbiotic hyphae within plant roots.

Genes involved in the synthesis of strigolactones play a pivotal role in promoting the germination of AMF spores and the directional growth of hyphae (Taulera et al., 2022). Specifically, strigolactone biosynthesis genes like CCD8 and MAX1 enable AMF to perceive signals secreted by plant roots, enhancing fungal movement toward the root zone (Boyno et al., 2023). During this process, genes involved in jasmonate and ethylene signaling, such as COI1 and EIN2, regulate the attachment of AMF to the root surface and the colonization process (Kadam et al., 2024). These hormonal signaling pathways also trigger metabolic changes in root cells, increasing the expression of phosphate transporter proteins (e.g., PT4) and optimizing nutrient transport (Ahmed et al., 2025).

Transcription factors induced by AMF play a critical role in regulating genetic networks that support the symbiotic relationship. For example, the RAM1 (Required for Arbuscular Mycorrhization 1) gene controls the expression of genes involved in arbuscule development and symbiotic nutrient transport (Ho-Plágaro et al., 2022). For example, NSP1 and NSP2 genes are responsible for lipid metabolism during mutualistic processes that are either required for the successful colonization of AMF in root cells (Hajiboland et al., 2024). Understanding the function of tolerance genes to abiotic stress conditions are critical with the AMF and plant symbiosis as well (Hajiboland et al., 2024). Key for ABF3 study are genes that improve tolerance to abiotic stress conditions Also in the symbiosis of AMF and plant differential gene expression analyses were used (Samanta et al., 2020).

In parallel with the increasing number of recent genomic and transcriptomic studies, that have highlighted more in detail the genetic networks underlying AMF's symbiosis (Kaya et al., 2023). Specifically, epigenetic regulation (i.e., histone modifications and DNA methylation) that regulate gene expression in root cells during AM colonization have been extensively studied and shed light on, manipulating the interaction dynamics (Kaya et al., 2023). Through these mechanisms gene expression is tailored to the specific symbiotic conditions with this allows both plant and fungus to gain benefits as maximally as possible.

4.2. Signal Transduction Pathways in Nutrient Cycling

The functionality of AMF in nutrient cycling is regulated by complex signal transduction pathways and genetic mechanisms between plants and fungi. This process is critical for the initiation, maintenance, and optimization of nutrient transport in the symbiotic relationship.

Strigolactones are key signaling molecules that promote the germination of AMF spores and the growth of hyphae (Boyno et al., 2023). Genes involved in strigolactone biosynthesis, such as CCD7, CCD8, and MAX1, regulate the production of these hormones and their secretion into the root zone (Selwal et al., 2023). AMF perceive strigolactones secreted by the plant and orient themselves toward the root surface, enhancing hyphal extension. During this process, signal molecules produced by AMF, such as Myc-LCOs (Mycorrhizal Lipochitooligosaccharides), activate early signaling pathways in plant root cells (Nasir et al., 2021). Plant root cells possess specialized receptor proteins to detect signals produced by AMF. For example, the SYMRK (Symbiosis Receptor-like Kinase) gene forms a receptor complex that perceives AMF signals and initiates intracellular signaling (Meena et al., 2023). This signaling triggers cellular responses through calcium signaling pathways regulated by DMI1, DMI2, and DMI3 genes, which control calcium release (Liu et al., 2022). Calcium signaling facilitates the remodeling of root cell walls and the entry of symbiotic hyphae into root cells.

The PT4 (Phosphate Transporter 4) gene, which is responsible for phosphate transport, becomes active during symbiosis and facilitates the transfer of phosphate to the plant via arbuscules (Luo et al., 2024). This process is supported by lipid metabolism regulated by RAM1 (Required for Arbuscular Mycorrhization 1) and RAM2 genes (Ho-Plágaro et al., 2022). Lipid secretion is a critical mechanism that strengthens AMF hyphal extensions and sustains the symbiotic relationship (Chauhan et al., 2022). Similarly, genes like AMT2, involved in nitrogen transport, are activated to optimize the transfer of nitrogen to plant roots (Rui et al., 2022).

Jasmonate and ethylene signaling also play essential roles in the symbiotic colonization process. Genes such as COI1 (Coronatine Insensitive 1) and EIN2 (Ethylene Insensitive 2) regulate the signaling of these hormones, promoting AMF attachment to the root surface and supporting colonization (Mierziak et al., 2024). These hormonal signaling pathways also induce metabolic changes in root cells, enhancing nutrient transport.

The recognition of mycorrhizal factors by plants occurs through bidirectional signaling mechanisms. For instance, Nod Factor-like signals secreted by the plant promote AMF root colonization, while Myc-LCOs and COs (Chitooligosaccharides) produced by AMF trigger the expression of symbiotic genes in plant root cells (Antar et al., 2021). During this process, genes such as CCaMK (Calcium and Calmodulin-dependent Kinase) and CYCLOPS regulate symbiotic signaling pathways, ensuring the continuity of the symbiosis (Bhupenchandra et al., 2024).

Lastly, interactions between AMF and the soil microbiota are also regulated by genetic-level signaling pathways. For example, in symbiotic interactions with rhizosphere bacteria, genes involved in phosphate solubilization and biological nitrogen fixation (e.g., the PHT1 phosphate transporter gene family and Nif genes) become active (Alzate Zuluaga et al., 2024). These signaling mechanisms enhance the synergy between AMF and rhizosphere bacteria, improving the efficiency of nutrient cycling and the sustainability of the symbiotic relationship. These genetic and mechanistic regulations enable the AMF-plant symbiosis to function as a complex network that not only supports plant nutrition but also contributes to ecosystem dynamics.

4.3. Genetic Regulation of AMF Networks

Genetic networks of AMF are essential to modulate a plant-fungi symbiotic interaction. Plant and fungal signaling molecules, as well as genetic mechanisms influencing those factors, are needed at every step in this process and genetics to start, maintain, and maximize symbiotic. The plant roots secrete strigolactones to stimulate the germination of AMF spores and attire hyphae on their surface . As these include genes CCD7, CCD8, and MAX1 involved in strigolactone biosynthesis (Verma et al., 2024). These signals are to be later recognized by AMF and induce hyphal growth of the fungal filaments merging with racemate (Fa et al., 2022). Mycorrhizal lipochitooligosaccharides (Myc-LCOs) and chitooligosaccharides (COs) produced by the fungi. The production of these signaling molecules could be perceived by SYN MBHRK (Yuan et al., 2022). These signals trigger the plant to produce calcium, which remodels cell walls and allows the symbiosis supplies to enter a root cell. This is further regulated intracellularly by CCaMK (Calcium and Calmodulin-dependent Kinase) gene-cytoplasmically controlled signaling pathways by CYCLOPS genes (Vergara et al., 2024).

During symbiosis the PT4, targets phosphate which is transported out during this process phytosymbiosis and into plant tissues via arbuscules There is support by lipid metabolism regulated by RAM 1 and ram 2 gene. It represents one of the most essential mechanisms to reinforce the AMF hyphal extensions and keep the symbiotic relationship. In parallel, other genes such as AMT2 involved in nitrate transport were responsive to have the best uptake of nitrate directly into plant roots in order to optimize an organ donor. The genetic control of AMF makes not only the transport of nutrients more efficient, but also makes it more resistant to environmental factors.

AMF's protection of the root system against pathogens were accomplished by enzymes and metabolic products produced by AMF that stimulate plant defense responses. For example, glucanases and chitinases are secreted by various fungi demonstrating the degree to which plant cell walls can be damaged, thereby preventing pathogen entry (Panicker et al., 2022). Moreover, jasmonic acid and ethylene signaling pathways up-regulate the expression of plant defense genes promoting plant defense-response machanism sustainability (Ahmed et al., 2025). Epigenetic modifications are seen as a major player in AMF-plant symbioses dynamic. These account for the regulation of gene expression to finely adjust fungal colonization of plant roots, and the promotion of nutrient translocation within plant cells.

The expression of symbiotic genes is regulated by epigenetic mechanisms such as histone modifications and DNA methylation (Kaya et al., 2023). For instance, increased level of H3K9 and H3K27 histone modifications improve the transcriptional activity of symbiotic genes thus helping their sustainability (Schardl et al., 2022).

In interactions with soil microbial communities, the genetic regulations of AMF play a central role. AMF act as a bridge among soil microorganisms, strengthening symbiotic relationships and preserving microbial diversity (Qiu et al., 2024). The synergy between rhizobacteria and AMF optimizes processes such as phosphate solubilization and biological nitrogen fixation (Janati et al., 2021). In this context, the phosphate transporter gene family (PHT1) and nitrogen fixation genes (NIF) become active (Altındal et al., 2024). The genetic mechanisms of AMF that enhance carbon sequestration and resilience to environmental stress factors are of great importance for agricultural sustainability.

5. Interactions with Soil Microbiota

5.1. The Central Role of AMF in Soil Microbial Communities

AMF are central actors in soil microbial communities and participate ecosystem functioning by having symbiotic hosts on roots (Khaliq et al., 2022). Colonized in plant roots, AMF delves into the fungal hyphae and establishes arbuscules structures between root tissues. They represent specialized interfaces for nutrient exchange between plant and fungus. AMF transport phosphorus, zinc as well other micronutrients to the plant, but they receive carbohydrates in exchange to the same parole of effect. This mutualistic interaction augments the solubility of phosphorus, which is essential to plant growth and more importantly beneficial to khaliqpoor soils.

Fungal hyphae form an extensive network in soil, gathering above the reach of plant root areas. This network also speeds up the degradation of organic matter and improves soil cycling of nutrients. Hence the physical structure that hyphae created reinforced soil aggregates, erosion protection and water holding capacity (Aminzadeh et al., 2025). This is one of the major components in rehabilitation of degraded or nutrient-poor soils through these effects. Additionally AMF is influenced with other microorganisms in the rhizosphere. Rhizobacteria form mutualisms with them and help process nitrogen fixation. Rhizobacteria convert atmospheric nitrogen into a plant-usable form while AMF increase phosphorus solubility, thus representing a synergistic effect (Kalamulla et al., 2024). These effects contribute a large part in making degraded or nutrient poor soils responsive by returning stress regenerative capacities on vegetation. AMF also interact with other soil microorganisms in the root-borne community (rhizosphere). In symbiosis with the rhizobacteria and assisting them carry out biological nitrogen fixation. NR is an ecological converter that converts atmospheric nitrogen to nitrate in a plant available form while AMF make phosphate more available, which creates a co-operation The collaboration between plants and the AMF make it easier for the former to leverage essential nutrients such as nitrogen and phosphorus from the soil. Therefor agricultural productivity increases, dependence on chemical fertilizers decreases and sustainability gets push.

Another key function of AMF in soil microbial communities is its effects on stabilizing the diversity and stability of soil microbial response. AMF more or less reconciles the competition among soil microorganisms, allowing their symbiotic relationships does not wither away (Wahab et al., 2023). Such diversity makes soil ecosystems resilient and help in mobilizing plant defense against both biotic and abiotic stresses. AMF establish resistance to strong environmental stresses such as drought, salinity and heavy metal contamination when plants get invaded by AMF (Riaz et al., 2021). AMF support detoxification processes by preventing the toxic ions from harming plants.

Overall, AFF is a comprehensive mechanism that contributes to both plant development and sustainability of soil ecosystems centered in AMF with multiple levels. These interactions are basic ecological processes that not only sustain healthy ecosystems in nature but also encourage sustainable agricultural practices. AMF contributions can contribute to the ecological balance and promote productivity in agricultural practice.

5.2. Synergistic Effects with Other Soil Microorganisms

AMF play a major role in the development of plant resistance with respect to both biotic and abiotic stresses through above soil microorganisms synergistic interactions. These exchanges take place with mechanisms such as pathogen suppression, microbial diversity inoculation and provision of plant resilience under stress conditions. Yet another is the pathogen suppression mechanism of AMF (Dey et al., 2022) affords defense in plants against diseases. AM fungi form a hyphal network in the root zone that provides a physical barrier and prevents access by pathogens to plant roots (Ebbisa et al., 2022). In addition, several of AMF produced metabolites provide direct inhibition on the growth of pathogens (Zhao et al., 2022). These also include antimicrobial compounds that inhibit the multiplication/activity of pathogens. Moreover, AMF could activate an endo systemic resistance in plants (Yu et al., 2022). This activation results not only in more robust defense responses of the whole plant than just the local site and at all tissues (systemic), but the ability to defend against pathogens is also enhanced. AMF are important in the maintenance of soil ecosystem balance by their ability to increase and maintain diversity of microbial life. AMF form tight symbiotic associations with a different group of rhizosphere beneficial microorganisms to assist its functions (Bao et al., 2022). For example, the rhizobacteria they partner with, enhance each other in processes such as nitrogen fixation and phosphorus solubilization. AMF furthermore ensure high nutrient turnover in soil, as well as a mode of populating beneficial microorganisms. Together these group factors increase the resistance of plants against stress and enables soil ecosystems to survive under varying condition.

The effects of AMF become even more pronounced under abiotic stress conditions such as drought, salinity, and heavy metal contamination (Ortas et al., 2021). Plants colonized by AMF exhibit greater resilience to drought stress because AMF enhance water retention in the soil and improve water uptake by the plant's root system (Madouh et al., 2023). Against salt stress, AMF protect plants by preventing the accumulation of salt ions in root tissues, thereby maintaining osmotic balance (Wang et al., 2022). In soils contaminated with heavy metals, AMF bind toxic ions, preventing them from reaching plants and supporting detoxification processes (Devi et al., 2021). This mechanism is associated with the ability of AMF to form a hyphal network that stabilizes heavy metals in the soil. As a result, plants are shielded from toxic effects and can sustain their growth.

In conclusion, the synergistic interactions between AMF and soil microorganisms are based on multifaceted mechanisms that support plant development and sustain the functioning of soil ecosystems. These mechanisms not only promote individual plant health but also contribute to the sustainability of agricultural production. The use of AMF as a biological fertilizer provides environmental and economic benefits, helping to preserve both natural ecosystems and agricultural systems.

6. Ecological and Environmental Impacts

6.1. Role of AMF in Carbon Sequestration

AMF play a significant ecological role in carbon sequestration and the regulation of the carbon cycle. The symbiotic relationship between AMF and plants facilitates the biological capture of atmospheric carbon dioxide (CO_2), contributing to the integration of carbon into soil ecosystems (Basiru et al., 2024; Boyno et al., 2024). This process supports environmental sustainability by reducing the carbon footprint in both natural ecosystems and agricultural areas.

Central to the role of AMF in carbon sequestration is the fact they form a symbiotic relationship with plant. Plants synthesize organic carbon compounds from atmospheric CO_2 through photosynthesis and plant roots then transfer some carbon to AMF (Solaiman et al., 2024). AMF metabolize this carbon and add to soil organic carbon accumulation by their respiration. For storing carbon in the soil, the hyphal network AMF establish is instrumental (Jeewani et al., 2021). Rhizosphere hyphal enhances soil structure by holding soil particles through a physical trapping of carbon. Furthermore, glomalin, an important soil protein to stabilize the long-term storage of carbon produced by AMF. Glomalin is one of the most efficient soil organic matter stabilizers providing an optimal carbon retention years (Hossain et al., 2021). It promotes carbonates volcanoes to create soil aggregates, thus protecting the carbon physically. This way it decreases the amount of soil carbon that is released to the atmosphere again resulting in less greenhouse gas emissions.

AMF filtering carbon from the atmospheric is only partly due to its carbon-saving. WoodWow IHF– hyphal network of AMFThe helper in helping to maintain soil organic matter levels and increase soil stability Table 1. Soil organic matter nourishes plant growth with nutrients and improves its water retention capability. This AMF effect is especially helpful under stress conditions like drought, for which the soil becomes a refuge (Benaffari et al., 2022). It will also bolster soil compactness and buildup of soil sovereignty by reducing erosion. We propose that the AMF hyphal network which are established between upper soil layers, facilitates more sustainable water transport in the soil (Wang et al., 2022). This is essential for soil health in farming and also wild systems.

The last but not least role of AMF in carbon sequestration is their reducing potential for green house gases emission (Basiru et al., 2024). By sequestering carbon into the soil, AMF lower atmospheric CO_2 levels (Agnihotri et al., 2022). This is one of the few strategies to tackle climate change. Agricultural areas also help the AMF as biofertilizer to substitute the chemical fertilizers depending on which much of agricultural caused greenhouse gas releases over the years. Overuse has the potential to produce high amounts of greenhouse gases, such as nitrous oxide (N_2O) in the soil.

Applications of AMF as a biofertilizer reduction of these emissions AMF plays a significant role in the carbon cycle for the sustainability of natural ecosystems and agricultural systems. In natural ecosystems, AMF networks increase soil carbon storage helping to maintain ecosystem resilience. In agro-ecosystems, the beneficial effects of AMF on carbon sequestration makes the crop production more sustainable (Park et al., 2024). AMF contributes to long-term agricultural productivity by increasing soil organic matter and soil physical properties. AM fungi not only sequester carbon in the soil, but they also increase soil health and reduce greenhouse gases whilst enhancing ecosystem resilience The use of AMF in natural ecosystems and agricultural areas is critical to solve climate change together with sustainable agricultural practices. AMF has great potentials for realizing sustainability both in the agricultural ecosystems and in the environment in future as a bioresource.

6.2. AMF Networks in Sustainable Agriculture

AMF networks are critical tools for the sustainability of agriculture in farmer perspective. Symbiosis with AMF allows plants to have better soil conditions AMF also directly benefits agricultural productivity and reduce dependency on chemical fertilizers. Excessive application of chemical fertilizers in water systems throws phosphorus and nitrogen into the sewers which mess with soil microbiome disrupting soils' overall productivity, consequently affecting fertility in the longer term. This is how AMF helps, which also aids the continuity of sustainable agricultural systems. AMF as Biological Fertilizer: Improvement in Soil Health and Preservation of Biodiversity in Agricultural Ecosystems (Kuila et al., 2022). It contributes to the soil biodiversity of diversity of these beneficial microorganisms that in many ways is also a reflection on the resilience of an ecosystem.

AMF bring higher level of soil organic matter in agricultural lands, thus helping to improve physical, chemical and biological properties of the soil. Over-retention

Table 1. Genes involved in AMF and plant symbiosis, their functions and related processes

Gene name	Function	Related process	Plant/Fungus
CCD7, CCD8, MAX1	Regulates strigolactone biosynthesis, promotes AMF spore germination.	Strigolactone production and hyphal growth	Plant
SYMRK	Forms a receptor complex to perceive AMF signals.	Signal perception and symbiosis initiation	Plant
DMI1, DMI2, DMI3	Regulates calcium signaling, activates symbiotic genes.	Intracellular signal transduction	Plant
CCaMK	Detects calcium signals and activates symbiotic genes.	Signal transduction and gene expression	Plant
CYCLOPS	Works with CCaMK to regulate the activation of symbiotic genes.	Signal transduction and symbiosis regulation	Plant
PT4	Facilitates phosphate transport through arbuscules.	Phosphate transport and nutrient cycling	Plant
RAM1, RAM2	Regulates lipid metabolism, supports hyphal growth in symbiosis.	Lipid secretion and symbiotic interaction	Plant
AMT2	Regulates nitrogen transport.	Nitrogen transport	Plant
PHT1	Phosphate transporter gene family, facilitates phosphate uptake by plants.	Phosphate solubilization and transport	Plant
NIF genes	Involved in nitrogen fixation.	Biological nitrogen fixation	Fungus/ Bacteria
Myc-LCOs	Signal molecules produced by AMF, initiate signaling in plant cells.	Signal transduction and symbiosis regulation	Fungus
COs	Produces signal molecules that reshape plant cell walls.	Cell wall remodeling	Fungus
H3K9, H3K27	Histone modifications enhance transcription of symbiotic genes.	Epigenetic regulation	Plant/Fungus
EIN2	Regulates ethylene signaling, supports symbiotic colonization.	Ethylene signaling	Plant
COI1	Regulates jasmonate signaling, optimizes plant defense responses.	Jasmonate signaling	Plant

and deficient nutrient from soil organic matter may lead to contamination of essential plant nutrients and restrict the accessibility of them to plants. It also helps to enhance the soil structure, which works in reducing erosion. AMF develops the network of hyphae that links soil particles together, improving soil physical stability and water flow within the soil (Agnihotri et al., 2022). This effect is especially advantageous in stress conditions like drought since it supplies the plant with water and defends agricultural production.

The above AMF impacts have been shown to in agricultural ecosystems so that they are able to help ensure long-term sustainability of patterns of land use and critical natural resource base. Furthermore the AMF use as bio-fertilizer provides economical advantages too. Decreasing the amount of chemical fertilizers they apply, and therefore lessening expenses and environmental burdens from chemical fertilizers. Overuse of the chemical fertilizers causes pollution in water and also increase greenhouse gas emissions that accelerate climate change. AMF facilitate natural isolation of nutrients that require sustainable agriculture, which may avoid the detrimental effect of these practices. The implementation of AMF in soil ecosystems could be such a way to increase agricultural outputs and sustainable environmental actions. Hence, the widespread availability of AMF networks in agricultural systems has considerable potential for both economic and environmental advantages.

6.3. AMF Response to Environmental Stressors

AMF play an important function in the strengthening of plant tolerance to stressors that cause agricultural productivity and ecosystem health. Drought, salinity and heavy metal contamination are just some of the many types of abiotic stress that can negatively affect plant growth and yield. AMF alleviate the impacts of these stress factors by inducing multiple physiological and biochemical defense mechanisms inside plants which further enhance their resilience to the adverse environmental stresses (Nie et al., 2024).

AMF act to help plants recover from drought stress firstly by improving their water uptake. By constructing a hyphal network in the root system that amends soil water, AMF allow plants to access water which would be unavailable to them. Hyphae penetrate soil areas not colonized by plant roots, evetually moving the water towards roots. Water loss reduction is the primary way that the plants are able to conserve water during these periods of drought. AMF also regulate water transport channels in plant root cells, increasing efficiency in water use (Tang et al., 2022). Thus, plants show an enhanced water use efficiency and become more drought-stress tolerant. This effect of AMF is especially important in water limited regions contributing to the sustainability of agricultural production under periods of drought.

Salinity stress: AMF help regulate soil ion homeostasis under AMF stress (Abbaspour et al., 2021). Salt water can lead to the accumulation of Na^+ and Cl^- ions in the root system of plants causing salt toxicity (Alharbi et al., 2022). AMF prevent the accumulation of these ions in plant root zone and conserve or restore ion homeostasis at plant cell level by xylem loading (Wang et al., 2022). Furtheron more AMF activate the synthesis of proline and various osmolytes which help in osmoregulation of plants (Han et al., 2024). It helps preserve the equilibrium of water within cells and increases cell tolerance to salinity stress These salinity-stress-reducing capabilities may be used as an alternative tool for agriculture to ease the high demand on agricultural productivity in such environment conditions, increasing tolerance to salinity (Vasilachi et al., 2023).

AMF reduce the bioavailability of heavy metals in soil as plants do not have access to these toxic elements (Riaz et al., 2021) the metal chelation and detoxification mechanisms of AMF decrease the metal accumulation process to plant roots (Liu et al., 2022). AMF enhance the activity of antioxidant enzymes, and protect the plant from oxidative stress (Yilmaz et al., 2023). In the above contaminated field conditions, AMF stimulate the biosynthesis of phytochelatin and Glutathione in plants that reduces the toxicity elements of toxic chemicals and enhances plant's tolerance to heavy metals (Ahammed et al., 2023). This is especially favorable in agricultural fields near industrial areas or soils polluted from mining activities, promoting plant survival rates and conserving soil productivity. COISAggravate heavy metal contamination can cause toxic effects on plants even inhibiting growth and development (Vasilachi et al., 2023). AMF reduce the bioavailability of heavy metals a sub soil environment/r by helping plants to avoid the toxic elements (Riaz et al., 2021). The metal chelation and detoxification mechanisms of AMF will facilitate to reduce the metal transfer to roots from soil (Liu et al., 2022). Furthermore amFase regulators can evoke proline enhancement and expression of sorbance or other osmolyte to maintain plants osmotic homeostasis in heavy metal stressed conditions.

The AMF responses to environmental stressors do not only mitigate the adverse effect on agricultural productivity but look to keep biotic integrity protected as well. The AMF symbiosis with plants enhances plant survival under stress and soil health and root rooting behavior also contribute natural resources conservation. Consequently, the functioning of AMF for environmental stressors is potentially quite meaningful to agricultural sustainability and ecosystem resiliency. On a larger scale biological solution to environmental stressors is available in the application of AMF into agricultural systems that are extremely widespread.

7. Biotechnological Applications of AMF

AMF have a mutualistic association with plants and play important roles in several biotechnological areas like agriculture, environmental biotechnology and forestry where they are amply used for various purposes. Fungi of this sort provide the necessary services to fertilize plants, enhance their growth and increase tolerance to droughts and salinity as well as soil health, and participate in bioremediation processes (Tomer et al., 2021). AMF have crucial role in plant production as they allow plants to grow better with less phosphorus and other nutrients. AMF manipulate plant metabolism that results in the production of phytohormones and secondary metabolites which include phenolic compounds, alkaloids and flavonoids that help plants to withstand abiotic stresses due to biotic and environmental factors (Amani Machiani et al., 2022). That is also why it contributes actively to raise the concentration of active components in medicinal and aromatic plants.

AMF + genetic engineering studies are among the opportunities for integrating stress tolerance mechanisms of plant species AMF-infected plant's responses and genetic/physiological adaptations that AM fungi mediated can be combined with use of Genetically Modified Organisms (GMO) allowed agriculture to generate more robust plant type. Also transcription factors that affect gene expression and when exposed to stress-specific genes are activated by AMF cooperating genes. AMF signals induce the activation of plant genes that could increase the resistance of the plant to abiotic stresses; i.e., droughts salinity and heavy-metal contamination (Sharma et al., 2021). More so, the application of genetic engineering techniques to increase the symbiotic ability of AMF has been advanced for plant species able to be acclimated into different and different environments.

Techniques including CRISPR-Cas9 or genetic modification aimed at the optimization of genetically profiles of plants spatially interacting with AMF, reliably increase water and nutrient use efficiencies (Chauhan et al., 2024). Besides this, observing epigenetic modifications in induced plants inoculated with AMF plant can help us to better understand and orient adaptations of plants to environmental conditions. All these approaches have tremendous potential for a sustainable agricultural production in the future. Additionally essential for agricultural production potential, some AMF species produce antioxidants and antimicrobial compounds that ameliorate the yield of medicinal plants (Zhao et al., 2022). That is why AMF are largely used in pharmaceutical biotechnology.

AMF is distributed over a wide range of applications in the biotech sector, from farming (agricultural) to environmental sustainability. The benefits that AMF provide, for instance by diminishing chemical fertilizer usage, increasing plant-stress resilience and sparing environmental restorative processes are enormous. To better understand and harness AMF for use in genetic engineering as well as biotechnology in the future, more research is necessary.

7.1. Utilizing AMF for Enhanced Nutrient Use Efficiency

AMF absorb nutrients from the soil through their hyphal networks and transfer them to plants. This process occurs at the molecular level through the coordination of various genes and cellular mechanisms. Hyphae, with their extensive surface area, enable the uptake of nutrients more efficiently and make nutrients from regions inaccessible to plant roots available for use (Das et al., 2022). The roots of plants colonized by AMF exhibit more efficient growth due to this nutrient transfer. The transfer of nutrients occurs through several key mechanisms:

Phosphorus Transport: Phosphorus (P) is essential for vital plant functions such as DNA synthesis, energy transfer, and photosynthesis. However, it is often found in insoluble forms in the soil. AMF convert phosphorus into soluble forms that plants can utilize. They collect inorganic phosphorus (Pi) from the soil via hyphae and transfer it to plants through phosphorus transporter proteins (PHT1 family) located in plant root cells (Srivastava et al., 2021). Phosphorus uptake is a critical process for plant growth and energy metabolism. During the interaction of AMF with root cells, the PT4 gene regulates phosphorus transport, thereby optimizing this process (Shao et al., 2021). Efficient phosphorus transport supports photosynthesis and cellular respiration mechanisms in plants, contributing to the maintenance of ecosystem balance.

Nitrogen Uptake: Nitrogen (N) is vital for protein synthesis and photosynthesis. It exists in the soil in forms such as nitrate (NO_3^-) and ammonium (NH_4^+), but its uptake by plants can sometimes be limited. AMF extract nitrogen forms like nitrate and ammonium from the soil environment through hyphae and transfer them to plant roots (Monika et al., 2022). Nitrogen plays a crucial role in biochemical processes such as protein synthesis and DNA replication. The AMT (Ammonium Transporter) gene family is a critical factor in regulating this process (Hui et al., 2022). The nitrogen provided by AMF can enhance plant growth, particularly in nitrogen-deficient soils, thereby increasing agricultural productivity.

Water Use Efficiency: AMF facilitate for plants to absorb water enabling drought tolerance. Their hyphal network allows water to be taken in from a greater soil volüme (Kakouridis et al., 2022) through which they also improve hydraulic conductivity, increase transport of water to roots. Especially in regions of water scarcity, this mechanism is of primary importance to ensure sustainable agricultural productivity.

Use of Micronutrient: Micronutrients such as Zn, Fe (chelated), Cu (expressed) are necessary for the growth and development in plant. AMF can absorb soil elements (zinc, iron and copper) to increase the bioavailability of these elements for plant growth. This mechanism could in turn, help to conserve plant physiology and increase productivity on agricultural lands where there are deficiency of micronutrients.

Carbon Transfer: AMF receive their energy and the carbon-based sugars that are produced from plants through photosynthesis (Bell et al., 2024). They at the same time increase plant tolerance to stress (e.g., salinity) Characterizing the root-soil interface and root architecture by micrometeorite soil respiration: Such mechanisms improve the efficiency of the carbon cycle and therefore represent an important contribution for ecosystem homeostasis (Carbon). AMF also help in sequestering Carbon for long term storage in soil ecosystems as well by soil organisms and in doing so they may be shaping resistance to climate change.

AM fungi take part in nutrient cycling and also represent ecosystem services like soil health, carbon storage, soil reinforcement, disease resistance by plants through them. AMF in agriculture IS an alternative to synthetic fertilizers and moreover supports for organic farming, AM within their capabilities of being suit up produced. AM holds a lot of hope, especially for the promotion of biofertilizers and soil microbiota. Application of bio-fertilizers with AMF at agricultural field not only increase the crops' yield but also soil safe.

7.2. Genetic Engineering Approaches to Enhance AMF Symbiosis

Several strategies for improving AMF coexistence have already been developed in the context of genetic engineering. They are designed to improve the efficiency of symbiosis by changing both plants and AMF's genetic and physiological characteristics.

Mycorrhizal gene inoculation – to another level, it is the implantation of genes that increase the root colonization and symbiotic relationship i.e. produce more mycorrhizal and are inserted into a plant. Selectively manipulating the expression of specific AMF–plant interaction genes can improve symbiotic efficacy by identifying specific genes associated with AM fungi. For example, the PT4 phosphate transporter genes or transcription factors such as RAM1, which are central to AM root interactions can be upregulated and thus promote the symbiosis (Kiddee et al., 2024).

Alterations in Plant-AMF Signal Transduction Networks: Signal transduction is central in the plant—AM symbiosis. AMF are the key fungus that plants communicate with by releasing strigolactones from their root tip in order to attract AMF and initiate symbiosis. The development of agricultural crops is achieved by genetically engineering these signaling molecules either to be increased in production/sensitivity, which cause AMF-plant root colonization. Moreover, modifying receptor proteins tethered to the plant cell wall can tether AMF – plant root cell interactions. Targeting genes through regulating these signaling pathways with gene-editing tools (Boyno et al., 2023).

Fungal Genome Editing: Genetically modifying AMF species can increase their root colonization capacity and establish a more effective symbiotic relationship with plants. Although working on AMF genetics is currently challenging, newly developed genome-editing tools (e.g., CRISPR-Cas9) can target genes that enhance AMF symbiotic abilities (Boyno et al., 2023). Through these advancements:

- The development of AMF mycelium can be accelerated,
- Their utilization of nutrients obtained from plants can be made more efficient,
- AMF species can be developed to be more resistant to environmental stress conditions.

Fungal genome engineering can produce specially developed symbiotic fungal species for agricultural use. These fungi can be optimized to better adapt to specific soil types or plant species.

Epigenetic Modifications: Epigenetic mechanisms regulate gene expression without altering the DNA sequence, influencing plant responses to environmental conditions and symbiotic organisms. To enhance AMF symbiosis, the epigenetic regulators of plants can be targeted (Kaya et al., 2023). For example:

- Activation of AMF-related genes can be increased through histone modifications or DNA methylation,
- MicroRNAs involved in establishing symbiosis with AMF can be regulated to optimize plant responses to AMF.

Such epigenetic modifications, combined with genetic engineering, can offer long-term solutions to enhance AMF symbiosis (Kaya et al., 2023). Additionally, advanced molecular biology and bioinformatics techniques are being used to better understand the symbiotic interactions between plants and AMF. RNA sequencing (RNA-Seq) and proteomic analyses can identify the genes activated in plant cells during AMF colonization, allowing these genes to be regulated through engineering approaches.

8. Challenges and Future Perspectives

8.1. Knowledge Gaps in AMF-Driven Nutrient Cycling

Arbuscular mycorrhizal fungi (AMF) colonizes the roots of plants and makes it possible for these organisms to take up phosphorus and nitrogen as well as other nutrients from the soil they may need for themselves. The process is important for plant productivity and ecosystem health. AMF in nutrient cycling and how this mutualism occurs remains poorly understood despite its substantial importance for agriculturally as well as environmentally applied work. These issues put AMF in one step out from being efficiently used in agricultural and environmental interactions.

The molecular and biochemical processes by which AMF provide nutrients to plants, especially necessary ones like phosphorus and nitrogen, still not entirely known. Unanswered still are queries on the genetic components involved in this symbiotic process and how environmental conditions affect these mechanisms. For example, different soil types or stress conditions might affect the symbiotic interaction between AMF and plants. More study is required to clarify the causes and effects of these differences. Nutrient cycling is also greatly influenced by AMF's interactions with soil microbial communities (Wang et al., 2021). Plant health and soil quality may be significantly affected by AMF's competition or cooperation with other microorganisms. We still, though, know too little about the results at the ecosystem level of these interactions. In particular, the variations between AMF's functions in natural ecosystems and agricultural ecosystems have to be better understood. Although AMF in agricultural systems can boost production, its role in natural ecosystems may be more complicated and have long-term effects. More thorough research on AMF's contributions to ecosystem services is therefore essential in this setting.

Ultimately, depending on plant species and soil type, AMF's involvement in nutrient cycling can differ greatly. For instance, some plant species have more pronounced symbiotic connections with AMF than others and so use this link to a lesser degree. More study is required to investigate the causes of these variations and determine how to maximize this symbiotic interaction. Knowing completely AMF's function in nutrient cycling will help to more efficiently use these organisms in environmental and agricultural uses.

8.2. Emerging Technologies for AMF Research

New technologies offer significant potential for improved knowledge of AMF and their effective application in agricultural uses. These technologies let one completely investigate AMF's genetic structure, biological functions, and symbiotic mechanisms. Advancing AMF research, thus, calls for handling specific issues these technologies generate.

Examining and changing the genetic composition of AMF using precise gene-editing technologies like CRISPR appears possible (Boyno et al., 2023). CRISPR technology can be used to investigate the functions of particular genes and boost the symbiotic capacity of AMF. But the use of these technologies is constrained by AMF's dual-nucleus genomic structure and their lab culture challenges. If CRISPR is to be properly incorporated into AMF research, more technical innovations and optimization are still needed.

Techniques of bioinformatics and systems biology are progressively clarifying the genetic structure and symbiotic interactions of AMF. Big data analytics allows more thorough investigation of AMF genomic, transcriptomic, and proteomic data. Such studies determine AMF nutrient transfer mechanism knowledge and environmental stress adaptation knowledge. However, instruments for this sector have to be more easily accessible and simple to use. It is also quite crucial to construct the required infrastructure for big data analysis in AMF studies.

Omics technologies—especially their capacity to expose biological functions and symbiotic mechanisms—helps one to grasp AMF (Salvioli and Bonfante, 2013). Tools in genomic, transcriptomic, proteomic, and metabolomic allow investigation of AMF gene expression and metabolic pathways. These tools can point out possible goals to maximize the symbiotic interaction between AMF and plants. On the other hand, using omics technologies might be expensive and technically difficult. Thus, it is absolutely vital to create more reasonably priced and efficient techniques.

Field studies and long-term monitoring enable one to understand the effects of AMF in agricultural and environmental applications. Examining the long-term impacts of AMF within ecosystems could help guide their use in sustainable agriculture by means of symbiotic interactions. Field studies, on the other hand, can be costly and time-consuming. Advancing AMF research depends on supporting and carrying out such studies on a larger scale.

AMF study technologies ultimately offer fascinating analysis of the genetic structure and biological functions of these entities. Among the gene-editing tools, CRISPR is one of the most potent for investigating and optimizing AMF symbiotic mechanisms; bioinformatics and omics technologies are also powerful tools for this goal. Field studies and long-term monitoring also help to assess AMF's ecosystem-level effects. Future studies will be absolutely necessary to fully exploit AMF's possibilities in environmental and agricultural applications. How are things going there?

References

Abbaspour, H., Pour, F.S. & Abdel-Wahhab, M.A. (2021). Arbuscular mycorrhizal symbiosis regulates the physiological responses, ion distribution and relevant gene expression to trigger salt stress tolerance in pistachio. Physiology and Molecular Biology of Plants 27(8), 1765–1778.

Agnihotri, R., Sharma, M.P., Prakash, A., Ramesh, A., Bhattacharjya, S., Patra, A.K. & Kuzyakov, Y. (2022). Glycoproteins of arbuscular mycorrhiza for soil carbon sequestration: Review of mechanisms and controls. Science of the Total Environment 806, 150571.

Ahammed, G.J., Shamsy, R., Liu, A. & Chen, S. (2023). Arbuscular mycorrhizal fungi-induced tolerance to chromium stress in plants. Environmental Pollution 327, 121597.

Ahmed, N., Li, J., Li, Y., Deng, L., Deng, L., Chachar, M. & Tu, P. (2025). Symbiotic synergy: How Arbuscular Mycorrhizal Fungi enhance nutrient uptake, stress tolerance, and soil health through molecular mechanisms and hormonal regulation. IMA Fungus 16, e144989.

Alharbi, K., Al-Osaimi, A.A. & Alghamdi, B.A. (2022). Sodium chloride (NaCl)-induced physiological alteration and oxidative stress generation in *Pisum sativum* (L.): A toxicity assessment. ACS Omega 7(24), 20819-20832.

Altındal, D. & Altındal, N. (2024). Different interactions of plants in the rhizosphere: Mechanisms and their ecological benefits. *In*: Microbiome and Nano-Cross-Talk (pp. 189–200). Academic Press.

Alzate Zuluaga, M.Y., Fattorini, R., Cesco, S. & Pii, Y. (2024). Plant-microbe interactions in the rhizosphere for smarter and more sustainable crop fertilization: The case of PGPR-based biofertilizers. Frontiers in Microbiology 15, 1440978.

Amani Machiani, M., Javanmard, A., Habibi Machiani, R. & Sadeghpour, A. (2022). Arbuscular mycorrhizal fungi and changes in primary and secondary metabolites. Plants 11(17), 2183.

Aminzadeh, A., Dorostkar, V. & Asghari, H.R. (2025). Soil structural stability improvement using arbuscular mycorrhizal fungi and biochar in water repellent and non-water repellent soil. Soil Use and Management 41(1), e70024.

An, X., Liu, J., Liu, X., Ma, C. & Zhang, Q. (2022). Optimizing phosphorus application rate and the mixed inoculation of arbuscular mycorrhizal fungi and phosphate-solubilizing bacteria can improve the phosphatase activity and organic acid content in alfalfa soil. Sustainability 14(18), 11342.

Antar, M., Gopal, P., Msimbira, L.A., Naamala, J., Nazari, M., Overbeek, W. & Smith, D.L. (2021). Inter-organismal signaling in the rhizosphere. Rhizosphere Biology: Interactions Between Microbes and Plants 255–293.

Bano, S.A. & Uzair, B. (2021). Arbuscular mycorrhizal fungi (AMF) for improved plant health and production. *In*: Microbial Biotechnology in Crop Protection (pp. 147–169). Singapore: Springer Singapore.

Bao, X., Zou, J., Zhang, B., Wu, L., Yang, T. & Huang, Q. (2022). Arbuscular mycorrhizal fungi and microbes interaction in rice mycorrhizosphere. Agronomy 12(6), 1277.

Basiru, S. & Hijri, M. (2024). Trade-off between soil organic carbon sequestration and plant nutrient uptake in arbuscular mycorrhizal symbiosis. Fungal Biology Reviews 49, 100381.

Basyal, B. (2024). Plant-arbuscular mycorrhizal fungi association under drought stress. *In*: Arbuscular Mycorrhizal Fungi in Sustainable Agriculture: Nutrient and Crop Management (pp. 211–221). Singapore: Springer Nature Singapore.

Begum, N., Qin, C., Ahanger, M.A., Raza, S., Khan, M.I., Ashraf, M. & Zhang, L. (2019). Role of arbuscular mycorrhizal fungi in plant growth regulation: Implications in abiotic stress tolerance. Frontiers in Plant Science 10, 1068.

Bell, C.A., Magkourilou, E., Ault, J.R., Urwin, P.E. & Field, K.J. (2024). Phytophagy impacts the quality and quantity of plant carbon resources acquired by mutualistic arbuscular mycorrhizal fungi. Nature Communications 15(1), 801.

Benaffari, W., Boutasknit, A., Anli, M., Ait-El-Mokhtar, M., Ait-Rahou, Y., Ben-Laouane, R. & Meddich, A. (2022). The native arbuscular mycorrhizal fungi and vermicompost-based organic amendments enhance soil fertility, growth performance, and the drought stress tolerance of quinoa. Plants 11(3), 393.

Bhantana, P., Rana, M.S., Sun, X.C., Moussa, M.G., Saleem, M.H., Syaifudin, M. & Hu, C.X. (2021). Arbuscular mycorrhizal fungi and its major role in plant growth, zinc nutrition, phosphorous regulation and phytoremediation. Symbiosis 84, 19–37.

Bhattacharyya, S.S., Ros, G.H., Furtak, K., Iqbal, H.M. & Parra-Saldívar, R. (2022). Soil carbon sequestration – An interplay between soil microbial community and soil organic matter dynamics. Science of the Total Environment 815, 152928.

Bhupenchandra, I., Chongtham, S.K., Devi, A.G., Dutta, P., Sahoo, M.R., Mohanty, S. & Swapnil, P. (2024). Unlocking the potential of arbuscular mycorrhizal fungi: Exploring role in plant growth promotion, nutrient uptake mechanisms, biotic stress alleviation, and sustaining agricultural production systems. Journal of Plant Growth Regulation 1–39.

Bonassa, G., Bolsan, A.C., Hollas, C.E., Venturin, B., Candido, D., Chini, A. & Kunz, A. (2021). Organic carbon bioavailability: Is it a good driver to choose the best biological nitrogen removal process? Science of the Total Environment 786, 147390.

Boyno, G. & Demir, S. (2022). Plant-mycorrhiza communication and mycorrhizae in inter-plant communication. Symbiosis 86(2), 155–168.

Boyno, G., Demir, S. & Danesh, Y.R. (2022). Effects of some biological agents on the growth and biochemical parameters of tomato plants infected with Alternaria solani (Ellis & Martin) Sorauer. European Journal of Plant Pathology 162(1), 19–29.

Boyno, G., Rezaee Danesh, Y., Demir, S., Teniz, N., Mulet, J.M. & Porcel, R. (2023). The complex interplay between arbuscular mycorrhizal fungi and strigolactone: mechanisms, sinergies, applications and future directions. International Journal of Molecular Sciences 24(23), 16774.

Boyno, G., Yerli, C., Çakmakci, T., Sahin, U. & Demir, S. (2024). The effect of arbuscular mycorrhizal fungi on carbon dioxide (CO_2) emission from turfgrass soil under different irrigation intervals. Journal of Water and Climate Change 15(2), 541–553.

Cao, N., Zha, L., Li, Q., Hu, W., Pang, J., Zhou, Z. & Meng, Y. (2025). Simultaneously improving soil AMF community and phosphorus uptake by cotton plants by continuous straw incorporation and optimizing phosphorus management. Land Degradation & Development 36(8), 2639–2650.

Chauhan, P.K., Upadhyay, S.K., Rajput, V.D., Dwivedi, P., Minkina, T. & Wong, M.H. (2024). Fostering plant growth performance under drought stress using rhizospheric microbes, their gene editing, and biochar. Environmental Geochemistry and Health 46(2), 41.

Chauhan, S., Mahawar, S., Jain, D., Udpadhay, S.K., Mohanty, S.R., Singh, A. & Maharjan, E. (2022). Boosting sustainable agriculture by arbuscular mycorrhiza under stress condition: Mechanism and future prospective. BioMed Research International 2022(1), 5275449.

Chen, X., Bai, Y., Lin, Y., Liu, H., Han, F., Chang, H., ... & Liu, Q. (2024). Genome-wide ıdentification and characterization of the pht1 gene family and ıts response to mycorrhizal symbiosis in *Salvia miltiorrhiza* under phosphate stress. Genes 15(5), 589.

Chourasiya, D., Gupta, M.M., Sahni, S., Oehl, F., Agnihotri, R., Buade, R., ... & Sharma, M. P. (2021). Unraveling the AM fungal community for understanding its ecosystem resilience to changed climate in agroecosystems. Symbiosis 1–16.

Das, P. P., Singh, K. R., Nagpure, G., Mansoori, A., Singh, R. P., Ghazi, I. A., ... & Singh, J. (2022). Plant-soil-microbes: A tripartite interaction for nutrient acquisition and better plant growth for sustainable agricultural practices. Environmental Research 214, 113821.

Demir, S., Danesh, Y.R., Boyno, G. & Najafi, S. (2022). Arbuscular mycorrhizal fungi in biotic and abiotic stress conditions: Function and management in horticulture. Sustainable Horticulture 157–183.

Devi, R., Behera, B., Raza, M.B., Mangal, V., Altaf, M.A., Kumar, R., ... & Singh, B. (2021). An insight into microbes mediated heavy metal detoxification in plants: A review. Journal of Soil Science and Plant Nutrition 1–23.

Dey, M. & Ghosh, S. (2022). Arbuscular mycorrhizae in plant immunity and crop pathogen control. Rhizosphere 22, 100524.

Duan, S., Feng, G., Limpens, E., Bonfante, P., Xie, X. & Zhang, L. (2024). Cross-kingdom nutrient exchange in the plant–arbuscular mycorrhizal fungus–bacterium continuum. Nature Reviews Microbiology 22(12), 773–790.

Ebbisa, A. (2022). Arbuscular mycorrhizal fungi (AMF) in optimizing nutrient bioavailability and reducing agrochemicals for maintaining sustainable agroecosystems. *In*: Arbuscular Mycorrhizal Fungi in Agriculture – New İnsights. IntechOpen.

El Malahi, S., Touhami, D., Ganoudi, M., Ennami, M., Maach, M., Idrissi Hassani, L.M. & Jauregui, I. (2025). Harnessing the power of microbial allies: AMF and PGPR as biostimulants for sustainable bioeconomy development in the global south. *In*: Sustainable Bioeconomy Development in the Global South: Volume I Status and Perspectives (pp. 261–291). Singapore: Springer Nature Singapore.

Etesami, H., Jeong, B.R. & Glick, B.R. (2021). Contribution of arbuscular mycorrhizal fungi, phosphate–solubilizing bacteria, and silicon to P uptake by plant. Frontiers in Plant Science 12, 699618.

Evelin, H., Devi, T.S., Gupta, S. & Kapoor, R. (2019). Mitigation of salinity stress in plants by arbuscular mycorrhizal symbiosis: Current understanding and new challenges. Frontiers in Plant Science 10, 470.

Fall, A.F., Nakabonge, G., Ssekandi, J., Founoune-Mboup, H., Apori, S.O., Ndiaye, A., ... & Ngom, K. (2022). Roles of arbuscular mycorrhizal fungi on soil fertility: Contribution in the improvement of physical, chemical, and biological properties of the soil. Frontiers in Fungal Biology 3, 723892.

Gou, X., Reich, P.B., Qiu, L., Shao, M., Wei, G., Wang, J. & Wei, X. (2023). Leguminous plants significantly increase soil nitrogen cycling across global climates and ecosystem types. Global Change Biology 29(14), 4028–4043.

Hajiboland, R. & Ahammed, G.J. (2024). Signaling events during the establishment of symbiosis between arbuscular mycorrhizal fungi and plant roots. *In*: Arbuscular Mycorrhizal Fungi and Higher Plants: Fundamentals and Applications (pp. 67–97). Singapore: Springer Nature Singapore.

Hammer, E.C., Arellano-Caicedo, C., Mafla-Endara, P.M., Kiers, E.T., Shimizu, T., Ohlsson, P. & Aleklett, K. (2024). Hyphal exploration strategies and habitat modification of an arbuscular mycorrhizal fungus in microengineered soil chips. Fungal Ecology 67, 101302.

Han, S., Cheng, Y., Wu, G., He, X. & Zhao, G. (2024). Enhancing salt tolerance in poplar seedlings through arbuscular mycorrhizal fungi symbiosis. Plants 13(2), 233.

Hartmann, M. & Six, J. (2023). Soil structure and microbiome functions in agroecosystems. Nature Reviews Earth & Environment 4(1), 4–18.

He, T., Zhang, X., Du, J., Gilliam, F.S., Yang, S., Tian, M., ... & Zhou, Y. (2023). Arbuscular mycorrhizal fungi shift soil bacterial community composition and reduce soil ammonia volatilization and nitrous oxide emissions. Microbial Ecology 85(3), 951–964.

He, X., Wang, W., Wang, X., Zhang, B., Shi, F., Kurakov, A.V. & Song, F. (2023). The potential of mycorrhizal fungi to increase terrestrial ecosystem carbon sink: A review. Eurasian Soil Science 56(11), 1724–1738.

Holka, M., Kowalska, J. & Jakubowska, M. (2022). Reducing carbon footprint of agriculture—Can organic farming help to mitigate climate change? Agriculture 12(9), 1383.

Ho-Plágaro, T. & García-Garrido, J.M. (2022). Molecular regulation of arbuscular mycorrhizal symbiosis. International Journal of Molecular Sciences 23(11), 5960.

Hossain, M.B. (2021). Glomalin and contribution of glomalin to carbon sequestration in soil: A review. Turkish Journal of Agriculture-Food Science and Technology 9(1), 191–196.

Hui, J., An, X., Li, Z., Neuhäuser, B., Ludewig, U., Wu, X., ... & Yuan, L. (2022). The mycorrhiza-specific ammonium transporter ZmAMT3; 1 mediates mycorrhiza-dependent nitrogen uptake in maize roots. The Plant Cell 34(10), 4066–4087.

Janati, W., Benmrid, B., Elhaissoufi, W., Zeroual, Y., Nasielski, J. & Bargaz, A. (2021). Will phosphate bio-solubilization stimulate biological nitrogen fixation in grain legumes? Frontiers in Agronomy 3, 637196.

Jeewani, P.H., Luo, Y., Yu, G., Fu, Y., He, X., Van Zwieten, L., ... & Xu, J. (2021). Arbuscular mycorrhizal fungi and goethite promote carbon sequestration via hyphal-aggregate mineral interactions. Soil Biology and Biochemistry 162, 108417.

Ji, C., Chen, Z., Kong, X., Xin, Z., Sun, F., Xing, J., ... & Cao, H. (2022). Biocontrol and plant growth promotion by combined Bacillus spp. inoculation affecting pathogen and AMF communities in the wheat rhizosphere at low salt stress conditions. Frontiers in Plant Science 13, 1043171.

Justi, M., Silva, C.A. & Rosa, S.D. (2022). Organic acids as complexing agents for iron and their effects on the nutrition and growth of maize and soybean. Archives of Agronomy and Soil Science 68(10), 1369–1384.

Kadam, S.B. & Barvkar, V.T. (2024). COI1-mediated jasmonic acid signalling regulates mycorrhizal colonisation intensity and is an indispensable component of mycorrhiza-induced resistance. Physiologia Plantarum 176(6), e14643.

Kakouridis, A., Hagen, J.A., Kan, M.P., Mambelli, S., Feldman, L.J., Herman, D.J., ... & Firestone, M.K. (2022). Routes to roots: Direct evidence of water transport by arbuscular mycorrhizal fungi to host plants. New Phytologist 236(1), 210–221.

Kalamulla, R. & Yapa, N. (2024). Co-inoculation of AMF and other microbial biofertilizers for better nutrient acquisition from the soil system. *In*: Arbuscular Mycorrhizal Fungi in Sustainable Agriculture: Nutrient and Crop Management (pp. 99–111).

Kaya, C., Uğurlar, F. & Adamakis, I.D.S. (2023). Epigenetic and hormonal modulation in plant–plant growth-promoting microorganism symbiosis for drought-resilient agriculture. International Journal of Molecular Sciences 24(22), 16064.

Keymer, A. & Gutjahr, C. (2018). Cross-kingdom lipid transfer in arbuscular mycorrhiza symbiosis and beyond. Current Opinion in Plant Biology 44, 137–144.

Khaliq, A., Perveen, S., Alamer, K.H., Zia Ul Haq, M., Rafique, Z., Alsudays, I.M., ... & Attia, H. (2022). Arbuscular mycorrhizal fungi symbiosis to enhance plant–soil interaction. Sustainability 14(13), 7840.

Khan, F., Siddique, A.B., Shabala, S., Zhou, M. & Zhao, C. (2023). Phosphorus plays key roles in regulating plants' physiological responses to abiotic stresses. Plants 12(15), 2861.

Khan, Y., Shah, S. & Tian, H. (2022). The roles of arbuscular mycorrhizal fungi in influencing plant nutrients, photosynthesis, and metabolites of cereal crops: A review. Agronomy 12(9), 2191.

Kiddee, S., Wongdee, J., Piromyou, P., Songwattana, P., Greetatorn, T., Boonkerd, N., ... & Tittabutr, P. (2024). Unveiling the tripartite synergistic interaction of plant-arbuscular mycorrhizal fungus symbiosis by endophytic Bacillus velezensis S141 in Lotus japonicus. Symbiosis 92(3), 355–367.

Kuila, D. & Ghosh, S. (2022). Aspects, problems and utilization of Arbuscular Mycorrhizal (AM) application as bio-fertilizer in sustainable agriculture. Current Research in Microbial Sciences 3, 100107.

Lilleskov, E.A., Kuyper, T.W., Bidartondo, M.I. & Hobbie, E.A. (2019). Atmospheric nitrogen deposition impacts on the structure and function of forest mycorrhizal communities: A review. Environmental Pollution 246, 148–162.

Liu, H., Lin, J.S., Luo, Z., Sun, J., Huang, X., Yang, Y., ... & Xie, F. (2022). Constitutive activation of a nuclear-localized calcium channel complex in Medicago truncatula. Proceedings of the National Academy of Sciences 119(34), e2205920119.

Liu, R. & Chen, Y. (2024). Mycorrhizal symbiosis: Evolution, opportunities, challenges, and prospects. *In*: Arbuscular Mycorrhizal Fungi in Sustainable Agriculture: Inoculum Production and Application (pp. 1–35).

Liu, Y.N., Liu, C.C., Zhu, A.Q., Niu, K.X., Guo, R., Tian, L., ... & Wang, B. (2022). OsRAM2 function in lipid biosynthesis is required for arbuscular mycorrhizal symbiosis in rice. Molecular Plant-Microbe Interactions 35(3), 187–199.

Liu, Y., He, G., He, T. & Saleem, M. (2022). Signaling and detoxification strategies in plant-microbes symbiosis under heavy metal stress: A mechanistic understanding. Microorganisms 11(1), 69.

Luo, M., Moorhead, D.L., Ochoa-Hueso, R., Mueller, C.W., Ying, S.C. & Chen, J. (2022). Nitrogen loading enhances phosphorus limitation in terrestrial ecosystems with implications for soil carbon cycling. Functional Ecology 36(11), 2845–2858.

Luo, Y., Ma, L., Feng, Q., Luo, H., Chen, C., Wang, S., ... & Li, N. (2024). Influence and role of fungi, bacteria, and mixed microbial populations on phosphorus acquisition in plants. Agriculture 14(3), 358.

Ma, X., Geng, Q., Zhang, H., Bian, C., Chen, H.Y., Jiang, D. & Xu, X. (2021). Global negative effects of nutrient enrichment on arbuscular mycorrhizal fungi, plant diversity and ecosystem multifunctionality. New Phytologist 229(5), 2957–2969.

Madouh, T.A. & Quoreshi, A.M. (2023). The function of arbuscular mycorrhizal fungi associated with drought stress resistance in native plants of arid desert ecosystems: A review. Diversity 15(3), 391.

Matos, I.F., do Carmo Silva, B., de Souza, S.B., Bertolazi, A.A., de Souza Pedroni, N.L., Intorne, A., ... & Ramos, A.C. (2021). Ecophysiology of nitrogen in symbiotic relationships of plants and microorganisms. *In*: Soil Nitrogen Ecology (pp. 33–50). Cham: Springer International Publishing.

Meena, M., Nagda, A., Mehta, T., Yadav, G. & Sonigra, P. (2023). Mechanistic basis of the symbiotic signaling pathway between the host and the pathogen. *In*: Plant-Microbe Interaction – Recent Advances in Molecular and Biochemical Approaches (pp. 375–387). Academic Press.

Mierziak, J. & Wojtasik, W. (2024). Epigenetic weapons of plants against fungal pathogens. BMC Plant Biology 24(1), 175.

Mishra, A.K., Bhowmick, M.K., Das, R., Pidikiti, P., Maurya, P.K., Sharma, S., ... & Singh, S. (2024). Potentials and prospects of AMF for soil carbon sequestration and nutrient cycling in rice-based cropping system. *In*: Arbuscular Mycorrhizal Fungi in Sustainable Agriculture: Nutrient and Crop Management (pp. 113–129). Singapore: Springer Nature Singapore.

Mishra, S., Mishra, A.K., Arya, R. & Mishra, V.C. (2024). Evaluating the ınfluence of ecological diversity on glomalin production and ıts ımplications for multifunctionality in ecosystem services. Innovation Discovery, 1.

Mitra, D., Nayeri, F.D., Sansinenea, E., Ortiz, A., Bhatta, B.B., Adeyemi, N.O., ... & Panneerselvam, P. (2023). Unraveling arbuscular mycorrhizal fungi interaction in rice for plant growth development and enhancing phosphorus use efficiency through recent development of regulatory genes. Journal of Plant Nutrition 46(13), 3184–3220.

Mohammed, A.E., Alotaibi, M.O. & Elobeid, M. (2022). Interactive influence of elevated CO_2 and arbuscular mycorrhizal fungi on sucrose and coumarin metabolism in Ammi majus. Plant Physiology and Biochemistry 185, 45–54.

Monika, Yadav, N., Mamta, Kumar, N., Kumar, A., Devi, S., ... & Arya, S.S. (2022). Arbuscular mycorrhizal fungi: A potential candidate for nitrogen fixation. In: Plant Stress Mitigators: Action and Application (pp. 217–234). Singapore: Springer Nature Singapore.

Morya, R., Bhargava, A., Lalitha, G.R., Vamshi, M., Verma, S., Rastogi, M. & Raj, S. (2023). Soil management practices to enhance carbon sequestration rates – A review. International Journal of Environment and Climate Change 13(11), 3762–3776.

Munene, R., Mustafa, O., Loftus, S., Banfield, C.C., Rötter, R.P., Bore, E.K., ... & Dippold, M.A. (2024). Contribution of Arbuscular Mycorrhiza and Exoenzymes to Nitrogen Acquisition of Sorghum Genotypes Under Drought. Available at SSRN 4681054.

Mythili, M. & Ramalakshmi, A. (2022). Unraveling the distribution of AMF communities and their metabolites associated with soils of minor millets. Rhizosphere 21, 100473.

Nasir, F., Bahadur, A., Lin, X., Gao, Y. & Tian, C. (2021). Novel insights into host receptors and receptor-mediated signaling that regulate arbuscular mycorrhizal symbiosis. Journal of Experimental Botany 72(5), 1546–1557.

Ndeko, A.B., Chuma, G.B., Kane, A. & Mushagalusa, G.N. (2024). Soil fertility and crop biofortification: Is there potential improvement through arbuscular mycorrhizal symbiosis in Sub-Saharan Africa? Discover Soil 1(1), 1–33.

Nie, W., He, Q., Guo, H., Zhang, W., Ma, L., Li, J. & Wen, D. (2024). Arbuscular mycorrhizal fungi: Boosting crop resilience to environmental stresses. Microorganisms 12(12), 2448.

Ortas, I., Rafique, M. & Çekiç, F.Ö. (2021). Do mycorrhizal fungi enable plants to cope with abiotic stresses by overcoming the detrimental effects of salinity and improving drought tolerance? Symbiotic Soil Microorganisms: Biology and Applications (pp. 391–428).

Pan, L. & Cai, B. (2023). Phosphate-solubilizing bacteria: Advances in their physiology, molecular mechanisms and microbial community effects. Microorganisms, 11(12), 2904.

Pang, F., Li, Q., Solanki, M.K., Wang, Z., Xing, Y.X. & Dong, D.F. (2024). Soil phosphorus transformation and plant uptake driven by phosphate-solubilizing microorganisms. Frontiers in Microbiology 15, 1383813.

Panicker, S. & Sayyed, R.Z. (2022). Hydrolytic enzymes from PGPR against plant fungal pathogens. *In*: Antifungal Metabolites of Rhizobacteria for Sustainable Agriculture (pp. 211–238). Cham: Springer International Publishing.

Parihar, M., Rakshit, A., Meena, V.S., Gupta, V.K., Rana, K., Choudhary, M., ... & Jatav, H.S. (2020). The potential of arbuscular mycorrhizal fungi in C cycling: A review. Archives of Microbiology 202, 1581–1596.

Park, S.H., Kang, B.R., Kim, J., Lee, Y., Nam, H.S. & Lee, T.K. (2024). Enhanced soil fertility and carbon dynamics in organic farming systems: The role of arbuscular mycorrhizal fungal abundance. Journal of Fungi 10(9), 598.

Poirier, Y. & Bucher, M. (2002). Phosphate transport and homeostasis in Arabidopsis. The Arabidopsis book/American Society of Plant Biologists, 1, e0024.

Priyadarshani, T.D.C., Weerasinghe, P.A., Law, S. & Zakeel, M.C.M. (2025). Hidden allies: Exploring the cutting-edge ınsights into the role of mycorrhizal fungi in plant health. *In*: Fungal Endophytes Volume II: Applications in Agroecosystems and Plant Protection (pp. 551–585). Singapore: Springer Nature Singapore.

Qiu, T., Peñuelas, J., Chen, Y., Sardans, J., Yu, J., Xu, Z., ... & Fang, L. (2024). Arbuscular mycorrhizal fungal interactions bridge the support of root-associated microbiota for slope multifunctionality in an erosion-prone ecosystem. Imeta 3(3), e187.

Qiu, Y., Guo, L., Xu, X., Zhang, L., Zhang, K., Chen, M., ... & Hu, S. (2021). Warming and elevated ozone induce tradeoffs between fine roots and mycorrhizal fungi and stimulate organic carbon decomposition. Science Advances 7(28), eabe9256.

Rawat, P., Das, S., Shankhdhar, D. & Shankhdhar, S.C. (2021). Phosphate-solubilizing microorganisms: Mechanism and their role in phosphate solubilization and uptake. Journal of Soil Science and Plant Nutrition 21(1), 49–68.

Riaz, M., Kamran, M., Fang, Y., Wang, Q., Cao, H., Yang, G., ... & Wang, X. (2021). Arbuscular mycorrhizal fungi-induced mitigation of heavy metal phytotoxicity in metal contaminated soils: A critical review. Journal of Hazardous Materials 402, 123919.

Rui, W., Mao, Z. & Li, Z. (2022). The roles of phosphorus and nitrogen nutrient transporters in the arbuscular mycorrhizal symbiosis. International Journal of Molecular Sciences 23(19), 11027.

Salmeron-Santiago, I.A., Martínez-Trujillo, M., Valdez-Alarcón, J.J., Pedraza-Santos, M.E., Santoyo, G., Pozo, M.J. & Chávez-Bárcenas, A.T. (2021). An updated review on the modulation of carbon partitioning and allocation in arbuscular mycorrhizal plants. Microorganisms 10(1), 75.

Salmeron-Santiago, I.A., Martínez-Trujillo, M., Valdez-Alarcón, J.J., Pedraza-Santos, M.E., Santoyo, G., López, P.A., ... & Chávez-Bárcenas, A.T. (2023). Carbohydrate and lipid balances in the positive plant phenotypic response to arbuscular mycorrhiza: Increase in sink strength. Physiologia Plantarum 175(1), e13857.

Salvioli, A. & Bonfante, P. (2013). Systems biology and "omics" tools: A cooperation for next-generation mycorrhizal studies. Plant Science 203, 107–114.

Samanta, S., Singh, A. & Roychoudhury, A. (2020). Involvement of sulfur in the regulation of abiotic stress tolerance in plants. Protective Chemical Agents in the Amelioration of Plant Abiotic Stress: Biochemical and Molecular Perspectives (pp. 437–466).

Sandhu, N., Sethi, M., Kumar, A., Dang, D., Singh, J. & Chhuneja, P. (2021). Biochemical and genetic approaches improving nitrogen use efficiency in cereal crops: A review. Frontiers in Plant Science 12, 657629.

Sarma, B., Kashtoh, H., Lama Tamang, T., Bhattacharyya, P.N., Mohanta, Y.K. & Baek, K.H. (2023). Abiotic stress in rice: Visiting the physiological response and its tolerance mechanisms. Plants 12(23), 3948.

Schardl, C.L., Afkhami, M.E., Gundel, P.E., Iannone, L.J., Young, C.A., Creamer, R., ... & Berry, D. (2022). Diversity of seed endophytes: Causes and implications. *In*: Plant Relationships: Fungal-Plant Interactions (pp. 83–132). Cham: Springer International Publishing.

Schimel, J.P. & Schaeffer, S.M. (2012). Microbial control over carbon cycling in soil. Frontiers in Microbiology 3, 348.

Selwal, N., Wani, A.K., Akhtar, N., Kaur, M. & Jassal, P.S. (2023). Molecular insights of Strigolactone biosynthesis, signalling pathways, regulatory roles, and hormonal crosstalks in plant systems. South African Journal of Botany 160, 9–22.

Shao, Y.D., Hu, X.C., Wu, Q.S., Yang, T.Y., Srivastava, A.K., Zhang, D.J., ... & Kuča, K. (2021). Mycorrhizas promote P acquisition of tea plants through changes in root morphology and P transporter gene expression. South African Journal of Botany 137, 455–462.

Sharma, K., Gupta, S., Thokchom, S.D., Jangir, P. & Kapoor, R. (2021). Arbuscular mycorrhiza-mediated regulation of polyamines and aquaporins during abiotic stress: Deep insights on the recondite players. Frontiers in Plant Science 12, 642101.

Sitoe, S.N.M. & Dames, J.F. (2022). Mitigating climate change: The influence of arbuscular mycorrhizal fungi on maize production and food security. *In*: Arbuscular Mycorrhizal Fungi in Agriculture – New Insights. IntechOpen.

Solaiman, Z. (2024). Role of AMF in organic matter decomposition, carbon sequestration and climate change mitigation. *In*: Arbuscular Mycorrhizal Fungi in Sustainable Agriculture: Nutrient and Crop Management (pp. 131–141). Singapore: Springer Nature Singapore.

Song, F., Song, G., Dong, A. & Kong, X. (2011). Regulatory mechanisms of host plant defense responses to arbuscular mycorrhiza. Acta Ecologica Sinica 31(6), 322–327.

Srivastava, R., Sirohi, P., Chauhan, H. & Kumar, R. (2021). The enhanced phosphorus use efficiency in phosphate-deficient and mycorrhiza-inoculated barley seedlings involves activation of different sets of PHT1 transporters in roots. Planta 254, 1–17.

Tahiri, A.I., Meddich, A., Raklami, A., Alahmad, A., Bechtaoui, N., Anli, M., ... & Oufdou, K. (2022). Assessing the potential role of compost, PGPR, and AMF in improving tomato plant growth, yield, fruit quality, and water stress tolerance. Journal of Soil Science and Plant Nutrition 1–22.

Tang, H., Hassan, M.U., Feng, L., Nawaz, M., Shah, A.N., Qari, S.H., ... & Miao, J. (2022). The critical role of arbuscular mycorrhizal fungi to improve drought tolerance and nitrogen use efficiency in crops. Frontiers in Plant Science 13, 919166.

Tang, H., Li, Q., Bao, Q., Tang, B., Li, K., Ding, Y., ... & Du, L. (2024). Interplay of soil characteristics and arbuscular mycorrhizal fungi diversity in alpine wetland restoration and carbon stabilization. Frontiers in Microbiology 15, 1376418.

Taulera, Q. (2022). Perception and Responses to Strigolactones in Arbuscular Mycorrhizal Fungi (Doctoral dissertation), Université Paul Sabatier-Toulouse III.

Thangavel, P., Anjum, N.A., Muthukumar, T., Sridevi, G., Vasudhevan, P. & Maruthupandian, A. (2022). Arbuscular mycorrhizae: Natural modulators of plant–nutrient relation and growth in stressful environments. Archives of Microbiology 204(5), 264.

Tharanath, A.C., Upendra, R.S. & Rajendra, K. (2024). Soil symphony: A comprehensive overview of plant–microbe interactions in agricultural systems. Applied Microbiology 4(4), 1549–1567.

Tirry, N., Ferioun, M., Kouchou, A., Laghmari, G., Asri, M., Zouitane, I., ... & El Ghachtouli, N. (2024). Plant growth-promoting rhizobacteria's (PGPR) effects on Medicago sativa growth, arbuscular mycorrhizal colonisation, and soil enzyme activities. International Journal of Environmental Studies 81(3), 1190–1208.

Tomer, A., Singh, R., Singh, S.K., Dwivedi, S.A., Reddy, C.U., Keloth, M.R.A. & Rachel, R. (2021). Role of fungi in bioremediation and environmental sustainability. Mycoremediation and Environmental Sustainability: Volume 3, 187–200.

Ullah, A., Gao, D. & Wu, F. (2024). Common mycorrhizal network: The predominant socialist and capitalist responses of possible plant–plant and plant–microbe interactions for sustainable agriculture. Frontiers in Microbiology 15, 1183024.

Vasilachi, I.C., Stoleru, V. & Gavrilescu, M. (2023). Analysis of heavy metal impacts on cereal crop growth and development in contaminated soils. Agriculture 13(10), 1983.

Vergara, C. & Araujo, K.E.C. (2024). Arbuscular mycorrhizal symbiosis: From infection signaling to bidirectional nutrient exchanges. Arbuscular Mycorrhizal Fungi in Sustainable Agriculture: Inoculum Production and Application, 401–418.

Verma, R., Sharma, B. & Rather, G.A. (2024). Strigolactones: Biosynthesis, regulation, signaling, roles, and response to stress. *In*: Strigolactones (pp. 147–188). Academic Press.

Wahab, A., Muhammad, M., Munir, A., Abdi, G., Zaman, W., Ayaz, A., ... & Reddy, S.P.P. (2023). Role of arbuscular mycorrhizal fungi in regulating growth, enhancing productivity, and potentially influencing ecosystems under abiotic and biotic stresses. Plants 12(17), 3102.

Wang, F., Zhang, L., Zhou, J., Rengel, Z., George, T.S. & Feng, G. (2022). Exploring the secrets of hyphosphere of arbuscular mycorrhizal fungi: Processes and ecological functions. Plant and Soil 481(1), 1–22.

Wang, J., Wang, J., He, J.Z., Zhu, Y.G., Qiao, N.H. & Ge, Y. (2021). Arbuscular mycorrhizal fungi and plant diversity drive restoration of nitrogen-cycling microbial communities. Molecular Ecology 30(16), 4133–4146.

Wang, J., Yuan, J., Ren, Q., Zhang, B., Zhang, J., Huang, R. & Wang, G.G. (2022). Arbuscular mycorrhizal fungi enhanced salt tolerance of Gleditsia sinensis by modulating antioxidant activity, ion balance and P/N ratio. Plant Growth Regulation 97(1), 33–49.

Wang, Y.J., He, X.H., Meng, L.L., Zou, Y.N. & Wu, Q.S. (2023). Extraradical mycorrhizal hyphae promote soil carbon sequestration through difficultly extractable glomalin-related soil protein in response to soil water stress. Microbial Ecology 86(2), 1023–1034.

Wang, Z., Lian, J., Liang, J., Wei, H., Chen, H., Hu, W. & Tang, M. (2024). Arbuscular mycorrhizal symbiosis modulates nitrogen uptake and assimilation to enhance drought tolerance of Populus cathayana. Plant Physiology and Biochemistry 210, 108648.

Wendel, A.S., Bauke, S.L., Amelung, W. & Knief, C. (2022). Root-rhizosphere-soil interactions in biopores. Plant and Soil 475(1), 253–277.

Wipf, D., Krajinski, F., van Tuinen, D., Recorbet, G. & Courty, P.E. (2019). Trading on the arbuscular mycorrhiza market: From arbuscules to common mycorrhizal networks. New Phytologist 223(3), 1127–1142.

Wu, S., Fu, W., Rillig, M.C., Chen, B., Zhu, Y.G. & Huang, L. (2024). Soil organic matter dynamics mediated by arbuscular mycorrhizal fungi – An updated conceptual framework. New Phytologist 242(4), 1417–1425.

Xiao, D., He, X., Xu, Z., Bai, S.H., Zhang, W., Hu, P., ... & Wang, K. (2023). Strong cooperations among diazotroph and arbuscular mycorrhizal fungi taxa promote free-living nitrogen fixation at soil-rock mixing layer. Geoderma 437, 116600.

Xu, F.Q., Meng, L.L., Kuča, K. & Wu, Q.S. (2024). The mechanism of arbuscular mycorrhizal fungi-alleviated manganese toxicity in plants: A review. Plant Physiology and Biochemistry 108808.

Yang, Y.M., Naseer, M., Zhu, Y., Wang, B.Z., Wang, S., Ma, Y., ... & Xiong, Y.C. (2023). Priming effects of nZVI on carbon sequestration and iron uptake are positively mediated by AM fungus in semiarid agricultural soils. Science of the Total Environment 882, 163632.

Yangyang, J.I.A., van der Heijden, M.G., Valzano-Held, A.Y., Jocher, M. & Walder, F. (2024). Mycorrhizal fungi mitigate nitrogen losses of an experimental grassland by facilitating plant uptake and soil microbial immobilization. Pedosphere 34(2), 399–410.

Yilmaz, A., Yildirim, E., Yilmaz, H., Soydemir, H.E., Güler, E., Ciftci, V. & Yaman, M. (2023). Use of arbuscular mycorrhizal fungi for boosting antioxidant enzyme metabolism and mitigating saline stress in sweet basil (*Ocimum basilicum* L.). Sustainability 15(7), 5982.

Yu, Y., Gui, Y., Li, Z., Jiang, C., Guo, J. & Niu, D. (2022). Induced systemic resistance for improving plant immunity by beneficial microbes. Plants 11(3), 386.

Yuan, P., Luo, F., Gleason, C. & Poovaiah, B.W. (2022). Calcium/calmodulin-mediated microbial symbiotic interactions in plants. Frontiers in Plant Science 13, 984909.

Zayed, O., Hewedy, O.A., Abdelmoteleb, A., Ali, M., Youssef, M.S., Roumia, A.F., ... & Yuan, Z.C. (2023). Nitrogen journey in plants: From uptake to metabolism, stress response, and microbe interaction. Biomolecules 13(10), 1443.

Zeng, J., Ma, S., Liu, J., Qin, S., Liu, X., Li, T., ... & Zhang, J. (2023). Organic materials and AMF addition promote growth of Taxodium 'zhongshanshan' by improving soil structure. Forests, 14(4), 731.

Zhang, C., Xiang, X., Yang, T., Liu, X., Ma, Y., Zhang, K., ... & Chu, H. (2024). Nitrogen fertilization reduces plant diversity by changing the diversity and stability of arbuscular mycorrhizal fungal community in a temperate steppe. Science of the Total Environment 918, 170775.

Zhao, Y., Cartabia, A., Lalaymia, I. & Declerck, S. (2022). Arbuscular mycorrhizal fungi and production of secondary metabolites in medicinal plants. Mycorrhiza 32(3), 221–256.

Zhou, M. & Li, Y. (2025). Effects of nitrogen addition on soil aggregate stability and mycorrhizal morphological characteristics: Differential responses of arbuscular mycorrhizal and ectomycorrhizal fungi. Forests 16(2), 282.

CHAPTER
5

Arbuscular Mycorrhizal Fungi: Applications, Management and Perspectives for Sustainable Systems

Fernando de Souza Buzo[1*], Camilla Santos Reis de Andrade da Silva[2], Julia Ferreira Xavier[2], Danilo Augusto Silvestre[2], Gabriela Cavalcanti Alves[2] and Maura Santos Reis de Andrade da Silva[3*]

[1] Agronomy Program, Universidade Brasil, Fernandópolis Campus, São Paulo, Brazil
[2] Department of Soil Science, Federal Rural University of Rio de Janeiro, Seropédica, Brazil
[3] Department of Agricultural Microbiology, São Paulo State University (UNESP), Brazil

1. Introduction

Arbuscular mycorrhizal fungi (AMF) establish symbiotic relationships with the roots of most plants, including species of agricultural importance. These fungi form a mutualistic association with plants, in which the plant supplies photoassimilates to the fungus, which enhances the plant's capacity to absorb water and nutrients from the soil (Berruti et al., 2016). Based on this, it has been found that AMFs are a strategy for increasing the sustainability and efficiency of production systems. This is because the implementation of management practices that favor the native AMF community and the inoculation of AMF can promote plant development and increase final productivity, while also improving fertilization efficiency and tolerance to water stresses (Chen et al., 2018).

They belong to the Glomeromycota phylum and form symbiosis with more than 80% of cultivated vascular plants (Ghorui et al., 2025; Ferreira Vilela & Barbosa, 2019; French, 2017), and they play an essential role in ecosystem services, especially in supporting plant growth, contributing to disease resistance, and improving soil quality (Singh et al., 2022). These fungi are capable of colonizing plant roots even in metal-contaminated soils (Saraswat & Rai, 2011) and play a key role in the bioremediation of contaminated areas (Ahammed & Hajiboland, 2024). Therefore, the inoculant formulation containing AMF is a biotechnology that can reduce the use of synthetic inputs in agriculture (Ghorui et al., 2025) (Figure 1).

*Corresponding authors: fsbuzo@gmail.com; maura@ufrrj.br

DOI: 10.1201/9781003669777-5

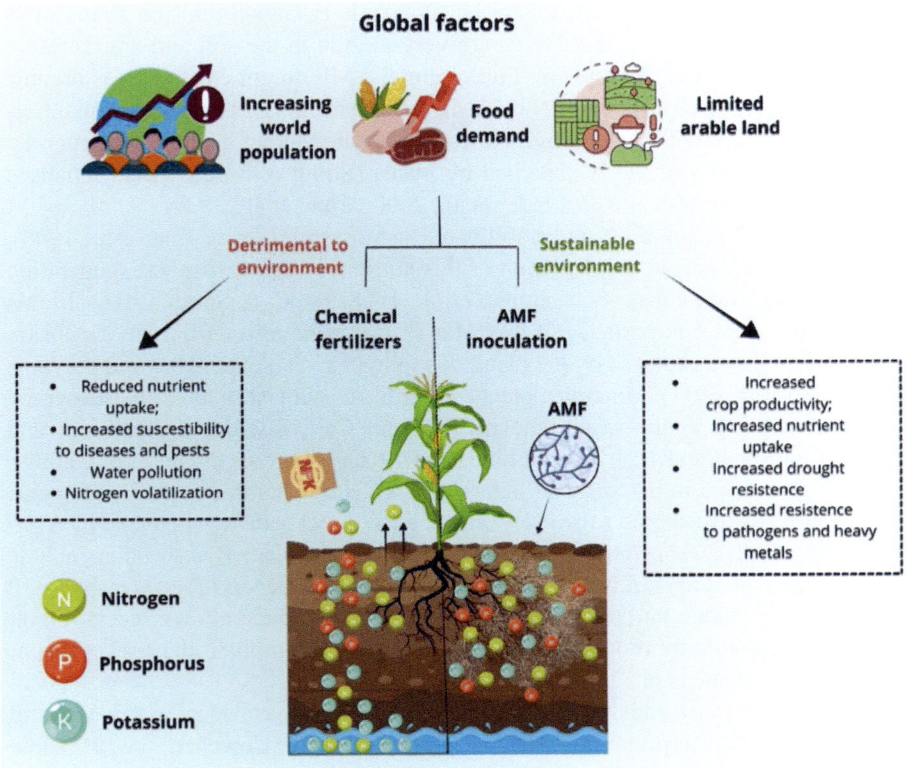

Fig. 1. Global factors have driven the use of chemical fertilizers, which increase productivity but cause environmental damage. AMFs improve nutrient absorption efficiency, confer greater tolerance to abiotic and biotic stresses and contribute to agricultural sustainability.

2. Applications of Arbuscular Mycorrhizal Fungi in Agriculture

AMF offer a range of benefits that can positively transform agricultural practices. One of their most significant effects is increased efficiency in nutrient absorption, particularly phosphorus (P). The benefit lies in the ability of AMF to colonize the roots and send out hyphae that spread through the soil. By effectively expanding the roots' reach, this mechanism improves soil exploration, resulting in greater absorption of water and nutrients, especially phosphorus (P) (Smith & Read, 2010). In addition, they benefit plants by absorbing forms of elements that are not available to roots (Moreira & Siqueira, 2006). Estimates state that the hyphal length density in the soil associated with plants can be hundreds of times greater than the root length density (Li et al., 2008), and they can absorb nutrients at distances up to 25 cm from the root (Jansa et al., 2003).

The literature suggests that AMF aid the plant in absorbing nitrogen (N), since the 15N content is consistently higher in mycorrhizal plants than in non-mycorrhizal ones (Ames et al., 1983; Hawkins et al., 2000). It is known that the forms of N absorbed by plants (NO_3^- and NH_4^+) are very mobile in the soil and are therefore rarely depleted in the rhizosphere. Furthermore, AMF do not easily access organic N, and thus, while absorption of N by AMF hyphae occurs and benefits the plant, its importance is lower than that observed for P, varying among studies, but generally not exceeding 50% of the N absorbed by plants (Smith & Smith, 2011; Johansen et al., 1996; Wang et al., 2017; Mäder et al., 2000; Wang et al., 2020).

Due to the low solubility and mobility of P in most soils (Schachtman et al., 1998), root absorption leads to the depletion of this nutrient in the rhizosphere, generating so-called depletion zones (Schachtman et al., 1998; Smith & Smith, 2011). In this context, the greater importance of AMF for P absorption arises from the expansion of the soil volume explored by the plant, due to hyphae that extend beyond the roots (Smith & Smith, 2011). Moreover, plants in symbiosis with AMF have been observed to exude organic acids, which chelate Al^{3+} and Ca^{2+}, dissolving aluminum and calcium phosphates (Klugh & Cumming, 2007; Klugh-Stewart & Cumming, 2009). Thus, P absorption in mycorrhizal plants occurs via two complementary pathways: the direct pathway (via roots themselves) and the indirect pathway mediated by AMF (Bücking et al., 2012). AMF mechanisms involved in plant P absorption include a higher rate of P uptake by hyphae, hyphal expansion beyond the rhizosphere, P storage by the fungi, and the smaller hyphal diameter, which allows access to soil spaces unreachable by roots (Sharif & Claassen, 2011; Smith et al., 2011; Garg & Pandey, 2015; Pepe et al., 2020; Jakobsen et al., 2001).

Compared to N and P, the relationship between mycorrhizal symbiosis and potassium (K) absorption has received less attention. However, the available literature indicates that the symbiosis is also beneficial for K uptake in plants (Alizadeh et al., 2025). Mycorrhizal bean plants showed greater absorption and storage of K. Similarly, Liu et al. (2019) found that AMF increased K acquisition in tomato plants, while Perner et al. (2007) and Baslam et al. (2013), observed higher K levels in mycorrhizal *Pelargonium peltatum* and *Lactuca sativa* plants, respectively. Several studies have reported that AMF extraradical hyphae can absorb K from the soil and supply it to host plants (George et al., 1992; Scheloske et al., 2004; Pallon et al., 2007; Olsson et al., 2008). However, there is still a gap in understanding the interaction between AMF and plant K supply (Garcia & Zimmermann, 2014).

Regarding calcium (Ca), magnesium (Mg), and sulfur (S), mycorrhizal symbiosis can enhance their absorption by plants, although few studies have demonstrated these effects. Studies on sulfur (S) indicate that its absorption by fungal hyphae and subsequent transport to the roots elevates the levels of this nutrient in plants (Gray & Gerdemann, 1973; Guo et al., 2007; Banerjee et al., 2003; Cavagnaro et al., 2006).

Regarding micronutrients, AMF can also absorb and transfer them to plants (Lehmann & Rillig, 2015; Ferrol et al., 2016), which is attributed to the greater soil volume explored by mycorrhizal plants (Chandrasekaran, 2020), since the availability of metallic micronutrients is limited by diffusion (Rengel & Marschner, 2005; Degryse et al., 2009). For example, AMF symbiosis has been found to improve Cu nutrition generally; for Fe, positive effects were observed only for a limited period within the plant cycle, while improvements in Mn nutrition were noted only in herbaceous plants, not in other plant types (Lehmann & Rillig, 2015).

Table 1. Results of the last 5 years' work on AMF inoculation in agricultural plants

Crop	Condition	AMF species	Results	Reference
Glycine max	• Pots • Soils contaminated with heavy metals	*Rhizophagus intraradices*	• Increased growth, P absorption and grain yield • Less transport of heavy metals towards the above-ground biomass (aerial part).	Adeyemi et al. (2021)
Glycine max	• Pots • Fertilization with selenite	*Claroideoglomus etunicatum* *Funneliformis mosseae*	• Increased amount of Se in seeds • Increase in root growth	Zhang et al. (2022)
Glycine max	• Vegetation house and field • Coinoculation with *Bradyrhizobium japonicum* • Three soybean varieties	*Gigaspora rosea* + *Rhizophagus clarus*	• Coinoculation generated better plant yields in the greenhouse • Greater dry biomass due to AMF in the greenhouse • Increase in height, dry biomass, number of plant pods^{-1} and plant yield^{-1} in the field	Dobo (2022)
Glycine max	• Field • Plants under water stress	*Glomus clarum* + *Funneliformis mosseae* + *Gigaspora margarita*	• Increase in plant and root length, leaf area and dry weight • Increased chlorophyll content and leaf N, P and K • Increased grain yield • Benefits were greater when co-inoculated with bacteria	Nader et al. (2024)
Glycine max	• Field • Two consecutive years • Doses of P applied	*Funneliformis mosseae* *Rhizophagus intraradices* *Claroideoglomus etunicatum*	• Inoculation increased shoot biomass, P absorption and grain yield of soybeans • Concluded that the optimal condition for increasing grain yield and P nutrition in Nigerian soybeans is achieved by inoculating with *R. intraradices* or *F. mosseae* along with a moderate rate of P (20 kg ha^{-1}).	Adeyemi et al. (2022)
Zea mays	• Field • Two consecutive years • Combined P and AMF doses • Response surface analysis	*Rhizoglomus intraradices*	• Higher foliar P content in the combination of 100% P_2O_5 with 120-140 g ha^{-1} of AMF • Higher Mg content in grains at maximum doses of P_2O_5 and inoculant combined • AMF increased ear diameter and number of rows • Did not influence grain yield	Buzo et al. (2021 and 2022a)

(Contd.)

(Contd.)

Crop	Condition	AMF species	Results	Reference
Zea mays	• Vegetation house • Plant growth-promoting rhizobacteria • Excessive concentrations of B	*Rhizophagus intraradices*	• Improved maize productivity under B stress by up to 20 mg B kg^{-1} • Applying mycorrhiza (in single or joint inoculation) increased the uptake of Boron (B) by the plant as a whole, with this increase observed in both the above-ground biomass and the root system.	Abdar et al. (2023)
Zea mays	• Field • Organic sweet corn • Combination of tillage practices	Native AMF *Rhizophagus irregularis* *Funneliformis mosseae*	• Increased concentrations of phosphorus (P) and potassium (K) and vitamins B6 and C in the grains • Increased concentrations of aspartic acid and glycine-potential of AMF to increase crop resilience to stress • Colonization by native AMF may have masked the effects of supplementary inoculation	Zinati et al. (2025)
Phaseolus vulgaris	• Pots	*Claroideoglomus claroideum; C. etunicatum; Funneliformis mosseae; Gigaspora rosea; Scutellospora calospora; Septoglomus constrictum; Rhizophagus intraradices; R. irregularis;* Native AMF community	• Inoculation did not affect productivity • Increase in the concentration of P, K and Cu in the seeds, depending on the species inoculated	Carrara et al. (2024)
Phaseolus vulgaris	• Field • Rhizobium strain • Reduced fertilization: only 70% of the conventional recommendation	*Rhizophagus intraradices* + *R. aggregatus* + *Septoglomus viscosum* + *Claroideoglomus etunicatum* + *C.* claroideum	• Inoculation did not change the results in terms of yield and fiber content. • The accumulation of Mg, K and Zn in the seeds was positively affected by the AMF	Massa et al. (2020)

Phaseolus vulgaris	• Pots • Water stress • Bacterium *Bacillus amyloliquefaciens*	*Glomus mosseae*	• Inoculation promoted a higher number of pods, pod length, pod weight and 100-seed weight in plants with little water restriction • Coinoculation promoted an increase in nutrient content (N, K, P, Mg and Ca), vitamin B1, folic acid and crude fiber in plants with little water restriction	Al-Amri (2021)
Phaseolus vulgaris	• Pots • Combination with different fungicides in seed treatment	*Rhizoglomus intraradices*	• It was observed that inoculation proportionally increased the dehydrogenase activity, as well as the root mass and the total plant mass. • The combination of the two highest doses of inoculant with metalaxyl + fludioxonil + thiabendazole or pyraclostrobin + methylthiophanate + fipronil in the seeds increased the biomass of the aerial part and the contents of K, Ca, Mg, S and Mn.	Buzo et al. (2022b)
Oryza sativa	• Field • Consortium with *Chrysopogon zizanioides* L. • 2 Management systems: conventional/chemical and organic	Native AMF inoculum	• AMF-colonization of rice roots low and not influenced by treatments • Increase in grain yield • No influence on plant growth	Madhushan et al. (2023)
Oryza sativa	• Pots • AMF combined with poultry manure, humified compost and Cd	*Claroideoglomus* sp. + *Microkamienskia peruviana* + *Microkamienskia* sp. + *Diversispora* sp. + *Claroideoglomus* etunicatum	• Greater root length, plant height, shoot and root biomass with combined AMF and compost • AMF + compost in pots contaminated with Cd resulted in improved plant growth • Higher levels of P and K in the root with AMF and compost combined	Vallejos-Torres et al. (2023)

(*Contd.*)

(Contd.)

Crop	Condition	AMF species	Results	Reference
Oryza sativa	• Field • Seedling production and nursery • Flooded cultivation	*Rhizophagus irregularis* *Funneliformis mosseae*	• AMF increases seedling growth, P nutrition and *Magnaporthe oryzae* fungus in rice seedlings grown aerobically in low-fertility soil • The application of *R. irregularis* results in an increase in grain yield and greater resistance to the blast (brusone) disease. • Effects depended on the rice variety	Martín-Cardoso et al. (2025)

In general, it is well known that AMF symbiosis increases plant tolerance to abiotic stresses. There are reports that AMF symbiosis enhances drought resistance, promoting greater water and nutrient absorption under such conditions (Maya & Matsubara, 2013). This increased tolerance to water deficit results from several AMF-induced effects on plants, including: hormonal balance changes (Asensio et al., 2012; Ludwig-Müller, 2010); improved plant water status and greater water uptake by hyphae (Porcel & Ruiz-Lozano, 2004; Gholamhoseini et al., 2013); increased antioxidant activity (Goicoechea et al., 2005; Zou et al., 2014; Amiri et al., 2015); and enhanced nutrient uptake (Al-Karaki, 2006; Zhao et al., 2015; Zhang et al., 2014).

AMF spore germination, hyphal elongation, and root colonization are reduced under salt stress (Porcel et al., 2012; Sheng et al., 2008). Nonetheless, AMF symbiosis provides plants with greater salt stress tolerance by promoting nutritional, physiological, and biochemical changes (Razvi et al., 2023), such as increased activity of antioxidant enzymes including catalase, and peroxidase, which eliminate reactive oxygen species and help alleviate salt stress (Hajiboland et al., 2010; He et al., 2007).

Research on AMF in agricultural crops still focuses mainly on pot experiments (Zhang et al., 2018; Ryan & Graham, 2018). Field results show contrasting outcomes, ranging from reduced productivity to no effect, to productivity increases (Erdinç et al., 2017; Zhang et al., 2018; Thirkell et al., 2017). In Table 1, we list studies from the last five years evaluating mycorrhizal inoculation in agricultural crops under field conditions and the results regarding their effects on productivity.

3. Management of the Native AMF Population

There are two main approaches to harnessing the benefits of AMF in agriculture: inoculation with selected strains and management of agricultural practices that favor the native AMF community in the soil. Inoculation involves introducing specific AMF spores or propagules into the soil or substrate, usually during planting or seeding. This practice can be particularly useful in degraded soils or where the native AMF population is low or inefficient (Wright, 2005).

However, the effectiveness of inoculation requires compatibility between the fungus and the host plant, soil conditions and competition with native microorganisms. On the other hand, management of the native AMF community focuses on creating environmental conditions that promote the diversity and activity of these fungi through practices such as no-till farming, crop rotation, cover crop use, and reduced chemical input use. These agricultural activities can increase the abundance and effectiveness of AMF in agriculture (Gosling et al., 2006).

The colonization and survival of AMF in the soil are directly related to the conditions provided by soil management practices, making it necessary to evaluate the resilience of AMF in different land use scenarios to guide more effective agricultural practices and conservation strategies (Zhang et al., 2018; Lehmann & Rillig, 2015).

A comparative study by Curaqueo et al. (2010) in a Mediterranean agroecosystem showed that the no-till system favored AMF establishment compared to conventional tillage. Similarly, research conducted under different edaphoclimatic conditions has reported that AMF dynamics are favored in systems with minimal soil disturbance

(Brito et al., 2017; Liu et al., 2022; Yang et al., 2024). This pattern had already been described by Kabir (2005), who pointed out that soil disturbance through ploughing and harrowing tends to reduce mycorrhizal colonization by breaking hyphal networks, decreasing inoculum potential, and exposing fungal structures such as hyphae, spores, and colonized roots to high temperatures, excessive oxidation, and predation, thus compromising their survival in the soil. However, crop rotation and the presence of mulch allows the stability of the mycorrhizal network, improving colonization, glomalin content and microbial resilience (Bowles et al., 2017).

Although soil management is an important factor, it is not the only determinant of AMF activity. Cultivation with non-host plant species in crop rotations, pastures, or agroforestry systems can also negatively affect AMF populations, as noted by Lekberg & Koide (2005). Furthermore, there are variations in mycorrhizal affinity among host species. Some plants stimulate AMF colonization and activity more intensely. According to Miranda, Vilela & Miranda (2005), in a crop rotation study in the Cerrado biome, maize showed greater root colonization by AMF compared to soybean. Similarly, Espindola et al. (1998) observed that the highest level of sweet potato mycorrhizal colonization was achieved after pre-cultivation with maize inoculated with *Glomus clarum*, surpassing results obtained with pre-cultivation of *Crotalaria juncea* under field conditions in southeastern Brazil. These discrepancies may be due to the inherent characteristics of these plant species, such as greater phosphorus demand, type of exudate or root architecture. In this regard, Šmilauer et al. (2020) highlight that host species characteristics govern the intensity and pattern of AMF colonization, directly influencing fungal community structure and functionality in the soil.

Regarding abiotic factors, it is common for mycorrhizal colonization to vary between dry and rainy seasons, directly affecting symbiosis and nutrient absorption efficiency (Miranda et al., 2005). Under temperate climate conditions, Jansa et al. (2002) found in a field study with maize that cold weather limits the recovery of mycorrhizal fungal communities after soil disturbances such as conventional tillage, reducing root reinfection capacity.

AMF colonization and diversity increase in systems with vegetation cover and reduced cultivation, with more pronounced positive effects in tropical climates, where microorganisms exhibit greater biological activity and regeneration speed (Bowles et al., 2017). Conversely, the study (Duarte & Maherali, 2022) analyzing AMF mutualistic responses under different climatic regimes showed that a moderate temperature increase of around 2-3°C tends to strengthen mycorrhizal symbiosis, especially in native plants, due to greater carbon allocation to roots and increased nutrient absorption efficiency. These findings indicate that even soils with similar physicochemical characteristics can vary significantly in AMF colonization, diversity, and resilience depending on climate, particularly temperature and humidity. Nevertheless, AMF demonstrates high adaptability and survival capacity. However, to express their functional potential, management strategies adapted to local conditions are required (Duarte & Maherali, 2022).

Organic fertilization through the use of manure increases the fungal community, making it a strategy to promote biological activity (Liu et al., 2022). Yang et al.

(2021) found that chemical fertilizer application reduced soil AMF diversity compared to the use of organic manure. In turn, Liu et al. (2016) indicated that AMF community structure correlated positively with phosphorus fertilization. Thus, appropriate phosphorus fertilization management through economical dosage is essential to preserve AMF functional diversity and increase the ecological benefits of mycorrhizal symbiosis, especially in intensive agricultural systems (Peng et al., 2022).

The appropriate selection of host crops in rotation programs, reduced soil disturbance, and economical fertilizer application are combined actions capable of conserving and stimulating AMF activity in the soil, promoting greater agricultural sustainability and contributing to the quality of fragile soils as well as plant nutrient efficiency (Figure 2).

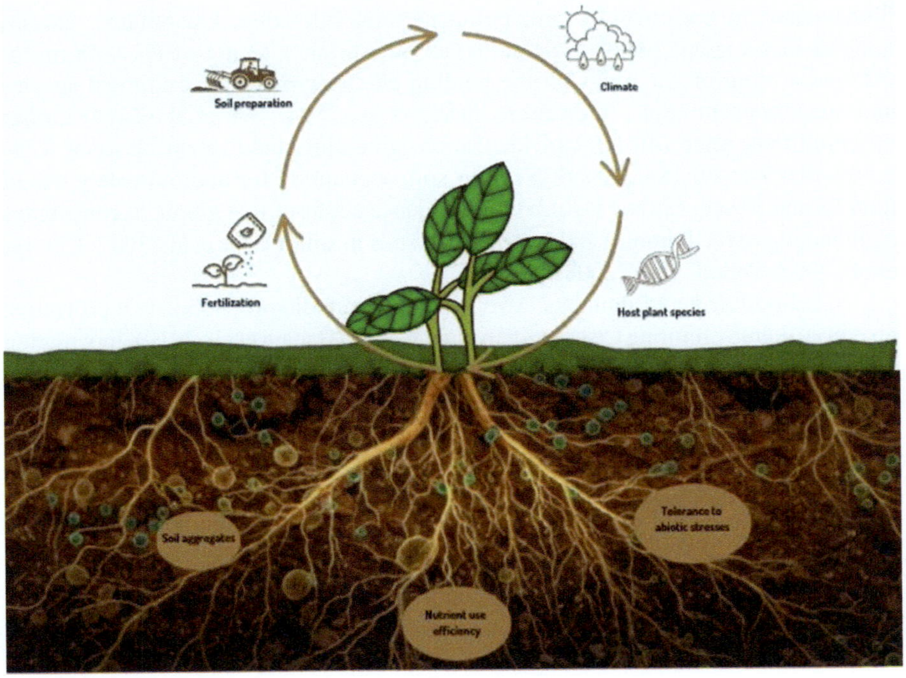

Fig. 2. Main management factors that favor the activity of AMF, as well as the environmental benefits provided to the soil and plants.

4. Use in Areas Contaminated with Heavy Metals

The application of arbuscular mycorrhizal fungi (AMF) in phytoremediation has emerged as a promising and environmentally sustainable strategy for the recovery of areas affected by heavy metals (Cabral et al., 2015; Morar et al., 2018). These fungi can retain and compartmentalize toxic metals within their structures, thereby reducing the bioavailability of these elements to plants. For instance, Chen et al. (2018) demonstrated through SR-μXRF imaging that cadmium (Cd) tends to accumulate within arbuscules and intercellular hyphae of AMF, while González-Guerrero et al. (2008) used synchrotron-based techniques to show that Cd, Cu, and Zn are mainly localized in the fungal spores and extra-radical mycelia, particularly in cell walls and vacuoles. Similarly, Wu et al. (2018), using proton-induced X-ray emission, suggested that Zn is sequestered by fungal thiol compounds within roots, whereas Pb and Cu precipitate as Fe-associated compounds in the rhizodermis of mycorrhizal roots.

AMF mitigate metal toxicity in soils primarily through two mechanisms: (i) direct adsorption of heavy metals onto fungal hyphae and (ii) immobilization of metals mediated by glomalin, a glycoprotein produced by AMF that contributes to soil aggregation and metal sequestration (De Andrade et al., 2008; Ferreira Vilela & Barbosa, 2019; Singh et al., 2022). Glomalin, a heat-resistant and insoluble protein present in the walls of hyphae and spores (Wang et al., 2019; Wu et al., 2014), is fundamental to soil stability and pollutant immobilization. Quantitative studies indicate that a single gram of glomalin can bind up to 1.12 mg of Pb, 0.08 mg of Cd, and 4.3 mg of Cu, although the binding efficiency depends on fungal species and metal concentrations (González-Chávez et al., 2004). Wu et al. (2014) further observed that, after 140 days of incubation, glomalin could retain between 0.21-1.78% of Pb and 0.38-0.98% of Ni from soil, accounting for approximately 4% of total Pb and Ni sequestration. Additional evidence confirms that glomalin contributes significantly to the immobilization of toxic metals in soil (Chern et al., 2007; Citterio et al., 2005; Vodnik et al., 2008).

In Cd-polluted environments, AMF increase glomalin production as a protective response, thereby limiting Cd transfer to plant tissues (Babadi et al., 2019). Inoculation with *Glomus aggregatum* has been shown to markedly decrease Cd, Pb, Zn, and Cu concentrations in plant shoots due to enhanced glomalin secretion and metal chelation in the rhizosphere (Nafady & Elgharably, 2018). Likewise, associations between *Rhizophagus intraradices*, *Glomus versiforme* (syn. *Diversispora epigaea*), and *Funneliformis mosseae* were reported to elevate total glomalin levels and soil pH, while reducing Pb and Cd uptake in maize (Hu et al., 2019).

Mycorrhizal symbiosis also promotes the synthesis of phytochelatins, peptides crucial for heavy-metal detoxification within plant cells. These molecules chelate metal ions, forming stable complexes that diminish cytotoxicity and facilitate sequestration into vacuoles (Garg & Chandel, 2012; Zhao et al., 2024). A well-documented case is the complexation of Cd ions by phytochelatins, which mitigates Cd toxicity in plant tissues (Han et al., 2021).

Experiments conducted with a mixed inoculum of *Gigaspora margarita*, *Glomus decipens*, *Scutellospora gilmori*, *Acaulospora* spp., and *Glomus* spp. isolated from decommissioned copper mines revealed an approximate 40% increase in the biomass of *Elsholtzia splendens*. This treatment also enhanced phosphorus content and promoted phytoextraction, with 98%, 68%, 68%, and 51% increases in Cu, Zn, Pb, and Cd accumulation in shoots, respectively (Wang et al., 2005, 2007). In field trials on Pb/Zn mining tailings, AMF inoculation increased *Chrysopogon zizanioides* (vetiver grass) dry matter by 8.1-13.8%, though metal reduction occurred mainly in roots rather than shoots (Wu et al., 2010). In addition, tree species established in mining-affected areas frequently show mycorrhizal colonization rates above 60%, with molecular analyses identifying species from *Acaulospora*, *Claroideoglomus*, *Diversispora*, *Glomus*, and *Sclerocystis* as key genera for phytoremediation initiatives in contaminated soils (Spruyt et al., 2014).

5. Recovery of Degraded Areas

AMF contribute to ecological restoration at different scales. These microorganisms are important mediators in disturbed communities and ecological succession (Asmelash et al., 2016; Neuenkamp et al., 2019). The first stages of ecological recovery involve successful colonization by pioneer plant species and the improvement of soil properties (Walker, Walker & Moral, 2007). A single mycorrhizal fungus can connect multiple plants, creating an interconnected mycorrhizal network (Simard & Austin, 2010). These networks support seedling regeneration, modify species interactions, and shift plant community dynamics, promoting greater plant diversity (Simard & Austin, 2010). AMF also prevent invasive plants by directly inhibiting them or indirectly enhancing the survival, establishment, and diversity of native plants, thus reducing available space for invaders (Schnitzer et al., 2011). In return, plants can allocate about 4–20% of their photosynthates to AMF (Lerat et al., 2003).

Vegetation and soil attributes are major factors influencing the composition and structure of AMF communities (Pereira et al., 2018). Common AMF families have distinct survival strategies to adapt to different environmental conditions (Chagnon et al., 2013; Figure 3). Glomerales species, characterized as ruderals, display rapid growth, efficient regeneration of fragmented hyphae, early investment in glomerospore formation and rapidly colonize roots (Chagnon et al., 2013; García de León et al., 2016). Meanwhile, stress-tolerant AMF, such as those from the Acaulosporaceae family, exhibit slow growth and produce mycelium that is highly resistant to abiotic stressors (Chagnon et al., 2013). Germination and early survival represent the final phases in the establishment of initial plant populations (Walker et al., 2007). Ruderal and stress-tolerant AMF characteristics predominate in early stages of ecological succession, improving seedling survival and establishment under stress conditions (García de León et al., 2016; Figure 3). After plants have survived their initial growing stage, different factors begin to play a central role in restoration efforts (Walker et al., 2007). Herbaceous plants often dominate pioneer communities because of their high reproductive capacity and the production of carbon-rich litter (Walker et al., 2007). This carbon availability, in turn, influences the composition of the soil microbial community (Li et al., 2017). AMF from the order Gigasporales, considered strong competitors, rely on greater carbon inputs from their plant hosts to support extensive mycelial growth for soil exploration (Chagnon et al., 2013; Figure 3). However, they tend to produce fewer glomerospores, prioritizing biomass expansion over reproduction (Chagnon et al., 2013).

Programs for restoring degraded ecosystems often focus on controlling the sequence in which species arrive and altering environmental filters to accelerate and/or initiate ecological succession (Hobbs et al., 2007). Several studies have shown that AMF provide potential benefits in land restoration (Pouyu-Rojas & Siqueira,

Fig. 3. Schematic representation of ecological succession in degraded environments, highlighting the interactions between plants and arbuscular mycorrhizal fungi (AMF). In degraded soils, AMF are mainly present as spores; in the pioneer stage, they associate with stress-tolerant ruderal plants; and in the intermediate stage, AMF function as competitors, supporting the establishment of more complex and shade-intolerant plant species.

2000; Schnitzer et al., 2011; Leifheit et al., 2014). However, the effectiveness of AMF inoculation in ecological restoration depends on the specific environmental context (Neuenkamp et al., 2019).

Hoeksema et al. (2010) observed that C4 grasses, compared to C3 grasses, have a stronger growth response to AMF inoculation. Plant nutrient demand can be a factor in growth response to mycorrhizal inoculation; C4 grasses are more dependent on AMF associations due to their relatively inefficient nutrient uptake (Hetrick et al., 1990; Wilson & Hartnett, 1998). Additionally, pioneer plants from tropical forests were more dependent on AMF inocula (Kiers et al., 2000). Inoculation with *Glomeromycota* fungi enhanced the growth and establishment of woody nitrogen-fixing plants used to initiate ecological restoration in degraded semi-arid areas (Caravaca et al., 2004; Barea, Azcón & Azcón-Aguilar, 2017). In restoration programs where the goal is rapid establishment of vegetation cover, mycorrhizal inoculation can be a beneficial tool for pioneer species with low nutrient uptake efficiency and high nutrient demand (Kiers et al., 2000; Neuenkamp et al., 2019). Soils with low nutrient contents, mainly total N and available P, demonstrated better plant responses to AMF inoculation than soils with high nutrient availability (Koide, 1991). In degraded forests, AMF inoculation improves seedling survival and establishment by increasing P uptake (Habte et al., 2001).

Degraded lands, particularly under harsh environmental conditions, exhibit low abundance and diversity of AMF (Cardozo-Junior et al., 2012). Disturbed soils benefit greatly from AMF inoculation, making it valuable for ecological restoration (Prach & Hobbs, 2008). Degraded grasslands have infertile soils, reducing vegetation cover and deteriorating soil structure and quality (Zhang et al., 2017). Xu et al. (2024) observed that mycorrhizal inoculation promoted growth in grass species used in degraded grassland restoration by increasing nutrient absorption. In degraded grasslands, AMF inoculation increased nutrient availability, mainly P and total N, which enhanced plant community productivity (Li et al., 2024). Mining-degraded sites have severe infertility problems and high concentrations of toxic metals (Li et al., 2014). AMF inoculation improves plant survival, growth, and nutrition in mining-impacted sites (Wang, 2017). Garcia et al. (2024) reported that AMF inoculation mitigates manganese soil toxicity and enhances plant growth. Hence, AMF inoculation is a potential tool for ecological recovery, particularly in early stages, benefiting pioneer plants and improving soil quality.

6. Challenges and Limitations

The cultivation of AMFs presents considerable difficulties since they are obligate biotrophs and therefore only grow in intimate association with host plant root systems (Brundrett & Tedersoo, 2018). These plants can be more efficiently colonized depending on the AMF species, owing to the capacity of certain species to suppress host defenses and optimize nutrient exchange processes (Lanfranco & Young, 2012). Field propagation is the most economical method for producing AMFs. In this system, host plants are inoculated and cultivated in sandy substrates, enabling AMFs to proliferate naturally. At the end of the growth cycles, mycorrhizal roots and propagule-bearing soil are collected, dried, and used as inoculum. Despite its simplicity, the approach has drawbacks, including variable yields, challenges in spore recovery, and a high risk of contamination of the inoculum by pests, pathogens, and weeds. Several of these limitations can be mitigated by producing AMF inoculum in soilless greenhouse systems, using sterile substrates such as vermiculite to cultivate host plants (Rouphael et al., 2015).

Another difficulty is the taxonomic identification of AMFs, which is based on microscopic characteristics such as spore color and size (Souza, 2015), requiring experienced professionals. In addition, non-traditional cultivation approaches, for instance, in vitro systems or using transformed root cultures, also require specialized laboratory facilities (Salomon et al., 2022).

Mass production of AMFs remains a challenge, but advances in cultivation techniques and the selection of efficient species have improved its feasibility. Their application in sustainable agriculture justifies continued efforts to overcome these limitations.

AMFs are considered generalists because they associate with a broad array of host plants (Davison et al., 2011). However, some plant–fungus combinations are more effective than others due to the plant's genetic selectivity, its biochemical capacity to produce exudates, and its evolutionary history.

AMF species lack strict host specificity; nevertheless, their performance is shaped by the environmental conditions under which the host plant grows; thus, environmental variability strongly affects AMF–plant symbiosis (Winagraski et al., 2019). Abiotic factors such as low phosphorus availability, pH between 5.5 and 7.5, and temperatures between 20°C and 30°C favor this relationship, as do biotic factors like the existing microbiome and management practices. AMF applications need to be harmonized with overall conservative field management practices, such as minimizing soil disturbance and incorporating cover crops. Jiang et al. (2021) observed that organic fertilizers and pesticides increase AMF biomass and impose a smaller negative effect on their diversity than mineral fertilization alone.

Therefore, precise selection of compatible plant–AMF combinations should be proposed for each ecosystem, supported by local studies and the adoption of beneficial agricultural practices through favorable management (Regvar et al., 2003; Rouphael et al., 2015).

Realizing the greatest gains from AMF activity requires the adoption of beneficial agricultural practices (Gosling et al., 2006), inoculating efficient AMF strains, and selecting suitable host plant–fungus combinations (Rouphael et al., 2015).

Another significant obstacle to the production and commercialization of AMF-based inoculants is the absence of standardization, which compromises their efficacy and reliability (Salomon et al., 2022). The main problems arising from this lack of standardization include inconsistency in inoculant composition, absence of standardized cultivation protocols, lack of regulations and studies under different conditions, and limited storage duration (Salomon et al., 2022b).

To change this scenario, it is necessary to establish technical standards for production, including inoculant identification and quality control, use standardized methods for spore counting and purity verification, invest in research to develop more stable and efficient inoculants, and educate producers on how to choose and apply commercial inoculants, as described by Salomon et al. (2022).

The adoption of bioinputs such as AMFs still faces resistance from many farmers, despite the proven benefits of their use. This resistance is due to factors such as historical dependence on synthetic pesticides and inorganic fertilizers (Liu et al., 2015), inconsistent results, lack of knowledge, and the initial cost of quality inoculants. Storage logistics also pose challenges, as products require specific temperature and humidity conditions to maintain viability, and there is a lack of government incentives to promote the use of bioinputs. Rillig et al. (2019) mention that there will be no significant change until AMFs (and other beneficial soil biota) are integrated into management decisions.

Changing this scenario will require joint efforts among governments, researchers, industry, and farmers to prove efficacy through scientific data and practical results, make bioinputs more accessible and user-friendly, and overcome conventional agribusiness paradigms.

With the advancement of sustainable agriculture and growing pressure for practices less dependent on chemical inputs, bioinputs are anticipated to assume a progressively more prominent role. Continued research and development are necessary to further improve the quality and efficiency of commercial AMF inoculants (Salomon et al., 2022).

7. Considerations

The commercialization of mycorrhizal-based inoculants has advanced in recent years. However, their high production cost still limits widespread adoption. Improving production processes and validating their effectiveness in field conditions are essential steps. In this regard, continuous technical support can contribute to identifying strengths and aspects that need improvement. The presence of counterfeit products in the marketplace weakens farmers' trust, slowing down sector growth. Furthermore, many producers remain unaware of bio-input technologies or face difficulties accessing reliable information, encountering socio-economic obstacles that hinder their adoption.

Despite these challenges, the mycorrhizal inoculant sector is projected to expand annually by about 14.3%, potentially reaching close to USD 1.087 billion by 2027 (Ghorui et al., 2025). Mycorrhizal symbiosis promotes increased nutrient uptake by plants, improves soil structure and quality, and enhances plant tolerance to environmental stresses. Nevertheless, it must be recognized that elements such as crop species, adopted management practices, soil and climate conditions, and soil microbial diversity can influence inoculation efficiency. Therefore, continued cooperation among researchers, governments, industry, and farmers is essential to overcome current limitations and consolidate AMFs as a viable approach for agriculture.

References

Abdar, N., Zarei, M., Shahriari, A.G. & Mirmazloum, I. (2023). The effect of arbuscular mycorrhizal inoculation and plant growth-promoting rhizobacteria on maize (*Zea mays* L.) under boron toxicity stress. Notulae Botanicae Horti Agrobotanici Cluj-Napoca 51(4), 13473–13473.

Adeyemi, N.O., Atayese, M.O., Sakariyawo, O.S. & Azeez, J.O. (2022). Mycorrhizal growth and phosphorus responses of tropical soybean (*Glycine max* L.) cultivars differing with arbuscular mycorrhizal fungi isolates and phosphorus application rates in a derived-savanna zone of Nigeria. Journal of Plant Nutrition 45(8), 1223–1239.

Adeyemi, N.O., Atayese, M.O., Sakariyawo, O.S., Azeez, J.O., Olubode, A., Ridwan, M. et al. (2021). Growth and phosphorus uptake of soybean (*Glycine max* L.) in response to arbuscular mycorrhizal fungus Rhizophagus intraradices inoculation in heavy metal-contaminated soils. Soil and Sediment Contamination: An International Journal 30(6), 698–713.

Ahammed, G.J. & Hajiboland, R. (2024). Introduction to arbuscular mycorrhizal fungi and higher plant symbiosis: characteristic features, functions, and applications. pp. 1–17. *In*: Ahammed, G.J. & Hajiboland, R. (Eds.). Arbuscular Mycorrhizal Fungi and Higher Plants: Fundamentals and Applications. Singapore: Springer Nature.

Al-Amri, S.M. (2021). Application of bio-fertilizers for enhancing growth and yield of common bean plants grown under water stress conditions. Saudi Journal of Biological Sciences 28(7), 3901–3908.

Al-Karaki, G.N. (2006). Nursery inoculation of tomato with arbuscular mycorrhizal fungi and subsequent performance under irrigation with saline water. Scientific Horticulture 109, 1–7.

Alizadeh, Z., Heidari, P. & Asghari, H.R. (2025). Exploring the influence of symbiosis between arbuscular mycorrhizal fungi and beans on potassium uptake and the activity of AKT and HKT genes. Scientific Reports 15(1), 1–12.

Ames, R.N., Reid, C.P.P., Porter, L.K. & Cambardella, C. (1983). Hyphal uptake and transport of nitrogen from two 15N-labeled sources by *Glomus mosseae*, a vesicular-arbuscular mycorrhizal fungus. New Phytologist 95(3), 381–396.

Amiri, R., Nikbakht, A. & Etemadi, N. (2015). Alleviation of drought stress on rose geranium [*Pelargonium graveolens* (L.) Herit.] in terms of antioxidant activity and secondary metabolites by mycorrhizal inoculation. Scientific Horticulture 197, 373–380.

Asensio, D., Rapparini, F. & Peñuelas, J. (2012). AM fungi root colonization increases the production of essential isoprenoids vs. nonessential isoprenoids especially under drought stress conditions or after jasmonic acid application. Phytochemistry 77, 149–161.

Asmelash, F., Bekele, T. & Birhane, E. (2016). The potential role of arbuscular mycorrhizal fungi in the restoration of degraded lands. Frontiers in Microbiology 7, 1095.

Babadi, M., Zalaghi, R. & Taghavi, M. (2019). A non-toxic polymer enhances sorghum-mycorrhiza symbiosis for bioremediation of Cd. Mycorrhiza 29(4), 375–387.

Banerjee, R., Evande, R., Kabil, Ö., Ojha, S. & Taoka, S. (2003). Reaction mechanism and regulation of cystathionine β-synthase. Biochimica et Biophysica Acta 1647, 30–35.

Barea, J.M., Palenzuela, J., Cornejo, P., Sánchez-Castro, I., Navarro-Fernández, C.M., Lopéz-García, A. et al. (2011). Ecological and functional roles of mycorrhizas in semi-arid ecosystems of Southeast Spain. Journal of Arid Environments 75(12), 1292–1301.

Barea, J.M., Pozo, M.J., Azcón, R. & Azcón-Aguilar, C. (2005). Microbial co-operation in the rhizosphere. Journal of Experimental Botany 56(417), 1761–1778.

Baslam, M., Garmendia, I. & Goicoechea, N. (2013). The arbuscular mycorrhizal symbiosis can overcome reductions in yield and nutritional quality in greenhouse-lettuces cultivated at inappropriate growing seasons. Scientific Horticulture 164, 145–154.

Berruti, A., Lumini, E., Balestrini, R. & Bianciotto, V. (2016). Arbuscular mycorrhizal fungi as natural biofertilizers: Let's benefit from past successes. Frontiers in Microbiology 6, 1559.

Bowles, T.M., Jackson, L.E., Loeher, M. & Cavagnaro, T.R. (2017). Ecological intensification and arbuscular mycorrhizas: A meta-analysis of tillage and cover crop effects. Journal of Applied Ecology 54, 1785–1793. doi: 10.1111/1365-2664.12815

Brito, L.F., Galvão, J.C.C., Giehl, J., Campos, S. de A. & Coelho, S.P. (2017). Agronomic traits and yield of organic maize under no-tillage system. Pesq. Agropec. Trop. 47, 72–79. doi: 10.1590/1983-40632016v4742774

Brundrett, M.C. & Tedersoo, L. (2018). Evolutionary history of mycorrhizal symbioses and global host plant diversity. New Phytology 220(4), 1108–1115. https://doi.org/10.1111/nph.14976.

Bücking, H., Liepold, E. & Ambilwade, P. (2012). The role of the mycorrhizal symbiosis in nutrient uptake of plants and the regulatory mechanisms underlying these transport processes. Plant Science 4, 108–132.

Buzo, F.S., Garé, L.M., Garcia, N.F.S., Silva, M.S.R.A., Martins, J.T., Silva, P.H.G. et al. (2023). Effect of mycorrhizae on phosphate fertilization efficiency and maize growth under field conditions. Scientific Reports 13, 3527.

Buzo, F.D.S., Garcia, N.F.S., Garé, L.M., Gato, I.M.B., Martins, J.T., Martins, J.O.M. et al. (2022a). Phosphate fertilization and mycorrhizal inoculation increase corn leaf and grain nutrient contents. Agronomy 12(7), 1597.

Buzo, F.S., Satin Mortinho, E., Lobo Santana, T.L., Militão Garcia, I., de Carvalho, C.L.M. & Filho, M.C.M.T. (2022b). Bean nutrition and development in the function of reduced phosphorus doses and inoculation with arbuscular mycorrhizal fungus. Journal of Plant Nutrition 45(13), 1942–1952.

Buzo, F.S., Garé, L.M., Garcia, N.F.S., de Andrade da Silva, M.S.R., da Silva, P.H.G., de Souza Morita, P.R. et al. (2022c). Chemical seed treatment and mycorrhizal inoculation provide better development and nutrition of common bean plants. Pest Management Science, 78(7), 2985–2994.

Cabral, L., Soares, C.R.F.S., Giachini, A.J. & Siqueira, O. (2015). Arbuscular mycorrhizal fungi in phytoremediation of contaminated areas by trace elements: Mechanisms and major benefits of their applications. World Journal of Microbiology and Biotechnology 31(11), 1655–1664.

Caravaca, F., Alguacil, M.M., Azcón, R., Diaz, G. & Roldán, A. (2004). Comparing the effectiveness of mycorrhizal inoculation and amendment with sugar beet, rock phosphate and *Aspergillus niger* to enhance field performance of the leguminous shrub *Dorycnium pentaphyllum* L. Applied Soil Ecology 25(2), 169–180.

Cardozo-Junior, F.M., Carneiro, R.F.V., Goto, B.T., Bezerra, A.A.C., Araújo, A.S.F. & Nunes, L.A.P.L. (2012). Arbuscular mycorrhizal fungi in degraded lands in northeast Brazil. African Journal of Microbiology Research 6, 7198–7205. doi: 10.5897/AJMR12.1132

Carrara, J.E., Reddivari, L. & Heller, W.P. (2024). Inoculation of black turtle beans (*Phaseolus vulgaris*) with mycorrhizal fungi increases the nutritional quality of seeds. Plant-Environment Interactions 5(1), e10128.

Cavagnaro, T.R., Jackson, L.E., Six, J., Ferris, H., Goyal, S., Asami, D. et al. (2006). Arbuscular mycorrhizas, microbial communities, nutrient availability, and soil aggregates in organic tomato production. Plant and Soil 282, 209–225.

Chagnon, P.L., Bradley, R.L., Maherali, H. & Klironomos, J.N. (2013). A trait-based framework to understand life history of mycorrhizal fungi. Trends in Plant Science 18, 484–491.

Chandrasekaran, M. (2020). A meta-analytical approach on arbuscular mycorrhizal fungi inoculation efficiency on plant growth and nutrient uptake. Agriculture 10(9), 370.

Chen, B., Nayuki, K., Kuga, Y., Zhang, X., Wu, S. & Ohtomo, R. (2018). Uptake and intraradical immobilization of cadmium by arbuscular mycorrhizal fungi as revealed by a stable isotope tracer and synchrotron radiation μx-ray fluorescence analysis. Microbes and Environments 33(3), 257–263.

Chen, M., Arato, M., Borghi, L., Nouri, E. & Reinhardt, D. (2018). Beneficial services of arbuscular mycorrhizal fungi-from ecology to application. Frontiers in Plant Science 9, 1270.

Chern, E.C., Tsai, D.W. & Ogunseitan, O.A. (2007). Deposition of glomalin-related soil protein and sequestered toxic metals into watersheds. Environmental Science & Technology 41(10), 3566–3572.

Citterio, S., Prato, N., Fumagalli, P., Aina, R., Massa, N., Santagostino, A. et al. (2005). The arbuscular mycorrhizal fungus Glomus mosseae induces growth and metal accumulation changes in *Cannabis sativa* L. Chemosphere 59(1), 21–29.

Curaqueo, G., Acevedo, E., Cornejo, P., Seguel, A., Rubio, R. & Borie, F. (2010). Tillage effect on soil organic matter, mycorrhizal hyphae and aggregates in a mediterranean agroecosystem. Revista de la ciencia del suelo y nutrición vegetal 10, 12–21. doi: 10.4067/S0718-27912010000100002

Davison, J., Opik, M., Daniell, T., Moora, M. & Zobel, M. (2011). Arbuscular mycorrhizal fungal communities in plant roots are not random assemblages. FEMS Microbiology Ecology 78, 103–115. 10.1111/j.1574-6941.2011.01103.x.

De Andrade, S.A.L., Silveira, A., Jorge, R.A. & Abreu, M.F. (2008). Cadmium accumulation in sunflower plants influenced by arbuscular mycorrhiza. International Journal of Phytoremediation 10(1), 1–13.

Degryse, F., Smolders, E., Zhang, H. & Davison, W. (2009). Predicting availability of mineral elements to plants with the DGT technique: A review of experimental data and interpretation by modelling. Environmental Chemistry, 6(3), 198–218.

Dobo, B. (2022). Effect of arbuscular mycorrhizal fungi (AMF) and rhizobium inoculation on growth and yield of *Glycine max* L. varieties. International Journal of Agronomy, 2022(1), 9520091.

Duarte, A.G. & Maherali, H. (2022). A meta-analysis of the effects of climate change on the mutualism between plants and arbuscular mycorrhizal fungi. Ecology and Evolution 12, e8518. doi: 10.1002/ece3.8518

Erdinç, C., Durak, E.D., Ekincialp, A., Şensoy, S. & Demir, S. (2017). Variations in response of determinate common bean (*Phaseolus vulgaris* L.) genotypes to arbuscular mycorrhizal fungi (AMF) inoculation. Turkish Journal of Agriculture and Forestry, 41(1), 1–9.

Espindola, J.A.A., Almeida, D.L., Guerra, J.G.M., Silva, E.M.R. & Souza, F.A. (1998). Influence of green manuring on mycorrhizal colonization and production of sweet potato. Pesquisa Agropecuária Brasileira 33, 339–347.

Ferreira Vilela, L.A. & Barbosa, M.V. (2019). Chapter 21 – Contribution of arbuscular mycorrhizal fungi in promoting cadmium tolerance in plants. pp. 553–586. *In*: Hasanuzzaman, M., Vara, P., Majeti, N., Nahar, K. (Eds.). Cadmium Tolerance in Plants [S.l.]. Academic Press.

Ferrol, N., Tamayo, E. & Vargas, P. (2016). The heavy metal paradox in arbuscular mycorrhizas: From mechanisms to biotechnological applications. Journal of Experimental Botany erw403.

French, K.E. (2017). Engineering mycorrhizal symbioses to alter plant metabolism and improve crop health. Frontiers in Microbiology 8, 1403.

García de León, D., Moora, M., Öpik, M., Neuenkamp, L., Gerz, M., Jairus, T. et al. (2016). Symbiont dynamics during ecosystem succession: Co-occurring plant and arbuscular mycorrhizal fungal communities. FEMS Microbiology Ecology 92(7), fiw097.

Garcia, K. & Zimmermann, S. (2014). The role of mycorrhizal associations in plant potassium nutrition. Frontiers in Plant Science 5, 337.

Garcia, K.G.V., de Souza Oliveira Filho, J., de Araújo Pereira, A.P. & Mendes Filho, P.F. (2024). Can inoculation of native arbuscular mycorrhizal fungi from a mining area attenuate stress of *Acacia mangium* Willd. to excess manganese? Journal of Soils and Sediments 24(9), 3252–3264.

Garg, N. & Chandel, S. (2012). Role of arbuscular mycorrhizal (AM) fungi on growth, cadmium uptake, osmolyte, and phytochelatin synthesis in *Cajanus cajan* (L.) Millsp. under NaCl and cd stresses. Journal of Plant Growth Regulation 31(3), 292–308.

Garg, N. & Pandey, R. (2015). Effectiveness of native and exotic arbuscular mycorrhizal fungi on nutrient uptake and ion homeostasis in salt-stressed *Cajanus cajan* L. (Millsp.) genotypes. Mycorrhiza 25, 165–180.

George, E., Häussler, K.U., Vetterlein, D., Gorgus, E. & Marschner, H. (1992). Water and nutrient translocation by hyphae of Glomus mosseae. Canadian Journal of Botany 70, 2130–2137.

Gholamhoseini, M., Ghalavand, A., Dolatabadian, A., Jamshidi, E. & Khodaei-Joghan, A. (2013). Effects of arbuscular mycorrhizal inoculation on growth, yield, nutrient uptake and irrigation water productivity of sunflowers grown under drought stress. Agricultural Water Management 117, 106–114.

Ghorui, M., Chowdhury, S. & Burla, S. (2025). Recent advances in the commercial formulation of arbuscular mycorrhizal inoculants. Frontiers in Industrial Microbiology 3, 1553472.

Goicoechea, N., Merino, S. & Sánchez-Díaz, M. (2005). Arbuscular mycorrhizal fungi can contribute to maintain antioxidant and carbon metabolism in nodules of *Anthyllis cytisoides* L. subjected to drought. Journal of Plant Physiology 162, 27–35.

González-Chávez, M.C.A., Gonzáles, R.C., Wright, S.F. & Nichols, K.A. (2004). The role of glomalin, a protein produced by arbuscular mycorrhizal fungi, in sequestering potentially toxic elements. Environmental Pollution 130(3), 317–323.

González-Guerrero, M., Melville, L.H., Ferrol, N., Lott, J.N.A., Azcon-Aguillar, C. & Peterson, R.L. (2008). Ultrastructural localization of heavy metals in the extraradical mycelium and spores of the arbuscular mycorrhizal fungus Glomus intraradices. Canadian Journal of Microbiology 54(2), 103–110.

Gosling, P., Hodge, A., Goodlass, G. & Bending, G.D., (2006). Arbuscular mycorrhizal fungi and organic farming. Agric. Ecosyst. Environ. 113, 17–35.

Gray, L.E. & Gerdemann, J.W. (1973). Uptake of sulphur-35 by vesicular-arbuscular mycorrhizal. Plant and Soil 39(3), 687–689.

Guo, T., Zhang, J., Christie, P. & Li, X. (2007). Pungency of spring onion as affected by inoculation with arbuscular mycorrhizal fungi and sulfur supply. Journal of Plant Nutrition 30(7), 1023–1034.

Habte, M., Miyasaka, S.C. & Matsuyama, D.T. (2001). Arbuscular mycorrhizal fungi improve early forest-tree establishment. *In*: Plant Nutrition: Food Security and Sustainability of Agro-Ecosystems through Basic and Applied Research (pp. 644–645). Dordrecht: Springer Netherlands.

Hajiboland, R., Aliasgharzadeh, N., Laiegh, S.F. & Poschenrieder, C. (2010). Colonization with arbuscular mycorrhizal fungi improves salinity tolerance of tomato (*Solanum lycopersicum* L.) plants. Plant and Soil 331, 313–327.

Han, Y., Zveuche, O., Dong, F., Ling, Q., Chen, Y., Sajid, S. et al. (2021). Unraveling the effects of arbuscular mycorrhizal fungi on cadmium uptake and detoxification mechanisms in perennial ryegrass (*Lolium perenne*). Science of The Total Environment 798, 149222.

Hawkins, H.J., Johansen, A. & George, E. (2000). Uptake and transport of organic and inorganic nitrogen by arbuscular mycorrhizal fungi. Plant and Soil 226, 275–285.

He, Z.Q., He, C.X., Zhang, Z.B., Zou, Z.R. & Wang, H.S. (2007). Changes of antioxidative enzymes and cell membrane osmosis in tomato colonized by arbuscular mycorrhizae under NaCl stress. Colloids and Surfaces B: Biointerfaces 59, 128–133.

Hetrick, B.D., Wilson, G.W.T. & Todd, T.C. (1990). Differential responses of C3 and C4 grasses to mycorrhizal symbiosis, phosphorus fertilization, and soil microorganisms. Canadian Journal of Botany 68(3), 461–467.

Hobbs, R.J., Jentsch, A. & Temperton, V.M. (2007). Restoration as a process of assembly and succession mediated by disturbance. *In*: Linking Restoration and Ecological Succession. (pp. 150–167). New York, NY: Springer New York.

Hoeksema, J.D., Chaudhary, V.B., Gehring, C.A., Johnson, N.C., Karst, J., Koide, R.T. et al. (2010). A meta-analysis of context-dependency in plant response to inoculation with mycorrhizal fungi. Ecology Letters 13(3), 394–407.doi: 10.1111/j.1461-0248.2009.01430.x

Hu, Z.-H., Zhuo, F., Jing, S.H., Li, X., Yan, T.-X., Lei, L.-L. et al. (2019). Combined application of arbuscular mycorrhizal fungi and steel slag improves plant growth and reduces Cd, Pb accumulation in *Zea mays*. International Journal of Phytoremediation 21(9), 857–865.

Jakobsen, I., Gazey, C. & Abbott, L.K. (2001). Phosphate transport by communities of arbuscular mycorrhizal fungi in intact soil cores. New Phytologist 149, 95–103.

Jansa, J., Mozafar, A., Anken, T., Ruh, R., Sanders, I. & Frossard, E. (2002). Diversity and structure of AMF communities as affected by tillage in a temperate soil. Mycorrhiza 12, 225–234. doi: 10.1007/s00572-002-0163-z

Jansa, J., Mozafar, A. & Frossard, E. (2003). Long-distance transport of P and Zn through the hyphae of an arbuscular mycorrhizal fungus in symbiosis with maize. Agronomie 23(5–6), 481–488.

Jiang, S., An, X., Shao, Y., Kang, Y., Chen, T., Mei, X. et al. (2021). Responses of arbuscular mycorrhizal fungi occurrence to organic fertilizer: A meta-analysis of field studies. Plant Soil 469, 89–105.

Johansen, A., Finlay, R.D. & Olsson, P.A. (1996). Nitrogen metabolism of external hyphae of the arbuscular mycorrhizal fungus Glomus intraradices. New Phytologist 133(4), 705–712.

Kabir, Z. (2005). Tillage or no-tillage: Impact on mycorrhizae. Canadian Journal of Plant Science 85, 23–29. doi: 10.4141/P03-160

Kiers, E.T., Lovelock, C.E., Krueger, E.L. & Herre, E.A. (2000). Differential effects of tropical arbuscular mycorrhizal fungal inocula on root colonization and tree seedling growth: Implications for tropical forest diversity. Ecology Letters 3(2), 106–113.

Klugh, K.R. & Cumming, J.R. (2007). Variations in organic acid exudation and aluminum resistance among arbuscular mycorrhizal species colonizing Liriodendron tulipifera. Tree Physiology, 27, 1103–1112.

Klugh-Stewart, K. & Cumming, J.R. (2009). Organic acid exudation by mycorrhizal *Andropogon virginicus* L. (broomsedge) roots in response to aluminum. Soil Biology and Biochemistry 41, 367–373.

Koide, R.T. (1991). Nutrient supply, nutrient demand and plant response to mycorrhizal infection. New Phytologist 117(3), 365–386.

Lanfranco, L. & Young, J.P. (2012). Genetic and genomic glimpses of the elusive arbuscular mycorrhizal fungi. Current Opinion in Plant Biology 15, 454–461. 10.1016/j.pbi.2012.04.003.

Lehmann, A. & Rillig, M.C. (2015). Arbuscular mycorrhizal contribution to copper, manganese and iron nutrient concentrations in crops–a meta-analysis. Soil Biology and Biochemistry 81, 147–158.

Leifheit, E.F., Verbruggen, E. & Rillig, M.C. (2014). Arbuscular mycorrhizal fungi reduce decomposition of woody plant litter while increasing soil aggregation. Soil Biology and Biochemistry 81, 323–328. doi:10.1016/j.soilbio.2014. 12.003

Lekberg, Y. & Koide, R.T. (2005). Arbuscular mycorrhizal fungi, rhizobia, available soil P and nodulation of groundnut (*Arachis hypogaea*) in Zimbabwe. Agriculture, Ecosystems & Environment 110, 143–148. doi: 10.1016/j.agee.2005.03.011

Lerat, S., Lapointe, L., Gutjahr, S., Piche, Y. & Vierheilig, H. (2003). Carbon partitioning in a split-root system of arbuscular mycorrhizal plants is fungal and plant species dependent. New Phytologist 157, 589–595. doi:10.1046/j.1469–8137.2003.00691.x

Li, H., Smith, F.A., Dickson, S., Holloway, R.E. & Smith, S.E. (2008). Plant growth depressions in arbuscular mycorrhizal symbioses: Not just caused by carbon drain? New Phytologist 178(4), 852–862.

Li, H., Zhang, J., Hu, H., Chen, L., Zhu, Y., Shen, H. & Fang, J. (2017). Shift in soil microbial communities with shrub encroachment in Inner Mongolia grasslands, China. European Journal of Soil Biology 79, 40–47.

Li, J., Xu, Z., Zhang, W., Yang, X., Struik, P.C., Jiang, S. et al. (2024). Soil microbiome mediates plant community productivity in grass-legume mixtures. Plant and Soil 149–168.

Liu, J., Liu, J., Liu, J., Cui, M., Huang, Y., Tian, Y. et al. (2019). The potassium transporter SlHAK10 is involved in mycorrhizal potassium uptake. Plant Physiology 180(1), 465–479.

Liu, W., Ma, K., Wang, X., Wang, Z. & Negrete-Yankelevich, S. (2022). Effects of no-tillage and biologically-based organic fertilizer on soil arbuscular mycorrhizal fungal communities in winter wheat field. Applied Soil Ecology 178, 104564. doi: 10.1016/j.apsoil.2022.104564

Liu, W., Zhang, Y., Jiang, S., Deng, Y., Christie, P., Murray, P.J. et al. (2016). Arbuscular mycorrhizal fungi in soil and roots respond differently to phosphorus inputs in an intensively managed calcareous agricultural soil. Scientific Reports 6(1), 24902. doi: 10.1038/srep24902

Liu, Y., Pan, X. & Li, J. (2015). A 1961–2010 record of fertilizer use, pesticide application and cereal yields: A review. Agronomy for Sustainable Development 35, 83–93. https://doi.org/10.1007/s13593-014-0259-9.

Ludwig-Müller, J. (2010). Hormonal responses in host plants triggered by arbuscular mycorrhizal fungi. (pp. 169–190). *In*: Koltai, H. & Kapulnik, Y. [eds.], Arbuscular Mycoorrhizas: Physiology and Function. Springer Netherlands.

Mäder, P., Vierheilig, H., Streitwolf-Engel, R., Boller, T., Frey, B., Christie, P. et al. (2000). Transport of 15N from a soil compartment separated by a polytetrafluoroethylene membrane to plant roots via the hyphae of arbuscular mycorrhizal fungi. The New Phytologist 146(1), 155–161.

Madhushan, K.W.A., Karunarathna, S.C., Dissanayake, D.M.D., Priyadarshani, T.D.C., Stephenson, S.L., Elgorban, A.M. et al. (2023). Effect of arbuscular mycorrhizal fungi on lowland rice growth and yield (*Oryza sativa* L.) under different farming practices. Agronomy 13(11), 2803.

Martín-Cardoso, H., Castillo, L., Busturia, I., Bücker, G., Marqués, L., Pla, E. et al. (2025). Arbuscular mycorrhizal fungi increase blast resistance and grain yield in Japonica rice cultivars in flooded fields. Rice 18(1), 1–17.

Massa, N., Cesaro, P., Todeschini, V., Capraro, J., Scarafoni, A., Cantamessa, S. et al. (2020). Selected autochthonous rhizobia, applied in combination with AM fungi, improve seed quality of common bean cultivated in reduced fertilization condition. Applied Soil Ecology 148, 103507.

Maya, M.A. & Matsubara, Y. (2013). Influence of arbuscular mycorrhiza on the growth and antioxidative activity in cyclamen under heat stress. Mycorrhiza 23, 381–390.

Miranda, J.C.C., Vilela, L. & Miranda, L.N. (2005). Dynamics and contribution of arbuscular mycorrhiza in production systems with crop rotation. Pesquisa Agropecuária Brasileira 40, 1005–1014. doi: https://doi.org/10.1590/S0100-204X2005001000009

Morar, F., Iantovics, L.B. & Gligor, A. (2018). Analysis of phytoremediation potential of crop plants in industrial heavy metal contaminated soil in the upper Mures River Basin. Journal of Environmental Informatics 31(1), 1–14.

Moreira, F.M.S. & Siqueira, J.O. (2006). Microbiologia e bioquímica do solo (2ª ed., pp. 407–447). Minas Gerais: Editora da UFLA.

Nader, A.A., Hauka, F.I., Afify, A.H. & El-Sawah, A.M. (2024). Drought-tolerant bacteria and arbuscular mycorrhizal fungi mitigate the detrimental effects of drought stress induced by withholding irrigation at critical growth stages of soybean (*Glycine max*, L.). Microorganisms 12(6), 1123.

Nafady, N.A. & Elgharably, A. (2018). Mycorrhizal symbiosis and phosphorus fertilization effects on Zea mays growth and heavy metals uptake. International Journal of Phytoremediation 20(9), 869–875.

Neuenkamp, L., Prober, S.M., Price, J.N., Zobel, M. & Standish, R.J. (2019). Benefits of mycorrhizal inoculation to ecological restoration depend on plant functional type, restoration context and time. Fungal Ecology 40, 140–149.

Olsson, P.A., Hammer, E.C., Wallander, H. & Pallon, J. (2008). Phosphorus availability influences elemental uptake in the mycorrhizal fungus *Glomus intraradices*, as revealed by particle-induced X-ray emission analysis. Applied and Environmental Microbiology 74(13), 4144–4148.

Pallon, J., Wallander, H., Hammer, E., Marrero, N.A., Auzelyte, V., Elfman, M. & Olsson, P.A. (2007). Symbiotic fungi that are essential for plant nutrient uptake investigated with NMP. Nuclear Instruments and Methods in Physics Research Section B: Beam Interactions with Materials and Atoms 260, 149–152.

Peng, S., Ban, M., Xing, W., Ge, Z. & Mao, L. (2022). Effects of nitrogen addition and seasonal change on arbuscular mycorrhizal fungi community diversity in a poplar plantation. Frontierts in Ecology and Evolution 10. doi: 10.3389/fevo.2022.1101698

Pepe, A., Giovannetti, M. & Sbrana, C. (2020). Appressoria and phosphorus fluxes in mycorrhizal plants: Connections between soil- and plant-based hyphae. Mycorrhiza 30, 589–600.

Pereira, C.M.R., Silva, D.K.A., Goto, B.T., Rosendahl, S. & Maia, L.C. (2018). Management practices may lead to loss of arbuscular mycorrhizal fungal diversity in protected areas of the Brazilian Atlantic Forest. Fungal Ecology 34, 50–58. https://doi.org/10.1016/j.funeco.2018.05.001

Perner, H., Schwarz, D., Bruns, C., Mäder, P. & George, E. (2007). Effect of arbuscular mycorrhizal colonization and two levels of compost supply on nutrient uptake and flowering of pelargonium plants. Mycorrhiza 17, 469–474.

Piwowarska, D., Kiedrzyńska, E. & Jaszczyszyn, K. (2024). A global perspective on the nature and fate of heavy metals polluting water ecosystems, and their impact and remediation. Critical Reviews in Environmental Science and Technology 54(19), 1436–1458.

Porcel, R. & Ruiz-Lozano, J.M. (2004). Arbuscular mycorrhizal influence on leaf water potential, solute accumulation, and oxidative stress in soybean plants subjected to drought stress. Journal of Experimental Botany 55, 1743–1750.

Porcel, R., Aroca, R. & Ruiz-Lozano, J.M. (2012). Salinity stress alleviation using arbuscular mycorrhizal fungi: A review. Agronomy for Sustainable Development 32, 181–200.

Pouyu-Rojas, E. & Siqueira, J.O. (2000). Arbuscular mycorrhiza land soil fertilization on post-transplant evelopment of out plants of seven forest species. Pesquisa Agropecuária Brasileira 35, 103–114. doi:10.1590/S0100-204X2000000100013

Prach, K. & Hobbs, R.J. (2008). Spontaneous succession versus technical reclamation in the restoration of disturbed sites. Restoration Ecology 16(3), 363–366.

Razvi, S.M., Singh, N., Mushtaq, A., Dar, S. & Hussain, S. (2023). Arbuscular mycorrhizal fungi for salinity stress: Anti-stress role and mechanisms. Pedosphere 33(1), 212–224.

Regvar, M., Vogel, K., Irgel, N., Wraber, T., Hildebrandt, U., Wilde, P. & Bothe, H., (2003). Colonization of pennycresses (*Thlaspi* spp.) of the Brassicaceae by arbuscular mycorrhizal fungi. Journal of Plant Physiology 160, 615–626.

Rengel, Z. & Marschner, P. (2005). Nutrient availability and management in the rhizosphere: Exploiting genotypic differences. New Phytologist 168(2), 305–312.

Riaz, M., Kamran, M., Fang, Y., Wang, Q., Cao, H., Yang, G. et al. (2021). Arbuscular mycorrhizal fungi-induced mitigation of heavy metal phytotoxicity in metal contaminated soils: A critical review. Journal of Hazardous Materials 402, 123919.

Rillig, M., Aguilar, T.C., Camenzind, T., Cavagnaro, T., Degrune, F., Hohmann, P. et al. (2019). Why farmers should manage the arbuscular mycorrhizal symbiosis. New Phytologist 222, 10.1111/nph.15602.

Rouphael, Y., Franken, P., Schneider, C., Schwarz, D., Giovannetti, M., Agnolucci, M. et al. (2015). Arbuscular mycorrhizal fungi act as biostimulants in horticultural crops. Scientia Horticulturae 196, 91–108. 10.1016/j.scienta.2015.09.002.

Ryan, M.H. & Graham, J.H. (2018). Little evidence that farmers should consider abundance or diversity of arbuscular mycorrhizal fungi when managing crops. New Phytologist 220(4), 1092–1107.

Saleem, M.H., Ali, S., Rehman, M., Rana, M.S., Rizwan, M., Kamran, M. et al. (2020). Influence of phosphorus on copper phytoextraction via modulating cellular organelles in two jute (*Corchorus capsularis* L.) varieties grown in a copper mining soil of Hubei Province, China. Chemosphere 248, 126032.

Salomon, M., Watts-Williams, S., McLaughlin, M., Bücking, H., Singh, B., Hutter, I. et al. (2022). Establishing a quality management framework for commercial inoculants containing arbuscular mycorrhizal fungi. iScience 25, 104636. 10.1016/j.isci.2022.104636.

Salomon, M.J., Demarmels, R., Watts-Williams, S.J., McLaughlin, M.J., Kafle, A., Ketelsen, C. et al. (2022b). Global evaluation of commercial arbuscular mycorrhizal inoculants under greenhouse and field conditions. Applied Soil Ecology 169, 104225. https://doi.org/10.1016/j. apsoil.2021.104225.

Saraswat, S. & Rai, J. (2011). Mechanism of metal tolerance and detoxification in mycorrhizal fungi. pp. 225–240. *In*: Khan, M.S., Zaidi, A., Goel, R. & Musarrat, J. (Eds.). Biomanagement of Metal-contaminated Soils. Dordrecht: Springer Netherlands.

Schachtman, D.P., Reid, R.J. & Ayling, S.M. (1998). Phosphorus uptake by plants: From soil to cell. Plant Physiology 116, 447–453.

Scheloske, S., Maetz, M., Schneider, T., Hildebrandt, U., Bothe, H. & Povh, B. (2004). Element distribution in mycorrhizal and nonmycorrhizal roots of the halophyte Aster tripolium determined by proton induced X-ray emission. Protoplasma 223, 183–189.

Schnitzer, S.A., Klironomos, J.N., Lambers, J.H.R., Kinkel, L.L., Reich, P.B., Xiao, K. et al. (2011). Soil microbes drive the classic plant diversity-productivity pattern. Ecology 92(2), 296–303. doi:10.1890/10-0773.1

Sharif, M. & Claassen, N. (2011). Action mechanisms of arbuscular mycorrhizal fungi in phosphorus uptake by *Capsicum annuum* L. Pedosphere 21, 502–511.

Sheng, M., Tang, M., Chen, H., Yang, H. & Huang, Y. (2008). Influence of arbuscular mycorrhizae on photosynthesis and water status of maize plants under salt stress. Mycorrhiza 18, 287–296.

Shetty, B.R., Jagadeesha, P.B. & Salmataj, S.A. (2025). Heavy metal contamination and its impact on the food chain: Exposure, bioaccumulation, and risk assessment. CyTA – Journal of Food 23(1), 2438726.

Shukla, S., Das S., Phutela, S., Tripathi, A., Kumari, C. & Shukla, S.K. (2024). Heavy metal stress in plants: Causes, impact and effective management. pp. 187–215. *In*: Kumar, N. (Ed.). Heavy Metal Toxicity: Human Health Impact and Mitigation Strategies. Switzerland, Cham: Springer Nature.

Simard, S.W. & Austin, M.E. (2010). The role of mycorrhizas in forest soil stability with climate change. pp. 275–302. *In*: Simard, S.W. [ed.]. Climate Change and Variability. Rijeka: Intech Open access.

Singh, H.V., Singh, U.B., Sahu, P.K., Malviya, D., Singh, S. & Saxena, A. (2022). Arbuscular mycorrhizal fungal symbiosis for mutual benefit: More than expectation. pp. 105–128. *In*: Singh, U.B., Rai, J.P. & Sharma, A.K. (Eds.). Re-visiting the Rhizosphere Eco-system for Agricultural Sustainability. Singapore: Springer Nature.

Šmilauer, P., Košnar, J., Kotilínek, M. & Šmilauerová, M. (2020). Contrasting effects of host identity, plant community, and local species pool on the composition and colonization levels of arbuscular mycorrhizal fungal community in a temperate grassland. New Phytologist 225, 461–473. doi: 10.1111/nph.16112

Smith, S.E., Jakobsen, I., Grønlund, M. & Smith, F.A. (2011). Roles of arbuscular mycorrhizas in plant phosphorus nutrition: Interactions between pathways of phosphorus uptake in arbuscular mycorrhizal roots have important implications for understanding and manipulating plant phosphorus acquisition. Plant Physiology 156, 1050–1057.

Smith, S.E. & Read, D.J. (2010). Mycorrhizal Symbiosis (3rd ed.). Academic Press.

Smith, S.E. & Smith, F.A. (2011). Roles of arbuscular mycorrhizas in plant nutrition and growth: New paradigms from cellular to ecosystem scales. Annual Review of Plant Biology 62(1), 227–250.

Souza, T. (2015). Handbook of Arbuscular Mycorrhizal Fungi. Switzerland: Springer International Publishing. https://doi.org/10.1007/978-3-319-24850-9.

Spruyt, A., Buck, M.T., Mia, A. & Straker, C.J. (2014). Arbuscular mycorrhiza (AM) status of rehabilitation plants of mine wastes in South Africa and determination of AM fungal diversity by analysis of the small subunit rRNA gene sequences. South African Journal of Botany 94, 231–237.

Thirkell, T.J., Charters, M.D., Elliott, A.J., Sait, S.M. & Field, K.J. (2017). Are mycorrhizal fungi our sustainable saviors? Considerations for achieving food security. Journal of Ecology 105, 921–929.

Umar, A.W., Naeem, M., Hussain, H., Ahmad, N. & Xu, M. (2025). Starvation from within: How heavy metals compete with essential nutrients, disrupt metabolism, and impair plant growth. Plant Science 353, 112412.

Ur Rahman, S., Han, J.C., Ahmad, M., Ashraf, M.N., Khaliq, M.A., Yousaf, M. et al. (2024). Aluminum phytotoxicity in acidic environments: A comprehensive review of plant tolerance and adaptation strategies. Ecotoxicology and Environmental Safety 269, 115791.

Vallejos-Torres, G., Gaona-Jimenez, N., Ordoñez-Sánchez, L., Lozano, C., Arevalo, A.A., Lozano, A. et al. (2023). Effects of arbuscular mycorrhizal fungi, poultry manure compost, and cadmium on plant growth and nutrient absorption of *Oryza sativa*. Chilean Journal of Agricultural Research 83(6), 656–665.

Vishwakarma, S., Ilyas, T., Malviya, D., Shafi, Z., Shahid, M., Yadav, B. et al. (2022). Arbuscular mycorrhizal fungi (AMF) as potential biocontrol agents. *In*: Rhizosphere Microbes: Biotic Stress Management (pp. 197–222). Singapore: Springer Nature Singapore.

Vodnik, D., Groman, H., Macek, I., van Elteren, J.T. & Kovacevic, M. (2008). The contribution of glomalin-related soil protein to Pb and Zn sequestration in polluted soil. Science of The Total Environment 392(1), 130–136.

Walker, L.R., Walker, J. & Moral, R.D. (2007). Forging a new alliance between succession and restoration. *In*: Linking Restoration and Ecological Succession (pp. 1–18). New York, NY: Springer New York.

Wang, F., Lin, X. & Yin, R. (2005). Heavy metal uptake by arbuscular mycorrhizas of Elsholtzia splendens and the potential for phytoremediation of contaminated soil. Plant and Soil 269(1), 225–232.

Wang, F.Y., Lin, X.G. & Yin, R. (2007). Role of microbial inoculation and chitosan in phytoextraction of Cu, Zn, Pb and Cd by *Elsholtzia splendens* – A field case. Environmental Pollution 147(1), 248–255.

Wang, Q., Mei, D., Chen, J., Lin, Y., Liu, J., Lu, H. & Yan, C. (2019). Sequestration of heavy metal by glomalin-related soil protein: Implication for water quality improvement in mangrove wetlands. Water Research 148, 142–152.

Wang, S., Chen, A., Xie, K., Yang, X., Luo, Z., Chen, J. et al. (2020). Functional analysis of the OsNPF4.5 nitrate transporter reveals a conserved mycorrhizal pathway of nitrogen acquisition in plants. Proceedings of the National Academy of Sciences of the United States of America 117(28), 16649–16659.

Wang, W., Shi, J., Xie, Q., Jiang, Y., Yu, N. & Wang, E. (2017). Nutrient exchange and regulation in arbuscular mycorrhizal symbiosis. Molecular Plant 10(9), 1147–1158.

Wilson, G.W. & Hartnett, D.C. (1998). Interspecific variation in plant responses to mycorrhizal colonization in tallgrass prairie. American Journal of Botany 85(12), 1732–1738.

Winagraski, E., Kaschuk, G., Monteiro, P.H.R., Auer, G. & Higa, A.R. (2019). Diversity of arbuscular mycorrhizal fungi in forest ecosystems of brazil: A review. Cerne 25(1), 25–35.

Wright, S.F. (2005). Management of arbuscular mycorrhizal fungi. *In*: Zobel, R.W., Wright, S.F. [eds.]. Roots and Soil Management: Interactions between Roots and the Soil. Book Series: Agronomy Monographs, Vol. 48, EUA: American Society of Agronomy, Inc. Crop Science Society of America, Inc. Soil Science Society of America, Inc.

Wu, S.C., Wong, C.C., Shu, W.S., Khan, A.G. & Whong, M.H. (2010). Mycorrhizo-remediation of lead/zinc mine tailings using vetiver: A field study. International Journal of Phytoremediation 13(1), 61–74.

Wu, S., Vosátka M., Vogel-Mikus, K., Kavcic, A., Kelemen, M., Sepec, L. et al. (2018). Nano zero-valent iron mediated metal (loid) uptake and translocation by arbuscular mycorrhizal symbioses. Environmental Science & Technology 52(14), 7640–7651.

Wu, Z., Mcgrouther, K., Huang, J., Wu, P., Wu, W. & Wang, H. (2014). Decomposition and the contribution of glomalin-related soil protein (GRSP) in heavy metal sequestration: Field experiment. Soil Biology and Biochemistry 68, 283–290.

Xu, H., Shi, Y., Chen, C., Pang, Z., Zhang, G., Zhang, W. & Kan, H. (2024). Arbuscular mycorrhizal fungi selectively promoted the growth of three ecological restoration plants. Plants 13(12), 1678.

Yang, B., Banerjee, S., Herzog, C., Ramírez, A.C., Dahlin, P. & van der Heijden, M.G.A. (2021). Impact of land use type and organic farming on the abundance, diversity, community composition and functional properties of soil nematode communities in vegetable farming. Agriculture, Ecosystems & Environment 318, 107488. doi: 10.1016/j.agee.2021.107488

Yang, H., Wang, G., Wang, J., Xiao, Q., Li, Z., De Clerck, C. et al. (2024). No-tillage facilitates soil organic carbon sequestration by enhancing arbuscular mycorrhizal fungi-related soil proteins accumulation and aggregation. Catena 245, 108323. doi: 10.1016/j.catena.2024.108323

Zhang, S., Lehmann, A., Zheng, W., You, Z. & Rillig, M.C. (2018). Arbuscular mycorrhizal fungi increase grain yields: A meta-analysis. New Phytologist 222(1), 543–555.

Zhang, Y., Yao, Q., Li, J., Hu, Y. & Chen, J. (2014). Growth response and nutrient uptake of Eriobotrya japonica plants inoculated with three isolates of arbuscular mycorrhizal fungi under water stress condition. Journal of Plant Nutrition 37, 690–703.

Zhang, Z., Li, B., Liu, Y., He, L., Pang, T., Chen, Z. et al. (2022). Arbuscular mycorrhizal fungal inoculation increases organic selenium accumulation in soybean (*Glycine max* (Linn.) Merr.) growing in selenite-spiked soils. Toxics 10(10), 565.

Zhao, R., Guo, W., Bi, N., Guo, J., Wang, L., Zhao, J. & Zhang, J. (2015). Arbuscular mycorrhizal fungi affect the growth, nutrient uptake and water status of maize (*Zea mays* L.) grown in two types of coal mine spoils under drought stress. Applied Soil Ecology 88, 41–49.

Zhao, S., Yan, L., Kamran, M., Liu, S.-S. & Riaz, M. (2024). Arbuscular mycorrhizal fungi-assisted phytoremediation: A promising strategy for cadmium-contaminated soils. Plants (Basel, Switzerland) 13(23), 3289, Nov. 22.

Zinati, G., Carrara, J.E., Das, S., Caetani, R., Kalra, A., Carr, E.A. et al. (2025). Impact of tillage practices and arbuscular mycorrhizal fungi inoculation on organic sweet corn yield and nutritional quality. Soil and Tillage Research 251, 106545.

Zou, Y.N., Huang, Y.M., Wu, Q.S. & He, X.H. (2014). Mycorrhiza-induced lower oxidative burst is related with higher antioxidant enzyme activities, net H_2O_2 effluxes, and Ca^{2+} influxes in trifoliate orange roots under drought stress. Mycorrhiza 25, 143–152.

CHAPTER
6

Brassinosteroids (BRs) Promote Growth and Development, Improving Heat Stress Tolerance

Sucharita Mohapatra[1], Soumya Shephalika Dash[2*], Lellapalli Rithesh[3], Jinisha Blessie JP[3], Shimi Jose[4], Juel Debnath[5], Gyanisha Nayak[6], Maksud Hasan Shah[7*], Rashadin Md Bodeul Ddoza[8] and Riyaz Sayyed[9]

[1] College of Agriculture, Odisha University of Agriculture and Technology, Bhubaneswar – 751003, India
[2] Palli-Siksha Bhavana, Visva-Bharati, Sriniketan – 731236, India
[3] Kerala Agricultural University, Vellayani – 695522, India
[4] Western Sydney University, Richmond-2753, Australia
[5] Kerala Agricultural University, Thrissur – 680656, India
[6] ICAR-Central Rice Research Institute, Cuttack – 753006, India
[7] Bidhan Chandra Krishi Viswavidyalaya, Nadia – 741252, India
[8] Bangladesh Agriculture University, Maimansingh – 2202, Bangladesh
[9] Department of Biological Sciences and Chemistry, College of Arts and Science, University of Nizwa, Nizwa, Oman

1. Introduction

Plants, being sessile organisms, are continually exposed to a various range of environmental stresses that adversely affect their growth, development, and productivity (Ferioun et al., 2025; Eswaran et al., 2024). Among these, heat stress is one of the most detrimental abiotic stresses, significantly impairing photosynthesis, respiration, membrane stability, reproductive development, and overall yield (Zahra et al., 2023). With the rising global temperatures attributed to climate change, crop productivity is increasingly threatened, especially in tropical and subtropical regions where temperatures often exceed optimal physiological thresholds. Consequently, enhancing thermotolerance in plants has become a critical area of research to ensure food security and agricultural sustainability. Phytohormones play a central role in mediating plant responses to environmental cues (Tanvere et al., 2023; Rafique et al., 2023), and among them, brassinosteroids (BRs) crucial phytohormones that

*Corresponding authors: soumyadashshephalika@gmail.com; maksudhasanshah@gmail.com

DOI: 10.1201/9781003669777-6

significantly influence plant growth, development, and stress resilience (Manghwar et al., 2022).

Brassinosteroids are a class of polyhydroxylated steroidal phytohormones structurally similar to animal steroid hormones (Khan et al., 2022; Usmani et al., 2025). Since their discovery in *Brassica napus* pollen in 1979, BRs have been broadly studied for their diverse physiological roles. They profoundly influence various aspects of plant growth and development. BR drive cell elongation and proliferation via upregulation of expansins, cyclins, and cellulose synthases, coordinate vascular differentiation, regulate reproductive organogenesis and seed germination (Manghwar et al., 2022). The biological activity of BRs is mediated through a well-characterized signaling pathway, primarily involving the receptor kinase *BRI1* (Brassinosteroid Insensitive 1) and its co-receptor *BAK1* (*BRI1*-Associated Receptor Kinase 1), followed by a cascade of phosphorylation and dephosphorylation events that regulate the activity of downstream transcription factors like *BES1* (*BRI1*-EMS-Suppressor 1) and *BZR1* (Brassinazole Resistant 1). These transcription factors modulate the expression of numerous genes associated with growth, development, and stress adaptation (Gruszka, 2013).

Brassinosteroids play a pivotal role in enhancing plant tolerance to heat stress by modulating physiological and molecular responses. Heat stress leads to reactive oxygen species (ROS) accumulation, protein denaturation, and membrane damage (Medina et al., 2021). BRs counteract these effects by boosting antioxidant enzymes like superoxide dismutase (SOD), Catalase (CAT), and ascorbate peroxidase (APX), thereby minimizing oxidative damage. They also upregulate heat shock proteins (HSPs), aiding in protein stabilization and refolding (Kothari and Lachowiec 2021). Studies in *Arabidopsis* and *Oryza sativa* confirm BR-induced HSP expression and thermotolerance (Anwar et al., 2018, Albertos et al., 2022). BRs improve chlorophyll content, photosynthesis, and membrane stability under heat. In crop like rice, BR foliar application enhances yield by regulating stomatal conductance, water retention, and osmoprotectants like proline and glycine betaine, maintaining cellular function during heat stress (Chandrakala et al., 2013, Sonjaroon et al., 2017).At the molecular level, thermo tolerance mediated by BRs involves several interconnected mechanisms. *BZR1* accumulates and dephosphorylates under warming conditions, forming an amplifying loop with *PIF4* to drive thermo-morphogenic growth, such as hypocotyl elongation. *BZR1* and *PIF4* co-activate gene networks regulating cell elongation via Ebox binding, aided by *ARF6* interactions (Lu et al., 2021). Additionally, BR signaling alleviates heat-induced damage by modulating heat shock transcription factors (*HSFA1s*), likely through *BIN2* (Brassinosteroid-Insensitive 2) and *BZR1*/*BES1* activity, driving expression of HSPs that safeguard protein homeostasis (Chen et al., 2022). BRregulated *ERF49*, a member of the ethyleneresponsive factor family, was identified as a repressor of thermotolerance; its inhibition by *BZR1* enhances *HSP* and antioxidant gene expression, improving survival at elevated temperatures (Chen et al., 2022). BRs also interact with other phytohormones such as abscisic acid (ABA), ethylene, jasmonic acid (JA), and salicylic acid (SA) to fine-tune stress responses. For instance, BR induced ROS can trigger ABA mediated stomatal closure to preserve leaf water content, while BR ethylene interactions via *BZR1–EIN3* govern cell elongation and developmental adaptations (Guo et al., 2024). In tomato, BR signaling enhances ABA accumulation under chilling, which parallels heat responses via upregulation of *NCED1*, suggesting similar regulation in heat via hormone interaction (An et al., 2023). Crosstalk with gibberellins through interaction with DELLAs refines BRinduced growth under stress (Gruszka 2018). Such crossregulatory dynamics form a complex web wherein BRs balance growth and defense under thermal stress. BRs are environmentally safe, biologically active at low concentrations, and increasingly explored as biostimulants under integrated stress management.

This chapter provides an in-depth overview of brassinosteroids, covering their discovery, occurrence, chemical structure, and biosynthetic pathways. It highlights their pivotal roles in key physiological processes, including cell elongation, vascular differentiation, and reproductive processes. The chapter also explores the BR signaling mechanism and its interaction with other hormonal pathways during heat stress. Emphasis is placed on their function in enhancing thermotolerance through antioxidant regulation, osmoprotection, and heat shock protein expression. Supported by recent studies, the chapter compiles species-specific responses to exogenous BR application and examines BR-regulated gene networks, offering insights into their potential for climate-resilient crop improvement.

2. Brassinosteroids

BRsare a class of phytohormones that play a vital role in plant growth and development. These polyhydroxysteroids are widely distributed throughout the plant kingdom and are structurally similar to steroid hormones found in animals and insects. BRs were first discovered in the pollen of *Brassica napus* and were subsequently named "brassins." Mitchell et al. (1970), reported that brassins significantly promoted cell elongation and division in the bean second-internode bioassay and enhanced yields when applied to young seedlings of radish, leafy vegetables, and potatoes. The most active BR, brassinolide, a lactone steroid, was isolated in 1979. A major breakthrough in BR research was the identification of the *BRI1* gene. *BRI1* encodes a ubiquitously expressed leucine-rich receptor-like kinase that plays a key role in BR signaling through the phosphorylation of serine and threonine residues (Li et al., 1997, Friedrichsen et al., 2000).

Brassinosteroids are plant growth-promoting molecules found in low concentrations across the plant kingdom and are widely distributed in both lower and higher plants. Young, actively growing tissues generally contain higher levels of BRs compared to mature tissues. Among plant organs, pollen and immature seeds are the richest sources, containing BRs in the range of 1–100 $\mu g\ kg^{-1}$ fresh weight, whereas shoots and leaves typically have lower concentrations, around 0.01–0.1 $\mu g\ kg^{-1}$ fresh weight. Since the discovery of brassinolide in 1979, numerous BR compounds have been identified from a wide variety of plant species, including angiosperms, gymnosperms, pteridophytes, bryophytes, and algae (Bajguz 2011). Among the BRs studied so far, castasterone, typhasterol, brassinolide, 6-deoxocastasterone (6-deoxoCS), 28-norcastasterone (28-norCS), and teasterone are the most commonly found in diverse plant species across different environmental conditions (Manghwar et al., 2022).

BRs are polyhydroxylated phytosterols synthesized from sterols such as campesterol, sitosterol, and cholesterol, which are metabolized through multiple intermediates within plant cells. All naturally occurring BRs identified to date share a common 5α-cholestane skeleton, with structural diversity arising from the type and orientation of oxygen-containing functional groups, particularly in rings A and B. Based on the number of carbon atoms in their steroid backbone, BRs are classified into three major types: C27, C28, and C29. Specifically, C27-BRs are based on a 5α-cholestane skeleton, C28-BRs on 5α-ergostane, and C29-BRs on 5α-stigmastane. In general, BRs are grouped into two forms: free (64 known compounds) and conjugated (5 known compounds). Depending on alkyl substitutions in the side chain, BRs can be further divided into C27 (10 compounds), C28 (38 compounds), and C29 (16 compounds). Moreover, based on modifications in the cholesterol side chain, BRs are categorized by different substituents at positions C-23, C-24, and C-25. These include forms such as 23-oxo, 24S-methyl, 24R-methyl, 24-methylene, 24S-ethyl, 24-ethylidene, 24-methylene-25-methyl, and 24-methyl-25-methyl, as well as those lacking substitutions at C-23, C-24, or both. Additionally, BRs can form conjugates with glucose and fatty acids (Bajguz et al., 2020).

2.1. Biosynthesis of BRs

A comprehensive understanding of the regulation of endogenous brassinosteroid levels through both biosynthesis and metabolic turnover is essential for constructing a molecular framework of BR function. The elucidation of the BR biosynthetic pathway has been primarily achieved through the use of gas chromatography-mass spectrometry (GC-MS), which enabled the identification of metabolic intermediates and the sequence of enzymatic reactions. This was accomplished by administering isotopically labeled precursors to *Catharanthus roseus* suspension cultures or *Arabidopsis* seedlings, followed by detailed metabolic profiling (Clouse 2002, Fujioka et al., 2003).

BRs including C27, C28, and C29 types, are synthesized through three distinct biochemical routes. The initial steps of biosynthesis are shared among these types and proceed via either the mevalonate (MVA) or the non-mevalonate (non-MVA) pathway. The divergence in biosynthetic routes occurs in the later stages, involving either cycloartenol- or cycloartanol-dependent pathways. The process begins with cycloartenol, a key sterol precursor, which undergoes a series of methylation, reduction, and desaturation reactions to produce canola sterol, a BR-specific intermediate. One of the principal sterols involved in BR biosynthesis is campesterol, which plays a dual role by contributing to membrane fluidity and serving as a precursor for BR production. Following a C-5 reduction, CR enters the late C-22 oxidation pathway. In this route, campesterolis first converted into campestanol (CN), which subsequently leads to the production of castasterone, a direct precursor of brassinolide, through two parallel oxidation sequences: C-6 early and C-6 late oxidation steps—collectively termed the CN-dependent pathway. Alternatively, campesterol may undergo hydroxylation at the C-22 position, initiating the C-22 early oxidation pathway, also known as the CN-independent route. castasterone is then transformed into brassinolide through BR-specific enzymatic reactions. Among these pathways, the C-22 early oxidation and the C-6 late oxidation sequences are predominantly active in most plant species for brassinolide biosynthesis. Several key enzymes facilitate these conversions, including DWARF 4 (DWF4), constitutive photomorphogenesis and dwarf (CPD), DEETIOLATED 2 (DET2), ROTUNDIFOLIA 3 (ROT3), and brassinosteroid-6-oxidase 1 and 2 (BR6ox1/2) (Zhao and Li, 2012, Gao et al., 2024).

2.2. Role of BRs in Growth and Development of Plants

Brassinosteroids are steroidal phytohormones that play an essential role in plant growth and development (Fig. 1) (Fang et al., 2019, Lin, 2020). Initially, their growth-promoting effects were attributed primarily to cell elongation; however, subsequent research revealed that BRs also influence cell division. It is now well established that BRs regulate a wide array of developmental and physiological processes beyond cell elongation and division, including xylem differentiation, photomorphogenesis, reproductive development, and responses to both abiotic and biotic stresses. BRs are involved in key functions such as stem cell maintenance, vascular development, floral transition, and the elongation of various cell types. They also play crucial roles in processes like hypocotyl elongation, root and shoot growth, stomatal patterning, pollen tube development, seed germination, xylem and tracheary element differentiation, senescence, and overall plant reproduction.

As polyhydroxylated steroidal compounds, BRs regulate several critical biological processes: they control the proliferation of leaf cells, and can induce

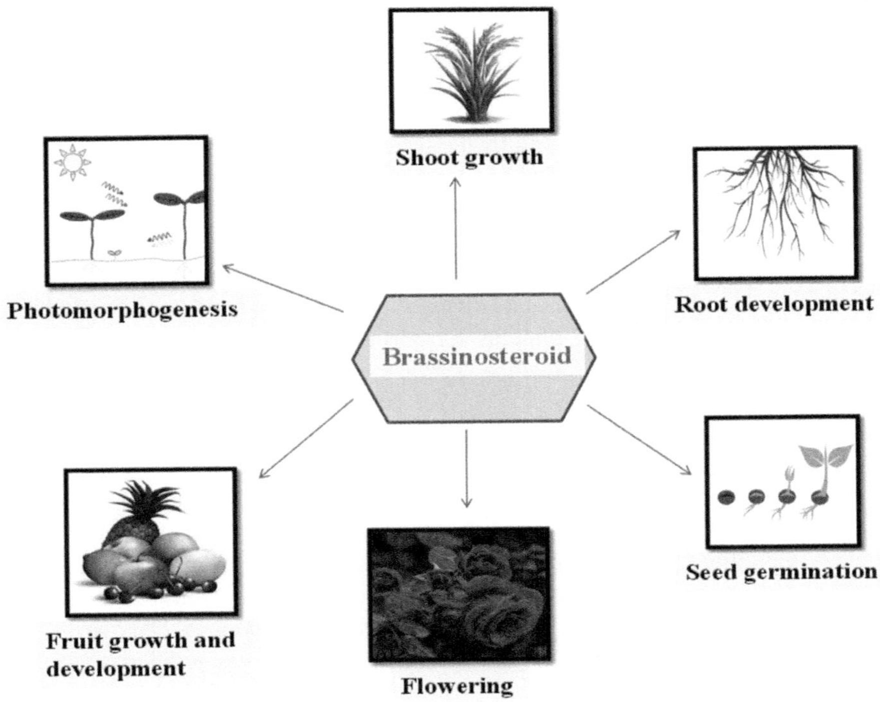

Fig. 1. Role of BRs in plant growth and development.

excessive growth in hydroponically grown plants. Furthermore, BRs enhance plant tolerance to abiotic and biotic stresses by modulating stomatal development and improving stress resilience mechanisms. Notably, they contribute to crop fertility by regulating both male and female reproductive development (Ahammed et al., 2020, Manghwar et al., 2022).

Under cold stress, BRs have been shown to increase the activity of antioxidant enzymes and enzymes involved in the Calvin cycle, thereby safeguarding the photosynthetic machinery (Jiang et al., 2013). In *Arabidopsis thaliana*, BRs upregulate genes such as *WRKY*46, *WRKY*54, and *WRKY*70, which positively influence BR-mediated growth responses (Chen et al., 2017). BRs also regulate seed development in *A. thaliana* by modulating gene expression related to ovule and seed formation, influencing seed size, weight, and number.

In rice, BRs are pivotal in determining grain yield and plant architecture. Genes such as *BRD1* and *D11*, involved in BR biosynthesis, influence plant height, while *OsDwarf1/OsDwarf2* encode C-6 oxidases essential for BR production and affect second internode elongation and seed size (Manghwar et al., 2022).

Recently, the application of BRs has gained attention as a promising strategy for improving crop productivity under both normal and stress conditions. BRs have been shown to mitigate the adverse effects of drought, salinity, heat, cold, disease, and nutrient deficiencies (Zhang et al., 2014). Key components of the BR signaling pathway have been identified, including *BZR1*, a positive regulator that activates BR responses and also mediates the negative feedback regulation of BR biosynthesis.

3. Signal Transduction of BRs

In plants, brassinosteroid (BR) signal transduction begins at the plasma membrane where BR molecules are perceived by the receptor kinase *BRI1*. Upon BR binding, BRI1 forms a complex with its co-receptor *BAK1*, triggering auto- and trans-phosphorylation events that initiate the signaling cascade (Fig. 2). Activated *BRI1* subsequently phosphorylates downstream signaling components such as BR-signaling kinase 1 (*BSK1*). *BSK1* then activates the *BSU1* (*BRI1* Suppressor 1) phosphatase, which inactivates BIN2, a GSK3-like kinase that negatively regulates BR signaling. In the absence of BRs, *BIN2* remains active and phosphorylates the transcription factors *BZR1* and *BZR2* (also known as *BES1*), preventing their nuclear translocation by promoting their interaction with 14-3-3 proteins and subsequent retention in the cytoplasm. Additionally, BIN2-mediated phosphorylation marks *BZR1/BZR2* for proteasomal degradation. However, once *BIN2* is inactivated by *BSU1*, Protein Phosphatase 2A (*PP2A*) dephosphorylates *BZR1* and *BZR2*, facilitating their release from 14-3-3 proteins and allowing them to accumulate in the nucleus. Inside the nucleus, these transcription factors bind to BR-responsive elements in the promoter regions of target genes, thereby regulating the expression of numerous genes involved in plant growth, development, and stress adaptation (Khan et al., 2022).

Furthermore, the activation of *BZR1* and *BZR2* also influences the expression of genes related to other stress-response pathways, such as those involved in osmotic regulation and membrane stability. Under heat stress, BRs enhance the fluidity of the plasma membrane by modulating its lipid composition, thereby preventing

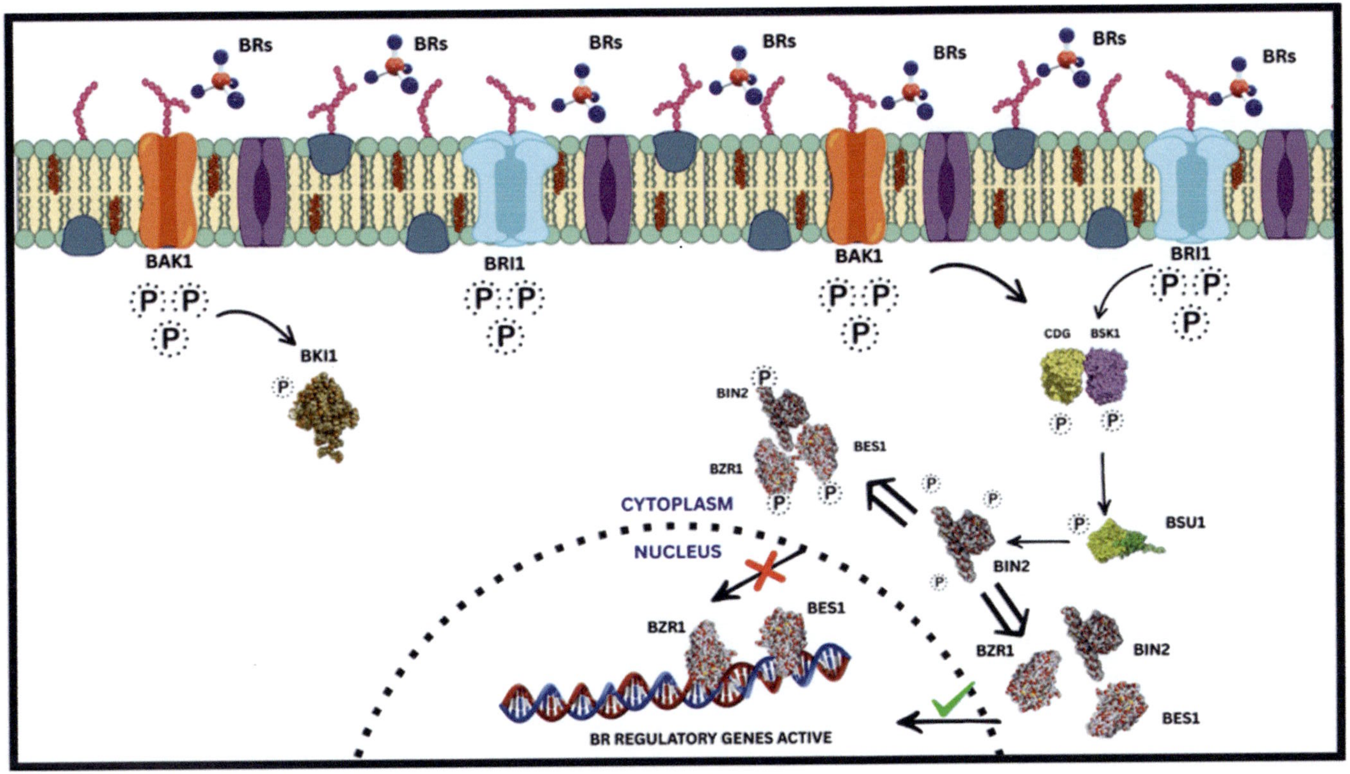

Fig. 2. BR signaling begins at the plasma membrane, where the hormone binds to the *BRI1* receptor and its co-receptor BAK1, activating their kinase domains. This activation triggers a phosphorylation cascade involving proteins like *BSK1* and *CDG1*. The cascade inhibits *BIN2*, a negative regulator kinase. Inhibition of *BIN2* allows transcription factors *BZR1* and *BES1* to accumulate and move into the nucleus, where they activate or repress BR-responsive genes, regulating plant growth and development.

membrane damage and maintaining cellular function (Rudolphi-Szydlo et al., 2022). The coordinated regulation of these genes by BRs ensures that the plant can efficiently manage the various physiological challenges posed by elevated temperatures, thus enhancing heat stress tolerance at the molecular level. In a gist, the activation of *BZR1* and *BZR2* by BR signalling plays a central role in regulating gene expression that governs both growth and stress tolerance. These transcription factors modulate the expression of genes involved in heat shock protein synthesis, ROS scavenging, and other stress-related pathways, providing plants with a robust mechanism to cope with heat stress. This genetic regulation by BRs is crucial for enhancing the plant's ability to withstand high-temperature conditions and maintain cellular integrity during periods of heat-induced stress.

3.1. BR-mediated Downstream Responses under Heat Stress

3.1.1. Expression of Heat Shock Proteins

HSPs are a family of highly conserved proteins that function as molecular chaperones, facilitating the proper folding of other proteins and preventing the aggregation of misfolded proteins under stress conditions, particularly heat (Malkeyeva et al., 2025). The expression of HSPs is tightly regulated at the genetic level, and BRs play a critical role in enhancing the synthesis of these proteins during heat stress (Kothari and Lachowiec, 2021). Upon the activation of *BZR1* and *BZR2* by BR signaling, the transcription of HSP genes, including *HSP70*, *HSP90*, and small HSPs (sHSPs), is induced. In heat-stressed plants, *HSP70* helps refold proteins that have been denatured due to elevated temperatures, ensuring that their functional conformation is restored. Similarly, *HSP90* stabilizes and activates other key regulatory proteins involved in cellular stress responses, maintaining protein function under extreme conditions. Small HSPs (sHSPs) are present in high quantities in stressed plants and act by preventing the misfolding and aggregation of proteins, thereby protecting cellular integrity (Cao et al., 2024). The upregulation of these essential chaperones by BRs ensures that plant cells can withstand extreme temperatures, maintain protein homeostasis, and continue vital cellular functions. By promoting the expression of HSPs, BRs enable plants to survive heat-induced stress and recover from protein damage, increasing resilience to high-temperature environments.

3.1.2. Antioxidant Defense and ROS Scavenging Genes

Heat stress induces the accumulation of ROS, which are highly reactive molecules that can cause extensive oxidative damage to cellular components, including lipids, proteins, and DNA (Juan et al., 2022). To mitigate this damage, plants have evolved sophisticated antioxidant defense mechanisms. These mechanisms involve the expression of antioxidant enzymes such as SOD, CAT, APX, and glutathione reductase (GR), which work synergistically to neutralize ROS and prevent cellular harm (Rajput et al., 2021). BRs enhance the expression of these antioxidant genes by activating the transcription factors *BZR1* and *BZR2*. Upon activation, *BZR1* and *BZR2* increase the transcription of genes involved in ROS scavenging, particularly those encoding SOD, CAT, APX, and GR. SOD plays a critical role in breaking down superoxide radicals, while CAT is essential for the detoxification of hydrogen peroxide. By upregulating the expression of these key enzymes, BRs help reduce ROS levels, thereby mitigating oxidative damage and enhancing the plant's ability to tolerate heat stress.

In addition to these enzymes, BRs also regulate the glutathione-ascorbate cycle, a vital pathway involved in ROS detoxification. The upregulation of GR and APX, which are involved in maintaining the cellular pool of antioxidants, further strengthens the plant's ability to scavenge ROS during heat stress. By boosting these protective antioxidant systems, BRs prevent oxidative damage, and improve overall function under elevated temperatures. This coordination of antioxidant defense mechanisms is essential for improving heat stress tolerance and supporting plant survival in high-temperature environments.

3.1.3. Regulation of Metabolic Pathways

BRs also play a crucial role in regulating metabolic pathways that are essential for maintaining energy balance and cellular homeostasis during heat stress. One key pathway influenced by BRs is the synthesis of compatible solutes, such as proline and trehalose, which help stabilize proteins and cellular structures under stressful conditions. These solutes act as osmotic regulators, ensuring that cells retain water and maintain turgor pressure, even during drought or heat stress. At the genetic level, BRs induce the expression of specific genes responsible for the biosynthesis of these compatible solutes. For example, proline dehydrogenase (PDH) and trehalose-6-phosphate synthase (TPS) are enzymes that regulate the synthesis of proline and trehalose, respectively (Sarkar and Sadhukhan, 2022). By activating these genes, BRs promote the accumulation of these solutes, which help plants cope with heat stress by stabilizing proteins, cellular membranes, and enzymes that may otherwise denature under high temperatures. Additionally, these compatible solutes help mitigate the effects of dehydration, further contributing to cellular protection during heat-induced stress. The accumulation of compatible solutes enhances the plant's ability to endure heat stress, improving overall heat tolerance and supporting long-term survival in environments where temperatures are elevated.

3.1.4. Gene Expression Profiling and BR-induced Heat Stress Tolerance

Recent advances in genomic technologies, particularly RNA sequencing (RNA-Seq), have provided valuable insights into the global changes in gene expression induced by BRs under heat stress conditions. These tools allow for the identification of a broad array of genes that are differentially expressed in response to BR treatment during heat stress. Notably, these genes are involved in various essential processes such as cell wall modification, protein folding, ROS scavenging, and metabolic regulation. Gene expression profiling of BR-treated plants under heat stress has revealed that BRs activate a distinct set of stress-responsive genes that enhance the plant's tolerance to high temperatures. This profiling highlights the importance of BR-induced transcription factors, HSPs, and antioxidant enzymes in improving heat stress resilience. The activation of these stress-related genes helps plants manage cellular damage caused by heat, stabilize proteins, and reduce oxidative stress.

Moreover, this profiling has deepened our understanding of how BRs modulate complex genetic networks that govern plant responses to heat stress. By identifying key genes and regulatory pathways involved in stress tolerance, this research provides valuable insights into the molecular mechanisms that underpin BR-induced heat stress resilience. These findings have important implications for the development of heat-tolerant crop varieties, which can be crucial for sustaining agricultural productivity in the face of climate change and increasing global temperatures.

4. Crosstalk between BRs and Other Plant Growth Regulators during Heat Stress

Rising global temperatures frequently subject plants to heat stress, significantly disrupting their physiological and biochemical processes, including growth, development, and productivity. BRs, a crucial group of steroidal phytohormones, participate in numerous vital plant functions. Among them, brassinolide is the most bioactive form, initially discovered in 1979 from the pollen of *Brassica napus*, with its molecular structure determined through crystal diffraction techniques (Grove et al., 1979). BRs are recognized as the sixth major class of endogenous plant hormones, following auxins, gibberellins, cytokinins, abscisic acid, and ethylene (Li et al., 2020). These hormones, including BRs, play a fundamental role in enhancing a plant's ability to cope with environmental stressors. BRs serve as central regulators, interacting with other hormonal pathways to influence gene expression, stress signaling mechanisms, and the plant's antioxidant defense during episodes of elevated temperature. Heat stress prompts widespread changes in the plant proteome aimed at minimizing cellular damage and ensuring survival. A hallmark of this response is the upregulation of HSPs, which are vital for protein folding, intracellular transport, and degradation under both stress and normal conditions (Wahid et al., 2007, Qu et al., 2013). These proteins contribute to thermotolerance by preserving the functionality of key physiological systems such as photosynthesis, water conservation, and membrane integrity (Ahn et al., 2006, Momcilovic et al., 2007).

While HSP expression increases under heat stress, the translation of various other genes such as those encoding transporters, detoxification enzymes, and regulatory proteins tends to be downregulated, redirecting cellular resources toward survival (Bita et al., 2013, Giri et al., 2017). BRs have been shown to elevate the production of multiple HSP families, highlighting their protective function in heat stress adaptation. When plants are treated with BRs during heat exposure, two major gene expression trends emerge: (1) BRs restore the expression of growth-related genes that are otherwise suppressed by heat stress, and (2) BRs amplify the expression of defense-related proteins beyond the levels induced by heat stress alone.

4.1. Crosstalk between BRs and Auxin

Auxin is a fundamental plant hormone that facilitates growth and development throughout a plant's life cycle (Zhang et al., 2022). The core components of auxin signaling within the nucleus include the SCFTIR1/AFB complex—comprised of TRANSPORT INHIBITOR RESPONSE 1/AUXIN SIGNALING F-BOX (*TIR1/*

AFB) proteins alongside AUXIN/INDOLE-3-ACETIC ACID (Aux/IAA) repressors and AUXIN RESPONSE FACTORs (*ARFs*), which are transcription factors (Korasick et al., 2014). When auxin levels are low, Aux/IAA proteins bind to ARFs and suppress their activity (Wang and Estelle, 2014). Conversely, elevated auxin levels promote its binding to the *SCF^TIR1/AFB* complex, which enhances the interaction with Aux/IAA repressors, thereby enabling ARFs to activate gene expression (Salehin et al., 2015). The interplay between BRs and auxin in regulating plant growth and stress responses involves both cooperative and opposing interactions (Hirano et al., 2017). BRs can influence auxin by altering its synthesis, directional transport, or signaling pathways (Bao et al., 2004). In return, auxin also modulates the production and signal transduction of BRs. For instance, external application of auxin upregulates genes like *CPD* and *DWF4*, which results in elevated BR levels (Chung et al., 2011). In rice, auxin-responsive transcription factors such as *OsARF11* can bind to the promoter region of Os*BRI1*, thereby enhancing its expression and promoting BR signaling. However, an excessive concentration of auxin can suppress BR signaling by inducing *BKI1*, a known inhibitor of the *BRI1* receptor (Devi et al., 2022).

4.2. Crosstalk between BRs and Abscisic Acid

Abscisic acid is recognized as a key hormone involved in plant stress responses, playing an essential role in regulating growth and adaptation during environmental stress (Yang et al., 2022). When plants are exposed to stress, ABA levels rise quickly. This increase is detected by specific receptors—PYRABACTIN RESISTANCE1 (*PYR1*), PYR1-LIKE (*PYL*), and REGULATORY COMPONENT OF ABA RECEPTOR (*RCAR*). The binding of ABA to these receptors promotes their interaction with PROTEIN PHOSPHATASE 2C (*PP2C*) proteins, which inhibits the phosphatase activity of *PP2C*s (Park et al., 2009). Under non-stressful conditions, brassinosteroids generally oppose the actions of ABA. Application of brassinolide, a type of BR, has been shown to suppress ABA levels by downregulating the expression of genes involved in its biosynthesis (Ha et al., 2018). Additionally, *PP2Cs*, known inhibitors of ABA signaling, can deactivate the kinase *BIN2* through dephosphorylation, thereby promoting BR signaling (Jiang et al., 2019). Moreover, *PP2C* enzymes like ABI1 can also activate *BES1* via dephosphorylation, enhancing BR-mediated signaling during heat stress (Albertos et al., 2022).

4.3. Crosstalk between BRs and Ethylene

Ethylene, a gaseous hormone composed of two carbon and four hydrogen atoms, is produced throughout various tissues and organs of plants. It plays a vital role in regulating physiological processes and in mediating responses to environmental stresses. During abiotic stress, the enzyme alternative oxidase (AOX) helps in mitigating excessive ROS induced by brassinolide, thereby protecting the photosynthetic machinery and improving stress tolerance, as observed in *Nicotiana benthamiana* (Deng et al., 2015). In *Cucumis sativus* L., brassinolide was found to stimulate both ethylene production and AOX gene expression under drought, salinity, and cold stress conditions (Wei et al., 2015). BRs influence ethylene-related processes by modulating genes and enzymes involved in ethylene biosynthesis and signaling pathways. Conversely, ethylene can also affect plant development and stress responses through mechanisms dependent on BR signaling. It further impacts BR biosynthesis by altering the expression of critical enzymes in its pathway. For example, the external application of 1-aminocyclopropane-1-carboxylic acid (ACC), a precursor to ethylene, enhances the expression of the BR biosynthesis gene *CPD*, leading to increased BR levels and signaling activity (De Grauwe et al., 2005). In rice, the ethylene-responsive factor SUBMERGENCE 1A (*SUB1A*) boosts BR production by upregulating BR biosynthetic genes such as *DWF4* and *DWF1* (Schmitz et al., 2013). The interaction between BRs and ethylene can be either cooperative or antagonistic, enabling plants to fine-tune their growth and resilience to stress conditions.

4.4. Crosstalk between BRs and Gibberellin

Gibberellins are essential for numerous plant developmental events, including the initiation of seed germination, elongation of stems, maturation of pollen, and the growth of leaves and flowers (Achard and Genschik, 2009). In GA-deficient conditions, DELLA proteins act as central repressors of GA signaling by binding to and inhibiting the activity of specific transcription factors within the pathway, thus obstructing the signaling cascade (Silverstone et al., 1998). Upon GA availability, the hormone binds to its nuclear receptor, GIBBERELLIN INSENSITIVE DWARF 1 (*GID1*), which promotes a stronger interaction with DELLA proteins, leading to their degradation and enabling the transcriptional activity of GA-responsive genes (Murase et al., 2008). BRs and GAs often act in coordination to regulate various physiological functions, such as promoting seed germination, shaping the apical hook, and improving tolerance to zinc stress (Xiong et al., 2022, Ren et al., 2023). However, BRs can also suppress GA biosynthesis by downregulating *SlGA20ox1* expression in tomato ovaries (Barro-Trastoy et al., 2020). Moreover, BR-responsive transcription factors like CESTA (*AtCES*) in *Arabidopsis* and its homolog SlCES in tomato are known to facilitate the breakdown of GAs by inducing genes involved in their catabolism (Shuai et al., 2022).

4.5. Crosstalk between BRs and Cytokinin

Cytokinins (CKs) are vital for a broad range of plant physiological processes, including cell division, tissue differentiation, organ formation, delay of leaf aging, uptake of nutrients, and responses to environmental stress (Li et al., 2021). The external application of 24-epibrassinolide (EBL), a brassinosteroid, enhances cytokinin accumulation by suppressing the expression of the CYTOKININ OXIDASE 3 (*CKX3*) gene, which is involved in cytokinin degradation (Yuldashev et al., 2012). The PROTEIN PHOSPHATASE WITH KELCH-LIKE DOMAINS 1 (*PPKL1*), a component implicated in brassinosteroid signaling, competes with *OsARR21* for binding with *OsAHP2*, thereby interrupting the *OsAHP2*-to-*OsARR21* phosphorylation cascade and reducing cytokinin-related responses (Liu et al., 2022). In Colombia, research has demonstrated that foliar sprays containing brassinosteroids or cytokinin-based biostimulants alleviated heat stress in rice by improving chlorophyll content, proline levels, and the efficiency of leaf gas exchange (Quintero-

Table. 1. Effects of exogenous application of BRs in ameliorating the heat stress on different plant species

Plant species	BR type & application method	Heat stress conditions	Observed effects	Reference
Soybean (*Glycine max*)	0.25 µM 24-epiBR (foliar spray)	High temperature stress	Increased relative water content, decreased MDA & ROS, Increased antioxidant enzymes (SOD, CAT, POD), improved PSII photochemistry (Fv/Fm), photosynthesis, and biomass.	Wang et al., 2022
Rice (*Oryza sativa*)	0.005 mg L^{-1} BR spray on seedlings	High temperature	Increased SOD & POD, decreased MDA & electrolyte leakage; Increased photosynthesis, chlorophyll, yield components.	Thussagunpanit et al., 2015, Pantoja-Benavides et al., 2021
Tomato (*Solanum lycopersicum*)	24-epiBR, 0.2 µM (foliar)	Heat shock/high temperature	Increased antioxidant enzymes, reduced lipid peroxidation, improved pollen thermotolerance, photosynthesis, chlorophyll content, PSII yield, and stomatal conductance.	Singh et al., 2005, Ogweno et al., 2008
Cucumber (*Cucumis sativus*)	24-epiBR foliar spray	Heat stress	Delayed chlorophyll degradation; Increased water use efficiency, antioxidant enzymes (APX, POD, SOD), AsA–GSH cycle enzyme activity, AsA/GSH levels, increased SPS, SS and AI activities.	Yu et al., 2004
Brassica spp.	EBR application	High temperature	Enhanced the activities of SOD, CAT, APOX, PPO, and auxinases.	Kumar et al., 2012
Wheat (*Triticum aestivum*)	EBR or homolog BR	High temperature	Enhanced antioxidant enzyme activities, chlorophyll, growth, membrane stability, and heat tolerance.	Wang et al., 2025
Banana (*Musa* spp.)	Homobrassinolide, ~0.1% in vitro	40°C exposure	Increased shoot growth & fresh weight; reduced heat damage.	Nassar, 2004
Leymus chinensis	Brassinolide, foliar	High-temperature stress	Enhanced antioxidant enzyme activities (SOD, POD, CAT, APX, GR), Accumulation of proline content, soluble protein and sugar content	Niu et al., 2016

(Contd.)

(*Contd.*)

Plant species	BR type & application method	Heat stress conditions	Observed effects	Reference
Elegant fig (*Ficus concinna*)	EBR pretreatment	35-40°C stress	Increased AsA–GSH cycle enzymes (APX, DHAR, GR, MDHAR), AsA/GSH content; decreased MDA, improved oxidative defense.	Jin et al., 2015
Grapevine (*Vitis* spp.)	Root-applied BR + salicylic acid	Heat stress	Improved photosynthetic efficiency, membrane stability, fruit development, and quality.	Sun et al., 2025
Kiwi (*Actinidia deliciosa*)	5 µmol L^{-1} EBR	Heat stress, 20°C for 20 days	Inhibited superoxide anion and hydrogen peroxide accumulation, enhancement of antioxidant activity, and delay of senescence.	Wang et al., 2020
Pepper (*Capsicum annuum*)	24-epibrassinolide	High-temperature exposure	Improved chlorophyll retention, protection against salt stress, and increased enzymatic antioxidative activity.	Abbas et al., 2013
Arabidopsis thaliana	10–100 nM 2,4epibrassinolide foliar spray	45-54°C	~92 % survival; induced expression of BR-regulated heat genes (e.g., *ERF49*)	Chen et al., 2022
Maize (*Zea mays*)	1 µM 24-epibrassinolide foliar	48°C, 14 h/day	Balanced CAT/SOD; improved post-heat growth recovery	Yadava et al., 2016
Fenugreek (*Trigonella foenum-graecum*)	4-16 µM 24-epibrassinolide	42°C, 6-24 h	Reduced MDA/EL; Increased antioxidants as well as diosgenin content via ABA/NO/H_2O_2 pathways	Sheikhi et al., 2023

EBR: 24-epibrassinolide; MDA: Malondialdehyde; ROS: Reactive Oxygen Species; SOD: Superoxide Dismutase; CAT: Catalase; POD: Peroxidase; Fv/Fm: Maximum Quantum Efficiency of Photosystem II; APX: Ascorbate Peroxidase; AsA–GSH: Ascorbate–Glutathione Cycle; AsA: Ascorbate; GSH: Glutathione; SPS: Sucrose Phosphate Synthase; SS: Sucrose Synthase; AI: Acid Invertase; MDHAR: Monodehydroascorbate Reductase; DHAR: Dehydroascorbate Reductase; GR: Glutathione Reductase

Calderon et al., 2021). According to Thussagunpanit et al. (2015), brassinosteroid application in rice plants at the reproductive stage protected the photosynthetic system during heat stress. This protection was evidenced by enhanced photosynthetic activity, increased stomatal conductance, and reduced photoinhibition. In wheat, the application of BRs was found to hasten the adaptation to high-temperature conditions and contributed to improved seedling survival under heat stress. Similarly, Hairat and Khurana (2015) reported that BR treatment under elevated temperatures led to improved photosynthesis and reduced membrane damage compared to untreated plants. Furthermore, Sonjaroon et al. (2018) found that applying 7,8-dihydro-8a-20-hydroxyecdysone (aDHECD), a synthetic BR analogue, at a concentration of 0.0001 μM significantly enhanced photosynthetic traits in heat-stressed rice plants compared to untreated controls.

Table 1 presents the impact of exogenous brassinosteroid application on various plant species under heat stress. It summarizes key physiological and biochemical improvements, including enhanced antioxidant activity, photosynthetic efficiency, and yield performance.

5. BRs Regulated Genes in Plant Growth and Development during Heat Stress

Under heat stress, plants experience extensive cellular damage and metabolic imbalance, but BRs mitigate these effects by activating a specific set of stress-responsive and growth-regulating genes. A major group of BR-regulated genes during heat stress includes HSPs such as *HSP70*, *HSP90*, and small HSPs (e.g., *HSP17.6*) (Dundar et al., 2025). These genes are transcriptionally activated by *BZR1*/ (*BES1: BRI1*-EMS-Suppressor 1) under heat stress and function as molecular chaperones to protect protein structure and prevent aggregation (Kelly-Bellow et al., 2023, Percio et al., 2025). BR application or enhanced BR signalling leads to a significant upregulation of these genes, thereby stabilizing cellular proteins and maintaining developmental integrity under high temperatures.

In addition to chaperones, BRs regulate transcription factors like *HSFA2* (Heat shock factor A2), which coordinates downstream expression of multiple HSPs (Haider et al., 2022). Studies show that *BZR1* directly or indirectly promotes *HSFA2* expression, enhancing thermotolerance without completely halting growth (Luo et al., 2022). Another critical gene is *DREB2A* (Dehydration responsive element binding 2A), which is upregulated under heat and drought conditions (Yu et al., 2021). Though primarily linked to ABA signalling, *DREB2A* is also influenced by BRs during combined stress and developmental stages, supporting transcriptional crosstalk between stress and growth programs (Chen et al., 2022).

BR signalling under heat stress also modulates antioxidant enzyme genes such as CAT2 (Catalase 2) and APX1 (Ascorbate peroxidase 1), which detoxify ROS generated during thermal stress (Wang et al., 2022). By enhancing ROS-scavenging capacity, BRs reduce cellular oxidative damage and support continued cell division and elongation (Halaji et al., 2024). Concurrently, Expansin (e.g., *EXPA1, EXPA8*) and XTH (Xyloglucan endotransglucosylase/Hydrolase) genes, which are central to cell wall loosening and elongation, remain transcriptionally active under BR regulation even during heat episodes, albeit at moderated levels to balance growth with survival (Furuya et al., 2024).

BRs also suppress growth-inhibitory genes during heat stress to maintain developmental progression. One such gene is *RD26* (Responsive to desiccation 26), a NAC-domain transcription factor that represses growth and promotes stress tolerance (Felagari et al., 2023). Under normal BR signalling, *RD26* is repressed by *BZR1* to sustain development. However, during heat stress, fine-tuned regulation ensures that *RD26* is not overly activated, avoiding excessive growth inhibition (Kothari and Lachowiec, 2021). This selective repression is critical in maintaining reproductive development and preventing yield loss under prolonged high temperatures.

Furthermore, BRs regulate biosynthetic and signaling genes such as *DWF4* and *CPD* under stress through feedback loops, adjusting endogenous BR levels based on environmental demand (Dündar et al., 2025). The repression of these biosynthesis genes under heat stress prevents overstimulation of growth responses, aligning internal BR signalling with the plant's thermal tolerance strategies.

In summary, during heat stress, BRs orchestrate a distinct transcriptional program by upregulating protective genes (HSPs, antioxidants), modulating developmental regulators (*EXP*, *XTH*), and suppressing inhibitory signals (*RD26*), all while coordinating with other hormonal pathways. This dual role enables plants to maintain vital growth processes while enhancing thermotolerance, making BR-regulated gene networks central to developmental resilience under heat conditions.

6. Conclusion

Brassinosteroids have been thoroughly characterized from their structural diversity and tightly regulated biosynthesis to their finely tuned signaling pathways mediated by *BRI1–BAK1* and downstream transcription factors *BES1*/*BZR1*. Their multifaceted role in orchestrating plant development is reflected in their regulation of key transcriptional networks that govern cell elongation, vascular differentiation, seed germination, flowering, and senescence. Under heat stress conditions, BRs engage synergistically with other phytohormones including ABA, ethylene, jasmonic acid, and salicylic acid to enhance physiological resilience through antioxidant defense, osmoprotection, *HSP* induction, and thermomorphogenic adaptations. Field trials and controlled-environment studies utilizing exogenous BR application across species such as rice, wheat, tomato, and cucumber have consistently demonstrated improvements in photosynthetic performance, membrane stability, water retention, and grain yield under thermal stress. Transcriptomic analyses reveal that BRs regulate extensive gene sets, including HSPs and transcription factors like *DREB*, *WRKY*, *NAC*, and *MYB*, thereby coupling growth with stress tolerance pathways. Looking ahead, future research must emphasize translational integration of BR biology into agronomic strategies. Particularly, detailed field-level studies across diverse agroecological zones are needed to optimize BR dosage, timing, and formulation for effective stress mitigation while ensuring compatibility with crop developmental stages. The development and deployment of stable synthetic analogues and BR-mimicking biostimulants offer promising avenues for scalable application.

Elucidating the spatiotemporal dynamics of BR signaling especially the interplay between *BRI1* variants and hormonally regulated networks will enhance precision breeding targets. Advanced omics, gene-editing tools, and modeling approaches should be leveraged to dissect BR crosstalk mechanisms, enabling cultivar-specific interventions that enhance stress resilience without growth penalties. Ultimately, harnessing brassinosteroids in synergy with sustainable agronomic practices can contribute significantly to climate-resilient agriculture and global food security.

References

Abbas, S., Latif, H.H. & Elsherbiny, E.A. (2013). Effect of 24-epibrassinolide on the physiological and genetic changes on two varieties of pepper under salt stress conditions. Pak. J. Bot. 45(4), 1273–1284.

Achard, P. & Genschik, P. (2009). Releasing the brakes of plant growth: How GAs shutdown DELLA proteins. J. Exp. Bot. 60(4), 1085–1092.

Ahammed, G.J., Li, X., Liu, A. & Chen, S. (2020). Brassinosteroids in plant tolerance to abiotic stress. J. Plant Growth Regul. 39(4), 1451–1464.

Ahn, Y.J. & Zimmerman, J.L. (2006. Introduction of the carrot HSP17.7 into potato (*Solanum tuberosum* L.) enhances cellular membrane stability and tuberization in vitro. Plant Cell Environ. 29(1), 95–104.

Albertos, P., Dundar, G., Schenk, P., Carrera, S., Cavelius, P., Sieberer, T. & Poppenberger, B. (2022). Transcription factor BES1 interacts with HSFA1 to promote heat stress resistance of plants. EMBO J. 41(3), e108664.

An, S., Liu, Y., Sang, K., Wang, T., Yu, J., Zhou, Y. & Xia, X. (2023). Brassinosteroid signaling positively regulates abscisic acid biosynthesis in response to chilling stress in tomato. J. Integr. Plant Biol. 65(1), 10–24.

Anwar, A., Liu, Y., Dong, R., Bai, L., Yu, X. & Li, Y. (2018). The physiological and molecular mechanism of brassinosteroid in response to stress: a review. Biol. Res. 51.

Bajguz, A. (2011). Brassinosteroids – Occurrence and chemical structures in plants. *In*: Brassinosteroids: A Class of Plant Hormone, pp. 1–27.

Bajguz, A., Chmur, M. & Gruszka, D. (2020). Comprehensive overview of the brassinosteroid biosynthesis pathways: substrates, products, inhibitors, and connections. Front. Plant Sci. 11: 1034.

Bao, F., Shen, J., Brady, S.R., Muday, G.K., Asami, T. & Yang, Z. (2004). Brassinosteroids interact with auxin to promote lateral root development in *Arabidopsis*. Plant Physiol. 134(4), 1624–1631.

Barro-Trastoy, D., Carrera, E., Banos, J., Palau-Rodriguez, J., Ruiz-Rivero, O., Tornero, P. & Perez-Amador, M.A. (2020). Regulation of ovule initiation by gibberellins and brassinosteroids in tomato and *Arabidopsis*: Two plant species, two molecular mechanisms. Plant J. 102(5), 1026–1041.

Bita, C.E. & Gerats, T. (2013). Plant tolerance to high temperature in a changing environment: Scientific fundamentals and production of heat stress-tolerant crops. Front. Plant Sci. 4, 273.

Cao, X., Wei, Y., Shen, B., Liu, L. & Mao, J. (2024). Interaction of the transcription factors BES1/BZR1 in plant growth and stress response. Int. J. Mol. Sci. 25(13), 6836.

Chandrakala, J.U., Chaturvedi, A.K., Ramesh, K.V., Rai, P., Khetarpal, S. & Pal, M. (2013). Acclimation response of signalling molecules for high temperature stress on photosynthetic characteristics in rice genotypes. Indian J. Plant Physiol. 18, 142–150.

Chen, J., Nolan, T.M., Ye, H., Zhang, M., Tong, H., Xin, P. & Yin, Y. (2017). *Arabidopsis* WRKY46, WRKY54, and WRKY70 transcription factors are involved in brassinosteroid-regulated plant growth and drought responses. Plant Cell 29(6), 1425–1439.

Chen, X., Xue, H., Zhu, L., Wang, H., Long, H., Zhao, J. & Zhu, S. (2022). ERF49 mediates brassinosteroid regulation of heat stress tolerance in *Arabidopsis thaliana*. BMC Biol. 20(1), 254.

Chung, Y., Maharjan, P.M., Lee, O., Fujioka, S., Jang, S., Kim, B. & Choe, S. (2011). Auxin stimulates DWARF4 expression and brassinosteroid biosynthesis in *Arabidopsis*. Plant J. 66(4), 564–578.

Clouse, S.D. (2002). Brassinosteroids. *Arabidopsis* Book 1: e0009.

De Grauwe, L., Vandenbussche, F., Tietz, O., Palme, K. & Van Der Straeten, D. (2005). Auxin, ethylene and brassinosteroids: tripartite control of growth in the Arabidopsis hypocotyl. Plant Cell Physiol. 46(6), 827–836.

Deng, X.G., Zhu, T., Zhang, D.W. & Lin, H.H. (2015). The alternative respiratory pathway is involved in brassinosteroid-induced environmental stress tolerance in *Nicotiana benthamiana*. J. Exp. Bot. 66(20), 6219–6232.

Devi, L.L., Pandey, A., Gupta, S. & Singh, A.P. (2022). The interplay of auxin and brassinosteroid signaling tunes root growth under low and different nitrogen forms. Plant Physiol. 189(3), 1757–1773.

Dundar, G., Ramirez, V.E. & Poppenberger, B. (2025). The heat shock response in plants: New insights into modes of perception, signaling, and the contribution of hormones. J. Exp. Bot. 76(7), 1970–1977.

Eswaran, S.U.D., Sundaram, L., Perveen, K., Bukhari, N.A. & Sayyed, R.Z. (2024). Osmolyte-producing microbial biostimulants regulate the growth of *Arachis hypogaea* L. under drought stress. BMC Microbio. 4, 165. https://doi.org/10.1186/s12866-024-03320-6

Fang, P., Yan, M., Chi, C., Wang, M., Zhou, Y., Zhou, J. & Yu, J. (2019). Brassinosteroids act as a positive regulator of photoprotection in response to chilling stress. Plant Physiol. 180(4), 2061–2076.

Felagari, K., Bahramnejad, B., Siosemardeh, A., Mirzaei, K., Lei, X. & Zhang, J. (2023). A comparison of the physiological traits and gene expression of brassinosteroids signaling under drought conditions in two chickpea cultivars. Agronomy 13(12), 2963.

Ferioun, M., Zouitane, I., Bouhraoua, S., Elouattassi, Y., Belahcen, D., Errabbani, A. et al. (2025). Applying microbial biostimulants and drought-tolerant genotypes to enhance barley growth and yield under drought stress. Front. Plant Sci. 15, 1494987. https://doi.org/10.3389/fpls.2024.1494987.

Friedrichsen, D.M., Joazeiro, C.A., Li, J., Hunter, T. & Chory, J. (2000). Brassinosteroid-insensitive-1 is a ubiquitously expressed leucine-rich repeat receptor serine/threonine kinase. Plant Physiol. 123(4), 1247–1256.

Fujioka, S. & Yokota, T. (2003). Biosynthesis and metabolism of brassinosteroids. Annu. Rev. Plant Biol. 54(1), 137–164.

Fujioka, S., Noguchi, T., Watanabe, T., Takatsuto, S. & Yoshida, S. (2000). Biosynthesis of brassinosteroids in cultured cells of *Catharanthus roseus*. Phytochem. 53(5), 549–553.

Furuya, T., Ohashi-Ito, K. & Kondo, Y. (2024). Multiple roles of brassinosteroid signaling in vascular development. Plant Cell Physiol. 65(10), 1601–1607.

Gao, Y., Ma, X., Zhang, Z., Wang, X. & Wang, Y. (2024). Brassinolides signaling pathway: Tandem response to plant hormones and regulation under various abiotic stresses. Hortic. Adv. 2(1), 27.

Ghoname, A.A., AbdelMotlb, N.A., Abdel-Al, F.S., Abu El-Azm, N.A., Abd Elhady, S.A., Merah, O. & Abdelhamid, M.T. (2023). Brassinosteroids or proline can alleviate yield inhibition under salt stress via modulating physio-biochemical activities and antioxidant systems in snap bean. J. Hortic. Sci. Biotechnol. 98(4), 526–539.

Giri, A., Heckathorn, S., Mishra, S. & Krause, C. (2017). Heat stress decreases levels of nutrient-uptake and assimilation proteins in tomato roots. Plants 6(1), 6.

Grove, M.D., Spencer, G.F., Rohwedder, W.K., Mandava, N., Worley, J.F., Warthen Jr., J.D. & Cook Jr., J.C. (1979). Brassinolide, a plant growth-promoting steroid isolated from *Brassica napus* pollen. Nature 281(5728), 216–217.

Gruszka, D. (2013). The brassinosteroid signaling pathway—New key players and interconnections with other signaling networks crucial for plant development and stress tolerance. Int. J. Mol. Sci. 14(5), 8740–8774.

Gruszka, D. (2018). Crosstalk of the Brassinosteroid Signalosome with Phytohormonal and stress signaling components maintains a balance between the processes of growth and stress tolerance. Int. J. Mol. Sci. 19(9), 2675.

Guo, F., Lv, M., Zhang, J. & Li, J. (2024). Crosstalk between brassinosteroids and other phytohormones during plant development and stress adaptation. Plant Cell Physiol. 65(10), 1530–1543.

Ha, Y.M., Shang, Y., Yang, D. & Nam, K.H. (2018). Brassinosteroid reduces ABA accumulation leading to the inhibition of ABA-induced stomatal closure. Biochem. Biophys. Res. Commun. 504(1), 143–148.

Hafeez, M.B., Zahra, N., Zahra, K., Raza, A., Batool, A., Shaukat, K. & Khan, S. (2021). Brassinosteroids: Molecular and physiological responses in plant growth and abiotic stresses. Plant Stress 2, 100029.

Haider, S., Raza, A., Iqbal, J., Shaukat, M. & Mahmood, T. (2022). Analyzing the regulatory role of heat shock transcription factors in plant heat stress tolerance: A brief appraisal. Mol. Biol. Rep. 49(6), 5771–5785.

Hairat, S. & Khurana, P. (2015). Improving photosynthetic responses during recovery from heat treatments with brassinosteroid and calcium chloride in Indian bread wheat cultivars. Am. J. Plant Sci. 6(11), 1827.

Halaji, B., Haghighi, M., Kovacs, G.P., Mirmazloum, I. & Szego, A. (2024). The role of brassinosteroids and nano-encapsulated brassinosteroids in *Capsicum* pepper growth and physiological adaptations to high-temperature stress. Horticulturae 10(10), 1062.

Hirano, K., Yoshida, H., Aya, K., Kawamura, M., Hayashi, M., Hobo, T. & Matsuoka, M. (2017). Small Organ Size 1 and Small Organ Size 2/DWARF and low-tillering form a complex to integrate auxin and brassinosteroid signaling in rice. Mol. Plant 10(4), 590–604.

Jiang, H., Tang, B., Xie, Z., Nolan, T., Ye, H., Song, G.Y. & Yin, Y. (2019). GSK3-like kinase BIN2 phosphorylates RD26 to potentiate drought signaling in *Arabidopsis*. Plant J. 100(5), 923–937.

Jiang, Y.P., Huang, L.F., Cheng, F., Zhou, Y.H., Xia, X.J., Mao, W.H. & Yu, J.Q. (2013). Brassinosteroids accelerate recovery of photosynthetic apparatus from cold stress by balancing the electron partitioning, carboxylation and redox homeostasis in cucumber. Physiol. Plant. 148(1), 133–145.

Jin, S.H., Li, X.Q., Wang, G.G. & Zhu, X.T. (2015). Brassinosteroids alleviate high-temperature injury in *Ficus concinna* seedlings via maintaining higher antioxidant defence and glyoxalase systems. AoB Plants 7, plv009.

Juan, C.A., Perez de la Lastra, J.M., Plou, F.J. & Perez-Lebena, E. (2021). The chemistry of reactive oxygen species (ROS) revisited: Outlining their role in biological macromolecules (DNA, lipids and proteins) and induced pathologies. Int. J. Mol. Sci. 22(9), 4642.

Kelly-Bellow, R., Lee, K., Kennaway, R., Barclay, J.E., Whibley, A., Bushell, C. et al. (2023). Brassinosteroid coordinates cell layer interactions in plants via cell wall and tissue mechanics. Science 380(6651), 1275–1281.

Khan, I., Awan, S.A., Ikram, R., Rizwan, M., Akhtar, N., Yasmin, H. et al. (2020). Effects of 24-epibrassinolide on plant growth, antioxidants defense system, and endogenous hormones in two wheat varieties under drought stress. Physiol. Planta. 1-11. https://onlinelibrary.wiley.com/doi/abs/10.1111/ppl.13237

Khan, M.T.A., Yusuf, M., Akram, W. & Qazi, F. (2022). Signal transduction of brassinosteroids under abiotic stresses. *In*: *Brassinosteroids Signalling:* Intervention with Phytohormones and Their Relationship in Plant Adaptation to Abiotic Stresses (pp. 1–16) Springer Singapore, Singapore.

Korasick, D.A., Westfall, C.S., Lee, S.G., Nanao, M.H., Dumas, R., Hagen, G. & Strader, L.C. (2014). Molecular basis for auxin response factor protein interaction and the control of auxin response repression. Proc. Natl. Acad. Sci. USA 111(14), 5427–5432.

Kothari, A. & Lachowiec, J. (2021). Roles of brassinosteroids in mitigating heat stress damage in cereal crops. Int. J. Mol. Sci. 22(5), 2706.

Kumar, S., Sirhindi, G., Bhardwaj, R., Kumar, M. & Arora, P. (2012). Role of 24-epibrassinolide in amelioration of high temperature stress through antioxidant defense system in *Brassica juncea* L. Plant Stress 6(1), 55–58.

Li, J. & Chory, J. (1997). A putative leucine-rich repeat receptor kinase involved in brassinosteroid signal transduction. Cell 90(5), 929–938.

Li, S.M., Zheng, H.X., Zhang, X.S. & Sui, N. (2021). Cytokinins as central regulators during plant growth and stress response. Plant Cell Rep. 40, 271–282.

Li, T., Lei, W., He, R., Tang, X., Han, J., Zou, L. et al. (2020). Brassinosteroids regulate root meristem development by mediating BIN2-UPB1 module in *Arabidopsis*. PLoS Genet. 16(7), e1008883.

Lin, W.H. (2020). Designed manipulation of the brassinosteroid signal to enhance crop yield. Front. Plant Sci. 11: 854.

Liu, D., Zhao, H., Xiao, Y., Zhang, G., Cao, S., Yin, W. et al. (2022). A cryptic inhibitor of cytokinin phosphorelay controls rice grain size. Mol. Plant 15(2), 293–307.

Lu, H.P., Wang, J.J., Wang, M.J. & Liu, J.X. (2021). Roles of plant hormones in thermomorphogenesis. Stress Biol. 1, 1–10.

Luo, J., Jiang, J., Sun, S. & Wang, X. (2022). Brassinosteroids promote thermotolerance through releasing BIN2-mediated phosphorylation and suppression of HsfA1 transcription factors in *Arabidopsis*. Plant Commun. 3(6), 100387.

Malkeyeva, D., Kiseleva, E.V. & Fedorova, S.A. (2025). Heat shock proteins in protein folding and reactivation. Vavilov J. Genet. Breed. 29(1), 7.

Manghwar, H., Hussain, A., Ali, Q. & Liu, F. (2022). Brassinosteroids (BRs) role in plant development and coping with different stresses. Int. J. Mol. Sci. 23(3), 1012.

Medina, E., Kim, S.H., Yun, M. & Choi, W.G. (2021). Recapitulation of the function and role of ROS generated in response to heat stress in plants. Plants 10(2), 371.

Mitchell, J.W., Mandava, N., Worley, J.F., Plimmer, J.R. & Smith, M.V. (1970). Brassins—A new family of plant hormones from rape pollen. Nature 225(5237), 1065–1066.

Momcilovic, I. & Ristic, Z. (2007). Expression of chloroplast protein synthesis elongation factor, EF-Tu, in two lines of maize with contrasting tolerance to heat stress during early stages of plant development. J. Plant Physiol. 164(1), 90–99.

Murase, K., Hirano, Y., Sun, T.P. & Hakoshima, T. (2008). Gibberellin-induced DELLA recognition by the gibberellin receptor GID1. Nature 456(7221), 459–463.

Nassar, A.H. (2004). Effect of homobrassinolide on *in vitro* growth of apical meristems and heat tolerance of banana shoots. Int. J. Agric. Biol. 6(5), 771–775.

Niu, J.H., Ahmad Anjum, S., Wang, R., Li, J.H., Liu, M.R., Song, J.X. et al. (2016). Exogenous application of brassinolide can alter morphological and physiological traits of *Leymus chinensis* (Trin.) Tzvelev under room and high temperatures. Chil. J. Agric. Res. 76(1), 27–33.

Ogweno, J.O., Song, X.S., Shi, K., Hu, W.H., Mao, W.H., Zhou, et al. (2008). Brassinosteroids alleviate heat-induced inhibition of photosynthesis by increasing carboxylation efficiency and enhancing antioxidant systems in *Lycopersicon esculentum*. J. Plant Growth Regul. 27, 49–57.

Pantoja-Benavides, A.D., Garces-Varon, G. & Restrepo-Diaz, H. (2021). Foliar growth regulator sprays induced tolerance to combined heat stress by enhancing physiological and biochemical responses in rice. Front. Plant Sci. 12, 702892.

Park, S.Y., Fung, P., Nishimura, N., Jensen, D.R., Fujii, H., Zhao, Y. et al. (2009). Abscisic acid inhibits type 2C protein phosphatases via the PYR/PYL family of START proteins. Science 324(5930), 1068–1071.

Percio, F., Rubio, L., Amorim-Silva, V. & Botella, M.A. (2025). Crucial roles of brassinosteroids in cell wall composition and structure across species: New insights and biotechnological applications. Plant Cell Environ. 48(3), 1751–1767.

Qu, A.L., Ding, Y.F., Jiang, Q. & Zhu, C. (2013). Molecular mechanisms of the plant heat stress response. Biochem. Biophys. Res. Commun. 432(2), 203–207.

Quintero-Calderon, E.H., Sanchez-Reinoso, A.D., Chavez-Arias, C.C., Garces-Varon, G. & Restrepo-Diaz, H. (2021). Rice seedlings showed a higher heat tolerance through the foliar application of biostimulants. Not. Bot. Horti Agrobot. Cluj-Napoca 49(1), 12120.

Rafique, N., Aqeel, M., Raja, N.I., Shabbir, G., Ajaib, M., Sayyed, R.Z. et al. (Feb 2023). Interactive effects of melatonin and salicylic acid on *Brassica napus* under drought condition. Plant and Soil. https://doi.org/10.1007/s11104-023-05942-7

Rajput, V.D., Harish, R.K., Singh, K.K., Verma, L., Sharma, F.R., Quiroz-Figueroa & Mandzhieva, S. (2021). Recent developments in enzymatic antioxidant defence mechanism in plants with special reference to abiotic stress. Biology 10(4), 267.

Ren, Y., Li, X., Liang, J., Wang, S., Wang, Z., Chen, H. et al. (2023). Brassinosteroids and gibberellic acid actively regulate the zinc detoxification mechanism of *Medicago sativa* L. seedlings. BMC Plant Biol. 23(1), 75.

Rudolphi-Szydło, E., Dyba, B., Janeczko, A., Latowski, D., Sadura, I. & Filek, M. (2022). Brassinosteroid-lipid membrane interaction under low and high temperature stress in model systems. BMC Plant Biol. 22(1), 246.

Salehin, M., Bagchi, R. & Estelle, M. (2015). SCF^TIR1/AFB-based auxin perception: Mechanism and role in plant growth and development. Plant Cell 27(1), 9–19.

Sarkar, A.K. & Sadhukhan, S. (2022). Imperative role of trehalose metabolism and trehalose-6-phosphate signaling on salt stress responses in plants. Physiol. Plant. 174(1), e13647.

Schmitz, A.J., Folsom, J.J., Jikamaru, Y., Ronald, P. & Walia, H. (2013). SUB1A-mediated submergence tolerance response in rice involves differential regulation of the brassinosteroid pathway. New Phytol. 198(4), 1060–1070.

Sharma, N., Kour, S., Kumar, D., Kaur, R., Khajuria, A. & Ohri, P. (2022). Role of brassinosteroids (BRs) in modulating antioxidative defense mechanism in plants growing under abiotic and biotic stress conditions. *In*: Antioxidant Defense in Plants: Molecular Basis of Regulation (pp. 325–367). Singapore: Springer Nature Singapore.

Sheikhi, S., Ebrahimi, A., Heidari, P., Amerian, M.R., Rashidi-Monfared, S. & Alipour, H. (2023). Exogenous 24-epibrassinolide ameliorates tolerance to high-temperature by adjusting the biosynthesis of pigments, enzymatic, non-enzymatic antioxidants, and diosgenin content in fenugreek. Sci. Rep. 13(1), 6661.

Shuai, H., Chen, T., Wlk, T., Rozhon, W., Pimenta Lange, M.J., Sieberer, T. et al. (2022). *SlCESTA* is a brassinosteroid-regulated bHLH transcription factor of tomato that promotes chilling tolerance and fruit growth when over-expressed. Front. Plant Sci. 13: 930805.

Silverstone, A.L., Ciampaglio, C.N. & Sun, T.P. (1998). The *Arabidopsis* RGA gene encodes a transcriptional regulator repressing the gibberellin signal transduction pathway. Plant Cell 10(2), 155–169.

Singh, I. & Shono, M. (2005). Physiological and molecular effects of 24-epibrassinolide, A brassinosteroid on thermotolerance of tomato. Plant Growth Regul. 47: 111–119.

Sonjaroon, W., Jutamanee, K., Khamsuk, O., Thussagunpanit, J., Kaveeta, L. & Suksamrarn, A. (2018). Impact of brassinosteroid mimic on photosynthesis, carbohydrate content and rice seed set at reproductive stage under heat stress. Agric. Nat. Resour. 52(3), 234–240.

Sun, Y., Sun, S., Zahid, M.S., Qiu, Q., Wang, L. & Wang, S. (2025). Root-applied brassinosteroid and salicylic acid enhance thermotolerance and fruit quality in heat-stressed 'Kyoho' grapevines. Front. Plant Sci. 16: 1563270.

Tanvere, S., Akhtar, N., Ilyas, N., Sayyed, R.Z., Fitriatin, B.N., Parveen, K. & Bukhari, N.A. (2023). Interactive effects of *Pseudomonas putida* and salicylic acid for mitigating drought tolerance in Canola (*Brassica napus* L.), Heliyon 9(3), e14193, https://doi.org/10.1016/j.heliyon.2023.e14193

Thussagunpanit, J., Jutamanee, K., Sonjaroon, W., Kaveeta, L., Chai-Arree, W., Pankean, P. et al. (2015). Effects of brassinosteroid and brassinosteroid mimic on photosynthetic efficiency and rice yield under heat stress. Photosynthetica 53, 312–320.

Usmani, L., Shakil, A., Khan, I., Alvi, T., Singh, S. & Das, D. (2025). Brassinosteroids in micronutrient homeostasis: Mechanisms and implications for plant nutrition and stress resilience. Plants. 14(4), 598.

Wahid, A., Gelani, S., Ashraf, M. & Foolad, M.R. (2007). Heat tolerance in plants: An overview. Environ. Exp. Bot. 61(3), 199–223.

Wang, C., Cui, H., Jin, M., Wang, J., Li, C., Li, Y. & Wang, Z. (2025). Enhancement of wheat resistance to dry–hot wind stress during grain filling by 24-epibrassinolide: Optimization of hormone balance and improvement of flag leaf photosynthetic performance. Front. Plant Sci. 16, 1552617.

Wang, R. & Estelle, M. (2014). Diversity and specificity: Auxin perception and signaling through the TIR1/AFB pathway. Curr. Opin. Plant Biol. 21, 51–58.

Wang, W., Xie, Y., Liu, C. & Jiang, H. (2022). The exogenous application of brassinosteroids confers tolerance to heat stress by increasing antioxidant capacity in soybeans. Agriculture 12(8), 1095.

Wang, X., Lu, Z., Su, J., Li, Y., Cao, M. & Gao, H. (2020). 24-Epibrassinolide delays senescence in harvested kiwifruit through effects on mitochondrial membrane and antioxidant activity. LWT 118, 108833.

Wei, L.J., Deng, X.G., Zhu, T., Zheng, T., Li, P.X., Wu, J.Q. et al. (2015). Ethylene is involved in brassinosteroids induced alternative respiratory pathway in cucumber (*Cucumis sativus* L.) seedlings response to abiotic stress. Front. Plant Sci. 6, 982.

Xiong, M., Yu, J., Wang, J., Gao, Q., Huang, L., Chen, C. et al. (2022). Brassinosteroids regulate rice seed germination through the BZR1-RAmy3D transcriptional module. Plant Physiol. 189(1), 402–418.

Yadava, P., Kaushal, J., Gautam, A., Parmar, H. & Singh, I. (2016). Physiological and biochemical effects of 24-epibrassinolide on heat-stress adaptation in maize (*Zea mays* L.). Nat. Sci. 8(4), 171–179.

Yang, X., Jia, Z., Pu, Q., Tian, Y., Zhu, F. & Liu, Y. (2022). ABA mediates plant development and abiotic stress via alternative splicing. Int. J. Mol. Sci. 23(7), 3796.

Yaqoob, U., Jan, N., Raman, P.V., Siddique, K.H. & John, R. (2022). Crosstalk between brassinosteroid signaling, ROS signaling and phenylpropanoid pathway during abiotic stress in plants: Does it exist. Plant Stress 4, 100075.

Yu, J.Q., Huang, L.F., Hu, W.H., Zhou, Y. H., Mao, W.H., Ye, S.F. et al. (2004). A role for brassinosteroids in the regulation of photosynthesis in *Cucumis sativus*. J. Exp. Bot. 55(399), 1135–1143.

Yu, M.H. (2021). Mechanistic study of brassinosteroid-regulated drought tolerance in *Arabidopsis*. The Chinese Univ. of Hong Kong (PhD thesis).

Yuldashev, R., Avalbaev, A., Bezrukova, M., Vysotskaya, L., Khripach, V. & Shakirova, F. (2012). Cytokinin oxidase is involved in the regulation of cytokinin content by 24-epibrassinolide in wheat seedlings. Plant Physiol. Biochem. 55, 1–6.

Zahra, N., Hafeez, M.B., Kausar, A., Al Zeidi, M., Asekova, S., Siddique, K.H. et al. (2023). Plant photosynthetic responses under drought stress: Effects and management. J. Agron. Crop Sci. 209(5), 651–672.

Zhang, C., Bai, M.Y. & Chong, K. (2014). Brassinosteroid-mediated regulation of agronomic traits in rice. Plant Cell Rep. 33(5), 683–696.

Zhang, Y., Yu, J., Xu, X., Wang, R., Liu, Y., Huang, S. et al. (2022). Molecular mechanisms of diverse auxin responses during plant growth and development. Int. J. Mol. Sci. 23(20), 12495.

Zhang, Z., Chen, Z., Song, H. & Cheng, S. (2023). From plant survival to thriving: Exploring the miracle of brassinosteroids for boosting abiotic stress resilience in horticultural crops. Front. Plant Sci. 14, 1218229.

Zhao, B. & Li, J. (2012). Regulation of brassinosteroid biosynthesis and inactivation. J. Integr. Plant Biol. 54(10), 746–759.

CHAPTER

7

Natural Supporters: How Signalling Molecules Help Plants to Combat the Effects of Heat Stress

A. Mounika[1], Phanindra P. V.[2], Santosh Kumari[1], Ritu Mawar[1*] and Riyaz Sayyed[3]

[1] ICAR-CAZRI, Jodhpur, Rajasthan – 542003, India

[2] Department of Plant Pathology, Banaras Hindu University, Varanasi – 221005, Uttar Pradesh

[3] Department of Biological Science and Chemistry, University of Nizwa, Nizwa 616, OMAN

1. Introduction

Global warming has considerably intensified the incident and importance of heat stress scenes, considerably reducing crop productivity. In reaction, plants generate elaborate physiological and microscopic methods, generally governed by phytohormones (Guo et al., 2019; Wang et al., 2019a; Zhou et al., 2019). Instead of working alone, phytohormones undertake far-reaching cross-ideas with additional indicating fragments to fine-tune stress reactions (Choudhary et al., 2012; Planas-Riverola et al., 2019). Brassinosteroids (BRs), a class of steroidal phytohormones, have accumulate significant interest for their parts in enlightening control and stress adaptation. BR-associated chemical were first found in Brassica napus irritant (Grove et al., 1979), and more than 70 have because happened recorded. BRs maybe found in a expansive range of plant variety, containing algae, pteridophytes, gymnosperms, and angiosperms. Hayat et al. (2003) recognized them as free or associated with sugars and fatty acids. They are essential for plant improvement (González-García et al., 2011). Brassinolide (BL), 24-epibrassinolide (24-EBL), and 28-homobrassinolide (28-HBL) are the three most influential bioactive BRs, and they have been widely intentional for their belongings on progress and stress reduction (Vardhini and Anjum 2015). BR biosynthesis principally takes place in the endoplasmic mesh, when in fact their idea is facilitated at the plasma membrane by BRASSINOSTEROID INSENSITIVE1 (BRI1) in addition to its co-receptor BAK1 (Nolan et al., 2020; Northey et al., 2016). When BRs bind to BRI1, a

*Corresponding author: ritumawar1976@gmail.com

DOI: 10.1201/9781003669777-7

indicating cascade is begun through suppression of the negative manager BKI1 and incitement of downstream transcription factors such as BRI1-EMS SUPPRESSOR1 (BES1) and BRASSINAZOLE RESISTANT1 (BZR1), that sequential control the expression of BR-responsive genes (Li et al., 1997; Li et al., 2002; Tong and Chu 2018; Planas-Riverola et al., 2019). Although first identified as promoters of container extension, BRs are also involved in processes like cell division, vascular incident, photomorphogenesis, reproduction, and senescence (Nolan et al., 2020). Mutations or deficiencies in BR biosynthesis or indicating influence characteristics such as dwarf improvement, late flowering, male infertility, and de-ineffectiveness in darkness (Clouse 2015). On the other hand, overexpression of BR biosynthetic genes augments inner BR levels, through enhancing yield and stress fortitude in crops (Divi and Krishna 2009; Xia et al., 2018). BRs donate absolutely to plant responses against dryness, salinity, heat, cold, difficult metals, and natural contaminators (Kagale et al., 2007; Rajewska et al., 2016; Xia et al., 2018), though the exact mechanisms are still being surveyed.

The interplay of BRs accompanying added hormones, including auxin, abscisic acid (ABA), salicylic acid (SA), jasmonic acid (JA), and nitric group of chemical elements (NO), places ruling class as main managers of stress adaptation in plants (Luo et al., 2010; Fan et al., 2012; Oh et al., 2012). Such coordination accompanying supplementary phytohormones authorizes BRs to improve thermotolerance (Divi et al., 2010). A key few this rule includes copy factors BES1 and BZR1, that influence the verbalization of stress-aware genes (Anwar et al., 2018). Both inside BRs and externally practiced BRs have existed reported to decrease heat-inferred damage, exemplification, by improving photosynthetic capacity through raised Rubisco exercise, enhanced quantum adeptness of PSII, and embellished chlorophyll radiance (Bajguz and Hayat 2009; Hayat et al., 2010; Fariduddin et al., 2014; Jin et al., 2015). This guardianship is connected to modifications in the heat-sympathetic transcriptome along with inflated levels of heat shock proteins (HSPs), that comprise molecular chaperones that maintain basic forms under heat and additional stresses (Dhaubhadel et al., 2002; Mazorra et al., 2011; Samakovli et al., 2014; Shigeta et al., 2015; Divi et al., 2016). Furthermore, BRs toughen sheath honor, reinforce antioxidant defenses, and support photosynthetic efficiency (Hayat et al., 2010; Kaur et al., 2013), while BR analogs help counterbalance the harmful impacts of heat stress on growth and plant structure (González-Olmedo et al., 2005). Numerous studies establish these thermoprotective roles of BRs indifferent crops. Increased heat resistance has existed recorded in Bromus inermis (Wilen et al., 1995), Brassica napus (Dhaubhadel et al., 1999), hay (Janeczko et al., 2011), soiled bean (Hayat et al., 2010), tomato (Mazorra et al., 2011), and rice (Thussagunpanit et al., 2015).

In light of these remarks, BRs hold meaningful potential for heartening crop resilience regardless of growing worldwide hotnesses. This chapter inquires to present a inclusive survey of the role of brassinosteroids in managing plant development and checking stress, accompanying special importance on heat stress environments.

2. Brassinosteroids: Structure, Biosynthesis and Signalling

Natural brassinosteroids (BRs) maintain a characteristic 5α-cholestane steroid determination with a infuse A/B ring composition. These compounds mainly feature two hydroxyl substituents on the A-ring, together with either a 6-ketone or a 7-oxa-6-ketone structure on the B-ring. The structural diversity of BRs is generated by differences in the number, kind, and spatial orientation of oxygenated functional groups present on these rings (Taiz and Zeiger 2010).

These modifications are carried out through successive oxidation and reduction steps that take place during BR biosynthesis. Based on structural traits and extent of modification, BRs can be generally grouped into conjugated and unconjugated types. The unconjugated BRs are additionally divided into three principal categories—C_{27}, C_{28}, and C_{29}—depending on variations in alkyl substitution on their side chains (Bajguz et al., 2020). Within these, 11 structural subgroups are recognized according to substituents located at the C-23, C-24, and C-25 positions. These include: 23-oxo, 24S-methyl, 24R-methyl, 24-methylene, 24S-ethyl, 24-ethylidene, 24-methylene-25-methyl, 24-methyl-25-methyl, and variants without substituents at C-23, C-24, or both. Of these, the C_{28}-type BRs represent the most prevalent forms in nature (Bajguz and Tretyn 2003). The most biologically potent brassinosteroid identified so far is brassinolide, chemically defined as (22R, 23R, 24S)-2α,3α,22,23-tetrahydroxy-24-methyl-B-homo-7-oxa-5α-cholestan-6-one (Fig. 1, Tang et al., 2016). By contrast, conjugated BRs—usually esterified with fatty acids or sugars—are typically produced through BR metabolism and are regarded as biologically inactive (Fujioka and Yokota 2003). These conjugates are thought to participate in BR storage and homeostasis rather than active signaling.

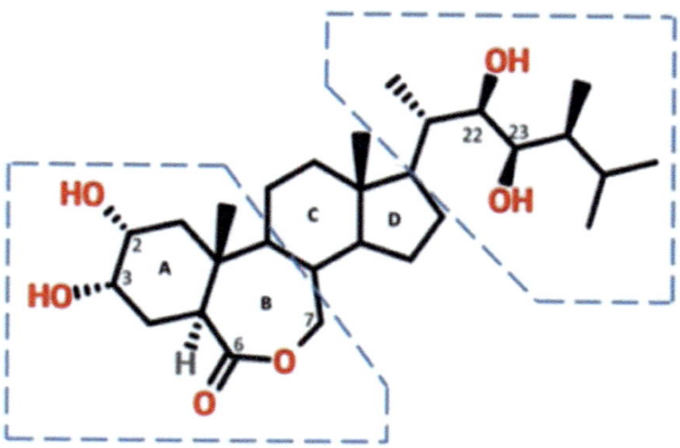

Fig. 1. The structure of Brassinolide (Tang et al., 2016).

Brassinosteroids (BRs) are produced through three distinct biosynthetic routes C_{27}, C_{28}, and C_{29} types—each originating from different sterol precursors. Despite these differences, all three routes share an initial common sequence of biosynthetic steps (Bajguz et al., 2020). BR formation starts with isopentenyl pyrophosphate (IPP), which arises either from the mevalonate pathway in higher plants or the non-mevalonate pathway in lower plants. IPP acts as the universal precursor of BRs and is converted first into squalene, which is subsequently cyclized into cycloartenol, a key branching molecule from which the C_{27}, C_{28}, and C_{29} biosynthetic paths diverge (Vriet et al., 2013). In the C_{27} branch, cholesterol derived from cycloartenol serves as the starting point, and both early and late C6 oxidation steps contribute to the production of 28-norbrassinolide (28-norBL) through a series of intermediates (Shimada et al., 2001). For C_{28} BRs, campesterol is the initial precursor, synthesized via 24-methylenelophenol. Two alternative oxidation pathways, one late C-22 oxidation route through campestanol, and an early C-22 oxidation pathway bypassing campestanol, lead to the generation of castasterone, a crucial intermediate that is eventually converted into brassinolide (BL), the most active C_{28} BR (Joo et al., 2012). Both early and late C6 oxidation reactions are part of these routes. The C_{29} branch originates from β-sitosterol, which undergoes successive modifications to yield 28-homobrassinolide (28-homoBL). Notably, these pathways are interconnected; for example, C_{28} BRs can also arise from either C_{27} or C_{29} precursors, reflecting the flexibility of BR biosynthesis (Joo et al., 2015). Ultimately, all pathways converge at castasterone, which acts as a common intermediate for the different BR types. Importantly, BR biosynthesis is tightly feedback-regulated, with biosynthetic gene expression adjusting according to endogenous BR concentrations. Additionally, plants utilize specific biosynthetic inhibitors to help maintain BR homeostasis (Rozhon et al., 2019). An overview of these pathways, including the principal enzymes, branching points, and known inhibitors, is shown in Fig. 2 (Bajguz et al., 2020).

The biosynthesis of BRs likely happens in the endoplasmic reticulum (ER), where the enzymes necessary have existed situated in Arabidopsis (Hu et al., 2025). BRs are moved from the ER to the apoplast by way of plasmodesmata by way of proteins ABCB1 and ABCB19 (Wei et al., 2024; Ying et al., 2024). BRs are discovered at the container surface by way of the skin sheath receptor BRASSINOSTEROID INSENSITIVE 1 (BRI1) and allure co-receptor SOMATIC EMBRYOGENESIS RECEPTOR KINASE3 (SERK3)/BRI1-ASSOCIATED KINASE1 (BAK1) (Fig. 3, Nolan et al., 2020). In the deficiency of BRs, inhibitors in the way that BKI1 and BIR3 block the receptor complex. Furthermore, by phosphorylating copy determinants containing BRI1-EMS SUPPRESSOR1 (BES1) and BRASSINAZOLE-RESISTANT1 (BZR1), BRASSINOSTEROID INSENSITIVE2 (BIN2) acts as a negative manager by restricting their project through various habits. Additionally, when BRZ-SENSITIVE-SHORT HYPOCOTYL1 (BSS1) is inactivated, it forms a complex among BES1 and BZR1 in the cytoplasm, while THIOREDOXIN H-TYPE5 decreases BZR1 in the core. This results in raised expression of BR-suppressed genes and dropped expression of BR-induced genes. When BRs are present, they start BR signaling via binding to BRI1 and its co-receptor BAK1. BKI1 and BIR3 separate from the receptor complex consequent to detection, enabling phosphorylation and activation of BRI1 and BAK1. Following phosphorylation of CONSTITUTIVE DIFFERENTIAL GROWTH1 (CDG1), the phosphatase BRI1-SUPPRESSOR1 (BSU1) is activated, inhibiting BIN2. Lastly, the activity is promoted by PROTEIN PHOSPHATASE2A (PP2A). Following that, the essential DIFFERENTIAL GROWTH1 (CDG1) phosphorylates and activates

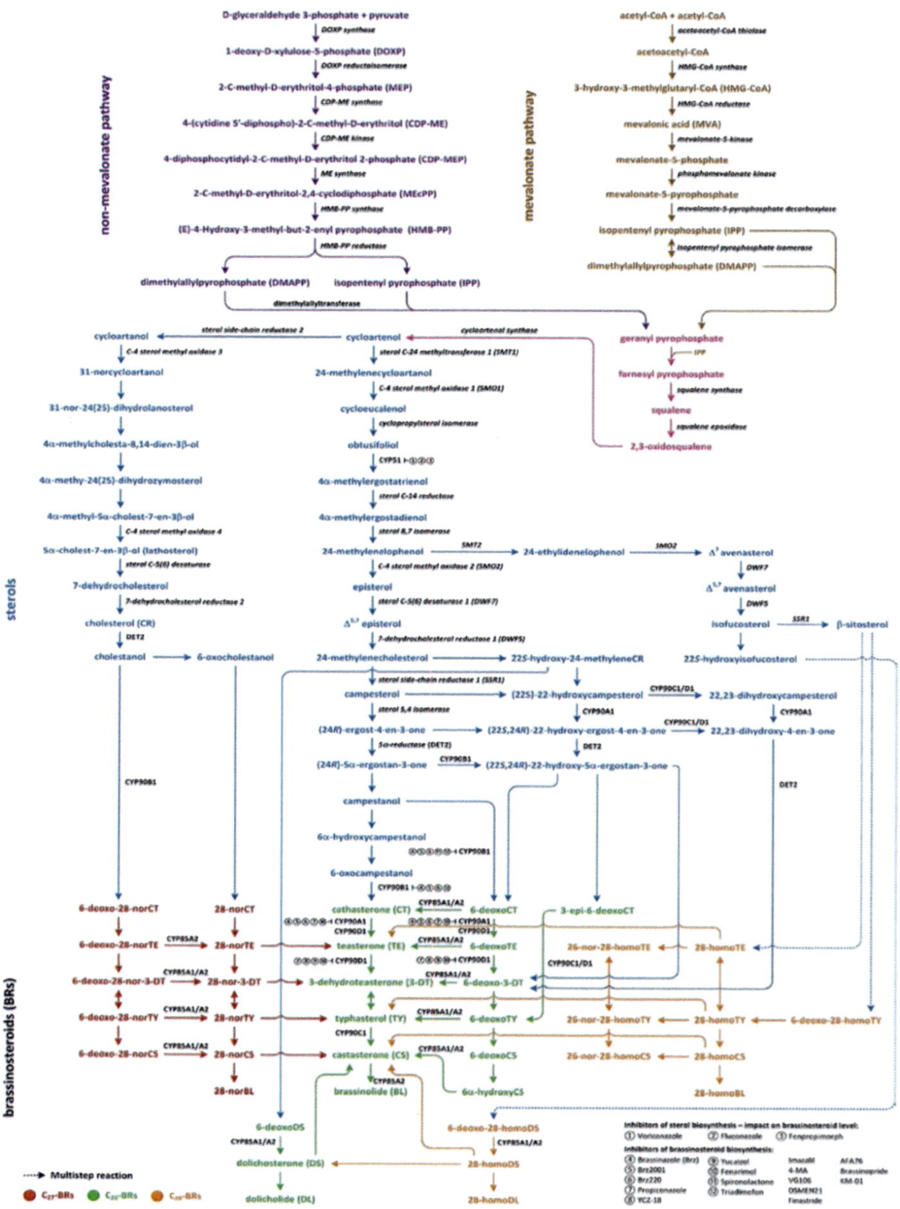

Fig. 2. The biosynthetic pathways of brassinosteroids (Bajguz et al., 2020).

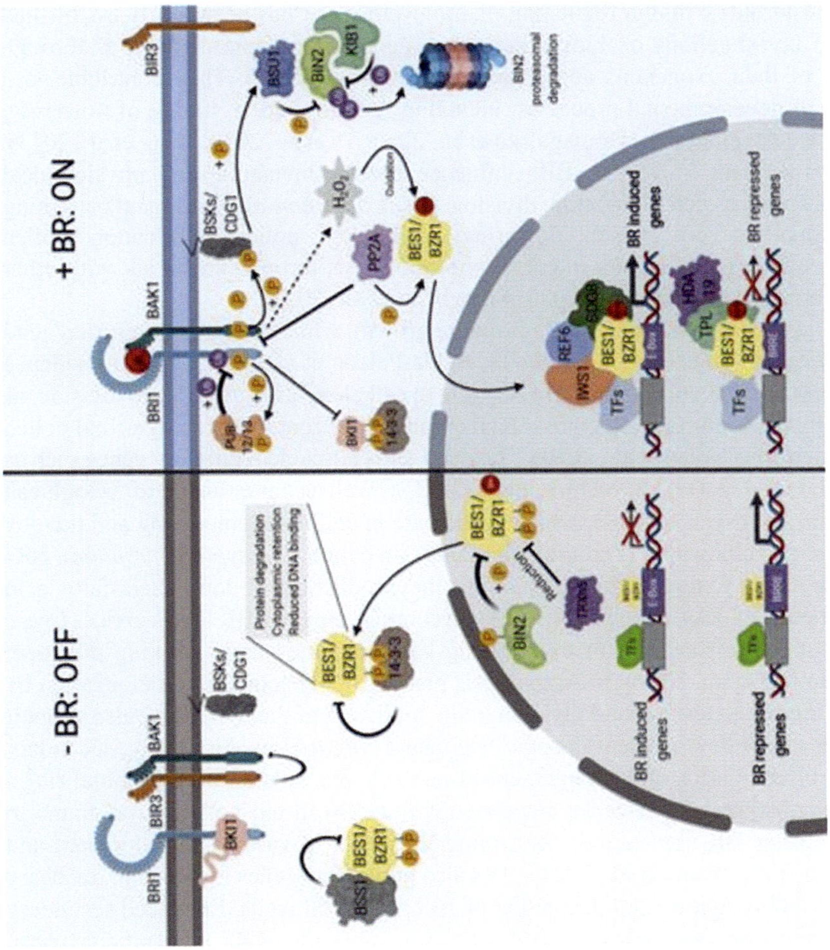

Fig. 3. The pathway of BR signalling (Nolan et al., 2020).

the phosphatase BRI1-SUPPRESSOR1 (BSU1), that suppresses BIN2. Finally, PROTEIN PHOSPHATASE2A (PP2A) improves the venture of BES1 and BZR1 in conjunction with additional transcription factors and cofactors, through stimulating BR-inferred genes and repressing BR-inhibited genes (Tang et al., 2016; Bajguz et al., 2020; Kim et al., 2020).

3. Role of BRs in Plant Development

Brassinosteroids's role in regulating plant development has been firmly established through investigations on biosynthetic and signaling mutants, as well as through studies of their exogenous application (Nolan et al., 2020). They contribute to a variety of developmental processes, including seed formation, timing of flowering, and pollen development (Domagalska et al., 2010; Ye et al., 2010; Jiang et al., 2013). Across the plant life cycle, BRs influence diverse metabolic and physiological functions such as cell elongation, division, tissue differentiation, stomatal patterning and formation, root growth, flowering, senescence, pollen germination, pollen tube elongation, and enhancement of photosynthesis through cross-talk with other phytohormones (Nolan et al., 2020; Manghwar et al., 2022).

At lower concentrations, BRs promote growth, while at higher levels they tend to suppress it (Clouse and Sasse 1998; Belda-Palzon et al., 2018). Recent evidence indicates that BRs contribute more strongly to cell elongation than to cell division, as shown by their influence on cytoskeletal organization, root meristem size, and pollen tube extension (Nolan et al., 2020). They activate cell-cycle regulatory genes such as CycB, CycA, CycD3.2, CycD3.1, and CDKB, as well as genes linked to Xyloglucan endotransglucosylase (XTR), which participate in cell wall remodeling and thereby facilitate cell elongation (Pereira-Netto 2009). In cotton (*Gossypium hirsutum*), BRs enhance fiber-elongation by modulating the production of long-chain-fatty acid molecules (VLCFAs) (Yang et al., 2023). In contrast, elevated BR levels in rice (*Oryza sativa*) suppress root growth by reducing cell elongation and shrinking meristem size (Jantapo et al., 2021). In Arabidopsis plants, BRs encourage root elongation by enhancing cell extension and division in the meristem of the root. They also support transverse root development in sugar beet (*Beta vulgaris*) by stimulating secondary xylem differentiation and enlargement of parenchyma cells between cambial rings. Moreover, expression of genes associated with cell wall biosynthesis was found to increase after BR application, confirming their role in vascular development and root thickening (Wang et al., 2021). BRs also upregulate genes involved in tracheary element differentiation and deposition of secondary cell walls. Enhanced secondary xylem production has been observed with overexpression of BR biosynthetic genes, highlighting the crucial role of this hormone in wood formation (Tang et al., 2020).

The timing of flowering is tightly regulated by the interaction of BR signaling with environmental cues. Mutants impaired in BR biosynthesis or signaling, such as *det2* and *bri1*, display significantly delayed flowering in *Arabidopsis*, indicating that BR contributes positively to floral transition (Li et al., 2010). BRs influence the activity of flowering integrator genes that drive the switch from vegetative to reproductive development, including SOC1 and FT (*FLOWERING LOCUS T*) (Li et al., 2018). Brassinosteroids enhance the activity of enzymes such as α-amylase, which mobilize stored nutrients required for early seedling establishment, thereby promoting germination. They also act synergistically with gibberellins and other phytohormones to overcome seed dormancy, particularly under stress conditions such as drought or salinity. BR-deficient mutants in *Arabidopsis* show delayed germination, further confirming their essential role in this process (Steber & McCourt 2001). Moreover, Kagale et al., reported that application of exogenous BRs boosts germination rates in several crop species by stimulating metabolic pathways essential for seed vigor (Kagale et al., 2007).

4. Heat Stress and its Impact on Plants

Climate change represents a serious challenge to global crop yields and food security. Rising temperatures driven by global warming lead to more frequent droughts and heat waves, which negatively affect both human and plant systems. In plants, heat stress refers to exposure to temperatures beyond a critical threshold for a duration sufficient to cause irreversible damage to growth and development. Short episodes of high temperature, often termed heat shocks, have emerged as major limiting factors for crop productivity (Mondal et al., 2023). Yield losses of about 15–35% in crop species are attributed to heat stress (Tayade et al., 2018). It also influences water availability, intensifies pest infestations, and accelerates disease outbreaks. Heat stress disrupts multiple physiological functions such as nutrient uptake, photosynthesis, respiration, and transpiration, ultimately reducing yield. Its impact depends on the crop type (field vs. horticultural) and growth stage, with reproductive phases being particularly vulnerable (Ostmeyer et al., 2020; Ravikiran et al., 2020). High temperatures severely damage cell membranes and the proteins associated with them (Mondal et al., 2023). They also deactivate enzymes, thereby disturbing metabolic pathways vital for normal plant growth. Suppression of major antioxidant enzymes, such as catalase, superoxide dismutase, and ascorbate peroxidase, as well as non-enzymatic antioxidants like phenols and ascorbic acid, leads to an excessive accumulation of reactive oxygen species (ROS), including free oxygen radicals (O_2^-), monooxygen (O_2), hydroxyl radicals (OH^-), and hydrogen peroxide (H2O2) (Chauhan et al., 2022). This excessive production of ROS disrupts the electron transport chains in mitochondria and chloroplasts, leading to oxidative damage to proteins, nucleic acids, and lipids, and inducing apoptosis (Hansanuzzaman et al., 2013; Sharma et al., 2019). High temperatures also inhibit the accumulation of solutes such as proline and soluble sugars, while inhibiting the synthesis of heat shock proteins (HSPs), which normally protect plants from heat stress (Saha et al., 2021).

Heat stress induces alterations in morphological, physiological, biochemical, and molecular processes, thereby hindering the growth and development of crops (Zhao et al., 2020). Extremely high temperatures can lead to premature leaf senescence, sunburn on leaves and twigs, inhibited growth, and blotching of leaves and fruits (Vollenweider and Gunthardt-Goerg 2005). For instance, in heat-sensitive rice varieties, Heat stress extended the flowering and ripening period by 50%, while reducing peak plant growth, the number of branches per plant, spike length, the number of spikelets per spike, and the number of grains per spike, thus decreasing grain yield (Kumar et al., 2023). In cereals, exposure to heat stress combined with reduced grain content leads to irritability and the onset of sterility. poor anther dehiscence, and reduced assimilate transport to the developing grain, producing small and shriveled kernels (Rahman et al., 2021). Heat stress also disrupts water and nutrient uptake, assimilation, and transport by altering uptake protein channels as well as root architecture and morphology (Mishra et al., 2023). In tomato, it reduces plant water and nutrient content, alters root biomass and structure, and increases transpiration (Duc et al., 2018). Within photosynthesis, photosystem II is considered the most heat-sensitive component (Li et al., 2020). Elevated temperature interferes with chlorophyll biosynthesis, electron transport, carbon dioxide fixation, and the efficiency of photosystems I and II (Abdelrahman et al., 2017; Sharma et al., 2019). In wheat, heat stress markedly lowers chlorophyll levels, RUBISCO activity, and photosynthetic rate (Almeselmani et al., 2012). It also disrupts source–sink relationships and reduces assimilate movement, ultimately contributing to yield decline (Sattar et al., 2020).

5. Brassinosteroids in Heat Stress Tolerance

Brassinosteroids help plants withstand a wide range of abiotic stresses such as drought, extreme temperatures, salinity, heavy metals, and others (Chaudhari et al., 2022). Whether synthesized endogenously or applied externally, BRs exert comparable effects in enhancing heat stress tolerance (Sadura and Janeczko, 2018). They regulate the expression of stress-responsive genes through diverse signaling networks (Kothari and Lachowiec, 2021). Firstly, they re-establish the production of proteins or metabolites crucial for different developmental functions. Secondly, they stimulate the expression of protective proteins that safeguard plants under heat stress. Under high temperature conditions, BRs modulate Ca^{2+} signaling, reduce lipid peroxidation, maintain cell membrane stability, preserve photosynthetic pigments, assist in CO_2 assimilation, regulate gas exchange, stabilize photosystems, and promote sugar accumulation (Sadura and Janeczko 2018). In plants, Ca^{2+} plays a central role in multiple signaling pathways associated with the synthesis of heat shock proteins (HSPs), transcription factors, and osmoprotectants such as γ-aminobutyric acid (GABA) (Kang et al., 2023). BRs influence the movement of these secondary messengers to initiate responses against heat stress. They bind to membrane receptors and increase cytoplasmic Ca^{2+} concentrations by opening Ca^{2+}-gated channels (Oh et al., 2012; Zhao et al., 2013). Application of 24-epibrassinolide and 24-epicastasterone, along with Ca^{2+} ions and antioxidants induced via ROS, was shown to protect cells of wheat coleoptiles from stress (Kolupaev et al., 2014). In maize, Ca^{2+} and a calcium-dependent protein kinase (CCaMK), termed ZmCCaMK, act together to promote BR-mediated antioxidant defense, with ZmCCaMK further enhancing BR-induced Ca^{2+} influx into the cytoplasm (Yan et al., 2015).

Based on current knowledge, brassinosteroids contribute to maintaining plasma membrane stability through multiple mechanisms (Sadura and Janeczko, 2018). They can alter the fatty acid profile of cell membranes by elevating the proportion of unsaturated fatty acids, since BRs participate in fatty acid biosynthesis, transport, and incorporation into membranes. A recent study demonstrated that nano-encapsulated BR application decreased the saturated-to-unsaturated fatty acid ratio in the membranes of bell pepper (Halaji et al., 2024). BRs are thought to embed themselves within membranes, thereby increasing the spacing between fatty acid chains and improving membrane performance. Additionally, they act indirectly on reactive oxygen species by lowering lipid peroxidation in membranes. In rice plants exposed to high day/night temperatures of 40/30 °C, treatment with 24-epibrassinolide and 7,8-dihydro-8α-20-hydroxyecdysone (DHECD), a BR mimic, enhanced lipid peroxidation in rice leaves (Thussagunpanit, 2015). Beyond this, BRs also influence the structure and composition of primary and secondary cell walls by promoting greater deposition of cellulose, hemicellulose, lignin, and pectin (Percio et al., 2025).

Brassinosteroids modulate both reactive-oxygen-species generation via NADPH oxidases and the ROS-scavenging machinery under different stress conditions (Bhatt et al., 2019). At low concentrations, BRs have been observed to enhance the production of ROS such as H_2O_2 and superoxide anions (O_2^-), which function as signaling messengers to trigger stress-responsive pathways. BR-induced H_2O_2 further activates stress-related regulators including transcription factors, Heat shock proteins (HSPs), aldehydes, and antioxidant enzymes scavenge harmful oxygen species (Zhu et al., 2013). Furthermore, nitric oxide (NO), which acts on the BR signaling pathway downstream of hydrogen peroxide (H_2O_2), plays a role in inhibiting harmful reactive oxygen species and promoting tolerance to BR-induced stress. Inhibition of nitric oxide biosynthesis leads to the complete silencing and suppression of the antioxidant gene expression and activity stimulated by BR (Zhang et al., 2011). The transcription factor BZR1 is considered a regulator of the BR signaling pathway in tomato heat stress responses through RBOH1-dependent reactive oxygen species signaling, which partially involves FER2 and FER3 (FERONIA homologs) (Yin et al., 2018). To counteract the accumulation of reactive oxygen species, plants also possess a variety of enzymatic antioxidants such as catalase (CAT), superoxide dismutase (SOD), ascorbate peroxidase (APX), glutathione reductase, guaiacol peroxidase, and glutathione peroxidase, and monodehydroascorbate reductase (Kothari and Lachowiec 2021). External application of 24-epibrassinolide (EBR) has been shown to enhance CAT, SOD, and APX activities and increase antioxidant compounds like ascorbic acid and reduced glutathione in soybean leaves, thereby alleviating growth inhibition caused by heat stress (Wang et al., 2022). In wheat, EBR treatment boosted SOD, CAT, and peroxidase (POD) activities under hot dry wind-induced stress (Wang et al., 2025) (Fig. 4).

Brassinosteroids help limit chlorophyll breakdown by either preventing its degradation or influencing the synthesis of enzymes responsible for chlorophyll catabolism under heat stress (Sadura and Janeczko 2018). They may also contribute to chlorophyll biosynthesis, as BR-deficient mutants typically show reduced chlorophyll levels (Janeczko et al., 2011). Application of 24-epibrassinolide has been reported to increase the concentrations of chlorophyll a, chlorophyll b, and carotenoids in several crops, including rice, eggplant, and melon (Wu et al., 2014; Zhang et al., 2014; Thussagunpanit et al., 2015). Similarly, root treatment with BR combined with salicylic acid enhanced chlorophyll content in heat-stressed 'Kyoho' grape leaves (Sun et al., 2025). BRs also influence gas exchange and stomatal conductance when plants are exposed to high temperature stress. In addition, they enhance photosystem performance, particularly PSII, by stimulating the enzymatic

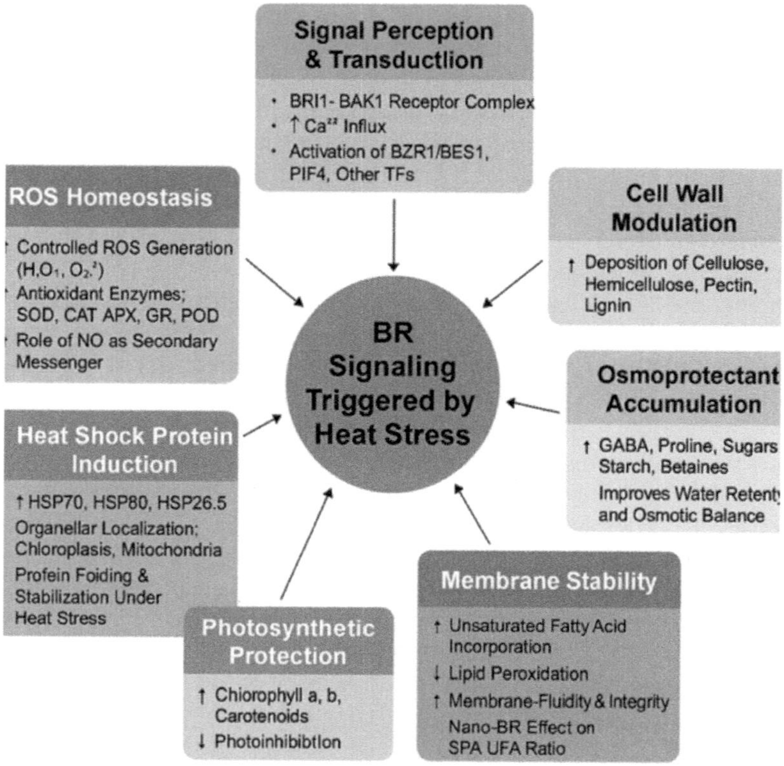

Fig. 4. Brassinosteroid-regulated mechanisms conferring heat stress tolerance.

machinery associated with photosynthesis (Xia et al., 2009; Janeczko et al., 2011). BRs further promote energy absorption, excitation energy trapping, and electron transport flux, thereby improving energy transfer within the PSII system of spring barley (Janeczko et al., 2011). Their beneficial effects on chloroplast and thylakoid membrane stability, enhanced pigment synthesis, improved CO_2 assimilation, and better water management in leaves collectively strengthen photosynthesis and ultimately boost crop yields (Sadura and Janeczko, 2018).

Brassinosteroids are known to enhance the levels of osmoprotectants such as γ-aminobutyric acid (GABA), soluble sugars, proline, betaines, and other amino acids to help maintain osmotic balance under high-temperature stress (Kothari and Lachowiec, 2021). External application of BR was shown to raise the total soluble content in grape leaves and played a key role in osmolyte accumulation (Lv et al., 2022). In rice, BR treatment improved the accumulation of soluble sugars, starch, sucrose, and total non-structural carbohydrates in leaves exposed to heat stress, thereby supporting better source–sink relationships (Nazir et al., 2023). During heat stress, heat shock proteins (HSPs) act as molecular chaperones, preventing injury to heat-sensitive proteins. They support proper folding, intracellular targeting, and stabilization of proteins required for critical physiological processes such as membrane integrity, water-use efficiency, and photosynthesis (Waters et al., 1996; Camejo et al.,

2005). HSPs build up in different organelles to maintain their functions; for example, increased mitochondrial HSPs stabilize electron transport chain complexes and protect translational machinery under heat stress (Downs and Heckathorn, 1998). Similarly, higher concentrations of HSPs in chloroplasts preserve the photosynthetic apparatus (Chauhan et al., 2012). The involvement of brassinosteroids in upregulating HSP expression under heat stress has been clearly demonstrated. In *Brassica napus* seedlings, application of 24-epibrassinolide enhanced the expression of *hsp70* and *hsp90* (Dhaubhadel et al., 2002). In *Arabidopsis*, BRs regulate the expression of *HSP26.5*, *HSP70*, and *HSP90.1* to confer thermotolerance (Chen et al., 2022).

The advancement of molecular techniques has enabled detailed insights into intracellular molecular interactions in plants. Mutant studies have made it possible to clarify the roles of specific genes or components in plant growth, development, and stress responses. The contribution of brassinosteroid signaling to thermotolerance has now been well demonstrated. BR receptors and their downstream components play critical roles in regulating plant development under elevated temperatures. During high-temperature stress, elevated BR levels activate the transcription factors BZR1 and BES1, which turn on BR-responsive genes and, either directly or through interaction with PHYTOCHROME INTERACTING FACTOR 4 (PIF4), regulate thermomorphogenesis (Ibanez et al., 2018; Martinez et al., 2018). BZR1 also represses ERF49, a negative regulator of thermotolerance, thereby promoting the expression of stress-inducible genes and heat shock proteins (Chen et al., 2022). In addition, BR signaling prevents BIN2-mediated phosphorylation of HEAT SHOCK TRANSCRIPTION FACTORS (HsfA1s), supporting their nuclear localization and DNA-binding capacity (Luo et al., 2022). Collectively, these findings underline the essential role of BRs in conferring thermotolerance to plants.

6. Interaction of BRs with Other Hormones and Signalling Molecules under Heat Stress

6.1. BR–ABA Cross-talk under Stress

One of the important hormones in combating biological stress, including heat, is abscisic acid. Abscisic acid increases the production of heat shock proteins (HSPs), which protect the cell from damage caused by reactive oxygen species (ROS) by accumulating on the plant cell wall (Hu et al., 2007; Rezaul et al., 2019). Its protective mechanism of action largely involves stomatal closure, modulation of antioxidant enzyme activity, and control of stress-responsive gene expression. The interaction between BR and ABA involves a delicate balance of both synergistic and antagonistic effects. While BRs are generally linked to promoting growth and development, ABA is often considered a stress-related hormone that restricts growth under unfavorable conditions to ensure survival. This contrasting relationship allows plants to balance the trade-off between sustaining growth and boosting stress tolerance (Divi et al., 2010).

During heat stress, the interaction between these hormones may alternate between inhibitory and cooperative modes. In some cases, ABA has been observed to suppress BR-mediated thermotolerance, likely by disrupting downstream BR signaling or diminishing its protective influence (Larkindale and Knight 2002; Divi et al., 2010). Conversely, under different conditions, BRs function as upstream regulators that promote ABA biosynthesis in a stress-dependent manner (Kurepin et al., 2008). This increase in ABA supports crucial defense mechanisms, such as boosting antioxidant capacity, modulating stomatal conductance, and activating stress-related genes (Kurepin et al., 2008). A reinforcing feedback cycle also exists, whereby BR-induced reactive oxygen species (ROS), especially hydrogen peroxide (H_2O_2), stimulate ABA synthesis, which in turn enhances ROS accumulation (Zhou et al., 2014). This feedback strengthens stress signaling and improves physiological resilience by maintaining membrane integrity, aiding osmotic adjustment, and enhancing detoxification processes (Zhou et al., 2014).

At the molecular scale, BR and ABA signaling intersect at several points, often through shared transcriptional regulators. For instance, BZR1, a key transcription factor in BR signaling, also governs genes responsive to ABA, indicating the presence of a co-regulated gene network (Sun et al., 2010). BRs additionally influence the expression and activity of heat shock transcription factors (HSFs), thereby coordinating with ABA to preserve cellular homeostasis during high-temperature stress (Mou et al., 2013). Overall, this carefully balanced hormonal interplay enables plants to adjust growth and defense strategies, activating either distinct or overlapping pathways depending on the type, severity, and duration of heat stress (Table 1).

6.2. BR Crosstalk with Salicylic Acid (SA) and Ethylene (ET) Under Heat Stress

Salicylic acid (SA) is widely known for its role in systemic acquired resistance (SAR) and for regulating plant responses to a range of biotic and abiotic stresses. In relation to heat stress, its cross-talk with brassinosteroids (BRs) has attracted increasing attention, with recent studies suggesting functional coordination between the two hormonal pathways. SA signaling appears essential for the effective expression of BR-mediated thermotolerance, since loss of SA sensitivity diminishes the protective effects of BRs under high-temperature conditions (Divi et al., 2010).

The cross-talk between BR and SA signaling during heat stress relies on shared regulators that link growth regulation with defense responses. A key hub in this interaction is NON-EXPRESSOR OF PATHOGENESIS-RELATED GENES 1 (NPR1), together with its downstream transcription factor WRKY70, which governs the expression of defense-related genes (Divi et al., 2010). The importance of this pathway is illustrated by the heat stress response of *npr1-1* mutants, which show heightened sensitivity and reduced induction of pathogenesis-related (PR) genes following SA treatment (Clarke et al., 2004; Larkindale et al., 2005; Clarke et al., 2009). When treated with exogenous BRs, these mutants displayed only a modest 2.4-fold improvement in survival compared to a ninefold increase observed in wild-type plants, underscoring the essential role of NPR1-mediated signaling in BR-induced thermotolerance (Divi et al., 2010). This interaction may also involve downstream signaling nodes such as BIN2 and BZR1, where NPR1 control could affect BR-related transcriptional outcomes under stress conditions.

Table 1. Summary of exogenous brassinosteroid (BR) applications under heat stress in various crops

Crop	BR type/Analog	Application method	Observed benefits under heat stress	References
Tomato	EBL, MH-5	Foliar spray	Increased mitochondrial sHSP, enhanced SOD activity, and protection of Rubisco	Singh and Shono (2005); Mazorra et al. (2002); Ogweno et al. (2008)
Arabidopsis	EBL	Pretreatment	Reduced bleaching and improved post-stress survival	Kagale et al. (2007)
Melon	EBL	Foliar spray	Improved CO_2 assimilation, pigment content, photochemical activity, and water-use efficiency	Zhang et al. (2013)
Eggplant	EBL (0.05–0.2 μM)	Foliar spray	Enhanced antioxidant defense and reduced ROS accumulation	Wu et al. (2014)
Wheat	EBL (0.01 μM)	Foliar spray	Increased biomass, photosynthetic efficiency, and antioxidant enzyme activity	Hussain et al. (2019)
Rice	αDHECD (0.0001 μM)	Applied at reproductive stage	Enhanced photosynthesis, carbohydrate content, seed setting, and seed weight	Sonjaroon et al. (2018)
Soybean	24-epibrassinolide	Foliar spray	Elevated SOD and CAT activity, reduced oxidative damage (MDA)	Wang et al. (2022)
Rice	Brassinosteroids	Foliar spray	Improved stomatal conductance, increased Fv/Fm ratio and panicle characteristics	Pantoja-Benavides et al. (2022)
Maize	100 ml of 0.25 mg/L BR (mainly composed of 24-epibrassinolide)	Foliar spray at post-silking stage	Improved accumulation, translocation, and distribution of biomass and nutrients	Xu et al. (2022)
Rice	24-epibrassinolide	Foliar spray at panicle initiation stage	Increased ABA production, enhance antioxidant activity	Yaanhua et al. (2023)
Wheat	24-epibrassinolide		Improved electron transport rate and efficiency of PSII	Jat et al. (2024)

(*Contd.*)

(*Contd.*)

Crop	**BR type/Analog**	**Application method**	**Observed benefits under heat stress**	**References**
Wheat	10 μmol·L−1 brassinosteroid (BRs),	Foliar spray at flowering period	increased wheat yield and improved starch quality	Wen et al., 2024
Grapevine (Kyoho)	200ppm 24-epibrassinolide (EBR)	Root application at flowering, young berry swelling and version stages	Mitigated PSII photoinhibition, lipid peroxidation, reduced malondialdehyde content, increased soluble protein content and fruit set	Sun et al., 2025
Maize	100 ml of 0.25 mg/L BR (mainly composed of 24-epibrassinolide)	Foliar spray at post-silking stage	Increased pasting viscosity, reduced gelatinization enthalpy and rapidly digestible starch content	Li et al., 2025

Physiological evidence also points toward a synergistic function of BRs and SA. Separate applications of each hormone have been reported to improve photosynthetic efficiency, mitigate oxidative damage, and enhance yield traits in crops exposed to heat stress. These comparable outcomes indicate that BR and SA may trigger overlapping or parallel signaling pathways, especially those linked to antioxidant enzyme activity and heat-responsive gene regulation (Sun et al., 2025). While the direct molecular interactions between BR and SA signaling are not yet fully clarified, the consistency of their individual impacts supports the idea of a coordinated role in thermotolerance.

Ethylene is an important regulator of plant adaptation to abiotic stress, functioning mainly through its control of antioxidant defense mechanisms and induction of heat shock proteins (HSPs). Although its interaction with brassinosteroids (BRs) has been observed under several stress conditions, the details of this relationship during heat stress are still not fully understood. Research suggests that BR signaling can influence ethylene biosynthesis and signaling by regulating the expression of ethylene-related genes, with transcription factors such as BZR1 playing a role in this regulation (Divi et al., 2010). This points to a wider hormonal network in which BRs interact with ethylene, abscisic acid (ABA), and other signaling molecules to orchestrate plant stress responses.

Under conditions of heat stress, BRs seem to act independently of ethylene. BR-mediated thermotolerance has been reported even in ethylene-insensitive mutants, indicating that ethylene signaling is not strictly required for BR-driven protection against high temperatures (Divi et al., 2010). This ethylene-independent role differs from its well-documented involvement in other stress situations, suggesting a selective and context-specific hormonal interaction. The capacity of BRs to enhance thermotolerance without dependence on ethylene highlights their distinct role in regulating defense mechanisms, including ROS detoxification, membrane stabilization, and activation of stress-responsive genes.

7. Genetic and Molecular Tools to Study BR-Mediated Heat Tolerance

The study of brassinosteroid (BR) signaling in relation to heat stress adaptation has advanced considerably through the use of genetic and molecular methods. Approaches such as mutant analysis, transgenic studies, and various omics technologies have proven particularly valuable in identifying the components and regulatory pathways involved in BR-mediated thermotolerance.

Mutants with defects in BR perception or signaling have provided valuable understanding of how plants use hormonal regulation to cope with heat stress. One of the first identified mutants was *bri1-4*, which carries a mutation in the BRASSINOSTEROID INSENSITIVE 1 (BRI1) receptor. This mutant displays a dwarf phenotype and accumulates high levels of endogenous brassinosteroids due to disrupted receptor-mediated feedback control. This phenotype emphasizes the essential role of BRI1 both in sensing BR signals and in preserving BR homeostasis under normal as well as stress conditions (Noguchi et al., 1999). The importance of other genes in signaling, including the BZR1 transcription factor and BR receptors, has also been revealed through the analysis of mutants that are either deficient in or overexpress these components (Chen et al., 2022).

Transgenic approaches that modify the expression of genes involved in BR biosynthesis and signaling have greatly improved our understanding of brassinosteroid (BR)-mediated responses to abiotic stress, especially heat. In rice, transgenic lines engineered to overexpress BR biosynthetic or signaling genes exhibited increased thermotolerance. These genetic modifications led to enhanced antioxidant enzyme activity and better photosynthetic performance, ultimately supporting higher survival and yield stability under high-temperature stress (Sharma et al., 2017). In *Arabidopsis*, the BRX gene has been shown to participate in crosstalk between BR and abscisic acid (ABA) signaling. Manipulation of BRX expression through genetic engineering altered ABA-responsive gene expression, suggesting that BR signaling influences stress adaptation partly through ABA-related pathways (Rodrigues, 2009). In potato, altering BR-associated genes through transgenic modification affected traits such as tuber dormancy and sprouting processes known to be sensitive to environmental cues like temperature (Di et al., 2024).

Omics-based approaches—including transcriptomics, proteomics, and metabolomics have collectively demonstrated the extensive regulatory role of BRs in shaping gene expression, protein activity, and metabolic adjustments under heat stress. Transcriptomic studies indicate that BRs upregulate a wide range of heat-responsive genes linked to antioxidant defenses, molecular chaperones, and hormonal interactions. In *Arabidopsis*, gene expression profiling revealed that BRs activate multiple defense pathways that contribute to thermotolerance (Divi et al., 2016). From a proteomic perspective, BRs increase the abundance of heat shock proteins and protein-folding enzymes, thereby safeguarding proteome stability during heat stress. These proteomic shifts are accompanied by metabolic reprogramming, which includes the accumulation of compatible solutes and secondary metabolites that promote osmoprotection and antioxidant activity (Siddiqui et al., 2019). In wheat, BR-driven transcriptional and proteomic modifications were shown to reinforce photosynthesis-related processes, aiding in the maintenance of cellular homeostasis under thermal stress (Margay et al., 2024). For example, BR application in *Gracilariopsis lemaneiformis* exposed to high-temperature stress upregulated genes involved in ROS scavenging and signaling, correlating with improved physiological tolerance (Chu et al., 2025).

Taken together, these strategies not only enhance our understanding of the mechanisms behind BR-mediated thermotolerance but also provide practical opportunities for improving crop resilience under heat stress driven by climate change.

8. Practical Applications and Agricultural Implications

Exogenous BR application supports plant resilience by stabilizing photosynthesis, improving antioxidant defenses, and protecting cellular integrity. In tomato, treatment with epibrassinolide (EBL) enhances mitochondrial heat shock proteins and photosynthetic efficiency (Singh and Shono 2005), while increased antioxidant enzyme activity and stabilization of Rubisco and Calvin cycle enzymes have been reported in multiple species (Mazorra et al., 2002; Ogweno et al., 2008). Arabidopsis seedlings treated with BRs exhibit reduced bleaching and improved survival under high temperatures (Kagale et al., 2007). Optimized formulations and integration with irrigation and nutrient practices have further improved BR performance under field conditions (Sharma et al., 2017). Similar effects have been observed in crops such as melon, eggplant, wheat, and rice, where Soybeans treated with BRs also display improved heat tolerance, marked by increased SOD and CAT activity and reduced oxidative damage (Wang et al., 2022). In wheat, BR application prior to heat stress helps maintain grain yield by preserving spike fertility (Kothari and Lachowiec, 2021).

Nevertheless, variable outcomes across studies highlight the importance of factors such as dosage, application method, BR compound, and timing (Kumari and Hemantaranjan, 2019). Excessive BR levels may inhibit growth, particularly root elongation, in multiple species (Clouse et al., 1996; Mussig et al., 2003; Trevisan et al., 2020). Additionally, the duration of BR effects may be transient, with limited persistence observed beyond one-week post-application (Nishikawa et al., 1994). Despite these challenges, BRs have gained commercial traction. For example, 2,4-epibrassinolide has been in agricultural use in Belarus and Russia since the early 1990s (Khripach et al., 2000). These compounds are also economically feasible and exhibit minimal toxicity to humans, animals, and ecosystems. BRs thus offer a promising, scientifically grounded approach to enhancing crop thermotolerance. Genetic approaches support these findings by confirming the role of BR signalling in thermotolerance. Manipulation of transcription factors such as BZR1 enhances heat response pathways via interaction with PIF4 (Oh et al., 2012). These traits, now accessible through gene editing technologies, can be incorporated into elite cultivars without affecting yield. Whether applied externally or introduced via genetic methods, brassinosteroids represent a vital component of modern strategies to sustain agricultural productivity in a warming climate.

9. Conclusion

Brassinosteroids (BRs) are key regulators in boosting plant tolerance to heat stress by coordinating growth-related and defense mechanisms. Their actions extend across hormonal interactions, regulation of gene expression, antioxidant systems, and metabolic adjustments—helping plants sustain physiological balance under high-temperature conditions. Both external application of BRs and genetic manipulation of BR signaling pathways have shown reliable advantages across diverse species in greenhouse and field studies. Although challenges such as variation in effectiveness and duration of action remain, BRs continue to represent a promising approach for climate-resilient crop production. Ongoing investigations into BR biosynthesis, signaling, and cross-talk with other hormones will enhance their application in breeding and developing heat-tolerant varieties to ensure future food security.

References

Abdelrahman, M., El-Sayed, M., Jogaiah, S., Burritt, D.J. & Tran, L. S. P. (2017). The "stay-green" trait and phytohormone signalling networks in plants under heat stress. Plant Cell Reports 36, 1009–1025. https://doi.org/10.1007/s00299-017-2119-y

Almeselmani, M., Deshmukh, P.S. & Chinnusamy, V. (2012). Effects of prolonged high temperature stress on respiration, photosynthesis and gene expression in wheat (*Triticum aestivum* L.) varieties differing in their thermotolerance. Plant Stress 6(1), 25–32.

Anwar, A., Liu, Y., Dong, R., Bai, L., Yu, X. & Li, Y. (2018). The physiological and molecular mechanism of brassinosteroid in response to stress: A review. Biological Research 51(1), 1–15. https://doi.org/10.1186/s40659-018-0195-2

Bajguz, A. & Hayat, S. (2009). Effects of brassinosteroids on the plant responses to environmental stresses. Plant Physiology and Biochemistry 47(1), 1–8. https://doi.org/10.1016/j.plaphy.2008.10.002

Bajguz, A. & Tretyn, A. (2003). The chemical structures and occurrence of brassinosteroids in plants. *In*: S. Hayat & A. Ahmad [eds.]. Brassinosteroids: Bioactivity and Crop Productivity (pp. 1–44). Springer. https://doi.org/10.1007/978-94-017-0948-4_1

Bajguz, A., Chmur, M. & Gruszka, D. (2020). Comprehensive overview of the brassinosteroid biosynthesis pathways: Substrates, products, inhibitors, and connections. Frontiers in Plant Science 11, 1034. https://doi.org/10.3389/fpls.2020.01034

Belda-Palazon, B., Gonzalez-Garcia, M.P., Lozano-Juste, J., Coego, A., Antoni, R., Julian, J., ... & Rodriguez, P.L. (2018). PYL8 mediates ABA perception in the root through non-cell-autonomous and ligand-stabilization–based mechanisms. Proceedings of the National Academy of Sciences 115(50), E11857–E11863. https://doi.org/10.1073/pnas.1815410115

Bhatt, D., Sharma, M., Nath, M., Bhatt, M.D. & Badoni, S. (2019). Brassinosteroid signalling and complex interplay of ROS, NADPH oxidase, and MAPK mediated biotic and abiotic stress acclimation in plants. *In*: Plant Signalling Molecules (pp. 407–416). Woodhead Publishing. https://doi.org/10.1016/B978-0-12-816451-8.00025-3

Camejo, D., Rodríguez, P., Morales, M.A., Dell'Amico, J.M., Torrecillas, A. & Alarcón, J.J. (2005). High temperature effects on photosynthetic activity of two tomato cultivars with different heat susceptibility. Journal of Plant Physiology 162(3), 281–289. https://doi.org/10.1016/j.jplph.2004.07.014

Chaudhuri, A., Halder, K., Abdin, M.Z., Majee, M. & Datta, A. (2022). Abiotic stress tolerance in plants: Brassinosteroids navigate competently. International Journal of Molecular Sciences 23(23), 14577. https://doi.org/10.3390/ijms232314577

Chauhan, H., Khurana, N., Nijhavan, A., Khurana, J.P. & Khurana, P. (2012). The wheat chloroplastic small heat shock protein (sHSP26) is involved in seed maturation and germination and imparts tolerance to heat stress. Plant, Cell & Environment 35(11), 1912–1931. https://doi.org/10.1111/j.1365-3040.2012.02525.x

Chauhan, J., Srivastava, J. P., Singhal, R. K., Soufan, W., Dadarwal, B. K., Mishra, U. N., ... & Sabagh, A. E. (2022). Alterations of oxidative stress indicators, antioxidant enzymes, soluble sugars, and amino acids in mustard [*Brassica juncea* (L.) Czern and Coss.] in response to varying sowing time, and field temperature. Frontiers in Plant Science 13, 875009. https://doi.org/10.3389/fpls.2022.875009

Chen, E., Yang, X., Liu, R., Zhang, M., Zhang, M., Zhou, F., ... & Li, C. (2022). *GhBEE3-like* gene regulated by brassinosteroids is involved in cotton drought tolerance. Frontiers in Plant Science 13, 1019146. https://doi.org/10.3389/fpls.2022.1019146

Choudhary, S.P., Yu, J.Q., Yamaguchi-Shinozaki, K., Shinozaki, K. & Tran, L.S.P. (2012). Benefits of brassinosteroid crosstalk. Trends in Plant Science 17(10), 594–605.

Chu, K., Dong, J., Li, H. & Zhao, C. (2025). Exogenous BR enhances host resistance to *Rhizoctonia solani* through reactive oxygen species homeostasis and photosynthesis improvement in *Oryza sativa*. Phytopathology. https://doi.org/10.1094/PHYTO-10-24-0322-R

Clarke, S.M., Cristescu, S.M., Miersch, O., Harren, F.J., Wasternack, C. & Mur, L.A. (2009). Jasmonates act with salicylic acid to confer basal thermotolerance in *Arabidopsis thaliana*. New Phytologist 182(1), 175–187. https://doi.org/10.1111/j.1469-8137.2008.02735.x

Clarke, S.M., Mur, L.A., Wood, J.E. & Scott, I.M. (2004). Salicylic acid dependent signalling promotes basal thermotolerance but is not essential for acquired thermotolerance in *Arabidopsis thaliana*. The Plant Journal 38(3), 432–447. https://doi.org/10.1111/j.1365-313X.2004.02054.x

Clouse, S. D. (2015). A history of brassinosteroid research from 1970 through 2005: Thirty-five years of phytochemistry, physiology, genes, and mutants. *Journal of Plant Growth Regulation, 34*, 828–844. https://doi.org/10.1007/s00344-015-9540-7

Clouse, S.D. & Sasse, J.M. (1998). Brassinosteroids: Essential regulators of plant growth and development. Annual Review of Plant Biology 49(1), 427–451. https://doi.org/10.1146/annurev.arplant.49.1.427

Clouse, S.D., Langford, M. & McMorris, T.C. (1996). A brassinosteroid-insensitive mutant in *Arabidopsis thaliana* exhibits multiple defects in growth and development. Plant Physiology 111(3), 671–678. https://doi.org/10.1104/pp.111.3.671

Dhaubhadel, S., Browning, K.S., Gallie, D.R. & Krishna, P. (2002). Brassinosteroid functions to protect the translational machinery and heat-shock protein synthesis following thermal stress. The Plant Journal 29(6), 681–691. https://doi.org/10.1046/j.1365-313X.2002.01257.x

Dhaubhadel, S., Chaudhary, S., Dobinson, K.F. & Krishna, P. (1999). Treatment with 24-epibrassinolide, a brassinosteroid, increases the basic thermotolerance of *Brassica napus* and tomato seedlings. Plant Molecular Biology 40, 333–342. https://doi.org/10.1023/A:1006283015582

Di, X., Wang, Q., Zhang, F., Feng, H., Wang, X. & Cai, C. (2024). Advances in the modulation of potato tuber dormancy and sprouting. International Journal of Molecular Sciences 25(10), 5078. https://doi.org/10.3390/ijms25105078

Divi, U.K. & Krishna, P. (2009). Brassinosteroid: A biotechnological target for enhancing crop yield and stress tolerance. New Biotechnology 26(3–4), 131–136. https://doi.org/10.1016/j.nbt.2009.07.006

Divi, U.K., Rahman, T. & Krishna, P. (2010). Brassinosteroid-mediated stress tolerance in *Arabidopsis* shows interactions with abscisic acid, ethylene and salicylic acid pathways. BMC Plant Biology 10, 1–14. https://doi.org/10.1186/1471-2229-10-151

Divi, U.K., Rahman, T. & Krishna, P. (2016). Gene expression and functional analyses in brassinosteroid-mediated stress tolerance. Plant Biotechnology Journal 14(1), 419–432. https://doi.org/10.1111/pbi.12396

Domagalska, M.A., Sarnowska, E., Nagy, F. & Davis, S.J. (2010). Genetic analyses of interactions among gibberellin, abscisic acid, and brassinosteroids in the control of flowering time in *Arabidopsis thaliana*. PLOS One 5(11), e14012. https://doi.org/10.1371/journal.pone.0014012

Duc, N.H., Csintalan, Z. & Posta, K. (2018). Arbuscular mycorrhizal fungi mitigate negative effects of combined drought and heat stress on tomato plants. Plant Physiology and Biochemistry 132, 297–307. https://doi.org/10.1016/j.plaphy.2018.09.011

Fan, X.Y., Sun, Y., Cao, D.M., Bai, M.Y., Luo, X.M., Yang, H.J., ... & Wang, Z.Y. (2012). BZS1, a B-box protein, promotes photomorphogenesis downstream of both brassinosteroid and light signalling pathways. Molecular Plant 5(3), 591–600. https://doi.org/10.1093/mp/sss041

Fariduddin, Q., Yusuf, M., Ahmad, I. & Ahmad, A. (2014). Brassinosteroids and their role in response of plants to abiotic stresses. Biologia Plantarum 58(1), 9–17. https://doi.org/10.1007/s10535-013-0374-5

Fujioka, S. & Yokota, T. (2003). Biosynthesis and metabolism of brassinosteroids. Annual Review of Plant Biology 54(1), 137–164. https://doi.org/10.1146/annurev.arplant.54.031902.134921

González-García, M.P., Vilarrasa Blasi, J., Zhiponova, M., Divol, F., Mora García, S., Russinova, E. & Caño Delgado, A.I. (2011). Brassinosteroids control meristem size by promoting cell cycle progression in *Arabidopsis* roots. Development 138(5), 849–859. https://doi.org/10.1242/dev.057331

González-Olmedo, J.L., Córdova, A., Aragón, C.E., Pina, D., Rivas, M. & Rodríguez, R. (2005). Effect of an analogue of brassinosteroid on FHIA-18 plantlets exposed to thermal stress. Infomusa 14(1), 18–20.

Grove, M.D., Spencer, G.F., Rohwedder, W.K., Mandava, N., Worley, J.F., Warthen, J.D.J. et al. (1979). Brassinolide, a plant growth-promoting steroid isolated from *Brassica napus* pollen. Nature 281, 216–217.

Guo, D.L., Wang, Z.G., Li, Q., Gu, S.C., Zhang, G.H. & Yu, Y.H. (2019). Hydrogen peroxide treatment promotes early ripening of Kyoho grape. Australian Journal of Grape and Wine Research 25(3), 357–362. https://doi.org/10.1111/ajgw.12399

Halaji, B., Haghighi, M., Kovács, G.P., Mirmazloum, I. & Szegő, A. (2024). The role of brassinosteroids and nano-encapsulated brassinosteroids in *Capsicum* pepper growth and physiological adaptations to high-temperature stress. Horticulturae 10(10), 1062. https://doi.org/10.3390/horticulturae10101062

Hasanuzzaman, M., Nahar, K., Alam, M.M., Roychowdhury, R. & Fujita, M. (2013). Physiological, biochemical, and molecular mechanisms of heat stress tolerance in plants. International Journal of Molecular Sciences 14(5), 9643–9684. https://doi.org/10.3390/ijms14059643

Haubrick, L.L. & Assmann, S.M. (2006). Brassinosteroids and plant function: Some clues, more puzzles. Plant, Cell & Environment 29, 446–457. https://doi.org/10.1111/j.1365-3040.2005.01481.x

Hayat, S., Ahmad, A. & Fariduddin, Q. (2003). Brassinosteroids: A regulator of 21st century. *In*: S. Hayat & A. Ahmad [eds.]. Brassinosteroids (pp. 231–246). Springer. https://doi.org/10.1007/978-94-017-0948-4_10

Hayat, S., Yadav, S., Ali, B. & Ahmad, A. (2010). Interactive effect of nitric oxide and brassinosteroids on photosynthesis and the antioxidant system of *Lycopersicon esculentum*. Russian Journal of Plant Physiology 57, 212–221. https://doi.org/10.1134/S1021443710020081

Heckathorn, S.A., Downs, C.A. & Coleman, J.S. (1999). Small heat shock proteins protect electron transport in chloroplasts and mitochondria during stress. American Zoologist 39(6), 865–876. https://doi.org/10.1093/icb/39.6.865

Hu, X., Jiang, M., Zhang, J., Zhang, A., Lin, F. & Tan, M. (2007). Calcium–calmodulin is required for abscisic acid-induced antioxidant defense and functions both upstream and downstream of H_2O_2 production in leaves of maize (*Zea mays*) plants. New Phytologist 173(1), 27–38. https://doi.org/10.1111/j.1469-8137.2006.01888.x

Hu, Z.L., Wei, H., Sun, L. & Russinova, E. (2025). Plant steroids on the move: Mechanisms of brassinosteroid export. Trends in Biochemical Sciences 50(6), 508–519. https://doi.org/10.1016/j.tibs.2025.03.003

Hussain, M., Khan, T.A., Yusuf, M. & Fariduddin, Q. (2019). Silicon-mediated role of 24-epibrassinolide in wheat under high-temperature stress. Environmental Science and Pollution Research 26, 17163–17172. https://doi.org/10.1007/s11356-019-04938-0

Ibañez, C., Delker, C., Martinez, C., Bürstenbinder, K., Janitza, P., Lippmann, R., ... & Quint, M. (2018). Brassinosteroids dominate hormonal regulation of plant thermomorphogenesis via BZR1. Current Biology 28(2), 303–310. https://doi.org/10.1016/j.cub.2017.11.077

Janeczko, A., Biesaga-Koscielniak, J., Dziurka, M., Oklestkova, J., Kocurek, M., Szarek-Lukaszewska, G. & Janeczko, Z. (2011). Response of Polish cultivars of soybean (*Glycine max* (L.) Merr.) to brassinosteroid application. Acta Scientiarum Polonorum. Agricultura 10(2).

Janeczko, A., Oklešťková, J., Pociecha, E., Kościelniak, J. & Mirek, M. (2011). Physiological effects and transport of 24-epibrassinolide in heat-stressed barley. Acta Physiologiae Plantarum 33, 1249–1259. https://doi.org/10.1007/s11738-010-0655-y

Jantapo, K., Wimonchaijit, W., Wang, W. & Chaiwanon, J. (2021). Supraoptimal brassinosteroid levels inhibit root growth by reducing root meristem and cell elongation in rice. Plants 10(9), 1962. https://doi.org/10.3390/plants10091962

Jat, M., Ray, M., Ahmad, M.A. & Prakash, P. (2024). Unravelling the photosynthetic dynamics and fluorescence parameters under ameliorative effects of 24-epibrassinolide in wheat (*Triticum aestivum* L.) grown under heat stress regime. Scientific Reports 14(1), 30745. https://doi.org/10.1038/s41598-024-79676-6

Jiang, Y.P., Huang, L.F., Cheng, F., Zhou, Y.H., Xia, X.J., Mao, W.H., ... & Yu, J.Q. (2013). Brassinosteroids accelerate recovery of photosynthetic apparatus from cold stress by balancing the electron partitioning, carboxylation and redox homeostasis in cucumber. Physiologia Plantarum 148(1), 133–145. https://doi.org/10.1111/j.1399-3054.2012.01696.x

Jin, S.H., Li, X.Q., Wang, G.G. & Zhu, X.T. (2015). Brassinosteroids alleviate high-temperature injury in *Ficus concinna* seedlings via maintaining higher antioxidant defence and glyoxalase systems. AoB Plants 7, plv009. https://doi.org/10.1093/aobpla/plv009

Joo, S.H., Jang, M.S., Kim, M.K., Lee, J.E. & Kim, S.K. (2015). Biosynthetic relationship between C28-brassinosteroids and C29-brassinosteroids in rice (*Oryza sativa*) seedlings. Phytochemistry 111, 84–90. https://doi.org/10.1016/j.phytochem.2014.11.006

Joo, S.H., Kim, T.W., Son, S.H., Lee, W.S., Yokota, T. & Kim, S.K. (2012). Biosynthesis of a cholesterol-derived brassinosteroid, 28-norcastasterone, in *Arabidopsis thaliana*. Journal of Experimental Botany 63(5), 1823–1833. https://doi.org/10.1093/jxb/err354

Kagale, S., Divi, U.K., Krochko, J.E., Keller, W.A. & Krishna, P. (2007). Brassinosteroid confers tolerance in *Arabidopsis thaliana* and *Brassica napus* to a range of abiotic stresses. Planta 225(2), 353–364. https://doi.org/10.1007/s00425-006-0361-6

Kang, X., Zhao, L. & Liu, X. (2023). Calcium signalling and the response to heat shock in crop plants. International Journal of Molecular Sciences 25(1), 324. https://doi.org/10.3390/ijms25010324

Kaur, R., Ohri, P. & Bhardwaj, R. (2013). Effect of 28-homobrassinolide on susceptible and resistant cultivars of tomato after nematode inoculation. Plant Growth Regulation 71, 199–205. https://doi.org/10.1007/s10725-013-9820-9

Khripach, V., Zhabinskii, V. & de Groot, A. (2000). Twenty years of brassinosteroids: Steroidal plant hormones warrant better crops for the XXI century. Annals of Botany 86(3), 441–447. https://doi.org/10.1006/anbo.2000.1227

Kim, Y., Park, S.U., Shin, D.M., Pham, G., Jeong, Y.S. & Kim, S.H. (2020). ATBS1-INTERACTING FACTOR 2 negatively regulates dark- and brassinosteroid-induced leaf senescence through interactions with INDUCER OF CBF EXPRESSION 1. Journal of Experimental Botany 71(4), 1475–1490. https://doi.org/10.1093/jxb/erz533

Kolupaev, Y.E., Vayner, A.A., Yastreb, T.O., Oboznyi, A.I. & Khripach, V.A. (2014). The role of reactive oxygen species and calcium ions in the implementation of the stress-protective effect of brassinosteroids on plant cells. Applied Biochemistry and Microbiology 50, 658–663. https://doi.org/10.1134/S0003683814060076

Kothari, A. & Lachowiec, J. (2021). Roles of brassinosteroids in mitigating heat stress damage in cereal crops. International Journal of Molecular Sciences 22(5), 2706. https://doi.org/10.3390/ijms22052706

Kumar, H., Chugh, V., Kumar, M., Gupta, V., Prasad, S., Kumar, S., … Kumar, M. (2023). Investigating the impact of terminal heat stress on contrasting wheat cultivars: A comprehensive analysis of phenological, physiological, and biochemical traits. Frontiers in Plant Science 14, 1189005. https://doi.org/10.3389/fpls.2023.1189005

Kumari, A. & Hemantaranjan, A. (2019). Brassinosteroids: The master regulators. *In*: A. Hemantaranjan [ed.]. Advances in Plant Physiology (Vol. 19, pp. 273). Scientific Publishers.

Kurepin, L.V., Qaderi, M.M., Back, T.G., Reid, D.M. & Pharis, R.P. (2008). A rapid effect of applied brassinolide on abscisic acid concentrations in *Brassica napus* leaf tissue subjected to short-term heat stress. Plant Growth Regulation 55, 165–167. https://doi.org/10.1007/s10725-008-9276-5

Larkindale, J. & Knight, M.R. (2002). Protection against heat stress-induced oxidative damage in *Arabidopsis* involves calcium, abscisic acid, ethylene, and salicylic acid. Plant Physiology 128(2), 682–695. https://doi.org/10.1104/pp.010320

Larkindale, J., Hall, J.D., Knight, M.R. & Vierling, E. (2005). Heat stress phenotypes of *Arabidopsis* mutants implicate multiple signalling pathways in the acquisition of thermotolerance. Plant Physiology 138(2), 882–897. https://doi.org/10.1104/pp.105.062257

Li, J., Nagpal, P., Vitart, V., McMorris, T.C. & Chory, J. (1996). A role for brassinosteroids in light-dependent development of *Arabidopsis*. Science 272(5260), 398–401. https://doi.org/10.1126/science.272.5260.398

Li, J. & Chory, J. (1997). A putative leucine-rich repeat receptor kinase involved in brassinosteroid signal transduction. Cell 90(5), 929–938.

Li, J., Wen, J., Lease, K.A., Doke, J.T., Tax, F.E. & Walker, J.C. (2002). BAK1, an *Arabidopsis* LRR receptor-like protein kinase, interacts with BRI1 and modulates brassinosteroid signalling. Cell 110(2), 213–222. https://doi.org/10.1016/S0092-8674(02)00812-7

Li, J., Guo, J., Liu, J., Qu, L., Li, G., Yang, H. & Lu, D. (2025). Exogenous brassinosteroids improve the structure and quality properties of waxy maize starch under post-silking heat and drought stress. Food Chemistry 490, 145093. https://doi.org/10.1016/j.foodchem.2025.145093

Li, L., Ye, H., Guo, H. & Yin, Y. (2010). *Arabidopsis* IWS1 interacts with transcription factor BES1 and is involved in plant steroid hormone brassinosteroid regulated gene expression. Proceedings of the National Academy of Sciences 107(8), 3918–3923. https://doi.org/10.1073/pnas.0909198107

Li, X., Wei, J.P., Ahammed, G.J., Zhang, L., Li, Y., Yan, P., … Han, W.Y. (2018). Brassinosteroids attenuate moderate high temperature-caused decline in tea quality by enhancing theanine biosynthesis in *Camellia sinensis* L. Frontiers in Plant Science 9, 1016. https://doi.org/10.3389/fpls.2018.01016

Luo, J., Jiang, J., Sun, S. & Wang, X. (2022). Brassinosteroids promote thermotolerance through releasing BIN2-mediated phosphorylation and suppression of HsfA1 transcription factors in *Arabidopsis*. Plant Communications 3(6), 100419. https://doi.org/10.1016/j.xplc.2022.100419

Luo, X.M., Lin, W.H., Zhu, S., Zhu, J.Y., Sun, Y., Fan, X.Y., … Wang, Z.Y. (2010). Integration of light- and brassinosteroid-signalling pathways by a GATA transcription factor in *Arabidopsis*. Developmental Cell 19(6), 872–883. https://doi.org/10.1016/j.devcel.2010.10.023

Lv, J., Dong, T., Zhang, Y., Ku, Y., Zheng, T., Jia, H. & Fang, J. (2022). Metabolomic profiling of brassinolide and abscisic acid in response to high-temperature stress. Plant Cell Reports 41(4), 935–946. https://doi.org/10.1007/s00299-022-02829-2

Manghwar, H., Hussain, A., Ali, Q. & Liu, F. (2022). Brassinosteroids (BRs) role in plant development and coping with different stresses. International Journal of Molecular Sciences 23(3), 1012. https://doi.org/10.3390/ijms23031012

Margay, A.R., Mehmood, A. & Bashir, L. (2024). Review on hormonal regulation of drought stress response in plants. International Journal of Plant & Soil Science, 36, 902–916. https://doi.org/10.9734/ijpss/2024/v36i84921

Martínez, C., Espinosa-Ruíz, A., de Lucas, M., Bernardo-García, S., Franco-Zorrilla, J.M. & Prat, S. (2018). PIF4-induced BR synthesis is critical to diurnal and thermomorphogenic growth. The EMBO Journal 37(23), e99552. https://doi.org/10.15252/embj.201899552

Mazorra, L.M., Holton, N., Bishop, G.J. & Núñez, M. (2011). Heat shock response in tomato brassinosteroid mutants indicates that thermotolerance is independent of brassinosteroid homeostasis. Plant Physiology and Biochemistry 49(12), 1420–1428. https://doi.org/10.1016/j.plaphy.2011.09.005

Mazorra, L.M., Nunez, M., Hechavarria, M., Coll, F. & Sánchez-Blanco, M.J. (2002). Influence of brassinosteroids on antioxidant enzymes activity in tomato under different temperatures. Biologia Plantarum 45, 593–596. https://doi.org/10.1023/A:1022390917656

Mishra, S., Spaccarotella, K., Gido, J., Samanta, I. & Chowdhary, G. (2023). Effects of heat stress on plant-nutrient relations: An update on nutrient uptake, transport, and assimilation. International Journal of Molecular Sciences 24(21), 15670. https://doi.org/10.3390/ijms242115670

Mondal, S., Karmakar, S., Panda, D., Pramanik, K., Bose, B. & Singhal, R.K. (2023). Crucial plant processes under heat stress and tolerance through heat shock proteins. Plant Stress 10, 100227. https://doi.org/10.1016/j.stress.2023.100227

Mou, L., Wang, Y., Li, H., Huang, Y., Jiang, T., Huang, W., … Gui, Y. (2013). A dominant-negative mutation of HSF2 associated with idiopathic azoospermia. Human Genetics 132, 159–165. https://doi.org/10.1007/s00439-012-1234-7

Müssig, C., Shin, G.H. & Altmann, T. (2003). Brassinosteroids promote root growth in *Arabidopsis*. Plant Physiology 133(3), 1261–1271. https://doi.org/10.1104/pp.103.028662

Nazir, F., Jahan, B., Kumari, S., Iqbal, N., Albaqami, M., Sofo, A. & Khan, M.I.R. (2023). Brassinosteroid modulates ethylene synthesis and antioxidant metabolism to protect rice (*Oryza sativa*) against heat stress-induced inhibition of source–sink capacity and photosynthetic and growth attributes. Journal of Plant Physiology 289, 154096. https://doi.org/10.1016/j.jplph.2023.154096

Nishikawa, N., Toyama, S., Shida, A. & Futatsuya, F. (1994). The uptake and the transport of 14C-labeled epibrassinolide in intact seedlings of cucumber and wheat. Journal of Plant Research 107, 125–130.

Noguchi, T., Fujioka, S., Choe, S., Takatsuto, S., Yoshida, S., Yuan, H., … Tax, F.E. (1999). Brassinosteroid-insensitive dwarf mutants of *Arabidopsis* accumulate brassinosteroids. Plant Physiology 121(3), 743–752. https://doi.org/10.1104/pp.121.3.743

Nolan, T.M., Vukasinovic, N., Liu, D., Russinova, E. & Yin, Y. (2020). Brassinosteroids: Multidimensional regulators of plant growth, development, and stress responses. The Plant Cell 32(2), 295–318. https://doi.org/10.1105/tpc.19.00335

Northey, J.G., Liang, S., Jamshed, M., Deb, S., Foo, E., Reid, J.B. et al. (2016). Farnesylation mediates brassinosteroid biosynthesis to regulate abscisic acid responses. Nature Plants 2, Article 16114. https://doi.org/10.1038/nplants.2016.114

Ogweno, J.O., Song, X.S., Shi, K., Hu, W.H., Mao, W.H., Zhou, Y.H., … Nogués, S. (2008). Brassinosteroids alleviate heat-induced inhibition of photosynthesis by increasing carboxylation efficiency and enhancing antioxidant systems in *Lycopersicon esculentum*. Journal of Plant Growth Regulation 27, 49–57. https://doi.org/10.1007/s00344-007-9030-7

Oh, E., Zhu, J.Y. & Wang, Z.Y. (2012). Interaction between BZR1 and PIF4 integrates brassinosteroid and environmental responses. Nature Cell Biology 14(8), 802–809. https://doi.org/10.1038/ncb2545

Ostmeyer, T., Parker, N., Jaenisch, B., Alkotami, L., Bustamante, C. & Jagadish, S.K. (2020). Impacts of heat, drought, and their interaction with nutrients on physiology, grain yield, and quality in field crops. Plant Physiology Reports 25(4), 549–568. https://doi.org/10.1007/s40502-020-00538-0

Pantoja-Benavides, A.D., Garces-Varon, G. & Restrepo-Díaz, H. (2022). Foliar cytokinins or brassinosteroids applications influence the rice plant acclimatization to combined heat stress. Frontiers in Plant Science 13, 983276. https://doi.org/10.3389/fpls.2022.983276

Percio, F., Rubio, L., Amorim-Silva, V. & Botella, M.A. (2025). Crucial roles of brassinosteroids in cell wall composition and structure across species: New insights and biotechnological applications. Plant, Cell & Environment 48(3), 1751–1767. https://doi.org/10.1111/pce.15258

Pereira-Netto, A.B., Roessner, U., Fujioka, S., Bacic, A., Asami, T., Yoshida, S. & Clouse, S.D. (2009). Shooting control by brassinosteroids: Metabolomic analysis and effect of brassinazole on *Malus prunifolia*, the Marubakaido apple rootstock. Tree Physiology 29(4), 607–620. https://doi.org/10.1093/treephys/tpn052

Planas-Riverola, A., Gupta, A., Betegón-Putze, I., Bosch, N., Ibañes, M. & Caño-Delgado, A.I. (2019). Brassinosteroid signalling in plant development and adaptation to stress. Development 146(5), dev169870. https://doi.org/10.1242/dev.169870

Rahman, M.S. & Rahman, M.S. (2021). Effects of elevated temperature on prooxidant-antioxidant homeostasis and redox status in the American oyster: Signalling pathways of cellular apoptosis during heat stress. Environmental Research 196, 110428. https://doi.org/10.1016/j.envres.2020.110428

Rajewska, I., Talarek, M. & Bajguz, A. (2016). Brassinosteroids and response of plants to heavy metals action. Frontiers in Plant Science 7, 629. https://doi.org/10.3389/fpls.2016.00629

Ravikiran, K.T., Gopala Krishnan, S., Vinod, K.K., Dhawan, G., Dwivedi, P., Kumar, P., … Singh, A.K. (2020). A trait-specific QTL survey identifies NL44, a NERICA cultivar as a novel source for reproductive stage heat stress tolerance in rice. Plant Physiology Reports 25, 664–676. https://doi.org/10.1007/s40502-020-00547-z

Rezaul, I.M., Baohua, F., Tingting, C., Weimeng, F., Caixia, Z., Longxing, T. & Guanfu, F. (2019). Abscisic acid prevents pollen abortion under high-temperature stress by mediating sugar metabolism in rice spikelets. Physiologia Plantarum 165(3), 644–663. https://doi.org/10.1111/ppl.12759

Rodrigues, A.D.P. (2009). Abscisic acid signal transduction: Regulation by HAB1 and interaction with brassinosteroids mediated by BRX. Doctoral Thesis, Polytechnic University of Valencia, Spain.

Rozhon, W., Akter, S., Fernandez, A. & Poppenberger, B. (2019). Inhibitors of brassinosteroid biosynthesis and signal transduction. Molecules 24(23), 4372. https://doi.org/10.3390/molecules24234372

Sadura, I. & Janeczko, A. (2018). Physiological and molecular mechanisms of brassinosteroid-induced tolerance to high and low temperature in plants. Biologia Plantarum 62(4), 601–616. https://doi.org/10.1007/s10535-018-0805-4

Saha, G., Mostofa, M.G., Rahman, M.M. & Tran, L.S.P. (2021). Silicon-mediated heat tolerance in higher plants: A mechanistic outlook. Plant Physiology and Biochemistry 166, 341–347. https://doi.org/10.1016/j.plaphy.2021.05.051

Samakovli, D., Margaritopoulou, T., Prassinos, C., Milioni, D. & Hatzopoulos, P. (2014). Brassinosteroid nuclear signalling recruits HSP 90 activity. New Phytologist 203(3), 743–757. https://doi.org/10.1111/nph.12843

Sattar, A., Sher, A., Ijaz, M., Ul-Allah, S., Rizwan, M.S., Hussain, M., ... & Cheema, M.A. (2020). Terminal drought and heat stress alter physiological and biochemical attributes in flag leaf of bread wheat. PLoS One 15(5), e0232974. https://doi.org/10.1371/journal.pone.0232974

Sharma, A., Boise, L.H. & Shanmugam, M. (2019). Cancer metabolism and the evasion of apoptotic cell death. Cancers 11(8), 1144. https://doi.org/10.3390/cancers11081144

Sharma, D., Singh, R., Tiwari, R., Kumar, R. & Gupta, V.K. (2019). Wheat responses and tolerance to terminal heat stress: A review. *In*: M. Hasanuzzaman, K. Nahar & M.A. Hossain [eds.]. Wheat Production in Changing Environments: Responses, Adaptation and Tolerance (pp. 149–173). Springer. https://doi.org/10.1007/978-981-13-6883-7_7

Sharma, I., Kaur, N. & Pati, P.K. (2017). Brassinosteroids: A promising option in deciphering remedial strategies for abiotic stress tolerance in rice. Frontiers in Plant Science 8, 2151. https://doi.org/10.3389/fpls.2017.02151

Shigeta, T., Zaizen, Y., Sugimoto, Y., Nakamura, Y., Matsuo, T. & Okamoto, S. (2015). Heat shock protein 90 acts in brassinosteroid signalling through interaction with BES1/BZR1 transcription factor. Journal of Plant Physiology 178, 69–73. https://doi.org/10.1016/j.jplph.2015.05.003

Shimada, Y., Fujioka, S., Miyauchi, N., Kushiro, M., Takatsuto, S., Nomura, T., ... & Yoshida, S. (2001). Brassinosteroid-6-oxidases from *Arabidopsis* and tomato catalyze multiple C-6 oxidations in brassinosteroid biosynthesis. Plant Physiology 126(2), 770–779. https://doi.org/10.1104/pp.126.2.770

Siddiqui, H., Sami, F., Juan, H.F. & Hayat, S. (2019). Brassinosteroid regulated physiological process: An omics perspective. *In*: Brassinosteroids: Plant Growth and Development (pp. 297–322). Springer. https://doi.org/10.1007/978-981-13-6058-9_10

Singh, I. & Shono, M. (2005). Physiological and molecular effects of 24-epibrassinolide, a brassinosteroid on thermotolerance of tomato. Plant Growth Regulation 47, 111–119. https://doi.org/10.1007/s10725-005-3252-0

Sonjaroon, W., Jutamanee, K., Khamsuk, O., Thussagunpanit, J., Kaveeta, L. & Suksamrarn, A. (2018). Impact of brassinosteroid mimic on photosynthesis, carbohydrate content and rice seed set at reproductive stage under heat stress. Agriculture and Natural Resources 52(3), 234–240. https://doi.org/10.1016/j.anres.2018.09.001

Steber, C.M. & McCourt, P. (2001). A role for brassinosteroids in germination in *Arabidopsis*. Plant Physiology 125(2), 763–769. https://doi.org/10.1104/pp.125.2.763

Sun, Y., Fan, X.Y., Cao, D.M., Tang, W., He, K., Zhu, J.Y., ... & Wang, Z.Y. (2010). Integration of brassinosteroid signal transduction with the transcription network for plant growth regulation in *Arabidopsis*. Developmental Cell 19(5), 765–777. https://doi.org/10.1016/j.devcel.2010.10.010

Sun, Y., Sun, S., Zahid, M.S., Qiu, Q., Wang, L. & Wang, S. (2025). Root-applied brassinosteroid and salicylic acid enhance thermotolerance and fruit quality in heat-stressed 'Kyoho' grapevines. Frontiers in Plant Science 16, 1563270. https://doi.org/10.3389/fpls.2025.1563270

Taiz, L. & Zeiger, E. (2010). Plant Physiology (5th ed.). Sinauer Associates.

Tang, J., Han, Z. & Chai, J. (2016). Q&A: What are brassinosteroids and how do they act in plants? BMC Biology 14, 1–5. https://doi.org/10.1186/s12915-016-0340-8

Tang, X., Wang, C., Liu, Y., He, G., Ma, N., Chai, G., ... & Zhou, G. (2020). Brassinosteroid signalling converges with auxin-mediated C3H17 to regulate xylem formation in *Populus*. Frontiers in Plant Science 11, 586014. https://doi.org/10.3389/fpls.2020.586014

Tang, Y., Liu, H., Guo, S., Wang, B., Li, Z., Chong, K. & Xu, Y. (2018). OsmiR396d affects gibberellin and brassinosteroid signalling to regulate plant architecture in rice. Plant Physiology 176(1), 946–959. https://doi.org/10.1104/pp.17.00964

Tayade, R., Nguyen, T., Oh, S.A., Hwang, Y.S., Yoon, I.S., Deshmuk, R., ... & Park, S.K. (2018). Effective strategies for enhancing tolerance to high-temperature stress in rice during the reproductive and ripening stages. Plant Breeding and Biotechnology 6(1), 1–18.

Thussagunpanit, J., Jutamanee, K., Sonjaroon, W., Kaveeta, L., Chai-Arree, W., Pankean, P. & Suksamrarn, A. (2015). Effects of brassinosteroid and brassinosteroid mimic on photosynthetic efficiency and rice yield under heat stress. Photosynthetica 53, 312–320. https://doi.org/10.1007/s11099-015-0106-5

Tong, H. & Chu, C. (2018). Functional specificities of brassinosteroid and potential utilization for crop improvement. Trends in Plant Science 23(11), 1016–1028. https://doi.org/10.1016/j.tplants.2018.08.007

Trevisan, S., Forestan, C., Brojanigo, S., Quaggiotti, S. & Varotto, S. (2020). Brassinosteroid application affects the growth and gravitropic response of maize by regulating gene expression in the roots, shoots and leaves. Plant Growth Regulation 92(1), 117–130. https://doi.org/10.1007/s10725-020-00626-z

Vardhini, B.V. & Anjum, N.A. (2015). Brassinosteroids make plant life easier under abiotic stresses mainly by modulating major components of antioxidant defense system. Frontiers in Environmental Science 2, 67. https://doi.org/10.3389/fenvs.2014.00067

Vollenweider, P. & Günthardt-Goerg, M.S. (2005). Diagnosis of abiotic and biotic stress factors using the visible symptoms in foliage. Environmental Pollution 137(3), 455–465. https://doi.org/10.1016/j.envpol.2005.01.032

Vriet, C., Russinova, E. & Reuzeau, C. (2013). From squalene to brassinolide: The steroid metabolic and signalling pathways across the plant kingdom. Molecular Plant 6(6), 1738–1757. https://doi.org/10.1093/mp/sst096

Wang, C., Cui, H., Jin, M., Wang, J., Li, C., Li, Y., ... & Wang, Z. (2025). Enhancement of wheat resistance to dry–hot wind stress during grain filling by 24-epibrassinolide: Optimization of hormone balance and improvement of flag leaf photosynthetic performance. Frontiers in Plant Science 16, 1552617. https://doi.org/10.3389/fpls.2025.1552617

Wang, D., Yang, Z., Wu, M., Wang, W., Wang, Y. & Nie, S. (2022). Enhanced brassinosteroid signalling via the overexpression of *SlBRI1* positively regulates the chilling stress tolerance of tomato. Plant Science 320, 111281. https://doi.org/10.1016/j.plantsci.2022.111281

Wang, F.H., Ahammed, G.J., Li, G.Y., Bai, P.T., Jiang, Y., Wang, S.X. & Chen, S.C. (2019). Ethylene is involved in red light-induced anthocyanin biosynthesis in cabbage (*Brassica oleracea*). International Journal of Agriculture and Biology 21, 955–963. https://doi.org/10.17957/IJAB/15.0980

Wang, L., Liu, J., Shen, Y., Pu, R., Hou, M., Wei, Q., ... & Wu, G. (2021). Brassinosteroids synthesised by *CYP85A/A1* but not *CYP85A2* function via a *BRI1*-like receptor but not via *BRI1* in *Picea abies*. Journal of Experimental Botany 72(5), 1748–1763. https://doi.org/10.1093/jxb/eraa557

Wang, W., Vinocur, B. & Altman, A. (2019a). Plant responses to drought, salinity and extreme temperatures: Towards genetic engineering for stress tolerance. Planta 218(1), 1–14.

Waters, E.R., Lee, G.J. & Vierling, E. (1996). Evolution, structure and function of the small heat shock proteins in plants. Journal of Experimental Botany 47(3), 325–338. https://doi.org/10.1093/jxb/47.3.325

Wei, L., Zhong, Y., Wu, X., Wei, S. & Liu, Y. (2024). Roles of nitric oxide and brassinosteroid in improving fruit quality during postharvest: Potential regulators? Journal of Agricultural and Food Chemistry 72(43), 23671–23688. https://doi.org/10.1021/acs.jafc.4c05680

Wen, J., Tang, Y., Li, J., He, T., Xiao, J., Nangia, V. & Liu, Y. (2024). Effects of exogenous brassinosteroids on the starch structure, physicochemical properties and digestibility of wheat under high-temperature stress at the early grain-filling stage. International Journal of Biological Macromolecules 283, 137690. https://doi.org/10.1016/j.ijbiomac.2024.137690

Wilen, R.W., Sacco, M., Gusta, L.V. & Krishna, P. (1995). Effects of 24-epibrassinolide on freezing and thermotolerance of bromegrass (*Bromus inermis*) cell cultures. Physiologia Plantarum 95(2), 195–202. https://doi.org/10.1111/j.1399-3054.1995.tb00827.x

Wu, X., Yao, X., Chen, J., Zhu, Z., Zhang, H. & Zha, D. (2014). Brassinosteroids protect photosynthesis and antioxidant system of eggplant seedlings from high-temperature stress. Acta Physiologiae Plantarum 36(2), 251–261. https://doi.org/10.1007/s11738-013-1406-7

Xia, X.J., Wang, Y.J., Zhou, Y.H., Tao, Y., Mao, W.H., Shi, K., ... & Yu, J.Q. (2009). Reactive oxygen species are involved in brassinosteroid-induced stress tolerance in cucumber. Plant Physiology 150(2), 801–814. https://doi.org/10.1104/pp.109.138230

Xia, X.J., Fang, P.P., Guo, X., Qian, X.J., Zhou, J., Shi, K., ... & Yu, J.Q. (2018). Brassinosteroid-mediated apoplastic H_2O_2-glutaredoxin 12/14 cascade regulates antioxidant capacity in response to chilling in tomato. Plant, Cell & Environment 41(5), 1052–1064. https://doi.org/10.1111/pce.13052

Xu, Y., Zhang, X., Yang, H. & Lu, D. (2022). Effects of exogenous brassinolide application at the silking stage on nutrient accumulation, translocation and remobilization of waxy maize under post-silking heat stress. Agriculture 12(5), 572. https://doi.org/10.3390/agriculture12050572

Yan, J., Guan, L., Sun, Y., Zhu, Y., Liu, L., Lu, R., ... & Zhang, A. (2015). Calcium and *ZmCCaMK* are involved in brassinosteroid-induced antioxidant defense in maize leaves. Plant and Cell Physiology 56(5), 883–896. https://doi.org/10.1093/pcp/pcv014

Yang, Z., Liu, Z., Ge, X., Lu, L., Qin, W., Qanmber, G., ... & Li, F. (2023). Brassinosteroids regulate cotton fiber elongation by modulating very-long-chain fatty acid biosynthesis. The Plant Cell 35(6), 2114–2131. https://doi.org/10.1093/plcell/koad060

Yanhua, C., Yaliang, W., Huizhe, C., Jing, X., Yikai, Z., Zhigang, W., ... & Yuping, Z. (2023). Brassinosteroids mediate endogenous phytohormone metabolism to alleviate high temperature injury at panicle initiation stage in rice. Rice Science 30(1), 70–86. https://doi.org/10.1016/j.rsci.2022.05.005

Ye, Q., Zhu, W., Li, L., Zhang, S., Yin, Y., Ma, H. & Wang, X. (2010). Brassinosteroids control male fertility by regulating the expression of key genes involved in *Arabidopsis* anther and pollen development. Proceedings of the National Academy of Sciences 107(13), 6100–6105. https://doi.org/10.1073/pnas.0912333107

Ying, W., Wang, Y., Wei, H., Luo, Y., Ma, Q., Zhu, H., ... & Sun, L. (2024). Structure and function of the *Arabidopsis* ABC transporter *ABCB19* in brassinosteroid export. Science 383(6689), eadj4591. https://doi.org/10.1126/science.adj4591

Zhang, A., Zhang, J., Zhang, J., Ye, N., Zhang, H., Tan, M. & Jiang, M. (2011). Nitric oxide mediates brassinosteroid-induced ABA biosynthesis involved in oxidative stress tolerance in maize leaves. Plant and Cell Physiology 52(1), 181–192. https://doi.org/10.1093/pcp/pcq187

Zhang, C., Bai, M.Y. & Chong, K. (2014). Brassinosteroid-mediated regulation of agronomic traits in rice. Plant Cell Reports 33, 683–696. https://doi.org/10.1007/s00299-014-1578-7

Zhang, Y.P., Zhu, X.H., Ding, H.D., Yang, S.J. & Chen, Y.Y. (2013). Foliar application of 24-epibrassinolide alleviates high-temperature-induced inhibition of photosynthesis in seedlings of two melon cultivars. Photosynthetica 51, 341–349. https://doi.org/10.1007/s11099-013-0031-4

Zhao, J., Lu, Z., Wang, L. & Jin, B. (2020). Plant responses to heat stress: Physiology, transcription, noncoding RNAs, and epigenetics. International Journal of Molecular Sciences 22(1), 117. https://doi.org/10.3390/ijms22010117

Zhao, L.N., Shen, L.K., Zhang, W.Z., Zhang, W., Wang, Y. & Wu, W.H. (2013). Ca^{2+}-dependent protein kinase11 and 24 modulate the activity of the inward rectifying K^+ channels in *Arabidopsis* pollen tubes. The Plant Cell 25(2), 649–661. https://doi.org/10.1105/tpc.112.103184

Zhou, J., Li, Y.S. & Chien, S. (2014). Shear stress–initiated signalling and its regulation of endothelial function. Arteriosclerosis, Thrombosis, and Vascular Biology 34(10), 2191–2198. https://doi.org/10.1161/ATVBAHA.114.303422

Zhou, Y., Guang, Y., Li, J., Wang, F., Ahammed, G.J. & Yang, Y. (2019). The *CYP74* gene family in watermelon: Genome-wide identification and expression profiling under hormonal stress and root-knot nematode infection. Agronomy 9(12), 872. https://doi.org/10.3390/agronomy9120872

Zhu, Y., Zuo, M., Liang, Y., Jiang, M., Zhang, J., Scheller, H.V., ... & Zhang, A. (2013). *MAP65-1a* positively regulates H_2O_2 amplification and enhances brassinosteroid-induced antioxidant defence in maize. Journal of Experimental Botany 64(12), 3787–3802. https://doi.org/10.1093/jxb/ert215

CHAPTER

8

Differential Plant Response to Biotic Stresses Via Arbuscular Mycorrhizal Fungi AMF

Ramila Fares[1,2,3*], Beatrice Farda[1], Rihab Djebaili[1], Marika Pellegrini[1], Younes Rezaee Danesh[4], Gökhan Boyno[4*], Necmettin Teniz[5], Rojbin Çevik[4] and Semra Demir[4]

[1] Department of Life, Health and Environmental Sciences, University of L'Aquila, L'Aquila, Italy

[2] Laboratory Biotechnology, Water, Environment and Health, Abbes Laghrour University of Khenchela, Khenchela, Algeria

[3] Laboratory of Mycology, Biotechnology and Microbial Activity, Brothers Mentouri University of Constantine 1, Constantine, Algeria

[4] Department of Plant Protection, Faculty of Agriculture, Van Yuzuncu Yil University, Van-Türkiye

[5] Department of Agricultural Biotechnology, Faculty of Agriculture, Van Yuzuncu Yil University, Van-Türkiye

1. Introduction

Plants, as sessile organisms, are constantly exposed to a variety of biotic stresses that can significantly affect their health, growth, and productivity (Mittal et al., 2023). These stresses arise from interactions with living organisms such as pathogens, herbivores, and pests, which can lead to substantial damage in both natural ecosystems and agricultural systems (Qasim et al., 2024). Over millions of years, plants have evolved complex defense mechanisms to cope with these biotic challenges (Das Laha et al., 2024). However, with the increasing global demand for food production, coupled with the rising frequency of biotic stress events, managing plant health has become a critical concern in modern agriculture. In recent years, there has been growing interest in the role of beneficial microorganisms in enhancing plant resilience to biotic stresses (Jeger et al., 2021). Among these, Arbuscular Mycorrhizal Fungi (AMF) stand out as key players in fostering plant health (Liu & Chen, 2024). AMF form symbiotic relationships with the roots of most terrestrial plants, providing numerous benefits, including improved nutrient uptake, enhanced

*Corresponding authors: ramila.fares@graduate.univaq.it; gokhanboyno@yyu.edu.tr

DOI: 10.1201/9781003669777-8

resistance to pathogens, and greater tolerance to herbivory (Bhantana et al., 2021). This chapter explores the intricate ways in which AMF mediate plant responses to biotic stresses, from altering plant physiological and biochemical pathways to enhancing systemic resistance. Through a detailed examination of the various biotic stresses plants face, including pathogen-induced stress, herbivore-induced stress, and competition with weeds, this chapter sheds light on the mechanisms by which AMF support plants in mitigating these challenges. Furthermore, it emphasizes the potential of AMF as a sustainable and eco-friendly approach to improving crop resilience, offering insights into future research directions and agricultural applications. Understanding the interplay between plants and AMF in the context of biotic stress will provide valuable knowledge for enhancing crop productivity and sustainability in an increasingly stressed agricultural environment.

2. Types of Biotic Stress

Biotic stress is a stress condition that induces negatively affecting plant health due to the attack of an external biotic agent (Sahu et al., 2022). Plants are naturally exposed to different types of biotic stress by getting attacked by pathogens including fungi, bacteria, and viruses, by getting infected by various types of pests, and by getting browsed by herbivores including insects, birds, and mammals (Bolaji Umar et al., 2022). Plants that are not able to enhance successful defense against pathogens and herbivores have difficulties expressing themselves, evolving, and producing offspring. In agriculture, the threat of crop yield by biotic factors being attacked on crops is an increasing challenge due to current environmental, economic, and regulatory limitations (Fig. 1) (Ali et al., 2024). In addition to these, the population is rapidly increasing, and thus, more crops and improved agriculture practices are required to meet the demand. Most pathogenic organisms are microbials which include bacteria, fungi, and viruses (El-Saadony et al., 2022). Some of the common examples of pests are aphids, beetles, grasshoppers, and lactritida; while caterpillars, beetles, and larva are common examples of insects. Thus plants have evolved two intrinsic defense systems known as innate immunity and acquired immunity (Gauthier et al., 2022). Biotic interactions that promote the health, growth, and survival of living organisms involving other living organisms are beneficial interactions. Plants are known to establish various beneficial partnerships with soil microorganisms. Among the soil microorganisms, certain bacterial communities specifically develop symbiotic relationships with plants that result in protection against soil-borne pathogens. This is known as induced systemic resistance. Pharmaceuticals and other industries depend heavily on natural plant species for their raw materials (Chiquito-Contreras et al., 2024). To date, crop breeding and genetic modifications are primarily used to develop stress-tolerant plant varieties. However, genetic modifications are subject to regulatory approvals and public resistance. On the contrary, modifying the plant microbiome using beneficial endophytes is a potential alternative for enhancing biotic stress tolerance in plants (Anand et al., 2023). Endophytic microorganisms help host plants more acclimated to the biotic conditions. However, the attack of various insect pests and pathogens impairs the yield of crops, ultimately leading

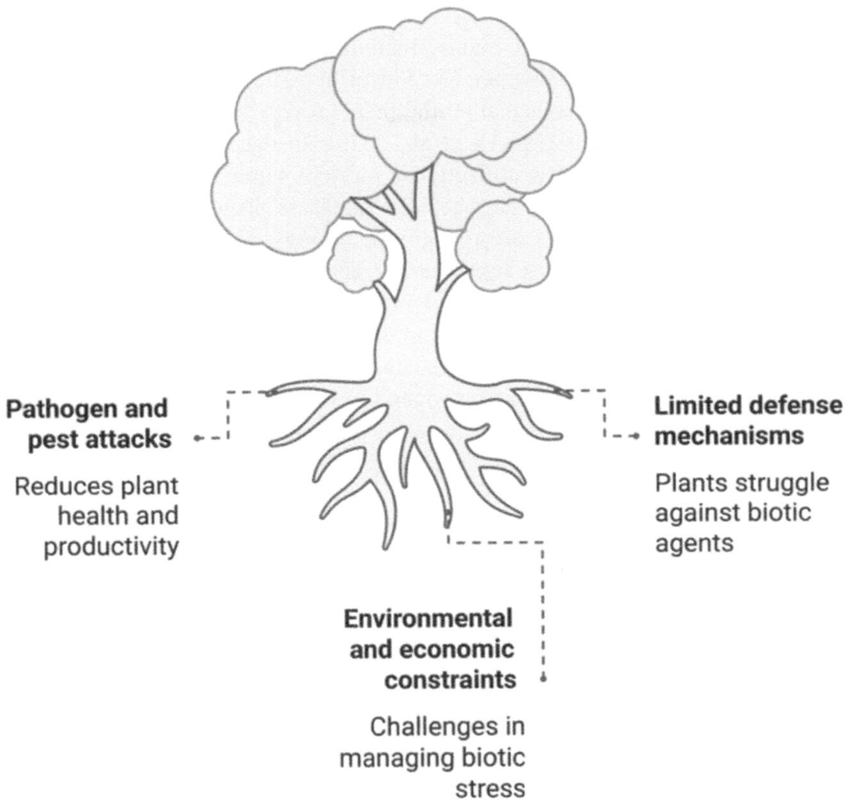

Fig. 1. The increasing threat to crop yields from biotic factors results from environmental, economic, and regulatory limitations.

to high economic losses (Skendžić et al., 2021). Major crops and economically important plants are frequently affected by various kinds of biotic stresses including fungal pathogens, insect pests, and nematodes (Kashyap et al., 2021). The infection significantly impacts the vascular system, leading to serious impairment of overall plant growth and development. Despite the advances in the field of biology, substantial research on biotic stresses remains elusive. Various studies have revealed the role of endophytic bacteria in conferring biotic stress tolerance in different plants. Among various endophytic bacteria, prolific infestation with specific bacterial strains can induce significant stress tolerance and promote plant growth (Flemer et al., 2022). This approach could provide an alternative tool for sustainable agriculture to manage various disease problems in crops without environmental hazards (Souto et al., 2021).

2.1. Pathogen-Induced Stress

Many plants in agricultural ecosystems are attacked by biotic factors such as pathogens, parasites, and herbivores. (Santoyo, 2022). Among herbivores, insects, nematodes, rodents, and birds are also major parasites of field crops. Many insect pests transmit viral diseases in plants, including the whitefly, the brown plant hopper, and the green plant hopper (X. Shi et al., 2021). Infection of plants by fungal parasites lead to diseases that inhibit plant growth, by either killing host cells or feed on living host cells (M. Ali et al., 2023). In necrotrophic fungi, the host is killed, and the dead tissue is utilized as a nutrient source, whereas in biotrophs, a healthy plant is needed since they actively suppress plant senescence (Salam & Khaidem, 2023). Plant viruses are recognized through disease symptoms such as leaf chlorosis, mosaic, ring pots, vein clearing, and stunted growth (Rajarammohan, 2021). Serious diseases caused by viral infection in tomatoes are also found in pepper and cucumber; seedlings growing in the affected nursery seed are commonly attacked by the fungal parasite, causing the development of a disease known as black root rot (Kumari et al., 2023). This pathogen has been identified as one of the most dangerous reported on soil-grown solanaceous vegetables; it also damages beans and various weeds, causing them to wilt. Diseased plants show a reduced root system and leaf yellowing due to N, Ca, and Fe nutrient deficiency. In the case of bent tomato seedlings, necrosis is also observed in the heart before the disease terminates those plants (El-Sayed et al., 2023). The virus has a high genetic variability favored by its lack of an epigenetic replication pathway; it consists of a single-stranded molecule of positive-sense RNA genome in the form of a rod, 8847 nt in size and crystallized as a rigid rod 18 nm in diameter with a helical axis 300 nm long.The genome is organized into ten open reading frames encoding proteins homologous to other known plant viruses. This virus is typically fed through the use of leaves (Gong et al., 2023). The onset of symptoms manifests within a few days, characterized by the emergence of diminutive, water-soaked lesions accompanied by necrotic tissue and the expulsion of aberrant root structures, all while mucilage exudes from the abaxial surface, resulting in the adhesion of adjacent leaves. Central to this phenomenon are the proteins present within root exudates, which orchestrate these complex interactions and contribute significantly to the overall plant response (Parasar et al., 2024). Cucumber mosaic virus, tobacco mosaic virus, and pepino mosaic virus are the most common viruses that infect tomatoes. Seeds and seedlings can also be infected with the virus, which can then be transmitted through seeds (Gonzalez et al., 2021). Systemic symptoms are easily seen in the new shoot, which shows a distorted and irregular appearance, leaf bunching, chlorotic spots, vein necrosis, etching, and low leaf size. Viral-infected tomato plants show stunted growth and leaf chlorosis. In heavily infested fields, the yield may reduce by up to 80% (Abdelkhalek et al., 2022).

2.2. Herbivore-Induced Stress

Biotic stresses, such as herbivory and pathogens, have been ubiquitous over the course of the evolution of plants. In contrast to abiotic stresses, biotic stress can be highly localized, but the perception of focal or systemic injuries by the leaf tissue is still not fully understood (Nawaz et al., 2023).When susceptible plants are challenged with herbivores, the herbivores injure the plants, generating the so-called herbivore-induced stress. In response to herbivory, plants emit substantial amounts of volatiles with isoprene, mono- and sesquiterpenes along with green-leaf volatiles. Despite an extensive source of information, there is relatively little information on the dynamic aspects of Volatile Organic Compounds (VOC) emissions (Pérez-Hedo et al., 2021). Stressed plants typically emit substantial amounts of volatile organic compounds including isoprenoids, benzenoids, green leaf volatiles, etc. Severe stresses triggering ROS formation result not only in the immediate damages at the stress site(s), but may also lead to extensive senescence of the entire organ (Bourtsoukidis et al., 2024). Relay Command emission of volatiles often takes place from the stressed organ, but the formation and remobilization of non-volatile storage pools in the healthy tissues of the stress site have also been implicated. There is extensive evidence that non-volatile stress signal molecules including phytohormones and volatile compounds are released from the stress site (Perez-Gil et al., 2024). Thus, it is likely that VOCs are involved in the systemic signaling of herbivore-induced stress. Given that the induced emissions are at maximum level under the conditions when the stress dose and ROS formation are at maximum level, and it may be seen that the involvement of induced isoprenoids in direct defense is plausible (Ngumbi et al., 2021). Smaller stress doses induced by herbivore feeding may also induce emission of volatiles, but these emissions are small and do not appear to be involved in direct defenses. Because of the mechanistically different mode of stress-induced ROS formation in response to biotic versus abiotic stress, there is currently no evidence of the involvement of induced emissions in direct defense against abiotic stresses, although such a role would be fully compatible with stress dose-dependent emissions of induced volatiles (Moreira et al., 2023).

2.3. Competition with Weeds

New corn chemical controls have been rapidly increasing, yet little is known about how these chemicals affect plant tolerance to living and even newly evolved stresses. Some suggest that the need for controls fades as agricultural practices advance (Iqbal et al., 2021). However, others convey that rapidly evolving types and frequencies of biotic stressors in agricultural fields will continue challenging the sustainability of farms and our food supply. In a two-year study, this work investigates the effects of 12 corn herbicides on the early physiological response of corn plants to three primary stresses that are living and currently dampen or can dampen the crop–weeds, shade, and reduced availability of NH_3NO_4 (Lo Presti et al., 2023; Zhang et al., 2022).

Weeds are a significant management problem for corn production in the Northern United States. Though present more infrequently, weeds, shade, and low N frequently co-occur and/or successively within corn fields.While this work underscores the need to sustain low biotic stress to obtain maximum yields from corn field agriculture, for the three corn stresses mentioned, a physiological basis for tradeoffs is discovered that may help predict when to expect them (Brown et al., 2022; Mcdonald et al., 2023). It is commonly assumed that plants have a shared pool of defense resources that can be generically allocated to resist most biotic stressors. Corn plant stress studies to date, however, have generally employed a single stress or a single type of stressor. This work utilizes physiological and transcriptomic assays to investigate how current herbicides affect corn plants' direct and typical early physiological response to newly-evolved living weed stress and how this response is similar or distinct to that elicited by existing non-weed stresses. Both weed pressure on corn resulting from herbicide resistance and the impact of herbicides on plant immunity are novel findings (Gill et al., 2022; Li et al., 2024; Sheoran et al., 2022).

3. Role of Arbuscular Mycorrhizal Fungi (AMF)

Arbuscular mycorrhizal fungi (AMF) are obligate biotrophic symbionts that colonize the roots of approximately 80% of land plant species. They can penetrate the root cortical cells, and a symbiosis is established (Basiru et al., 2023). AMF cause a dramatic change in the host plant's morphology, structure, and physiology, and branched structures are formed inside the root for nutrient exchange. The mutualistic association of AMF and plant hosts is among the most important biotic interactions in terrestrial ecosystems (Sun et al., 2022). The roles of mycorrhizal symbiosis and host-plant interactions in regulating the quantity and composition of AMF, improving agricultural productivity, maintaining plant community diversity, regulating carbon cycling and storage, and enhancing soil carbon accumulation have been summarized (Noceto et al., 2021). Consistent colonization of AMF can help mitigate PI, as demonstrated in several studies. In the model plant species L. sativa, AMF colonization significantly influenced the mycorrhization indices, including the percentage of root length colonized by AMF (M%) and root length of AMF (A%) (Kakabouki et al., 2023). AMF reduced leaf poisons to growing plants (Albqmi et al., 2023). AMF mitigate PI by regulating intracellular signalling pathways, enhancing symbiotic effectiveness, regulating root responses, increasing soil biogeochemical processes, and promoting beneficial rhizobacteria. In this study, Cladosporium sp. negatively affected Artemisia argyi growth but was found to improve biomass after AMF inoculation (Sonbol et al., 2025). Meanwhile, mycelium release and ammonium nitrogen from AMF might be the main reasons for affecting the host plant growth after the introduction of AMF, although further research is needed (Thangavel, Naser, et al., 2022).

3.1. Introduction to AMF

Arbuscular mycorrhizal fungi (AMF), members of Glomeromycota, can penetrate the vigorously growing materials of plant roots and reach the inner root cortical cells, accommodating a symbiosis with approximately 85% of land plant species (Díaz et al., 2025). There is increasing evidence that AMFs are essential for the evolution of terrestrial plants. After hyphate germination on soil or root exudates, AMF spores undergo mitotic division to form a short germ tube (0.5–1 mm long). This structure originates either specialized germ cells or long lateral hyphae, which begin to grow actively and approach the root surface. Several plant root factors play key roles about AMF, which induce hyphal branching of AMF (Malar C et al., 2021; Zhou et al., 2022).One of these factors is strigolactones. An ancient plant–AMF symbiosis signaling pathway evolved during early land colonization. During co-evolution, AMF has developed effector-like protein/peptides that hijack host plant signaling pathways of different plant families, reprogramming gene expression to modulate plant defenses to its advantage (Nasfi et al., 2025). On the plant side, gene families recognize AMF effectors via direct or indirect protein interactions to trigger diverse or compatible responses in. Various host plant species have captured this ancient mutualism, and although phylogenetically unrelated, they share conserved and/or convergent plant signaling pathways (Berger & Gutjahr, 2021). A significant and complex crosstalk on a primary metabolic pathway between the host and AMF is vital in AM symbiosis. Lignin, a product of phenylpropanoid precursor metabolism, is involved in plant responses to biotic or abiotic stress, modulating AM fungal colonization. This may hinder AM hyphal penetration and prevent AMF from establishing successful symbiosis (Lin et al., 2021; S. Yadav & Chattopadhyay, 2023). AMF strikingly alters the natural and agricultural soil communities and dramatically affects the plant-associated microbe communities, determining the stability of plant AM symbiosis and nutrient acquisition. Understanding AMF signaling networks will provide insights into the possible management of AM symbiosis to enhance nutrient acquisition and improve plant and soil health (Brar et al., 2024; Das et al., 2022).

3.2. Mechanisms of AMF Interaction with Plants

The extrinsic mycelium of Arbuscular Mycorrhizal Fungi (AMF) entering the epidermal cells of the host roots and the cortical tissue of the plants forms a dendritic arbuscular structure through a continuous bifurcation process (F. Wang et al., 2022). The tips of the branched structures form specialized compartments by invading the plant membrane. Inside this compartment, AMF structures include arbuscules, vesicles, and hyphal coils. This AMF colonization structure is the site for the exchange of nutrients and signals between the plant host and the fungi, establishing a new nutrient conduit parallel to the Apoplast and Plant Endocytosis Pathway (PEP)(Howard et al., 2022). AMF can significantly affect plants' development and growth as they regulate water utilization, nutrition uptake, and hormone balance from biological and abiotic stresses. The interplay between AMF and plants can be divided into three subclasses on the plant side: biotrophic interaction, symbiotic system, and factor-based interaction (Demir et al., 2022). The mycorrhizal fungus in the soil secretes strigolactones to attract nearby AMF. Hormones produced by AMF can positively regulate nutrient delivery and root growth. In order to form symbiotic relationships, AMF needs to recruit fatty acids from the host plant to facilitate mycorrhizal colonization (Andrino et al., 2021). Lipids are the most primitive and basic nutrients for fungi. The transcription factor RAM1 is crucial for arbuscular mycorrhiza symbiosis (AMS), as it regulates the activity of glycerol-3-phosphate acyltransferase RAM2, which regulates the transfer and delivery of fatty acids, the essential nutrient for AMF, from plants to AMF (Song et al., 2025). The transcriptional activators ERM1 and WRI5a can directly bind to the promoters and then activate the expression of FatM, STR, and STR2; these target genes play important roles in fatty acid biosynthesis and transfer (Ge, 2024). Furthermore, the transcriptional regulatory complex ERM1/WRI5a–ERF12–TOPLESS plays an important regulatory role in arbuscular-containing cells, which helps maintain a stable and beneficial symbiosis. In addition to obtaining fatty acids from host plants, AMF can also obtain sugars from host plants in the mature stage. After AMF successfully invade host plant root epidermal cells to form symbionts, they can expand the effective absorption range of host plants in the soil (Wang et al., 2023a).

4. Plant Responses to Biotic Stresses

In nature, plants live in association with beneficial microorganisms, such as those belonging to the phyla Glomeromycota, which form arbuscular mycorrhizae (AM). In return for an enhanced nutrient uptake, particularly phosphorus (P), the plants provide carbohydrates for their fungal partners. This interaction is perceived as a cooperative mutualism between plants and fungi (Bennett & Groten, 2022). Such associations have other benefits, including enhancing plant growth and productivity, fruit quality, nutrient content, stress tolerance, disease control, soil structure improvement, and soil nutrient cycling. These changes are made directly through various morphological, physiological, or biochemical changes in plants or indirectly through changes in the rhizosphere microbial community structure (Ullah et al., 2021). By bundling the roots and forming a physical barrier to the fungal plant pathogens, the protective effect of AM fungi may also impair the invasion potential of other necrotrophic soil-borne pathogens. The formation and development of the symbiotic structure are regulated by chemical signaling transmitted between both fungal and plant partners (Boyno et al., 2022). The primary fungal signal molecules, known as strigolactones (SLs) and lipochitin oligosaccharides (LCOs), are recognized by the plant and initiate a cascade of signaling events to activate a common signaling pathway of the plant's cellular factors. This pathway promotes AM fungi's colonization of the roots and leads to prospective and currently known signaling molecules such as cytokinins, polyamines, and reactive oxygen species (ROS) (Teyssier et al., 2024). SLs have also been reported to promote the functional activity of AM fungi and nitrogen-fixing bacteria. The defense mechanisms include the production of phytoalexins, accumulation of pathogenesis-related (PR) proteins, and the hypersensitive response involving a rapid programmed cell death of infected plant cells (Chourasiya et al., 2022). The ability of AM fungi to facilitate plant uptake of porous pesticides and produce plant defense-boosting mediated compounds has indicated a crucial role in combating soil-borne diseases. Taken together, it is non-exhaustive knowledge on plant responses to biotic stresses that tend to vary depending on the pathogen levels of infection and association with AM fungi (Yu et al., 2022).

4.1. Physiological Responses

Beneficial fungi have evolved extraradical structures, such as hyphal networks, to increase the foraging area for resource-vulnerable plants, which can establish physiological and developmental mechanisms. Moreover, arbuscular mycorrhizal fungi (AMF) positively influence the physiological responses of plants under various biotic and abiotic stress conditions (Wahab et al., 2023). Such responses allow elicitors or signals released by AMF, which are sensed and partially interpreted by the appropriate host plants, to trigger protective or adaptive pathways. Such pathways effectively regulate constitutive or systemic responses that are generally ineffective in controlling pathway-specific stresses (Sarkar & Sadhukhan, 2023). Other pathways cause the specific inhibition of the effectors or signals that recognize or transduce the former. A balance is struck among pathways for specific biotic stresses such that negatively modulated responses allow the establishment of physiological and developmental mechanisms. Enabling positive and negative regulatory feedback controls two waves of host plant responses (Panahi, 2024). AMF exerts multiple effects on the plant's responses to biotic stresses, which could be beneficial or detrimental. Such responses are mediated by signaling pathways stimulated following AMF establishment in either roots or leaves, including Phospholipase D, Apoplastic Reactive Oxygen Species, Ca^{2+}, and Mitogen-Activated Protein Kinase (Bhupenchandra et al., 2024). AMF increase the production of protective/antioxidative compounds, modulates defense genes, and alters xylem-to-phloem transportation, enhancing antifungal or antiherbivore plant defences. By enhancing nutrient supply, AMF increases energy and metabolic substrates for freshly activated host defense responses to benefit pathogens and herbivores. By modulating the second wave of host responses or suppressing pathogen growth, roots-localized AMF confer protection in healthy leaves (Tan et al., 2022). Herbivores or pathogens condone plants purified by AMF hyphae, and their protective effects spread from AMF-colonized organs to non-colonized organs. Interaction between AMF and herbivores/pathogens could occur, but the outcome is context-dependent. In enhancing the demands for alternate resources, AMF may release host plants from competition by herbivores and pathogens during co-infection precision. However, whether grass- or shrub-dominated systems are less common in interunit deserts is poorly understood (Elhamouly et al., 2022).

4.2. Biochemical Responses

AMFs are now widely recognized by a growing body of literature to enhance the resistance of host plants to biotic stress (Alrajhi et al., 2024). Many of these studies focus primarily on the mechanisms AMF confer biotic stress tolerance to host plants and the induced changes in several biochemical and physiological responses. Review articles have recently highlighted multiple mechanisms, including strengthening plant innate immunity, producing phytohormones, enhancing antioxidant activity, etc. Significant biochemical responses of host plants in response to fungal colonization under biotic stresses, such as increased activity of the peroxidase and polyphenol oxidase enzymes (Kumar et al., 2024). Comprehensive overviews of the effects of biotic stresses on host plant growth and physiology (Dixit et al., 2024). Additionally, the studies primarily focus on AM-inoculated plants under single biotic stress, such as pathogens or herbivores, but relatively few studies consider both arbuscular mycorrhizal fungi (AMF) and the simultaneous effect of two bio-stresses on the physiology of the host plants (Bhupenchandra et al., 2024). The AMF's positive effects on the plants' growth, physiology, and yield in various organic substrates and saline conditions have been presented. Several AMF species have been tested for their efficiency in alleviating plant osmotic stress. For example, another study shows the efficacy of Glomus mosseae in enhancing growth and metabolic activity in tomato plants under stress induced by osmotic agents (Chen et al., 2024). Similarly, *Glomus intraradices* have also induced a more efficient carbon mineralization in their rhizosphere under drought conditions. In other studies, *G. etunicatum* has been widely reported in heart-leaved ice plants for its protective role against decreased plant growth and metabolism under salinity and drought stress. Despite this, several studies have been conducted emphasizing the beneficial effects of AM symbiosis against biotic and abiotic stresses on plant physiology and biochemistry (Cerrato et al., 2024).

4.3. Morphological Changes

Even though AMF root invasions predominantly occur in the root cortex and are often hypogeous, they provoke dramatic changes in the entire root system structure, including the root and root hair morphology, root branching, and root architecture. All AMFs modify root morphology with differential responses among species (Mitra et al., 2021). In chickpeas, AMF promotes root elongation and development and increases root weight. Mycorrhizal plants showed thicker roots and higher root-to-shoot ratio with higher total root length and root surface area than did non-mycorrhizal plants in rice and tomato colonized with G. irregular and G. intraradices, respectively (Yonas & Zawar, 2024). Root encasement in hyphae of G. intraradices increased in *Trichoderma harzianum* colonized or PGPR-treated soybean seedlings over 36 days. With higher P and N levels or AMF inoculation, the probability of basal root number >3, >5, and > 7 increased in maize, resulting in significantly modified root systems with more basal roots (Lotfabadi et al., 2022). Mycorrhizal-dependent differences in root morphology near barriers between soil and hydrogel or between hydrogel and hydrogel were more significant than the number of roots within either substrate type, suggesting an increase in the proliferative and exploratory growth of arbuscular mycorrhizal (AM) roots in contrasting water availability conditions (Hammer et al., 2024). Root hairs are pivotal structures that effectively absorb water and nutrients from the soil, extending root establishment into the soil and promoting plant tolerance to water deficit. Under water deficit, AMF strongly stimulated root hair formation in tomatoes and soybeans and significantly downregulated root hair elongation (Al-Amri, 2021). Following an intense drought treatment-induced decline with a partial recovery in re-watered conditions, both drought treatment types increased mycorrhizal colonization in tomato roots. AMF-associated changes were slower but more sustained in tapering regions closer to the root apex (Felföldi et al., 2022).

5. AMF Mediated Resistance Mechanisms

AMF can maintain a chemically awake root zone network and physical barrier to prevent entry of microbes into roots via morphological and structural modifications, such as dry root patches, root hairs, thicker cortices, and short secondary roots clusters with longer lateral and deeper sunk root canals (Tilgam et al., 2022).

5.1. Enhanced Nutrient Uptake

The arbuscular mycorrhizal (AM) fungi were found to provide the host plants with enhanced nutrient uptake, especially for P, N, and K. This function seems to be a primary one for AM fungi during their coevolution with land plants (Bhantana et al., 2021). Almost all plants in terrestrial ecosystems entered into symbiotic interactions with AM fungi, indicating the evolutionarily preserved importance of AM fungi.

Beneficial effects of AM fungi on plant nutrition uptake are common responses to abiotic stresses. A recent meta-analysis revealed that AM fungi improved foliar N and P concentrations under drought conditions, which suggested that AM fungi can indirectly facilitate plant performance under water deficit by maintaining plant foliar nutrient status (Boutasknit et al., 2021). In many cases, improved foliar nutrient status by AM fungi will lead to enhanced photosynthesis and reduced oxidative stress in plants. Higher leaf photosynthetic capacity and better photosystem II (PS II) photochemical efficiency of plants colonized with AM fungi also play a crucial physiological role in the enhanced drought resistance (Wang et al., 2023a). The contribution of mycorrhizal fungal hyphae to soil aggregation and the inhibition of the return of soil organic carbon to the atmosphere, increase the longer-term carbon storage, and improve nutrient retention within soils are derived from enhanced plant growth. Therefore, AM fungi can contribute to the long-term relevance of ecosystem services in land plants during and after their coevolution (Wahab et al., 2023). AM fungi are obligate biotrophs, which rely entirely on the host plant for their carbon supply. However, host plants were found to exude higher amounts of CO_2 when colonized with AM fungi than when uncolonized (Oliveira et al., 2022). In addition, the exudates from AM-fungi-colonized plant roots also can stimulate the soil CO_2 respiration in the oxic and anoxic conditions, which indicates that the plant–AM fungal symbiosis has an effect on the flow of carbon substrate (Fiorilli et al., 2021). Products of the AM fungi, which are released from their hyphae and mycorrhizal roots into soils, can also stimulate soil microorganisms. The fungus-induced changes in the microbe-to-microbe interactions could alter the cascade of the fertility chain, ultimately affecting overall ecosystem services. Thus, AM fungi promote both plant growth and soil health as well as the ecological function of terrestrial ecosystems (Fall et al., 2022).

5.2. Induced Systemic Resistance

Recent studies provide enough evidence of a common signalling between abscisic acid (ABA), biotic stress signals and plant resistance conditioned by a mycorrhiza fungus, AMF. A testing of the priming nature of the defence response triggered by AMF is the approach to discover if it is an AMF-specific resistance mechanism (Hussain et al., 2024). Accordingly, plants colonised by a mycorrhizal fungus were resistant to a fungal pathogen attack when the colonization occurred at least 3 weeks before the onset of the stress (Lutz et al., 2023). However, a recovery of sensitivity to the attack by the pathogen as the time of colonization increased occurred in already mature tomato plants. Upon infection, a differential induction of Cinnamate-4-hydroxylase (C4H) and chalcone synthase (CHS) expression and an accumulation of phytoalexins occurred mainly in the non-mycorrhizal half of the plant surrounded by mycorrhiza. While delayed, the infection was then not restricted to the locally inoculated leaves. It could be detected far from the point of infection in the leaves through specific fungal and plant defence gene expression (Rostami et al., 2022). Biostimulants evoke metabolic changes enabling plants to mobilize a greater availability of nutrients from pre-existent soil pools or reserve pools within the plant (Ben Mrid et al., 2021). AMF enhance bioavailability of nutrients in both soil and root-nodules and can alter root architecture, growing a greater number of thin roots and root hairs. This provides greater access to nutrient pools, thereby increasing plant productivity in terms of growth, yield, quality and harvestability. It is likely that AMF and/or their rooting structures play a role in plant root recognition of beneficial microbes promoting Keystonism similar to that known for *Pseudomonas spp.*, some of which produce ACC-deaminase which may alter signaling pathways (Gough et al., 2022). Keystonism and pre-conditioning have been used to describe a range of plant preparedness signaling observed that leads to differential plant responses to subsequent biotic stresses. These events provide opportunities to enhance pathogen resistance with respect to a given pathogen and/or a given plant species or cultivar, and suggest there are a variety of factors and pathways involved in priming and acquisition of this enhanced performance (Leisner et al., 2022).

5.3. Alteration of Plant Defense Pathways

Normal plant growth depends on fruitful interaction of plants with beneficial mycorrhizal fungi as well as on the optimum efficiency of associated microbe species. These may include bacteria and fungi that are both beneficial as well as pathogenic (Thoms et al., 2021). Mycorrhizal fungi acquisition of nutrients along with protection against poorly functioning pathogens or parasitic fungi is a recognized mutualistic relationship. The consequences for plants with bud developed arbuscular mycorrhizal (AM) fungi capable of accelerating host development include early and healthy flowering in melons, increased seed weight in wheat, and increased phloem loading and sap flow in red maples (Phour & Sindhu, 2024). The mechanisms involved in altered plant performance were recently found to involve cytoplasmic enzyme production and significantly altered defense pathways in the arbuscular mycorrhizal (AM) association (Segal et al., 2025). Selective breeding for faster flowering in short-day annual species such as sunflower and radioactively dating herbarium specimens led to the contemporary hypothesis for crop domestication (Santoyo, 2022). Associated AM fungi could carbocycle cause host early flowering and increased mutualistic benefit. Microcosm studies with a broad array of microbes had been used for decades. Compared to non AM fungi/processes, enhancement of plant response can – via selection to respond to species in AM fungal communities with ecological similarity – by rapid failure to flower, reduced carbon allocation and thereby reduced edible yield (Ramírez-Serrano et al., 2022). Enhancing dose response fit with AM fungal treatment delivery could cause ASD enhancement of performance via increased phloem loading and expanded root architecture, and these responsive traits could co-vary with maximum benefit under conditions non conducive to co-varying microbes with mutualism promoting functions such as plastic nodulation/modification of root mass ratio. Quantitative gene knockout for colonized root trait targeting AM fungal biosynthetic gene clusters followed by mutagenesis could uncover generality behind this phenomenon (Primieri et al., 2022).

6. Case Studies of AMF in Biotic Stress Mitigation

Although investigations into addressing AMF degree in soil started first, research later occurred about the impact of soil AMF degree on plant development. The first case study was performed by with minimum disturbance to soil (Qin et al., 2022). This was a concept of AMF biofunctionality in soil. Finally, application of AMF with compost is the most effective. Soil physicochemical changes and AMF recovery were also monitored after 10 years of AMF bioengineering by periodic soil sampling. From the results of this case study, function-based AMF bioengineering is practical and applicable to ameliorate and restore more degraded landscapes. In another case study, AMF bioengineering by composting with psyllium residues was attempted as native AMF biotization, with special consideration for soil moisture reestablishment, soil rolling and compaction, and establishment of soil food webs (Ait Rahou et al., 2021; Tahiri et al., 2022). AMF response indicators and soil physicochemical changes were monitored after AMF bioengineering. Test items of soil compaction and food webs improved physiological properties mainly under biotization. These case studies neither have focused on the effects of AMF isolates on plant growth and stress mitigation abilities in pot experiments nor applied various AMF species against diseases in disease-prone areas (Begum et al., 2022). AMF provides systemic resistance, increases leaf firmness, balanced nutritional elements, enhancing endogenous chilling reaction, promote fruit growth and ripening, and greater production quality although 30% of fruits were infected by the pathogens. Plus, AMF may provide salinity tolerance via regulation of shoot- and root-based growth, transpiration, ion content and distribution, and osmotic features (Melini et al., 2023).*

6.1. AMF and Fungal Pathogens

The role of arbuscular mycorrhizal fungi (AMF) in promoting plant fertility is widely accepted. In contrast, the role of AMF in defense against fungal pathogens is more contentious despite research indicating that AMF can protect plants from fungal parasites (Brito et al., 2021). In addition to direct effects on pathogen populations, AMF can also help plants cope with parasitism by modifying the defence response, resulting in increased tolerance of infection rather than decreased parasitism. Recent work examining the interactions between AMF and fungal root pathogens under glasshouse conditions has identified significant variation in both the cost of AMF colonization and the ability of AMF to provide protection against fungi from plant to plant. This has implications for our understanding of the processes controlling AMF communities and the abiotic and biotic factors necessary for them to effectively protect their hosts (Bano & Uzair, 2021; Fabiyi et al., 2024). In an agricultural context, AMF have been suggested as a biocontrol agent of pest pathogens, including the soil-borne plant pathogen, which is increasingly important in the context of sustainable agriculture (Selvaraj & Thangavel, 2021). To maximize the efficacy of AMF biocontrol, it will be important to evaluate the trade-offs and mechanisms by which individual AMF suppress pathogen populations and the factors that select for these traits in agricultural soils. This leads to AMF distribution and in turn to the effect of AMF on grapevine health and yield in the vineyard setting (Velaz et al., 2025). An overview of the known effects of AMF on their plant hosts shows that the extent to which they support grapevine function, with respect to nutrient acquisition, water relations, defense against biotic stress and possibly abiotic stress. Fungal root pathogens belonging to the genus present serious threats to grapevine health and longevity, and a recent phylogenetic approach to predicting susceptibility to these pathogens is outlined (Cataldo et al., 2022).In glasshouse trials, members of AMF communities isolated from vineyard soils differ in the intensity of the benefit they confer on grapevine hosts. In doing so, they alter patterns of pathogen development in roots, rhizosphere, and in growth media (Giudice, 2022). Some members of AMF communities possess intrinsic traits that are responsible for the observed patterns of pathogen interference but themselves differ in the costs of being colonized. A theoretical model of the interactions between plant hosts, AMF and pathogens suggests that under some conditions a delicate balance of biotic interactions will exist (Smulders, 2022). New questions about the distribution and functioning of AMF in natural and agricultural systems are posed, including the control of dispersion and the impacts of rivers on inoculum potential and activity.

6.2. AMF and Insect Herbivores

Mycorrhizal associations may affect both direct and indirect plant defenses against herbivores. While some studies suggest AMF-mediated alterations in direct defenses, reports on the influence of these fungi on the release of herbivore-induced plant volatiles (HIPVs) or other indirect defenses are rare. Colonization of plants with AMF can potentially modify the plants' ability to defend themselves against insect herbivores via changes in the plant's nutritional quality or the release of HIPVs (Sardans et al., 2021; Stratton et al., 2021). Direct effects of AMF on the nutritional value of the plant are likely to influence herbivore preferences and performance. High nutrient uptake may positively affect growth and efficiency of herbivores, as they typically prefer nutrient-rich plants (Zeng, 2024). Alternatively, mycorrhizae-induced alterations in concentrations of nutrients that are frequently limiting for herbivores may negatively affect them. Reduced availability of such nutrients may increase feeding and oviposition preference of voracious herbivores, such as in the case of sodium . Herbivores located in low-nutrient areas may be especially prone to attack plant species harboring AMF with low translocation capacity (Bernaola et al., 2021). Another way AMF affect plants is via increases in observed concentrations of secondary metabolites that can have effects on leaf toughness. The reduced leaf toughness would explain observations of increased damage levels on mycorrhizal plants. Changes in the concentration of nitrogen-containing compounds affecting food quality and alterations in glucosinolate profiles may similarly influence herbivore resistance (Wu et al., 2021).Mycorrhization can also alter the timing of phenolic compound accumulation. Increased secondary metabolite levels can improve the plants' ability to resist herbivore attack but can cost the plant extra nitrogen, which may negatively affect the plant's ability to provide AMF with carbohydrates (Erb et al., 2021).

6.3. AMF and Viral Infections

Viral pathogens, which are primarily included in the group of biotrophic pathogens, are among the main reasons for decreased yield and quality of crops. With the introduction of new genotypes and cultivars in some crops, the number of virosis increased and negatively affected the crop production. Studies have shown that a wide range of plant viruses could be transmitted by AMF or infect mycorrhizal plants. Such studies are in limited number and mainly involved non-virulent viruses, indicating a gap of knowledge regarding the interaction of virulent viruses with AMF (Metwally et al., 2024). Recently, a preliminary test showed that AM fungi potentially decrease the titer of Tomato bushy stunt virus (TBSV) and some signs of the infection; however, the degree of control was dependent on the type of AMF . In general, viruses within the RNA and DNA groups could use AMF as vectors and this was documented for a wide range of non-virulent viruses. Viruses can prepare the host for infection by either enhancing or limiting the mycorrhizal symbiosis (Dutta et al., 2022; Pramanik et al., 2021). On the one side, viruses are able to disrupt mycorrhizal symbiosis and the development of arbuscules, which negatively affected cytokinin and auxin levels in tomato plants. On the other side, there are cases indicating that the degree of mycorrhizal colonization, expression of symbiotic genes, soil organic carbon and nitrogen increase in response to viral infection. These results prompted the investigation of the regulation of genes involved in the mycorrhizal symbiosis in the presence of TBSV and ToMV and mycorrhizal fungi (Deja-Sikora et al., 2023; Poudel et al., 2021). The results also showed that the AM fungi effect is not only on plant physiology, ameliorating important stress effects, but also on changing the viruses themselves to less virulent forms. These results suggested an interesting area of research concerning the relationship between AMF and viral infection in plants, urged by the need to re-evaluate and modify current paradigms that do not generally acknowledge the importance of AMF in managing viral infection in crops (Fiorilli et al., 2024).Further investigations are essential to dissect the highly complex interactions involved in differential plant response to AMF and viral infection. In summary, the plant response to the interaction with AM fungi and viral infection is multifaceted, indicated by differential regulation of PR proteins in response to TBSV and ToMV infection. The presented data may provide practical implications for healthier crops (Fabiyi et al., 2024).

7. Factors Influencing AMF Efficacy

The efficacy of AMF in colonizing roots and improving host characteristics and conditions depended not only on the fungal species and strains but also on host species, environment, and soil conditions. While the super colonization of AMF in roots could benefit the plant, it might impair host development through other parts (Grünfeld et al., 2022). These responses may vary depending on one or more of the many existing inoculum factors such as AMF communities, their genetic backgrounds, host ranges, soil conditions, moisture, or nutrients availability. The variability also extended to the symbiosis establishment such as spore germination, pre-colonization stages, and root response (Basiru et al., 2022). Within AMF inconsistent responses were observed in magnitudes and patterns of host colonization, increase in host parameters and conditions, and spore production. Some documented poor root colonization in phosphorous fertilizer responsive plant species (Sathyabhama et al., 2022). There were groups of crop species that were responsive to AMF inoculation, but their responses were reduced by chemical fertilization. Crops such as maize and sorghum were found among these species and these types of non-responder plant species could be used for bio-control of AMF in intentionally introduced ecosystems. Pastures and reforested sites with cotton, mustard, sunflower could have AMF parasite and less mycotropic crops such as faba beans were included in viable farming rotations. (El Omari & El Ghachtouli, 2021).The calcium phosphate treated spore extract of AMF at a proper dose increased root colonization but reduced shoot weight of higher efficacy AMF inoculation relative to lesser ones. Excluding AMF washed soils could lose a considerable fraction of AMF, and microfraction of soil or mixture of soil of several pathogens could conceal the availability of inoculum for beneficial AMF as well as rhizobium (Ul Haq et al., 2023).Within AMF significant responses were observed in host root colonization rates and root length but no effects were seen on host shoot height, degree of mycorrhizal colonization, and shoot weight depending on the improvement of soil conditions (Opitz et al., 2021). AMF also improved height and weight of pure Zamia compared with other species planted together, suggesting that they helped to eliminate inter-plant competition. Within isolated AMF species its ecological efficacy and amendment difference were observed in spore germination radial growth rates, width of mycelial growth inhibition zones, and hyphal bait test ratio (Soltangheisi et al., 2024). Comparing with the biotopes, AMF at one biotope was more effective in hyphal growth and parasitic impact to the host and this efficacy was limited at that biotope. Field applications of twelve species of AMF fungi were done at ten host plants and pot cultivation study showed that two species were the most effective symbionts (Ducousso-Détrez et al., 2022).

7.1. Soil Type and Nutrient Availability

A mutualistic family of fungi, the arbuscular mycorrhizal (AM) fungi, offers great promise for sustainable agricultural practices. The genetic versatility of AM fungi and their biotic and abiotic stress tolerance traits highlights the need for their application in different soils to enhance stress resistance in plants as well as soil health (Branco et al., 2022). AM fungi are some of the earliest known mutualists of land plants, and their co-evolution with plant hosts over hundreds of millions of years has shaped habitat for most plant species. In return, these fungi receive carbon-rich plant photosynthates and other nutrients that promote their reproduction and help build carbon pools in soils on a landscape scale (Rosling et al., 2024). The formation of symbiotic associations with AM fungi requires spatial investigation of plant root cells and the rhizosphere, in addition to cell-type/organ-level changes that improve host foraging. AM fungi increase soil and root surface area to volume ratios and pore connectivity, and modify soil structure and aggregation to eliminate barriers to growth (Hu et al., 2024). They remodel the plant root architecture to enable efficient capturing and sharing of nutrients in the soils. Fungi display remarkable phenotypic plasticity in changing their formations of structure, which alter the pathways of resource allocation and exchange in AM mutualisms (Ekanayaka et al., 2025). In considering short- and long-term benefits of AM symbiosis to their partners, several tradeoffs arise in the acquisition of different nutrients and water. Differences in soil types influence the levels of nutrients available for the plant (Scavo et al., 2022) Arbuscular mycorrhizal fungal responses to salt stress in plants are influenced by the availability of nutrients from the soil and temperature (Sousa et al., 2024). Moreover, different genera of AM fungi enhance the degree of salinity stress tolerance in plants differently due to the variations in their effectiveness to ameliorate ion accumulation. With the above understanding, it would be insightful to investigate how these factors alter plant-mycorrhizal fungal interactions and responses with stressors (Sousa et al., 2024).

7.2. Plant Species and Genotype

Depending on the plant species, the response to the application of an AMF varies greatly. Even using the same AMF isolate, some species may respond positively, while others may respond negatively or not at all. This response is relatively well-studied for nutrient enhancement, but starts being explored for resistance against herbivores (Berger & Gutjahr, 2021).Overall, given the vast amount of literature on different fungal genera, and the even larger amount of species belonging to those fungal genera, a systematic approach cannot be achieved. Nonetheless, it has been shown that several species of the same genus (e.g., *Glomus, Claroideoglomus, Entrophospora*) lead to a variety of different plant responses (Zhu et al., 2020). Even when closely related (consensus sequences show over 95% of identity), AMF may induce significantly different responses in the host plants (Seeliger et al., 2024). This may be due to the co-evolution of host and fungal species. It has been shown that AMF have switched hosts in evolution, with fungi from modern-day grasslands arising from C3 AMF ancestors of ancient C4 grasslands (Seeliger et al., 2024). Approximately two-thirds of AMF-generated transcripts were compatible with plant two-component systems, pointing to an important signaling role of AM fungi on plants. Numerous studies have analyzed the interaction of specific plant species with specific AMF in a biogeographic context, allowing identification of most of the variation in plant species response due to AMF type. Genotype effects of the AMF on host plant have also been seen, but have not been studied as well (Bourret et al., 2021; Jin et al., 2023).Understanding the molecular mechanisms of mycorrhization has involved identifying the genes involved in the mycorrhizal symbiosis using a forward genetics approach, isolating plant mutant lines with an impaired development of the fungus (Venice et al., 2021). Most of the work characterizing the genetic processes underlying the fungal colonization of the root system has focused on the early stages of the association. Genetic approaches have led to the identification of a number of essential genes for the colonization and the development of the fungus inside the plant (Branco et al., 2022). The mycorrhizal toolkit of plant genes was shown to confer compatibility with AMF, and it is highly conserved among plant species capable of forming symbiosis. The current knowledge supports the idea that most plants on Earth today have the ability to host mycorrhizal fungi (Prasad, 2021). However, and even using the same AMF taxa or isolate, a given plant species may produce a plethora of different responses. The characterization of the genetic components controlling the host plant response to mycorrhiza, combined with advances in high-throughput genomics, transcriptomics, and metabolomics, should soon clarify the currently obscure functions of the mycorrhizal phenotype (Sweeney et al., 2021).

7.3. Environmental Conditions

AMFs have evolved with terrestrial plants for more than 400 million years and are the most prevalent and diverse group of root endosymbiotic fungi within the soil ecosystem (Liu & Chen, 2024). AMF symbiosis predominates among terrestrial plants, including those adapted to extreme habitats such as alpine tundra and deserts. AMF and their plant hosts are co-adapted and co-evolved to cope with biotic and abiotic stresses in the environment (Wahab et al., 2023).The application of AMF biomass yielded a substantial increase in plant biomass and resulted in a significant alteration of soil physico-chemical and biological properties. AMF colonization positively affected plant height, stem diameter, and leaf area (Wang et al., 2023b). The formulation of AMF in peat-based substrates had a significant effect on enhancing root development, root surface area, and plant growth (Haines, 2023). AMF inoculation of tomato seedlings improved root architecture by enhancing the expression of the root hair-mediated Unusual Rotten Roots-1, which promotes root hair production in tomato plants in a concentration-dependent manner. The grafting of tomato seedlings onto the AMF-inoculated rootstock substantially enhanced the number of roots and root length developed by the scion plant. Two AMF and doses were studied to increase the tolerance of tomato to water stress. The number of effective mycorrhizal spores increased with G. mosseae treatment as the water-stress level increased (Balliu et al., 2021; W. Zhang et al., 2024).Water-stress reduced the number of leaves and leaf area in the plants compared to the respective control plants without fungi (Rehman et al., 2022). Foliar plant height and width were decreased under water-stress conditions; however, *G. mosseae* fungi ameliorate the effects of water-stress conditions (Santra & Banerjee, 2021).The donuts of pigments in the stems and branches were also decreased due to the growth of all fungi under water-stress conditions which ameliorated the water-stress effect. There were differences in root-nodulation numbers and weights by AMF and doses of fungal inoculation, and the highest nodulation was found to occur *by G. mosseae* (Cheema & Garg, 2022; Gough et al., 2022).

8. Implications for Agriculture

Some agricultural practices exert pressure on plant production systems, which disrupts the natural balance required for stable crop production (A. Rehman et al., 2022). In response, the agricultural community has focused on developing healthy soils and optimizing the soil biota, which regulates soil supply . Beneficial soil microorganisms are key promoters of sustainable and enhanced agricultural production systems (Streletskii et al., 2024). Plant mycorrhization is promoted and recommended as a sustainable approach. Currently, the presence of arbuscular mycorrhizal fungi (AMF) compositions might be suppressed due to agricultural practices, which reduce the abundance of AMF necessary for the establishment of mycorrhiza (El Omari & El Ghachtouli, 2021).This is supported by findings which indicate that intensive agricultural practices may cause biotic impoverishment of soils, including mycorrhizal fungi. Complementing the efforts of mycorrhization processes or the restoration of degraded agricultural soils with mycorrhizal fungi isolates may promote soil health (Amaral et al., 2023). Plant companionship is another natural practice, which has been effective in organic agriculture and is part of more sustainable crop production systems (Gamage et al., 2023) Companion plants might cue and promote AMF symbiosis as well, while intercropping systems with a diverse composition of crops might promote AMF composition and prevalence, elemental cycling, and plant productivity (Moreira et al., 2024).Various tools developed for genetic assisted breeding are promoting the studies of germplasm, functional analysis of genes affecting AMF function, or AMF root synaptic selection, and may also be transferred to plants' defense responses against pathogenic fungi (Kaur et al., 2022).The biotechnology of beneficial root symbionts is a genetic breeding or microbiome transplantation for agricultural systems which are effective in recovering or improving production after biotic attacks (Harman et al., 2021). However, mass propagation methods and applications are still limited by biotic constraints. Such systems might be attempted, but it is the core functions of growth control by AMF which should be screened before transplanting to any agricultural systems (Bhantana et al., 2021).

8.1. Sustainable Practices

The development and use of sustainable plant-compatible practices are critical for preventing and controlling the arrival of potentially dangerous phytopathogens (Schiavi et al., 2022). Current control measures rely primarily on chemical fungicides to prevent the start or slow the further spread of plant diseases. These applied control measures may be paralleled by different cultural measures that improve soil–plant health (McLaughlin et al., 2023). The application of AMF aids in the re-establishment of endogenous AMF on land where the ranch was abandoned and/or compromised by some other anthropogenic disturbance (Meddich, 2023). The role of AMF can be critical in the sense that they can ameliorate symptoms of most important root diseases. AM fungal inoculation pre-exposure to a virulent race of a nematodecan substantially lessen the onset of pests' symptoms on previously healthy plants (Samon et al., 2024). Several AMF species when tested either individually or in blends can ameliorate symptoms of both root diseases. The mechanism of action of AMF in lessening root disease development is still a subject of investigation. AMF can lessen the build-up of a biotic stress as a consequence of improved plant and/or soil health at the appropriate population densities of both the AMF and pest at the initiation of pest infestation(Cai et al., 2021; Wang et al., 2024). (Unfortunately, the above-mentioned sustainable plant-compatible practices have not been sufficiently integrated into wider pest management plans. The excessive and unplanned use of synthetic agrochemical fertilizers and pesticides is consigned to be an integral cause of soil and water pollution and of climatic shifts (Nath et al., 2023). Earth's carrying capacity is also being diminished rapidly as a result of intensification of modern day agriculture. Currently, biocontrol agents and biopesticides are being touted as alternatives to control and/or prevent weeds, phytopathogens and invertebrate crop pests (Hamrouni et al., 2024). Development and use of cost-effective plant-compatible biopesticide and biofertilizer products from locally obtained AMF and filamentous fungi, bacterial biocontrol agents, nematophagous fungi and environmentally sustainable plant species have recently added to the number of novel and ecologically compatible options available to regulators and users (Ferreyra-Suarez et al., 2024; Richa, 2023). Essential services needed for sustainable agriculture cannot be delivered by the large farming sector(s) of the economy alone. Investment is needed in Research and Development, and simple insurance-based land-use change options should be implemented to restore wildlife habitats (Prasad et al., 2023).

8.2. Crop Management Strategies

The commercial AM formulations should contain taxa that are adapted to a specific region of use and improve plant response to the target environmental conditions (Campos et al., 2023). The best-mycorrhizal crop species or varieties should be selected. When applied to the soil, mycorrhizal plants should be kept as "mycorrhization farms." The AM fungi isolate should be selected based on functional trait analysis, especially for the strains with more beneficial capacities in nutrient uptake as well as the resilience to soil biota and other edaphic factors (Davison et al., 2021).Terrestrial ecosystems for fungal diversity and the sprawling evolutionary development of the amoeboflagellates should also be reviewed (Baudy et al., 2021). For ameliorating heavy metal (HM) toxicity. Arbuscular mycorrhizal fungi (AMF), soil saprophytic fungi, and bioweathers are regarded as the best agents for remediation of contaminated heavy metals in both terrestrial and aquatic ecosystems (Bhantana et al., 2021). A sustainable way is to employ metal-resistant microorganisms for biotechnological application to remove them (or their toxicity) from contaminated sites. In-depth analysis of the evolutionary history of AMF will give some insight into early land colonization of plants and evolution of terrestrial ecosystems. Further research on their relevance is warranted (Montoliu-Nerin et al., 2021).By better understanding the molecular interactions, AMF could become an environmentally benign epidemic for sustainable crop production under environmental stresses (Thangavel, Anjum et al., 2022). Researches on AMF presently are too scattered without theoretical framework guiding future research endeavors for better defining AMF taxonomy, broadening their applications in crop production systems, or enhancing their scorching capacities for bioremediation of terrestrial and aquatic ecosystems (Sweeney et al., 2022). The commercial AM formulations should contain taxa that are adapted to a specific region of use and improve plant response to the target environmental conditions. The best-mycorrhizal crop species or varieties should be selected (Ricigliano et al., 2021).

9. Future Directions

The focus of future mycorrhizal studies will likely include identifying AMF species that can induce greater plant resistance to a broad spectrum of pathogens (Sarkar and Sadhukhan, 2023). Alterations in rhizosphere microbial communities due to mycorrhizal inoculation in response to soil-borne pathogens require further investigation to determine the responsible functional groups (Deng et al., 2024). Moreover, several unexplored areas may be investigated to acquire comprehensive knowledge of AMF biocontrol mechanisms. Such areas may include the roles of plant growth-promoting rhizobacteria and soil saprophytic microorganisms in plant biocontrol of soil-borne diseases. Addressing these challenges will provide opportunities for further research on the functional characterization of AMF (Chauhan et al., 2022).Understanding the mechanisms regulating plant functions under single and combined biotic and abiotic stresses using omics technologies needs further clarity (B. Yadav et al., 2022). Information on AMF-mediated alterations in the transcriptome, proteome, and metabolome of different plant species infected with different pathogens under variable environmental conditions is limited. Whole-genome sequencing of mycorrhizal fungi is still limited (Q. S. Wu et al., 2023). Genomic information of AMF will provide deeper insights into the complex interactions between plant and pathogen and the metabolic and signaling pathways involved in mycorrhizal symbiosis (J. Shi et al., 2023).The ability to decipher the ecological role of AMF will greatly benefit ecological studies. Aside from the direct effects of biotic stresses, the indirect effects of AMF on plant interactions with herbivores and pathogens through bottom-up and top-down control are challenges to be addressed in future experiments. Recent findings show that AMF-colonized plants exhibit lower susceptibility to insect herbivores and higher resistance to shoot-feeding herbivores. Biocontrol of foliar diseases by AMF is also a promising research area (Anandakumar et al., 2025).

9.1. Research Gaps

Over the past decade, a considerable amount of research has been devoted to exploring the potential benefits of AMF on host plants' performance during the exposure to biotic and abiotic stressors (Sonbol et al., 2025). Numerous AMF species and their corresponding biocontrol mechanisms have been examined in this light. Nevertheless, several crucial points remain to be addressed to define more accurate and effective AMF–host plant–stresses interactions (Moarrefzadeh et al., 2022). First, effects of AMF inoculation on stress severity through the entire post-inoculation period have been mostly overlooked. Understanding the developmental stage of plant to which AMF will respond more effectively to alleviate stress is crucial prior to application in agricultural practice (Dowarah et al., 2022).In addition, there is limited knowledge regarding whether and how various AMF species modulate host plant's performance during their consecutive exposure to different stressors. Whether similar or different stress alleviatory mechanisms are employed by AMF is unclear too (Zou et al., 2020). More accurate examination of biocontrol mechanisms by multiple AMF species in different stressful conditions will improve our understanding of AMF–host plant specialization under field conditions (Ahammed & Hajiboland, 2024). Second, more attention should be paid to selected ecologically relevant abiotic and biotic stresses in parallel to the fundamental information being generated mainly under laboratory conditions about AMF and their positive effects on plant responses to herbivory or disease (Orine et al., 2022).Current knowledge regarding AMF plant responses to soil salinity and heavy metal pollution, both of which represent the most important ecological constraints to crop production on a global scale, are especially lacking (Sagar et al., 2021). Recent advances in multi-omics and various bioinformatics tools have greatly improved our understanding of complex AMF–host plant interactions at multiple levels. These technologies should favor improvement of biocontrol mechanisms by AMF under eco-relevant conditions that have been mostly neglected (Jain et al., 2024).

9.2. Technological Innovations

Fungi from the phylum Glomeromycetes establish symbiosis with >80% of terrestrial plants, and their association with higher plants, termed arbuscular mycorrhiza, is one of the most common interactions in terrestrial ecosystems (Putzke, 2025).These mutualistic fungi promote plant growth by increasing phosphate uptake and by exerting important, phosphate-independent effects on plant physiology and biochemistry. These fungi are also regarded as biopesticides, which assist the plant in biotic stress by inducing systemic resistance (Dowarah et al., 2022). During the interaction between beneficial AM fungi and plant roots, the early signaling events and fungal surface compounds triggering the plant's saturation of the distinct colonization stages and active fungal growth remain largely unknown (Bhupenchandra et al., 2024). AM fungi and plant regulate each other's gene expression during the pre-symbiotic phase. Yet, evidence of which fungal signals activate downstream plant signaling pathways involved in the establishment of AM symbiosis remains poorly understood (Lidoy et al., 2023).Actively growing AM fungal hyphae are surrounded by a vast fungal cell surface area, much of which consists of polysaccharides, glycoproteins, and lipids exposed on the fungal cell surface. These fungal surface compounds interact with plant co-receptors to activate intracellular signaling pathways that may regulate plant responses to mycorrhizal colonization (Delaux & Schornack, 2021). These surface components of AM fungi were previously thought to act as molecular signals; several of them, especially polysaccharides and proteins, have been implicated in the successful establishment of AM symbiosis through physicochemical mechanisms (Etesami et al., 2021).

10. Conclusion

In conclusion, Arbuscular Mycorrhizal Fungi (AMF) represent a crucial biotic factor in enhancing plant resilience to various biotic stresses. Through complex interactions, AMF modulate plant defense mechanisms, increase nutrient uptake, and support plants in dealing with pathogens, herbivores, and competition from weeds. These fungi not only mitigate biotic stress but also improve soil health and support sustainable agricultural practices by reducing the need for chemical inputs like fertilizers and pesticides. As a promising alternative for increasing agricultural productivity, AMF contribute to sustainable food production systems. However, to fully harness their potential, more research is needed to unravel the specific mechanisms by which AMF interact with plants under different stress conditions and to identify the most effective AMF strains for diverse agricultural contexts. This will allow for more tailored and efficient use of AMF in both conventional and organic farming systems, offering a sustainable pathway for improving crop resilience in the face of increasing global agricultural challenges. Further studies on the ecological and genetic interactions between plants, AMF, and environmental stressors will pave the way for new agricultural innovations, advancing the integration of AMF into sustainable farming practices.

References

Abdelkhalek, A., Aseel, D.G., Király, L., Künstler, A., Moawad, H. & Al-Askar, A.A. (2022). Induction of systemic resistance to tobacco mosaic virus in tomato through foliar application of Bacillus amyloliquefaciens strain TBorg1 culture filtrate. Viruses 14(8), 1830. https://doi.org/10.3390/v14081830

Ahammed, G. & Hajiboland, R. (2024). Arbuscular Mycorrhizal Fungi and Higher Plants: Fundamentals and Applications. https://doi.org/10.1007/978-981-99-8220-2

Ait Rahou, Y., Ait-El-Mokhtar, M., Anli, M., Boutasknit, A., Ben-Laouane, R., Douira, A. et al. (2021). Use of mycorrhizal fungi and compost for improving the growth and yield of tomato and its resistance to Verticillium dahliae. Archives of Phytopathology and Plant Protection 54(13–14), 665–690. https://doi.org/10.1080/03235408.2020.1854938

Al-Amri, S.M. (2021). Application of bio-fertilizers for enhancing growth and yield of common bean plants grown under water stress conditions. Saudi Journal of Biological Sciences 28(7), 3901–3908. https://doi.org/10.1016/J.SJBS.2021.03.064

Albqmi, M., Selim, S., Al-Sanea, M.M., Alnusaire, T.S., Almuhayawi, M.S., Jaouni, S.K.A., ... & AbdElgawad, H. (2023). Interactive effect of arbuscular mycorrhizal fungi (AMF) and olive solid waste on wheat under arsenite toxicity. Plants 12(5), 1100.

Ali, J., Mukarram, M., Ojo, J., Dawam, N., Riyazuddin, R., Ghramh, H.A., ... & Bayram, A. (2024). Harnessing phytohormones: Advancing plant growth and defence strategies for sustainable agriculture. Physiologia Plantarum 176(3), e14307.

Ali, M.A., Abdellah, I.M. & Eletmany, M.R. (2023). Towards sustainable management of insect pests: Protecting food security through ecological intensification. International Journal of Chemical and Biochemical Sciences 24(4), 386–394.

Alrajhi, K., Bibi, S. & Abu-Dieyeh, M. (2024). Diversity, distribution, and applications of arbuscular mycorrhizal fungi in the Arabian Peninsula. Saudi Journal of Biological Sciences 31(2), 103911. https://doi.org/10.1016/J.SJBS.2023.103911

Amaral, L., Thomas, P., Amedi, A. & Striem-Amit, E. (2023). Longitudinal stability of individual brain plasticity patterns in blindness. *In* : biorxiv.org. https://doi.org/10.1101/2023.11.01.565196

Anand, U., Pal, T., Yadav, N., Singh, K., Tripathi, V., Choudhary, K. et al. (2023). Current scenario and future prospects of endophytic microbes: Promising candidates for abiotic and biotic stress management for agricultural and environmental. Microbial Ecology 86(3), 1455–1486. Springer. https://doi.org/10.1007/s00248-023-02190-1

Anandakumar, S., Kalaiselvi, T. & Sivakumar, U. (2025). Physio-biochemical response of mycorrhizal and non-mycorrhizal blackgram (Vigna mungo L.) plants to varying populations of herbivorous insect infestation (Spodoptera litura). International Journal of Tropical Insect Science. https://doi.org/10.1007/S42690-025-01471-1

Andrino, A., Guggenberger, G., Sauheitl, L., Burkart, S. & Boy, J. (2021). Carbon investment into mobilization of mineral and organic phosphorus by arbuscular mycorrhiza. Springer 57(1), 47–64. https://doi.org/10.1007/s00374-020-01505-5

Balliu, A., Zheng, Y., Sallaku, G., Fernández, J.A., Gruda, N.S. & Tuzel, Y. (2021). Environmental and cultivation factors affect the morphology, architecture and performance of root systems in soilless grown plants. Horticulturae 7(8), 243. https://doi.org/10.3390/horticulturae7080243

Bano, S.A. & Uzair, B. (2021). Arbuscular Mycorrhizal Fungi (AMF) for improved plant health and production. Microbial Biotechnology in Crop Protection 147–169. https://doi.org/10.1007/978-981-16-0049-4_6

Basiru, S., Ait Si Mhand, K. & Hijri, M. (2023). Disentangling arbuscular mycorrhizal fungi and bacteria at the soil-root interface. Mycorrhiza 33(3), 119–137. https://doi.org/10.1007/S00572-023-01107-7

Basiru, S. & Hijri, M. (2022). The potential applications of commercial arbuscular mycorrhizal fungal inoculants and their ecological consequences. Microorganisms 10(10), 1897.

Baudy, P., Zubrod, J.P., Konschak, M., Kolbenschlag, S., Pollitt, A., Baschien, C., ... & Bundschuh, M. (2021). Fungal–fungal and fungal–bacterial interactions in aquatic decomposer communities: bacteria promote fungal diversity. Ecology 102(10), e03471.

Begum, N., Wang, L., Ahmad, H., Akhtar, K., Roy, R., Khan, M.I. et al. (2022). Co-inoculation of arbuscular mycorrhizal fungi and the plant growth-promoting rhizobacteria improve growth and photosynthesis in tobacco under drought stress by up-regulating antioxidant and mineral nutrition metabolism. Microbial Ecology 83(4), 971–988.

Ben Mrid, R., Benmrid, B., Hafsa, J., Boukcim, H., Sobeh, M. & Yasri, A. (2021). Secondary metabolites as biostimulant and bioprotectant agents: A review. Science of The Total Environment 777, 146204. https://doi.org/10.1016/J.SCITOTENV.2021.146204

Bennett, A.E. & Groten, K. (2022). The costs and benefits of plant-arbuscular mycorrhizal fungal interactions. Annual Review of Plant Biology 73, 649–672. https://doi.org/10.1146/ANNUREV-ARPLANT-102820-124504

Berger, F. & Gutjahr, C. (2021). Factors affecting plant responsiveness to arbuscular mycorrhiza. Current Opinion in Plant Biology 59, 101994. https://doi.org/10.1016/J.PBI.2020.101994

Bernaola, L. & Stout, M.J. (2021). The effect of mycorrhizal seed treatments on rice growth, yield, and tolerance to insect herbivores. Journal of Pest Science 94(2), 375–392. Springer. https://doi.org/10.1007/s10340-020-01279-7

Bhantana, P., Rana, M.S., Sun, X. Cheng, Moussa, M.G., Saleem, M.H., Syaifudin, M. et al. (2021). Arbuscular mycorrhizal fungi and its major role in plant growth, zinc nutrition, phosphorous regulation and phytoremediation. Symbiosis 84(1), 19–37. https://doi.org/10.1007/S13199-021-00756-6

Bhupenchandra, I., Chongtham, S.K., Devi, A.G., Dutta, P., Sahoo, M.R., Mohanty, S. et al. (2024). Unlocking the potential of arbuscular mycorrhizal fungi: Exploring role in plant growth promotion, nutrient uptake mechanisms, biotic stress alleviation, and sustaining agricultural production systems. Journal of Plant Growth Regulation. https://doi.org/10.1007/s00344-024-11467-9

Bolaji Umar, O., Amudalat Ranti, L., Shehu Abdulbaki, A., Lukman Bola, A., Khadijat Abdulhamid, A., Ramat Biola, M. et al. (2022). Stresses in plants: Biotic and abiotic. *In*: Current Trends in Wheat Research. IntechOpen. https://doi.org/10.5772/intechopen.100501

Bourret, R.B., Foster, C.A. & Goldman, W.E. (2021). Predicted functional and structural diversity of receiver domains in fungal two-component regulatory systems. MSphere 6(5). https://doi.org/10.1128/MSPHERE.00722-21

Bourtsoukidis, E., Seco, R., Neri, L., Rapparini, F., Vinci, G., Gallo, F. & Balestreri, C. (2024). Biogenic volatile organic compound emissions in response to climate change-induced environmental stresses. Biogenic Volatile Organic Compounds and Climate Change 49–86. https://doi.org/10.1016/B978-0-12-821076-5.00002-1

Boutasknit, A., Baslam, M., Ait-El-Mokhtar, M., Anli, M., Ben-Laouane, R., Ait-Rahou, Y. et al. (2021). Assemblage of indigenous arbuscular mycorrhizal fungi and green waste compost enhance drought stress tolerance in carob (Ceratonia siliqua L.) trees. Scientific Reports 11(1), 22835. https://doi.org/10.1038/s41598-021-02018-3

Boyno, G. & Demir, S. (2022). Plant-mycorrhiza communication and mycorrhizae in inter-plant communication. Symbiosis 86(2), 155–168. Springer, https://doi.org/10.1007/s13199-022-00837-0

Branco, S., Schauster, A., Liao, H.L. & Ruytinx, J. (2022). Mechanisms of stress tolerance and their effects on the ecology and evolution of mycorrhizal fungi. New Phytologist 235(6), 2158–2175.

Brar, B., Bala, K., Saharan, B.S., Sadh, P.K. & Duhan, J.S. (2024). Bio-boosting agriculture: Harnessing the potential of fungi-bacteria-plant synergies for crop improvement. Discover Plants 1(1), 21. Springer. https://doi.org/10.1007/s44372-024-00023-0

Brito, I., Carvalho, M. & Goss, M.J. (2021). Managing the functional diversity of arbuscular mycorrhizal fungi for the sustainable intensification of crop production. Plants People Planet 3(5), 491–505. https://doi.org/10.1002/PPP3.10212

Brown, B., Gallandt, E.R., DiTommaso, A., Salon, P., Smith, R.G., Ryan, M.R. et al. (2022). Improving weed management based on the timing of emergence peaks: A case study of problematic weeds in Northeast USA. Frontiers in Agronomy 4. https://doi.org/10.3389/FAGRO.2022.888664/FULL

Cai, X., Zhao, H., Liang, C., Li, M. & Liu, R. (2021). Effects and mechanisms of symbiotic microbial combination agents to control tomato fusarium crown and root rot disease. Frontiers in Microbiology 12. https://doi.org/10.3389/FMICB.2021.629793/FULL

Campos, E.V.R., Pereira, A. do E.S., Aleksieienko, I., do Carmo, G.C., Gohari, G., Santaella, C. et al. (2023). Encapsulated plant growth regulators and associative microorganisms: Nature-based solutions to mitigate the effects of climate change on plants. Plant Science 331, 111688. https://doi.org/10.1016/J.PLANTSCI.2023.111688

Cataldo, E., Fucile, M. & Mattii, G.B. (2022). Biostimulants in viticulture: A sustainable approach against biotic and abiotic stresses. Plants 11(2), 162. https://doi.org/10.3390/plants11020162

Cerrato, M.D., Mir-Rosselló, P.M., Cortés-Fernández, I., Ribas-Serra, A., Douthe, C., Cardona, C., ... & Gil Vives, L. (2024). Insights on physiological, antioxidant and flowering response to salinity stress of two candidate ornamental species: The native coastal geophytes Pancratium maritimum L. and Eryngium maritimum L. Physiology and Molecular Biology of Plants 30(9), 1533–1549.

Chauhan, S., Mahawar, S., Jain, D., Udpadhay, S.K., Mohanty, S.R., Singh, A. & Maharjan, E. (2022). Boosting sustainable agriculture by arbuscular mycorrhiza under stress condition: Mechanism and future prospective. BioMed Research International 2022(1), 5275449.

Cheema, A. & Garg, N. (2022). Differential effectiveness of Arbuscular Mycorrhizae in improving rhizobial symbiosis by modulating sucrose metabolism and antioxidant defense in Chickpea under As stress. Symbiosis 86(1), 49–69. https://doi.org/10.1007/S13199-021-00815-Y

Chen, G., Yang, A., Wang, J., Ke, L., Chen, S. & Li, W. (2024). Effects of the synergistic treatments of arbuscular mycorrhizal fungi and trehalose on adaptability to salt stress in tomato seedlings. Microbiology Spectrum 12(3). https://doi.org/10.1128/spectrum.03404-23

Chiquito-Contreras, C.J., Meza-Menchaca, T., Guzmán-López, O., Vásquez, E.C. & Ricaño-Rodríguez, J. (2024). Molecular insights into plant–microbe interactions: A comprehensive review of key mechanisms. Frontiers in Bioscience – Elite 16(1). https://doi.org/10.31083/J.FBE1601009/HTM

Chourasiya, D., Gajghate, R., Bharti, A., Prakash, A. & Sharma, M.P. (2022). Deciphering the role of phytohormones in the regulation of arbuscular mycorrhizal fungal symbiosis and mechanisms involved. New and Future Developments in Microbial Biotechnology and Bioengineering: Sustainable Agriculture: Revisiting Green Chemicals 427–447. https://doi.org/10.1016/B978-0-323-85581-5.00022-7

Das Laha, S., Kundu, A. & Podder, S. (2024). Impact of biotic stresses on the Brassicaceae family and opportunities for crop improvement by exploiting genotyping traits. Planta 259(5). https://doi.org/10.1007/S00425-024-04379-1

Das, P.P., Singh, K.R., Nagpure, G., Mansoori, A., Singh, R.P., Ghazi, I.A. et al. (2022). Plant-soil-microbes: A tripartite interaction for nutrient acquisition and better plant growth for sustainable agricultural practices. Environmental Research 214, 113821. https://doi.org/10.1016/J.ENVRES.2022.113821

Davison, J., Moora, M., Semchenko, M., Adenan, S.B., Ahmed, T., Akhmetzhanova, A.A. et al. (2021). Temperature and pH define the realised niche space of arbuscular mycorrhizal fungi. New Phytologist 231(2), 763–776. https://doi.org/10.1111/nph.17240

Deja-Sikora, E., Werner, K. & Hrynkiewicz, K. (2023). AMF species do matter: Rhizophagus irregularis and Funneliformis mosseae affect healthy and PVY-infected Solanum tuberosum L. in a different way. Frontiers in Microbiology 14. https://doi.org/10.3389/FMICB.2023.1127278/FULL

Delaux, P.M. & Schornack, S. (2021). Plant evolution driven by interactions with symbiotic and pathogenic microbes. Science 371(6531). https://doi.org/10.1126/SCIENCE.ABA6605

Demir, S., Danesh, Y.R., Boyno, G. & Najafi, S. (2022). Arbuscular mycorrhizal fungi in biotic and abiotic stress conditions: function and management in horticulture. Sustainable Horticulture: Microbial Inoculants and Stress Interaction 157–183. https://doi.org/10.1016/B978-0-323-91861-9.00011-2

Deng, Y., Kong, W., Zhang, X., Zhu, Y., Xie, T., Chen, M. et al. (2024). Rhizosphere microbial community enrichment processes in healthy and diseased plants: Implications of soil properties on biomarkers. Frontiers in Microbiology 15. https://doi.org/10.3389/FMICB.2024.1333076/FULL

Díaz, V., Villalobos, M., Arriaza, K., Flores, K., Hernández-Saravia, L.P. & Velásquez, A. (2025). Decoding the dialog between plants and arbuscular mycorrhizal fungi: A molecular genetic perspective. Genes 16(2), 143. https://doi.org/10.3390/genes16020143

Dixit, S., Sivalingam, P.N., Baskaran, R.K.M., Senthil-Kumar, M. & Ghosh, P.K. (2024). Plant responses to concurrent abiotic and biotic stress: Unravelling physiological and morphological mechanisms. Plant Physiology Reports 29(1), 6–17. https://doi.org/10.1007/s40502-023-00766-0

Dowarah, B., Gill, S.S. & Agarwala, N. (2022). Arbuscular mycorrhizal fungi in conferring tolerance to biotic stresses in plants. Journal of Plant Growth Regulation 41(4), 1429–1444. https://doi.org/10.1007/S00344-021-10392-5

Ducousso-Détrez, A., Raveau, R., Fontaine, J., Hijri, M. & Lounès-Hadj Sahraoui, A. (2022). Glomerales dominate arbuscular mycorrhizal fungal communities associated with spontaneous plants in phosphate-rich soils of former rock phosphate mining sites. Microorganisms 10(12), 2406. https://doi.org/10.3390/microorganisms10122406

Dutta, P., Kumari, A., Mahanta, M., Biswas, K.K., Dudkiewicz, A., Thakuria, D. et al. (2022). Advances in nanotechnology as a potential alternative for plant viral disease management. Frontiers in Microbiology 13. https://doi.org/10.3389/FMICB.2022.935193/PDF

Ekanayaka, A.H., Karunarathna, S.C., Tibpromma, S., Dutta, A.K., Tennakoon, D.S., Karunarathna, A. et al. (2025). Species evolution: Cryptic species and phenotypic noise with a particular focus on fungal systematics. Frontiers in Cellular and Infection Microbiology 15. https://doi.org/10.3389/FCIMB.2025.1497085/FULL

El Omari, B. & El Ghachtouli, N. (2021). Arbuscular mycorrhizal fungi-weeds interaction in cropping and unmanaged ecosystems: A review. Symbiosis 83(3), 279–292. https://doi.org/10.1007/S13199-021-00753-9

Elhamouly, N.A., Hewedy, O.A., Zaitoon, A., Miraples, A., Elshorbagy, O.T., Hussien, S. et al. (2022). The hidden power of secondary metabolites in plant-fungi interactions and sustainable phytoremediation. Frontiers in Plant Science 13. https://doi.org/10.3389/FPLS.2022.1044896/FULL

El-Saadony, M.T., Saad, A.M., Soliman, S.M., Salem, H.M., Ahmed, A.I., Mahmood, M. et al. (2022). Plant growth-promoting microorganisms as biocontrol agents of plant diseases:

Mechanisms, challenges and future perspectives. Frontiers in Plant Science 13. https://doi.org/10.3389/FPLS.2022.923880/FULL

El-Sayed, E.S.R., Mohamed, S.S., Mousa, S.A., El-Seoud, M.A.A., Elmehlawy, A.A., & Abdou, D.A. (2023). Bifunctional role of some biogenic nanoparticles in controlling wilt disease and promoting growth of common bean. AMB Express 13(1), 41. Springer. https://doi.org/10.1186/s13568-023-01546-7

Elvira González, L., Peiró, R., Rubio, L. & Galipienso, L. (2021). Persistent southern tomato virus (STV) interacts with cucumber mosaic and/or pepino mosaic virus in mixed-infections modifying plant symptoms, viral titer and small RNA accumulation. Microorganisms, 9(4), 689.

Erb, M., Züst, T. & Robert, C.A.M. (2021). Using plant chemistry to improve interactions between plants, herbivores and their natural enemies: Challenges and opportunities. Current Opinion in Biotechnology 70, 262–265. https://doi.org/10.1016/J.COPBIO.2021.05.011

Etesami, H., Jeong, B.R. & Glick, B.R. (2021). Contribution of arbuscular mycorrhizal fungi, phosphate–solubilizing bacteria, and silicon to P uptake by plant. Frontiers in Plant Science 12. https://doi.org/10.3389/FPLS.2021.699618/FULL

Fabiyi, O.A., Ansari, R.A., Rizvi, R., Sumbul, A. & Mahmood, I. (2024). Influence of arbuscular mycorrhizal fungi on soil health amelioration and plant fitness under hostile environment. Mycorrhizal Symbiosis and Agroecosystem Restoration 227–248. https://doi.org/10.1007/978-981-99-5030-0_10

Fall, A.F., Nakabonge, G., Ssekandi, J., Founoune-Mboup, H., Apori, S.O., Ndiaye, A. et al. (2022). Roles of arbuscular mycorrhizal fungi on soil fertility: Contribution in the improvement of physical, chemical, and biological properties of the soil. Frontiers in Fungal Biology 3. https://doi.org/10.3389/FFUNB.2022.723892/FULL

Felföldi, Z., Vidican, R., Stoian, V., Roman, I.A., Sestras, A.F., Rusu, T. & Sestras, R.E. (2022). Arbuscular mycorrhizal fungi and fertilization influence yield, growth and root colonization of different tomato genotype. Plants 11(13), 1743. https://doi.org/10.3390/plants11131743

Ferreyra-Suarez, D., García-Depraect, O. & Castro-Muñoz, R. (2024). A review on fungal-based biopesticides and biofertilizers production. Ecotoxicology and Environmental Safety 283, 116945. https://doi.org/10.1016/J.ECOENV.2024.116945

Fiorilli, V., Martínez-Medina, A., Pozo, M.J. & Lanfranco, L. (2024). Plant immunity modulation in arbuscular mycorrhizal symbiosis and its impact on pathogens and pests. Annual Review of Phytopathology 62(1), 127–156. https://doi.org/10.1146/ANNUREV-PHYTO-121423-042014

Fiorilli, V., Salmeron-Santiago, I.A., Martínez-Trujillo, M., Valdez-Alarcón, J.J., Pedraza-Santos, M.E., Santoyo, G. et al. (2021). An updated review on the modulation of carbon partitioning and allocation in arbuscular mycorrhizal plants. Microorganisms. https://doi.org/10.3390/microorganisms10010075

Flemer, B., Gulati, S., Bergna, A., Rändler, M., Cernava, T., Witzel, K. et al. (2022). Biotic and abiotic stress factors induce microbiome shifts and enrichment of distinct beneficial bacteria in tomato roots. Phytobiomes Journal 6(4), 276–289. https://doi.org/10.1094/PBIOMES-10-21-0067-R

Gamage, A., Gangahagedara, R., Gamage, J., Jayasinghe, N., Kodikara, N., Suraweera, P. & Merah, O. (2023). Role of organic farming for achieving sustainability in agriculture. Farming System 1(1), 100005. https://doi.org/10.1016/J.FARSYS.2023.100005

Gauthier, A.E., Rotjan, R.D. & Kagan, J.C. (2022). Lipopolysaccharide detection by the innate immune system may be an uncommon defence strategy used in nature. Open Biology 12(10). https://doi.org/10.1098/RSOB.220146

Ge, Y. (2024). Root Nodule and Ectomycorrhizae Symbioses Recruit Mechanisms from the Ancestral Arbuscular Mycorrhizae Symbiosis. https://search.proquest.com/openview/615066c0b52b0a0c3596195ea7c793ad/1?pq-origsite=gscholar&cbl=2026366&diss=y

Gill, T., Gill, S.K., Saini, D.K., Chopra, Y., de Koff, J.P. & Sandhu, K.S. (2022). A comprehensive review of high throughput phenotyping and machine learning for plant stress phenotyping. Phenomics 2(3), 156–183. https://doi.org/10.1007/S43657-022-00048-Z

Giudice, G. (2022). New Plant Breeding Techniques and Priming as a Multiple Level Strategy for the Control of Downy Mildew Infection in Grapevine. https://air.unimi.it/handle/2434/924372

Gong, P., Shen, Q., Zhang, M., Qiao, R., Jiang, J., Su, L. et al. (2023). Plant and animal positive-sense single-stranded RNA viruses encode small proteins important for viral infection in their negative-sense strand. Molecular Plant 16(11), 1794–1810. https://doi.org/10.1016/j.molp.2023.09.020

Gough, E.C., Owen, K.J., Zwart, R.S. & Thompson, J.P. (2022). The role of nutrients underlying interactions among root-nodule bacteria (Bradyrhizobium sp.), arbuscular mycorrhizal fungi (Funneliformis mosseae) and root-lesion nematodes (Pratylenchus thornei) in nitrogen fixation and growth of mung bean (Vigna radiata). Plant and Soil 472(1–2), 421–449. https://doi.org/10.1007/s11104-021-05254-8

Grünfeld, L., Skias, G., Rillig, M.C. & Veresoglou, S.D. (2022). Arbuscular mycorrhizal root colonization depends on the spatial distribution of the host plants. Mycorrhiza 32(5), 387–395. https://doi.org/10.1007/s00572-022-01087-0

Haines, O.S. (2023). Evaluating Alternatives to Peat Based Growing Substrates for Use in Ornamental Horticulture. https://search.proquest.com/openview/2f8e76f2db821b1f986e1034be77b97a/1?pq-origsite=gscholar&cbl=2026366&diss=y

Hammer, E.C., Arellano-Caicedo, C., Mafla-Endara, P.M., Kiers, E.T., Shimizu, T., Ohlsson, P. & Aleklett, K. (2024). Hyphal exploration strategies and habitat modification of an arbuscular mycorrhizal fungus in microengineered soil chips. Fungal Ecology 67, 101302. https://doi.org/10.1016/J.FUNECO.2023.101302

Hamrouni, R., Regus, F., Farnet Da Silva, A.M., Orsiere, T., Boudenne, J.L., Laffont-Schwob, I. et al. (2024). Current status and future trends of microbial and nematode-based biopesticides for biocontrol of crop pathogens. Critical Reviews in Biotechnology. https://doi.org/10.1080/07388551.2024.2370370

Harman, G., Khadka, R., Doni, F. & Uphoff, N. (2021). Benefits to plant health and productivity from enhancing plant microbial symbionts. Frontiers in Plant Science 11. https://doi.org/10.3389/FPLS.2020.610065/FULL

Howard, N., Pressel, S., Kaye, R.S., Daniell, T.J. & Field, K.J. (2022). The potential role of Mucoromycotina 'fine root endophytes' in plant nitrogen nutrition. Physiologia Plantarum 174(3). https://doi.org/10.1111/PPL.13715

Hu, C., Xu, T., Wang, S., Bian, H. & Dai, H. (2024). Effect of aminating lignin loading with arbuscular mycorrhizal fungi on soil aggregate structure improvement. Polymers 16(12), 1701. https://doi.org/10.3390/polym16121701

Hussain, H.A., Uddin, S., Hussain, S., Rahman, A., Khalid, S., Hussain, W. et al. (2024). Crosstalk between phytohormones and environmental signals in the regulation of mycorrhizal symbiosis. Essential Minerals in Plant-Soil Systems: Coordination, Signaling, and Interaction under Adverse Situations 197–214. https://doi.org/10.1016/B978-0-443-16082-0.00007-2

Iqbal, Z., Iqbal, M.S., Hashem, A., Abd_Allah, E.F. & Ansari, M.I. (2021). Plant defense responses to biotic stress and its interplay with fluctuating dark/light conditions. Frontiers in Plant Science 12. https://doi.org/10.3389/FPLS.2021.631810/FULL

Jain, A., Sarsaiya, S., Singh, R., Gong, Q., Wu, Q. & Shi, J. (2024). Omics approaches in understanding the benefits of plant-microbe interactions. Frontiers in Microbiology 15. https://doi.org/10.3389/FMICB.2024.1391059/FULL

Jeger, M., Beresford, R., Bock, C., Brown, N., Fox, A., Newton, A. et al. (2021). Global challenges facing plant pathology: multidisciplinary approaches to meet the food security and environmental challenges in the mid-twenty-first century. CABI Agriculture and Bioscience 2(1), 20. https://doi.org/10.1186/s43170-021-00042-x

Jin, S., Hui, M., Lu, Y. & Zhao, Y. (2023). An overview on the two-component systems of Streptomyces coelicolor. World Journal of Microbiology and Biotechnology 39(3). https://doi.org/10.1007/S11274-023-03522-6

Kakabouki, I., Stavropoulos, P., Roussis, I. & Mavroeidis, A. (2023). The Influence of Salinity and Arbuscular Mycorrhizal Fungi (AMF) on the Growth and Yield Performances of Flax (*Linum usitatissimum* L.). https://www.preprints.org/frontend/manuscript/b90db233e4ba3a00d34c868f7ae3e707/download_pub

Kashyap, B. & Kumar, R. (2021). Sensing methodologies in agriculture for monitoring biotic stress in plants due to pathogens and pests. Inventions 6(2), 29. https://doi.org/10.3390/inventions6020029

Kaur, S., Samota, M.K., Choudhary, M., Pandey, A.K., Sharma, A. & Thakur, J. (2022). How do plants defend themselves against pathogens – Biochemical mechanisms and genetic interventions. Physiology and Molecular Biology of Plants 28(2), 485–504. https://doi.org/10.1007/s12298-022-01146-y

Kumar, A., Dutta, J., Chandrawanshi, N.K., Ekka, A. & Sethi, S.K. (2024). Arbuscular mycorrhizal fungi in alleviation of biotic stress tolerance in plants: A new direction in sustainable agriculture. Entrepreneurship with Microorganisms 355–369. https://doi.org/10.1016/B978-0-443-19049-0.00006-2

Kumari, N., Sharma, V., Patel, P. & Sharma, P.N. (2023). Pepper mild mottle virus: A formidable foe of capsicum production—A review. Frontiers in Virology 3. https://doi.org/10.3389/FVIRO.2023.1208853/FULL

Leisner, C.P., Potnis, N. & Sanz-Saez, A. (2023). Crosstalk and trade-offs: Plant responses to climate change-associated abiotic and biotic stresses. Plant, Cell & Environment 46(10), 2946–2963. https://doi.org/10.1111/pce.14532

Li, Y., Yang, K. & Wu, B. (2024). Feature selection and spectral indices for identifying maize stress types. Applied Spectroscopy. https://doi.org/10.1177/00037028241279328

Lidoy, J., López-García, Á., Amate, C., García, J.M., Flors, V., García-Garrido, J.M. et al. (2023). Regulation of mycorrhizal colonization under stress in tomato depends on symbiotic efficiency. Environmental and Experimental Botany 215, 105479. https://doi.org/10.1016/J.ENVEXPBOT.2023.105479

Lin, P., Zhang, M., Wang, M., Li, Y., Liu, J. & Chen, Y. (2021). Inoculation with arbuscular mycorrhizal fungus modulates defense-related genes expression in banana seedlings susceptible to wilt disease. Plant Signaling & Behavior 16(5). https://doi.org/10.1080/15592324.2021.1884782

Liu, R. & Chen, Y. (2024). Mycorrhizal Symbiosis: Evolution, opportunities, challenges, and prospects. Arbuscular Mycorrhizal Fungi in Sustainable Agriculture: Inoculum Production and Application 1–35. https://doi.org/10.1007/978-981-97-0296-1_1

Lo Presti, D., Di Tocco, J., Massaroni, C., Cimini, S., De Gara, L., Singh, S. et al. (2023). Current understanding, challenges and perspective on portable systems applied to plant monitoring and precision agriculture. Biosensors and Bioelectronics 222, 115005. https://doi.org/10.1016/J.BIOS.2022.115005

Lotfabadi, Z.E., Weisany, W., Nawroz, Tahir, A.-R., & Torkashvand, M. (2022). Arbuscular mycorrhizal fungi species improve the fatty acids profile and nutrients status of soybean cultivars grown under drought stress. Journal of Applied Microbiology 132, 2177. https://doi.org/10.1111/jam.15326

Lutz, S., Bodenhausen, N., Hess, J., Valzano-Held, A., Waelchli, J., Deslandes-Hérold, G. et al. (2023). Soil microbiome indicators can predict crop growth response to large-scale inoculation with arbuscular mycorrhizal fungi. Nature Microbiology 8. https://doi.org/10.1038/s41564-023-01520-w

Malar, C.M., Krüger, M., Krüger, C., Wang, Y., Stajich, J.E., Keller, J. et al. (2021). The genome of Geosiphon pyriformis reveals ancestral traits linked to the emergence of the arbuscular mycorrhizal symbiosis. Current Biology 31(7), 1570-1577. e4. https://doi.org/10.1016/j.cub.2021.01.058

Mcdonald, S.T., Sarangi, D., Rees, J.M. & Jhala, A.J. (2023). A follow-up survey to assess stakeholders' perspectives on weed management challenges and current practices in Nebraska, USA. Agrosystems, Geosciences & Environment 6(3). https://doi.org/10.1002/agg2.20425

McLaughlin, M.S., Roy, M., Abbasi, P.A., Carisse, O., Yurgel, S.N. & Ali, S. (2023). Why do we need alternative methods for fungal disease management in plants? Plants 12(22), 3822. https://doi.org/10.3390/plants12223822

Meddich, A. (2023). Biostimulants for resilient agriculture—Improving plant tolerance to abiotic stress: A concise review. Gesunde Pflanzen 75(4), 709–727. https://doi.org/10.1007/S10343-022-00784-2

Melini, F., Melini, V., Luziatelli, F., Abou Jaoudé, R., Ficca, A.G. & Ruzzi, M. (2023). Effect of microbial plant biostimulants on fruit and vegetable quality: Current research lines and future perspectives. Frontiers in Plant Science 14. https://doi.org/10.3389/FPLS.2023.1251544/FULL

Metwally, R.A., Taha, M.A., El-Moaty, N.M.A. & Abdelhameed, R.E. (2024). Attenuation of Zucchini mosaic virus disease in cucumber plants by mycorrhizal symbiosis. Plant Cell Reports 43(2), 54. https://doi.org/10.1007/s00299-023-03138-y

Mitra, D., Guerra Sierra, B.E., Khoshru, B., De Los Santos Villalobos, S., Belz, C., Chaudhary, P. et al. (2021). Impacts of arbuscular mycorrhizal fungi on rice growth, development, and stress management with a particular emphasis on strigolactone effects on root development. Communications in Soil Science and Plant Analysis 52(14), 1591–1621. https://doi.org/10.1080/00103624.2021.1892728

Mittal, U., Kumar, V., Kukreja, S., Singh, B., Pandey, N.K. & Goutam, U. (2023). Role of beneficial elements in developing resilience to abiotic and biotic stresses in plants: Present status and future prospects. Journal of Plant Growth Regulation 42(6), 3789–3813. https://doi.org/10.1007/S00344-022-10840-W

Moarrefzadeh, N., Khateri, H. & Sharifi, R. (2022). The effects of some arbuscular mycorrhizal fungi on plant growth and biocontrol of Ascochyta blight in two chickpea varieties. Journal of Microbial Biology. https://doi.org/10.22108/bjm2021.130031.1410

Montoliu-Nerin, M., Sánchez-García, M., Bergin, C., Kutschera, V.E., Johannesson, H., Bever, J.D. & Rosling, A. (2021). In-depth phylogenomic analysis of arbuscular mycorrhizal fungi based on a comprehensive set of de novo genome assemblies. Frontiers in Fungal Biology 2. https://doi.org/10.3389/FFUNB.2021.716385/FULL

Moreira, X., Martín-Cacheda, L., Quiroga, G., Lago-Núñez, B., Röder, G. & Abdala-Roberts, L. (2024). Testing the joint effects of arbuscular mycorrhizal fungi and ants on insect herbivory on potato plants. Planta 260(3), 66. https://doi.org/10.1007/s00425-024-04492-1

Moreira, X., Martín-Cacheda, L., Vázquez-González, C., Rasmann, S., Röder, G. & Abdala-Roberts, L. (2023). Volatile-mediated signalling between potato plants in response to insect herbivory is not contingent on soil nutrients. Journal of Chemical Ecology 49(9–10), 507–517. https://doi.org/10.1007/s10886-023-01445-y

Nasfi, S., Shahbazi, S., Bitterlich, K., Šečić, E., Kogel, K.-H. & Steinbrenner, J. (2025). A pipeline for validation of *Serendipita indica* effector-like sRNA suggests cross-kingdom communication in the symbiosis with Arabidopsis. Journal of Experimental Botany 76(6), 1811–1829. https://doi.org/10.1093/jxb/erae515

Nath, A., Bhuyan, P., Gogoi, N. & Deka, P. (2023). Pesticides and chemical fertilizers: Role in soil degradation, groundwater contamination, and human health. Xenobiotics in Urban Ecosystems: Sources, Distribution and Health Impacts 131–160. https://doi.org/10.1007/978-3-031-35775-6_7

Nawaz, M., Sun, J., Shabbir, S., Khattak, W.A., Ren, G., Nie, X. et al. (2023). A review of plants strategies to resist biotic and abiotic environmental stressors. Science of The Total Environment 900, 165832. https://doi.org/10.1016/J.SCITOTENV.2023.165832

Ngumbi, E., Dady, E. & B. Calla (2021). Flooding and herbivory: The effect of concurrent stress factors on plant volatile emissions and gene expression in two heirloom tomato varieties. BMC Plant Biology 22(1). https://doi.org/10.1186/s12870-022-03911-3

Noceto, P.-A., Bettenfeld, P., Boussageon, R., Hériché, M., Sportes, A., Diederik van Tuinen & Courty, P.-E. (2021). Arbuscular mycorrhizal fungi, a key symbiosis in the development of quality traits in crop production, alone or combined with plant growth-promoting bacteria. Mycorrhiza 1(6), 3. https://doi.org/10.1007/s00572-021-01054-1

Oliveira, T.C., Cabral, J.S.R., Santana, L.R., Tavares, G.G., Santos, L.D.S., Paim, T.P. et al. (2022). The arbuscular mycorrhizal fungus Rhizophagus clarus improves physiological tolerance to drought stress in soybean plants. Scientific Reports 12(1), 9044. https://doi.org/10.1038/s41598-022-13059-7

Opitz, M.W., Daneshkhah, R., Lorenz, C., Ludwig, R., Steinkellner, S. & Wieczorek, K. (2021). Serendipita indica changes host sugar and defense status in Arabidopsis thaliana: Cooperation or exploitation? Planta 253(3), 74. https://doi.org/10.1007/s00425-021-03587-3

Orine, D., Defossez, E., Vergara, F., Uthe, H., van Dam, N.M. & Rasmann, S. (2022). Arbuscular mycorrhizal fungi prevent the negative effect of drought and modulate the growth-defence trade-off in tomato plants. Journal of Sustainable Agriculture and Environment 1(3), 177–190. https://doi.org/10.1002/sae2.12018

Panahi, B. (2024). Transcriptome signature for multiple biotic and abiotic stress in barley (*Hordeum vulgare* L.) identifies using machine learning approach. Current Plant Biology 40, 100416. https://doi.org/10.1016/J.CPB.2024.100416

Parasar, B.J., Sharma, I. & Agarwala, N. (2024). Root exudation drives abiotic stress tolerance in plants by recruiting beneficial microbes. Applied Soil Ecology 198, 105351. https://doi.org/10.1016/J.APSOIL.2024.105351

Perez-Gil, J., Behrendorff, J., Douw, A. & Vickers, C.E. (2024). The methylerythritol phosphate pathway as an oxidative stress sense and response system. Nature Communications 15(1), 5303. https://doi.org/10.1038/s41467-024-49483-8

Pérez-Hedo, M., Alonso-Valiente, M., Vacas, S., Gallego, C., Pons, C., Arbona, V. et al. (2021). Plant exposure to herbivore-induced plant volatiles: A sustainable approach through eliciting plant defenses. Journal of Pest Science 94(4), 1221–1235. https://doi.org/10.1007/S10340-021-01334-X

Phour, M. & Sindhu, S.S. (2024). Arbuscular mycorrhizal fungi: An eco-friendly technology for alleviation of salinity stress and nutrient acquisition in sustainable agriculture. Arbuscular Mycorrhizal Fungi in Sustainable Agriculture: Nutrient and Crop Management, 223–267. https://doi.org/10.1007/978-981-97-0300-5_11

Poudel, M., Mendes, R., Costa, L.A.S., Bueno, C.G., Meng, Y., Folimonova, S.Y. et al. (2021). The role of plant-associated bacteria, fungi, and viruses in drought stress mitigation. Frontiers in Microbiology 12. https://doi.org/10.3389/FMICB.2021.743512/FULL

Pramanik, D., Shelake, R.M., Park, J., Kim, M.J., Hwang, I., Park, Y. & Kim, J.-Y. (2021). CRISPR/Cas9-mediated generation of pathogen-resistant tomato against tomato yellow leaf curl virus and powdery mildew. International Journal of Molecular Sciences 22(4), 1878. https://doi.org/10.3390/ijms22041878

Prasad, K. (2021). Arbuscular mycorrhizal fungi and plant collaborations influences ecology and environmental changes for global sustainable development. Journal of Ecology and Natural Resources 5(1), 1–16. https://doi.org/10.23880/jenr-16000231

Prasad, P.V.V., Bhatnagar, N., Bhandari, V., Jacob, G., Narayan, K., Echeverría, R. et al. (2023). Patterns of investment in agricultural research and innovation for the Global South, with a focus on sustainable agricultural intensification. Frontiers in Sustainable Food Systems 7. https://doi.org/10.3389/FSUFS.2023.1108949/FULL

Primieri, S., Magnoli, S.M., Koffel, T., Stürmer, S.L. & Bever, J.D. (2022). Perennial, but not annual legumes synergistically benefit from infection with arbuscular mycorrhizal fungi and rhizobia: A meta-analysis. New Phytologist 233(1), 505–514. https://doi.org/10.1111/NPH.17787

Putzke, J. (2025). The Diversity of the Fungal World. https://books.google.com/books?hl=fr&lr=&id=aEFLEQAAQBAJ&oi=fnd&pg=PP10&dq=Lopes,+C.+H.,+Galiano,+V.,+%26+Zorzi,+A.+F.+(2025).+and+Ecological+Importance.+The+Diversity+of+the+Fungal+World.+%5BHTML%5D&ots=0FXJVkfwSw&sig=oz_nTr1xf5mjwDz6EoJWIvKiJqc

Qasim, M., Islam, W., Rizwan, M., Hussain, D., Noman, A., Khan, K.A. et al. (2024). Impact of plant monoterpenes on insect pest management and insect-associated microbes. Heliyon 10(20), e39120. https://doi.org/10.1016/j.heliyon.2024.e39120

Qin, M., Li, L., Miranda, J.P., Tang, Y., Song, B., Oosthuizen, M.K. & Wei, W. (2022). Experimental duration determines the effect of arbuscular mycorrhizal fungi on plant biomass in pot experiments: A meta-analysis. Frontiers in Plant Science 13. https://doi.org/10.3389/FPLS.2022.1024874/FULL

Rajarammohan, S. (2021). Redefining plant-necrotroph interactions: The thin line between hemibiotrophs and necrotrophs. Frontiers in Microbiology 12. https://doi.org/10.3389/FMICB.2021.673518/FULL

Ramírez-Serrano, B., Querejeta, M., Minchev, Z., Gamir, J., Perdereau, E., Pozo, M.J. et al. (2022). Mycorrhizal benefits on plant growth and protection against Spodoptera exigua depend on N availability. Journal of Plant Interactions 17(1), 940–955. https://doi.org/10.1080/17429145.2022.2120212

Rehman, A., Farooq, M., Lee, D.J. & Siddique, K.H.M. (2022). Sustainable agricultural practices for food security and ecosystem services. Environmental Science and Pollution Research 29(56), 84076–84095. https://doi.org/10.1007/S11356-022-23635-Z

Rehman, M.M.U., Zhu, Y., Abrar, M., Khan, W., Wang, W., Iqbal, A. et al. (2022). Moisture- and period-dependent interactive effects of plant growth-promoting rhizobacteria and AM fungus on water use and yield formation in dryland wheat. Plant and Soil. https://doi.org/10.1007/S11104-022-05641-9

Richa. (2023). Liquid bio-fertilizers: Prospects and challenges. Metabolomics, Proteomes and Gene Editing Approaches in Biofertilizer Industry 77–99. https://doi.org/10.1007/978-981-99-3561-1_5

Ricigliano, V.A., Williams, S.T. & Oliver, R. (2021). Effects of different artificial diets on commercial honey bee colony performance, health biomarkers, and gut microbiota. BMC Veterinary Research 18(1). https://doi.org/10.1186/s12917-022-03151-5

Rosling, A., Eshghi Sahraei, S., Kalsoom Khan, F., Desirò, A., Bryson, A.E., Mondo, S.J. et al. (2024). Evolutionary history of arbuscular mycorrhizal fungi and genomic signatures of obligate symbiosis. BMC Genomics 25(1), 529. https://doi.org/10.1186/s12864-024-10391-2

Rostami, Z., Fazeli, A. & Hojati, Z. (2022). The isolation and expression analysis of cinnamate 4-hydroxylase and chalcone synthase genes of Scrophularia striata under different abiotic elicitors. Scientific Reports 12(1), 8128. https://doi.org/10.1038/s41598-022-12361-8

Sagar, A., Rathore, P., Ramteke, P.W., Ramakrishna, W., Reddy, M.S. & Pecoraro, L. (2021). Plant growth promoting rhizobacteria, arbuscular mycorrhizal fungi and their synergistic interactions to counteract the negative effects of saline soil on agriculture: Key macromolecules and mechanisms. Microorganisms 9(7), 1491. https://doi.org/10.3390/microorganisms9071491

Sahu, P.K., Jayalakshmi, K., Tilgam, J., Gupta, A., Nagaraju, Y., Kumar, A. et al. (2022). ROS generated from biotic stress: Effects on plants and alleviation by endophytic microbes. Frontiers in Plant Science 13. https://doi.org/10.3389/FPLS.2022.1042936/FULL

Salam, M.D. & Khaidem, S. (2023). 5 Microbial oxidative enzymes in alleviation of biotic stress in plants. *In*: Microbial Oxidative Enzymes (pp. 91–108). De Gruyter. https://doi.org/10.1515/9783111062235-005

Sam-On, M.F.S., Mustafa, S., Yusof, M.T., Hashim, A.M. & Aizuddin, K.N.A.K. (2024). Exploring the global trends of bacillus, trichoderma and entomopathogenic fungi for pathogen and pest control in chili cultivation. Saudi Journal of Biological Sciences 31(8), 104046.

Santoyo, G. (2022). How plants recruit their microbiome? New insights into beneficial interactions. Journal of Advanced Research 40, 45–58. https://doi.org/10.1016/J.JARE.2021.11.020

Santra, H.K. & Banerjee, D. (2021). Production, optimization, characterization and drought stress resistance by β-glucan-rich heteropolysaccharide from an endophytic fungi Colletotrichum alatae LCS1 isolated from Clubmoss (Lycopodium clavatum). Frontiers in Fungal Biology 2. https://doi.org/10.3389/FFUNB.2021.796010/FULL

Sardans, J., Gargallo-Garriga, A., Urban, O., Klem, K., Holub, P., Janssens, I.A. et al. (2021). Ecometabolomics of plant–herbivore and plant–fungi interactions: A synthesis study. Ecosphere 12(9). https://doi.org/10.1002/ECS2.3736

Sarkar, A.K. & Sadhukhan, S. (2023). Unearthing the alteration in plant volatiles induced by mycorrhizal fungi: A shield against plant pathogens. Physiologia Plantarum 175(1). https://doi.org/10.1111/PPL.13845

Sathyabhama, M., Viswanathan, R., Prasanth, C.N., Malathi, P. & Sundar, A.R. (2022). Differential host responses of sugarcane to Colletotrichum falcatum reveal activation of probable effector triggered immunity (ETI) in defence responses. Plant Cell Reports 41(6), 1461–1476. https://doi.org/10.1007/S00299-022-02870-1

Scavo, A., Fontanazza, S., Restuccia, A., Pesce, G.R., Abbate, C. & Mauromicale, G. (2022). The role of cover crops in improving soil fertility and plant nutritional status in temperate climates. A review. Agronomy for Sustainable Development 42(5). https://doi.org/10.1007/s13593-022-00825-0

Schiavi, D., Ronchetti, R., Di Lorenzo, V., Salustri, M., Petrucci, C. et al. (2022). Circular hazelnut protection by lignocellulosic waste valorization for nanopesticides development. Applied Sciences 12(5), 2604. https://doi.org/10.3390/app12052604

Seeliger, M., Hilton, S., Muscatt, G., Walker, C., Bass, D., Albornoz, F. et al. (2024). New fungal primers reveal the diversity of Mucoromycotinian arbuscular mycorrhizal fungi and their response to nitrogen application. Environmental Microbiome 19(1), 71.

Segal, E., Ecol, S. K.-Annu. Rev. & 2025, undefined. (2025). Determinants of Plant-Mycorrhizal Fungal Distributions and Function Under Global Change. Papers.Ssrn.Com https://doi.org/10.1146/annurev-ecolsys-102723-051821

Selvaraj, A. & Thangavel, K. (2021). Arbuscular Mycorrhizal Fungi: Potential Plant Protective Agent Against Herbivorous Insect and Its Importance in Sustainable Agriculture (pp. 319–337). https://doi.org/10.1007/978-3-030-51916-2_19

Sheoran, S., Gupta, M., Kumari, S., Kumar, S. & Rakshit, S. (2022). Meta-QTL analysis and candidate genes identification for various abiotic stresses in maize (Zea mays L.) and their implications in breeding programs. Molecular Breeding 42(5). https://doi.org/10.1007/S11032-022-01294-9

Shi, J., Wang, X. & Wang, E. (2023). Mycorrhizal symbiosis in plant growth and stress adaptation: From genes to ecosystems. Annual Review of Plant Biology 74, 569–607. https://doi.org/10.1146/ANNUREV-ARPLANT-061722-090342

Shi, X., Zhang, Z., Zhang, C., Zhou, X., Zhang, D. & Liu, Y. (2021). The molecular mechanism of efficient transmission of plant viruses in variable virus–vector–plant interactions. Horticultural Plant Journal 7(6), 501–508. https://doi.org/10.1016/J.HPJ.2021.04.006

Skendžić, S., Zovko, M., Živković, I.P., Lešić, V. & Lemić, D.L. (2021). The impact of climate change on agricultural insect pests. Insects. https://doi.org/10.3390/insects12050440

Smulders, L. (2022). Interactions of Soil Microbiome, Plant Defenses and Domestication in Tomato. https://digibug.ugr.es/handle/10481/77151

Soltangheisi, A., Hales-Henao, A., Pena, R. & Tibbett, M. (2024). Species-specific effects of mycorrhizal symbiosis on Populus trichocarpa after a lethal dose of copper. Ecotoxicology and Environmental Safety 272, 116112. https://doi.org/10.1016/J.ECOENV.2024.116112

Sonbol, H., Korany, S., Nhs, M., Abdi, I., Maridueña-Zavala, M.G., Alsherif, E.A. et al. (2025). Exploring the benefits of AMF colonization for improving wheat growth, physiology and metabolism, and antimicrobial activity under biotic stress from aphid infection. BMC Plant Biology 25(1). https://doi.org/10.1186/s12870-025-06196-4

Song, F., Ji, C., Wang, T., Zhang, Z., Duan, Y., Yu, M. et al. (2025). Genome-wide identification, expression, and protein interaction of GRAS family genes during arbuscular mycorrhizal symbiosis in Poncirus trifoliata. International Journal of Molecular Sciences 26(5), 2082. https://doi.org/10.3390/ijms26052082

Sousa, B., Soares, C., Sousa, F., Martins, M., Mateus, P., Rodrigues, F. et al. (2024). Enhancing tomato plants' tolerance to combined heat and salt stress – The role of arbuscular mycorrhizae and biochar. Science of The Total Environment 948, 174860. https://doi.org/10.1016/J.SCITOTENV.2024.174860

Souto, A.L., Sylvestre, M., Tölke, E.D., Tavares, J.F., Barbosa-Filho, J.M. & Cebrián-Torrejón, G. (2021). Plant-derived pesticides as an alternative to pest management and sustainable agricultural production: Prospects, applications and challenges. Molecules 26(16), 4835. https://doi.org/10.3390/molecules26164835

Stratton, C.A., Ray, S., Bradley, B.A., Kaye, J.P., Ali, J.G. & Murrell, E.G. (2021). Nutrition vs association: Plant defenses are altered by arbuscular mycorrhizal fungi association not by nutritional provisioning alone. BMC Plant Biology 22(1). https://doi.org/10.1186/s12870-022-03795-3

Streletskii, R.A., Astaykina, A.A., Belov, A.A., Cheptsov, V.S. & Vetrova, A.A. (2024). Beneficial soil microorganisms and their role in sustainable agriculture. Sustainable Agricultural Practices 293–333. https://doi.org/10.1016/B978-0-443-19150-3.00013-8

Sun, D., Yang, X., Wang, Y., Fan, Y., Ding, P., Song, X. et al. (2022). Stronger mutualistic interactions with arbuscular mycorrhizal fungi help Asteraceae invaders outcompete the phylogenetically related natives. New Phytologist 236(4), 1487–1496. https://doi.org/10.1111/NPH.18435

Sweeney, C.J., Bottoms, M., Ellis, S., Ernst, G., Kimmel, S., Loutseti, S. et al. (2022). Arbuscular mycorrhizal fungi and the need for a meaningful regulatory plant protection product testing strategy. Environmental Toxicology and Chemistry 41(8), 1808–1823. https://doi.org/10.1002/etc.5400

Sweeney, C.J., de Vries, F.T., van Dongen, B.E. & Bardgett, R.D. (2021). Root traits explain rhizosphere fungal community composition among temperate grassland plant species. New Phytologist 229(3), 1492–1507. https://doi.org/10.1111/nph.16976

Tahiri, A.-I., Meddich, A., Raklami, A., Alahmad, A., Bechtaoui, N., Anli, M. et al. (2022). Assessing the potential role of compost, PGPR, and AMF in improving tomato plant growth, yield, fruit quality, and water stress tolerance. Journal of Soil Science and Plant Nutrition 1(1), 3. https://doi.org/10.1007/s42729-021-00684-w

Tan, M., Li, Y., Xu, J., Yan, S. & Jiang, D. (2022). Effects of arbuscular mycorrhizal fungi-colonized Populus alba × P. berolinensis seedlings on the microbial and metabolic status of gypsy moth larvae. Insects 13(11), 1002. https://doi.org/10.3390/insects13111002

Teyssier, E., Grat, S., Landry, D., Ouradou, M., Rich, M.K., Fort, S. et al. (2024). LysM-RLK plays an ancestral symbiotic function in plants. *In*: BioRxiv. https://doi.org/10.1101/2024.01.16.575821

Thangavel, P., Anjum, N.A., Muthukumar, T., Sridevi, G., Vasudhevan, P. & Maruthupandian, A. (2022). Arbuscular mycorrhizae: Natural modulators of plant–nutrient relation and growth in stressful environments. Archives of Microbiology 204(5), 264. https://doi.org/10.1007/S00203-022-02882-1

Thoms, D., Liang, Y. & Haney, C.H. (2021). Maintaining symbiotic homeostasis: How do plants engage with beneficial microorganisms while at the same time restricting pathogens? Molecular Plant–Microbe Interactions 34(5), 462–469. https://doi.org/10.1094/MPMI-11-20-0318-FI

Tilgam, J., Sreeshma, N., Priyadarshini, P., Bhavyasree, R.K., Choudhury, S., Bharati, A. & Ashajyothi, M. (2022). Rhizosphere Engineering for Systemic Resistance/Tolerance to Biotic and Abiotic Stress. 271–300. https://doi.org/10.1007/978-981-19-4101-6_14

Ul Haq, J., Sharif, M., Akbar, W.A., Ur Rahim, H., Ahmad Mian, I., Ahmad, S. et al. (2023). Arbuscular mycorrhiza fungi integrated with single super phosphate improve wheat-nitrogen-phosphorus acquisition, yield, root infection activity, and spore density in alkaline-calcareous soil. Gesunde Pflanzen 75(3), 539–548. https://doi.org/10.1007/S10343-022-00718-Y

Ullah, A., Bano, A. & Khan, N. (2021). Climate change and salinity effects on crops and chemical communication between plants and plant growth-promoting microorganisms under stress. Frontiers in Sustainable Food Systems 5. https://doi.org/10.3389/FSUFS.2021.618092/PDF

Velaz, M., Santesteban, L.G. & Torres, N. (2025). Mycorrhizae and grapevines: The known unknowns of their interaction for wine growers' challenges. Journal of Experimental Botany. https://doi.org/10.1093/jxb/eraf081/8069425

Venice, F., Chialva, M., Domingo, G., Novero, M., Carpentieri, A., Salvioli di Fossalunga, A. et al. (2021). Symbiotic responses of Lotus japonicus to two isogenic lines of a mycorrhizal fungus differing in the presence/absence of an endobacterium. Plant Journal 108(6), 1547–1564. https://doi.org/10.1111/TPJ.15578

Wahab, A., Muhammad, M., Munir, A., Abdi, G., Zaman, W., Ayaz, A. et al. (2023). Role of arbuscular mycorrhizal fungi in regulating growth, enhancing productivity, and potentially influencing ecosystems under abiotic and biotic stresses. Plants 12(17), 3102. https://doi.org/10.3390/plants12173102

Wang, F., Zhang, L., Zhou, J., Rengel, Z., George, T.S. & Feng, G. (2022). Exploring the secrets of hyphosphere of arbuscular mycorrhizal fungi: Processes and ecological functions. Plant and Soil 481(1–2), 1–22. https://doi.org/10.1007/s11104-022-05621-z

Wang, M., Wang, Z., Guo, M., Qu, L. & Biere, A. (2023b). Effects of arbuscular mycorrhizal fungi on plant growth and herbivore infestation depend on availability of soil water and nutrients. Frontiers in Plant Science 14. https://doi.org/10.3389/FPLS.2023.1101932/FULL

Wang, M., Xiang, L., Tang, W., Chen, X., Li, C., Yin, C. & Mao, Z. (2024). Improving apple orchard health: The role of arbuscular mycorrhizal fungi in alleviating replant disease and strengthening soil microbial communities. Applied Soil Ecology 196, 105278. https://doi.org/10.1016/J.APSOIL.2024.105278

Wang, Q., Liu, M., Wang, Z., Li, J., Liu, K. & Huang, D. (2023a). The role of arbuscular mycorrhizal symbiosis in plant abiotic stress. Frontiers in Microbiology 14. https://doi.org/10.3389/FMICB.2023.1323881/FULL

Wu, Q.S., Silva, F.S.B., Hijri, M. & Kapoor, R. (2023). Editorial: Arbuscular mycorrhiza-mediated augmentation of plant secondary metabolite production. Frontiers in Plant Science 14. https://doi.org/10.3389/FPLS.2023.1150900/FULL

Wu, Y.H., Wang, H., Liu, M., Li, B., Chen, X., Ma, Y.T. & Yan, Z.Y. (2021). Effects of native arbuscular mycorrhizae isolated on root biomass and secondary metabolites of Salvia miltiorrhiza Bge. Frontiers in Plant Science 12. https://doi.org/10.3389/FPLS.2021.617892/FULL

Yadav, B., Kaur, V., Narayan, O.P., Yadav, S.K., Kumar, A. & Wankhede, D.P. (2022). Integrated omics approaches for flax improvement under abiotic and biotic stress: Current status and future prospects. Frontiers in Plant Science 13. https://doi.org/10.3389/FPLS.2022.931275/FULL

Yadav, S. & Chattopadhyay, D. (2023). Lignin: The building block of defense responses to stress in plants. Journal of Plant Growth Regulation 42(10), 6652–6666. https://doi.org/10.1007/s00344-023-10926-z

Yonas, M.W. & Zawar, S. (2024). Optimizing chickpea growth: Unveiling the interplay of arbuscular mycorrhizal fungi and rhizobium for sustainable agriculture. Soil Use and Management 40(2). https://doi.org/10.1111/SUM.13057

Yu, L., Zhang, W., Geng, Y., Liu, K. & Shao, X. (2022). Cooperation with arbuscular mycorrhizal fungi increases plant nutrient uptake and improves defenses against insects. Frontiers in Ecology and Evolution 10. https://doi.org/10.3389/FEVO.2022.833389/FULL

Zeng, M. (2024). The mutual effect of nutrients on plant–herbivore interactions. Plant Ecology 225(10), 1035–1045. https://doi.org/10.1007/s11258-024-01452-3

Zhang, P., He, E., Romero-Freire, A., Xia, B., Ying, R., Liu, Y. et al. (2022). Plant intelligence in a rapidly changing world: Implementation of plant-plant communications in managed plant systems. Sustainable Horizons 2, 100008. https://doi.org/10.1016/J.HORIZ.2022.100008

Zhang, W., Xia, K., Feng, Z., Qin, Y., Zhou, Y., Feng, G. et al. (2024). Tomato plant growth promotion and drought tolerance conferred by three arbuscular mycorrhizal fungi is mediated by lipid metabolism. Plant Physiology and Biochemistry 208, 108478. https://doi.org/10.1016/J.PLAPHY.2024.108478

Zhou, L., Zhou, X., He, Y., Fu, Y., Du, Z., Lu, M. et al. (2022). Global systematic review with meta-analysis shows that warming effects on terrestrial plant biomass allocation are influenced by precipitation and mycorrhizal association. Nature Communications 13(1), 4914. https://doi.org/10.1038/s41467-022-32671-9

Zhu, T., Li, L., Duan, Q., Liu, X. & Chen, M. (2020). Progress in our understanding of plant responses to the stress of heavy metal cadmium. Plant Signaling & Behavior 16(1). https://doi.org/10.1080/15592324.2020.1836884

Zou, Y.-N., Wu, Q.-S. & Kuča, K. (2020). Unravelling the role of arbuscular mycorrhizal fungi in mitigating the oxidative burst of plants under drought stress. Plant Biology 23(S1), 50–57. https://doi.org/10.1111/plb.13161

CHAPTER
9

Enhancing Pollutant Removal: The Role of Arbuscular Mycorrhizae in Soil-Plant Dynamic

Juan F. Saldarriaga[1*], Julián E. López[2] and Carolina Montoya-Ruiz[3]

[1] Dept. Civil and Environmental Engineering, Universidad de los Andes, Carrera 1Este #19A-40, 111711, Bogotá, Colombia
[2] Facultad de Arquitectura e Ingeniería, Institución Universitaria Colegio Mayor de Antioquia, Carrera 78 # 65-46, 050034, Medellín, Colombia
[3] Facultad de Ciencias de la Vida, Universidad Nacional de Colombia, Sede Medellín Calle 59A #63-20, 050034, Medellín, Colombia

1. Understanding Arbuscular Mycorrhizal Fungi

Arbuscular mycorrhizal fungi (AMF) are the most ubiquitous symbiotic microbes in terrestrial ecosystems, forming connections with about 70% of vascular plant species (Saia & Jansa, 2022). These fungi create mutualistic relationships by infiltrating the root cortical cells of their hosts and creating specialized branched structures called arbuscules, which promote efficient nutrient exchange (Wang et al., 2023). The highly regulated process of establishing this symbiosis starts with molecular recognition between the partners: strigolactones, which are released by plant roots and detected by AMF, are signaling molecules that cause the fungi to produce lipochitooligosaccharides (Myc factors), which the plant recognizes. After this reciprocal identification, AMF hyphae colonize the root cortex and creates a hyphopodium on the root epidermis (Wang et al., 2023). At the same time, they spread a large extraradical mycelium into the surrounding soil, which can make up 9–55% of the total biomass of soil microorganisms (Akter et al., 2025; Ahmed et al., 2025).

Arbuscular mycorrhizal fungi increase plant resistance to abiotic stresses like drought and salinity, improves the production and accumulation of significant bioactive compounds in plants, improves soil structure and fertility, and grow plant uptake of mineral nutrients, especially phosphorus, through these networks (Willis et al., 2013). Their function in tying together the root systems of various plants,

*Corresponding author: jf.saldarriaga@uniandes.edu.co, juanfelorza@gmail.com

DOI: 10.1201/9781003669777-9

promoting nutrient cycling, and sustaining soil microbial communities which enhance ecosystem resilience and productivity further emphasizes the ecological significance of AMF (Akter et al., 2025).

These play a pivotal role in maintaining soil health by participating in essential processes like organic matter decomposition, soil mineral and fertilizer nutrient uptake, and soil structure enhancement, these are crucial in preserving soil health (Wang et al., 2016b). Because of the vast hyphal networks that AMF creates, the root exploration zone is greatly expanded, increasing the amount of soil that plants can access to absorb nutrients and water (Wang et al., 2016b; Akter et al., 2025). Phosphorus, nitrogen, potassium, calcium, sulfur, zinc, copper, and other essential mineral nutrients are better absorbed by this dense extraradical mycelium, which frequently leads to higher plant biomass and better crop quality. Additionally, AMF helps to stabilize the structure of soil by generating glomalin, a glycoprotein that serves as a long-term soil binding agent and encourages the development and stabilization of soil macroaggregates (Wang et al., 2021; Saia and Jansa, 2022). These macroaggregates improve water infiltration, reduce erosion and nutrient losses, and enhance soil aeration and water retention, all of which are critical for crop productivity and resilience (Phour and Sindhu, 2024).

The continuous renewal and eventual decomposition of AMF hyphae further enriches soil organic matter and maintain aggregate stability, continuously contributing 3–20% of total soil organic carbon (SOC) in ecosystems (Swartz et al., 2022). Beyond these physical and chemical benefits, AMF also bolsters plant resistance to a range of environmental stresses, including drought, salinity, nutrient deprivation, extreme temperatures, heavy metals, pests, and diseases. Collectively, these mechanisms underscore the integral role of AMF in fostering friable, fertile soils that support sustainable plant growth and ecosystem function (Hao et al., 2022; Ahmed et al., 2025).

Understanding the diversity and function of these mycorrhizal partners is essential for developing effective bioremediation strategies. The taxonomy of AMF has undergone major changes, particularly with the advent of genetic methods that complement traditional morphological identification (Schüssler and Walker, 2010). Currently, AMF are categorized under the phylum Glomeromycota or, depending on the classification method, under the subphylum Glomeromycotina within Mucoromycota (Tedersoo et al., 2024). The monophyly of Glomeromycota is supported by recent phylogenomic analyses, which identify four major orders: Paraglomerales, Diversisporales, Archaeosporales, and Glomerales (Montoliu-Nerin et al., 2021; Tedersoo et al., 2024). Conventional taxonomy depended on spore morphology, which was helpful but limited by cryptic diversity and phenotypic plasticity. Molecular approaches, particularly DNA sequencing of ribosomal genes, have enabled more precise species delimitation and the discovery of new genera and species (Schüssler and Walker, 2010). As of recent estimates, approximately 300–343 AMF species have been described, grouped into 33 genera and 12 families (Montoliu-Nerin et al., 2021). Different AMF species have been associated with pollutant removal, with the most relevant species listed in (Table 1).

Table 1. Principal arbuscular mycorrhizal fungi (AMF) species with impact in the pollutant removal

AMF species	Pollutant type	Detoxification role	References
Rhizophagus irregularis	Heavy metals, Dioxins/Furans, Emerging pollutants	Immobilizes Cd, Pb, As, Cr; protects plants from dioxins/furans toxicity; assists in removal of emerging pollutants in wetlands	Jiang et al., 2016; Meglouli et al., 2020; Banerjee et al., 2024
Funneliformis mosseae	Heavy metals, Organic pollutants	Enhances plant tolerance and accumulation of Cd, Pb, Zn, As; aids in stabilization and removal of hydrocarbons and pesticides	Boorboori and Zhang, 2024; Li et al., 2024
Glomus versiforme	Heavy metals, Organic pollutants	Reduces plant uptake of toxic metals; improves growth in contaminated soils; supports degradation of hydrocarbons	Tan et al., 2015; Jiang et al., 2016
Claroideoglomus etunicatum	Heavy metals, Organic pollutants	Promotes soil health; assists in detoxification of both organic (e.g., PAHs) and inorganic pollutants	Hao et al., 2022
Acaulospora laevis	Heavy metals	Contributes to immobilization of metals and supports plant survival in polluted soils	Ma et al., 2022a
Rhizophagus intraradices	Heavy metals, Organic pollutants	Aids in phytoremediation of Cd, Pb, and organic pollutants such as PAHs and pesticides	Ma et al., 2022a; Colombo et al., 2024

2. Pollutant Removal Mechanisms

AMF play a crucial role in contaminant removal through several mechanisms, such as: immobilization and sequestration in fungal structures, chemical transformation of contaminants, degradation of organic pollutants, and enhancement of plant tolerance and detoxification (Table 1). Implementation of these mechanisms in the remediation field is covered below.

2.1. Immobilization and Sequestration in Fungal Structures and Chemical Transformation of Contaminants

One of the main ways that AMF aid in the bioremediation of polluted environments is by immobilization and sequestration in fungal structures. These procedures lessen the mobility and bioavailability of harmful contaminants, especially heavy metals. The main components of the fungal cell wall are chitin, glucans, proteins, and lipids. These substances offer a variety of functional groups, such as carboxyl, amino, and phosphate, that can bind metal ions and other contaminants through a process known as biosorption (Figure 1). This passive binding process immobilizes pollutants on the fungal surface, effectively reducing their concentration in soil solution (Kumar et al., 2021; Riaz et al., 2021). In addition to surface biosorption, AMF actively absorb specific metals into their cytoplasm, where they may be encapsulated in specialized vesicles or vacuoles. By serving as reservoirs, these intracellular structures keep toxic metals out of metabolic processes and shield both the host plant and the fungal cells from harm (Wu et al., 2015; Dinakarkumar et al., 2024).

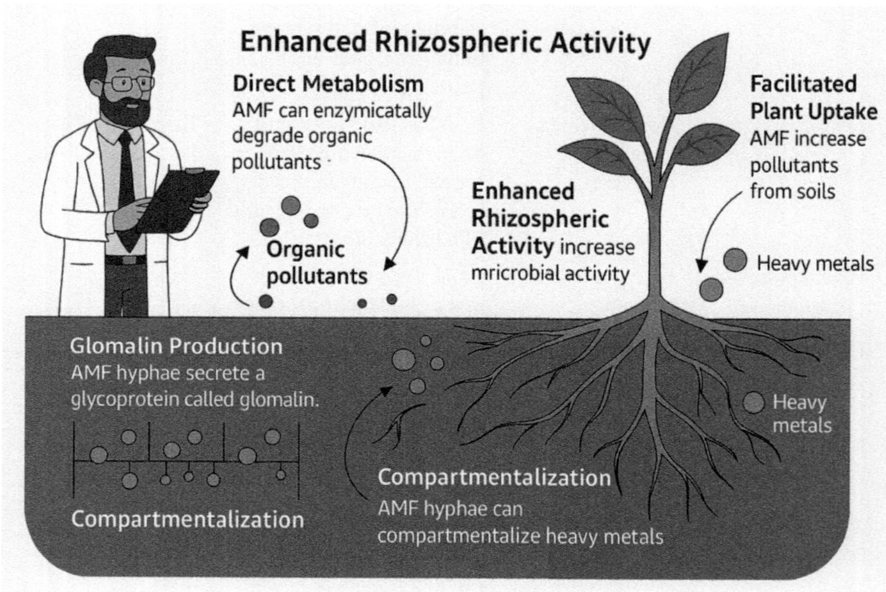

Fig. 1. Pollutant removal mechanisms

On the other hand, AMF have another immobilization mechanism associated with Glomalin production. Glomalin, a special glycoprotein, has a remarkable affinity for heavy metals like Cr, Pb, and Cd. In addition to immobilizing metals, glomalin promotes the stability and development of soil aggregates, which physically trap pollutants in soil matrix and lessens their potential to leach. This dual role makes glomalin a key factor in both chemical and physical immobilization of pollutants (Zhao et al., 2024; Dinakarkumar et al., 2024).

AMF can chemically transform certain contaminants, facilitating their immobilization. For example, AMF have been shown to reduce toxic hexavalent chromium Cr(VI) to the less toxic and less mobile trivalent chromium Cr(III), which precipitates as Cr(III)-phosphate analogues on fungal surfaces. This reduction and precipitation process significantly decrease Cr bioavailability and toxicity in the soil-plant system (Wu et al., 2015). Similarly, as taken up by plants is chemically transformed by AMF-associated processes: arsenate As(V) is enzymatically reduced to arsenite As(III) by arsenate reductase and then either sequestered in vacuoles or expelled, reducing its toxicity and translocation (Boorboori and Zhang, 2022). AMF can induce precipitation of metals in the rhizosphere by altering local chemical conditions. For example, secretion of organic acids and other metabolites can change pH and redox potential, leading to the formation of insoluble metal complexes or mineral precipitates (Figure 1). This immobilization reduces metal mobility and bioavailability (Wei et al., 2023; Yang et al., 2024).

2.2. Degradation of Organic Pollutants

In contaminated soils and sediments, AMF have a major indirect effect in the breakdown of organic contaminants. Through a variety of synergistic pathways, their presence improves the breakdown of complex organic pollutants, including phthalates, pesticides, petroleum hydrocarbons, and polycyclic aromatic hydrocarbons (PAHs). By promoting the development and activity of several microbial populations that are able to break down organic contaminants, AMF colonization modifies the rhizosphere environment. Increased root exudation and changes to the physicochemical characteristics of the soil cause this stimulation, which gives bacteria and fungi that break down pollutant's substrates and ideal growing conditions. Consequently, the rhizosphere's microbial consortia improve their ability to break down persistent organic chemicals, hastening the mineralization and detoxifying of such compounds in soils (Gao et al., 2011; Goswami et al., 2023; Vargas et al., 2024).

Also, the breakdown of complex organic compounds depends on the development of extracellular enzymes including laccases, peroxidases, and hydrolases, which are increased by AMF and related microbial communities. These enzymes depolymerize resistant contaminants, such as PAHs and chlorinated insecticides, into less hazardous, simpler molecules by catalyzing oxidative and hydrolytic processes. Enzymatic activity is frequently greater in mycorrhizal soils than in non-mycorrhizal controls, indicating that AMF indirectly increases the rhizosphere microbiome's capacity for biodegradation (Gao et al., 2011; Dagher et al., 2020).

Furthermore, AMF generates an increase in plant biomass and root exudation. Plant development and biomass accumulation are encouraged by AMF symbiosis, which results in a larger root system. Greater amounts and a wider variety of root exudates, such as sugars, amino acids, organic acids, and secondary metabolites, are released by larger root systems. Heterotrophic microorganisms use these exudates as sources of energy and nutrients, which promotes the microbial breakdown of organic contaminants. Thus, the increased rhizodeposition generates a positive feedback loop that promotes the biodegradation of pollutants and the restoration of soil health (Gao et al., 2011; George and Ray, 2023).

2.3. Plant Tolerance and Detoxification

Through a variety of physiological, biochemical, and molecular reactions, AMF considerably improve plant tolerance and detoxifying systems against a range of abiotic challenges, such as drought, heavy metal toxicity, and nutrient shortages. An important detoxifying mechanism brought about by AMF symbiosis is the increased production of metallothioneins and phytochelatins in host plants. In order to reduce their free ionic forms and lessen oxidative damage, these tiny, cysteine-rich peptides and proteins bind heavy metals intracellularly and create stable complexes that sequester harmful ions including Cd, Ni, and Cr. For instance, it has been demonstrated that Glomus intraradices colonization raises the amounts of phytochelatin in tomato and alfalfa, allowing these plants to store metals mostly in their roots while shielding their shoots from harm (Begum et al., 2019; Ahammed et al., 2023).

Another important issue in stress tolerance is Increased Antioxidant Enzyme Activities. Reactive oxygen species (ROS) are produced in plants by heavy metals and other pollutants, which causes oxidative stress. Antioxidant enzymes that detoxify ROS and shield biological components from oxidative damage, such as superoxide dismutase (SOD), catalase (CAT), and peroxidases (POD), are activated more when AMF symbiosis is present. Plant resistance to metal stress and other abiotic stressors, such as drought, is enhanced by this enzymatic overexpression. Comparing AMF-inoculated plants to non-inoculated controls under stress, studies have shown a considerable increase in antioxidant enzyme activity (Begum et al., 2019); (Leudo et al., 2020; Cruz et al., 2021; Sun et al., 2024).

3. Heavy Metal Remediation

The majority of terrestrial plants' roots create mutualistic symbiotic connections with AMF, which are crucial partners in the rehabilitation of heavy metal-contaminated soils. By serving as a biological filter and stabilizer within the soil-plant system, AMF not only improves the uptake of essential nutrients like phosphorus and micronutrients but also significantly lessens the phytotoxic effects of heavy metals (Wahab et al., 2023; khaliq et al., 2024). AMF's vast hyphal networks enhance the absorptive surface area of plant roots, which helps with the release of glomalin, a glycoprotein that binds heavy metals and enhances soil fertility and structure, as well as the sequestration and compartmentalization of hazardous metals inside root tissues (Reza Boorboori and Zhang, 2022; Etesami and Glick, 2023). This organic alliance promotes the growth of robust agroecosystems that can flourish in the face of heavy metal stress and provides an affordable, sustainable, and ecologically friendly way to restore damaged, polluted areas. The application of AMF-based techniques stands out as a viable strategy for ecological restoration and the advancement of sustainable agriculture in places impacted by heavy metal pollution, especially as research continues to unravel the physiological and molecular mechanisms underpinning these interactions. AMF have demonstrated remarkable potential for remediating soils contaminated with various heavy metals and metalloids. They enhance plants' ability to take up metals such as Cu, Ni, Pb, and Zn. This capability makes them particularly valuable in phytoremediation strategies targeting metal-contaminated soils (Sánchez-Castro et al., 2023).

3.1. Dual Mechanisms in Heavy Metal Remediation: Responses at Low Metal Concentrations

The removal of heavy metals from soil is a critical function of AMF, which also play a crucial role in soil-plant dynamics due to their adaptive responses to different metal concentrations. When heavy metal concentrations are low, AMF's hyphal networks extend outside the root zone, improving the uptake of important micronutrients like copper and zinc as well as some non-essential elements (Wahab et al., 2023; Zhao et al., 2024). These networks get access to soil micropores and help carry nutrients and metals to the roots of plants, a process that is controlled by specific fungal transporter proteins. In addition to providing sustenance for plants, this selective absorption facilitates the regulated phytoextraction of metals, which can help with biofortification and the slow remediation of contaminated soils (Begum et al., 2019; Khaliq et al., 2022). For instance, research has demonstrated that AMF, such as *Rhizophagus irregularis*, can enhance zinc uptake in maize, fostering robust plant development without being harmful. Apart from improved absorption at low metal concentrations, AMF is also essential for phytostabilization in highly contaminated soils (Moreira et al., 2021; Saboor et al., 2021). To decrease the mobility and bioavailability of heavy metals like lead, cadmium, and chromium, they release glomalin, a glycoprotein that forms stable compounds with these metals (Figure 2). The risk of heavy metals entering the food chain is reduced, and aboveground tissues are protected by this immobilization process, which stops excessive metal transfer to plant shoots. Additionally, AMF colonization increases the plant's antioxidant defenses and attracts metal-resistant bacteria to the rhizosphere, both of which contribute to the reduction of oxidative stress and the maintenance of plant health in the face of metal stress. AMF significantly improves ecosystem resilience and soil structure. In addition to immobilizing metals, the glomalin they generate also promotes better soil aggregation, less erosion, and improved water retention—all of which support a more robust and varied soil microbial population. Better nutrient cycling and increased ecological stability follow from this (Ghorui et al., 2024; Son et al., 2024). AMF is an essential partner in ecological restoration and sustainable agriculture because of its dual function in balancing metal intake and immobilization,

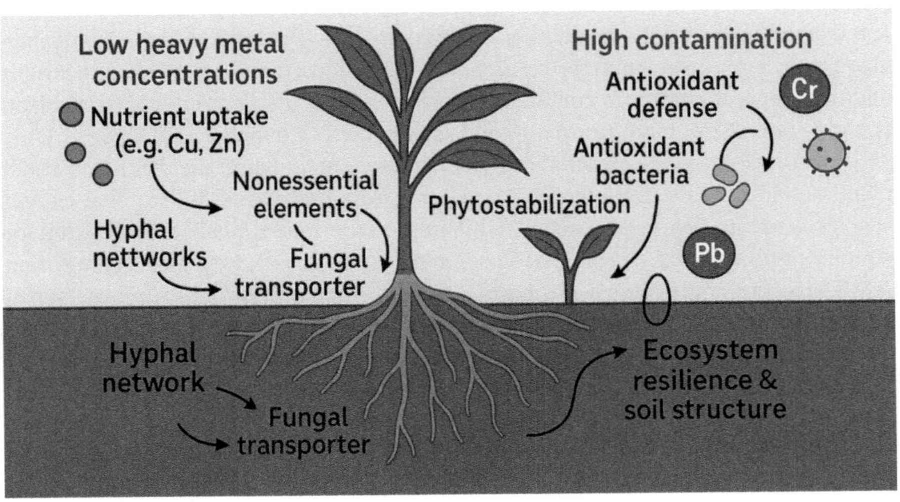

Fig. 2. Adaptive role of arbuscular mycorrhizal fungi in heavy metal management and soil-plant health.

which enables plants to flourish in both nutrient-deficient and contaminated situations. Their ecological significance and the need for more research to maximize their usage in field-scale remediation and soil health management are highlighted by their capacity to adaptively alter metal dynamics in the soil-plant system (Kalamulla et al., 2022; Wahab et al., 2023).

3.2. Dual Mechanisms in Heavy Metal Remediation: Responses at High Metal Concentrations

Particularly when metal concentrations are high, AMF have a major impact on soil-plant dynamics by removing and immobilizing heavy metals from the soil. AMF use a number of defense mechanisms to reduce toxicity for themselves and their host plants when exposed to high concentrations of heavy metals such as cadmium, lead, copper, and chromium (Leudo et al., 2020; Saldarriaga et al., 2023; Crişan et al., 2024; Zhao et al., 2024). The sequestration of heavy metals within fungal hyphae and vesicles in plant roots is one main tactic. This successfully compartmentalizes and immobilizes these toxic ions to prevent them from reaching sensitive plant tissues or interfering with vital nutrient intake. The metal-binding glycoprotein glomalin, which is also produced by AMF, is essential for stabilizing and ensnaring heavy metals in the soil matrix, hence decreasing their mobility and bioavailability. This helps to preserve the integrity of soil structure by encouraging aggregate stability in addition to preventing excessive metal accumulation in plant shoots (Yang et al., 2017).

As a biological filter, AMF absorbs and stores heavy metals in their structures at high metal concentrations, shielding the host plant from harmful impacts that would otherwise disrupt physiological and biochemical functions like respiration, photosynthesis, and nutrition metabolism (Dhalaria et al., 2020; Wahab et al., 2023). Further decreasing the reactivity and uptake of toxic metals by plants, AMF can also release organic acids and metallothioneins, which chemically change the metals into less dangerous forms or bind them in stable complexes (Figure 2). Mycorrhizal plants are able to detoxify and immobilize metals, which allows them to grow more healthily and with more tolerance in contaminated surroundings than non-mycorrhizal plants (Leudo et al., 2020; Reza Boorboori and Zhang, 2022; Zhao et al., 2024).

Furthermore, AMF strengthens the antioxidant defense mechanisms of the plant, assisting in the fight against oxidative stress brought on by reactive oxygen species generated by heavy metals, which otherwise hinder plant growth and harm cellular components. Assuring that heavy metal stress does not impair plant nutrition, AMF also promotes improved phosphorus and micronutrient absorption, even in contaminated soils. In general, AMF stabilize heavy metals, promote plant health, and preserve soil fertility, all of which aid in the rehabilitation of polluted soils. For this reason, they are an essential part of phytoremediation techniques in areas where heavy metal pollution is high (Hasanuzzaman et al., 2020; Mansoor et al., 2023).

3.3. Factors Influencing AMF Effectiveness in Remediation

The AMF in supporting heavy metal remediation depends on a complex interaction of biological, chemical, and environmental factors. On the biological side, the type of plant or ecotype involved can greatly influence the success of the process. This is due to differences in how various plants tolerate and accumulate metals, their root structures, and how well they interact with particular AMF species (Ai et al., 2022; Zhao et al., 2024). Likewise, AMF strains themselves vary in their resilience to metal stress, their ability to help plants absorb or block metals, and how efficiently they colonize plant roots. The overall success of this partnership hinges on how compatible the plant and fungal species are (Figure 3).

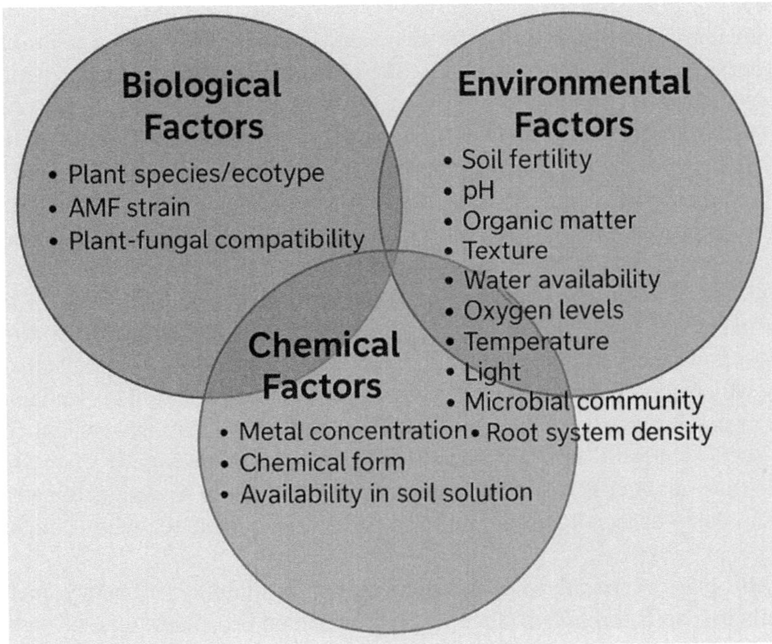

Fig. 3. AMF-supported heavy metal remediation.

From a chemical perspective, the characteristics of the heavy metal contaminant play a crucial role. These include its total concentration, chemical form, and especially its availability in the soil solution. Since each metal behaves differently in terms of toxicity and mobility, their interactions with AMF structures and secreted compounds also vary, which in turn affects remediation efficiency (Tchounwou et al., 2012; Cruz et al., 2019, 2021b; Kafle et al., 2022).

Environmental conditions further shape the outcome of AMF-assisted remediation. Factors such as soil fertility, pH, organic matter content, texture, water availability, oxygen levels, temperature, light, microbial community composition, and root system density all influence how well AMF can colonize plant roots, perform metabolic functions, and affect metal availability and uptake (Figure 3). These conditions ultimately determine the impact of AMF on plant health and the broader ecosystem (Ndeko et al., 2024; Laishram et al., 2025).

Studies have shown that native AMF strains from metal-contaminated environments tend to be especially well-suited for use in phytoremediation. Their long-term adaptation to stressful local conditions and co-evolution with native plant species make them particularly robust and effective partners in mitigating heavy metal pollution (Riaz et al., 2021; Zhao et al., 2024).

3.4. Applications in Environmental Restoration

Using AMF for environmental cleanup offers a flexible and effective approach, primarily through two key methods: phytostabilization and phytoextraction. Phytostabilization involves keeping heavy metals locked in the soil to prevent them from spreading or being absorbed by plants and animals. AMF play a central role in this process by producing compounds like organic acids and glomalin, which can bind metals and keep them from moving freely (Yan et al., 2020). Their extensive hyphal networks also help trap metals physically. This reduced mobility significantly lowers the risk of metals leaching into water sources or entering the food chain. On top of that, AMF supports healthier root development, which strengthens soil structure, improves stability, and reduces erosion—making this approach particularly useful for securing contaminated sites and limiting further environmental damage (George and Ray, 2023). Phytoextraction, on the other hand, focuses on pulling metals out of the soil using plants. When the level of contamination is moderate and conditions are well-controlled, AMF can enhance the ability of plants to absorb metals, which are then removed as the plant biomass is harvested. This technique has shown promising results with crops like wheat and even with high-value medicinal plants such as alfalfa. In such cases, AMF helps boost the uptake of metals into plant tissues, making it easier to clean up polluted soils in a sustainable and cost-effective way—especially in agricultural areas where land recovery is a priority (Kuang et al., 2025).

AMF play a crucial role in reducing the availability of heavy metals in contaminated soils through a range of interconnected processes. One of their most important functions is helping to trap heavy metals within plant roots and fungal hyphae, which prevents these toxic elements from moving into the upper parts of the plant and, ultimately, from entering the food chain. A key mechanism behind this is the production of glomalin-associated soil proteins (GASPs), which are known to bind metals like Pb, keeping them locked in the soil and less harmful to plants (Lu et al., 2023; Son et al., 2024). In addition, AMF can capture metals by chelating them within their fungal structures or attaching them to the surface of their hyphae. This action helps limit how much of these metals are taken up by plants and curbs their movement through the soil. Research, including both lab experiments and broader meta-analyses, has consistently shown that plants associated with AMF tend to accumulate more metals in their roots while keeping the levels in leaves and stems much lower. This kind of selective metal retention not only enhances the plant's ability to tolerate contamination but also helps protect the parts of the plant that are most vulnerable or most likely to be consumed. Beyond managing metals directly, AMF also contribute to better soil health. By supporting the formation of stable soil aggregates and improving moisture retention, they indirectly help reduce the movement and availability of heavy metals. On top of this, AMF foster more diverse and robust microbial communities in the soil, which can work alongside the fungi to further detoxify and stabilize contaminants. Altogether, these biological interactions make AMF vital allies in restoring polluted soils, offering a natural and sustainable path for reducing heavy metal risks in the environment (Yan et al., 2020; Muhammad et al., 2024).

4. Organic Pollutant Degradation

The symbiotic interaction between AMF and plant roots is used by AMB to accelerate degradation processes, making it a revolutionary method for cleaning up soils contaminated with organic contaminants. Using the fungi's capacity to promote microbial activity, enhance plant health, and either directly or indirectly digest pollutants, this technique makes money. Numerous groups of organic pollutants, such as pesticides, petroleum hydrocarbons, phthalate esters, and polycyclic aromatic hydrocarbons (PAHs), are effectively remedied by AMB; each class requires a different set of processes. As an example of the adaptability and promise of AMB in tackling soil contamination, we go deeply into these mechanisms below, backed by case examples and open-access scientific material (Khaliq et al., 2022; Wahab et al., 2023).

4.1. Polycyclic Aromatic Hydrocarbons (PAHs)

Persistent organic pollutants, or PAHs, are produced by oil spills, industrial operations, and incomplete burning of fossil fuels. They are resistant to natural degradation because of their complex molecular architectures and hydrophobic nature. AMF uses a variety of synergistic processes to improve PAH dissipation (Patel et al., 2020). First, AMF's hyphal networks spread out past the root zone, forming a "hyphosphere" that attracts bacteria that break down hydrocarbons, including Mycobacterium, Bacillus, and Pseudomonas. Catabolic genes, such as nidA and phnAc, are found in these microorganisms. These genes encode enzymes, such as dioxygenases, that degrade PAH rings into less hazardous intermediates (Iffis et al., 2014; Sun et al., 2023). According to different studies, higher microbial diversity and enzymatic activity in the rhizosphere caused Glomus mosseae-colonized clover (*Trifolium pratense*) to increase PAH breakdown by 20–40% when compared to non-mycorrhizal controls (Sun et al., 2014; Malicka et al., 2021).

Second, AMF directly oxidizes PAHs by secreting extracellular enzymes such as laccases and peroxidases. For instance, studies found that pyrene levels in Lolium multiflorum infected with *Rhizophagus irregularis* decreased by 30%, which was associated with fungal peroxidase activity (Cheng et al., 2021). Benzo[a]pyrene and other high-molecular-weight PAHs that are otherwise resistant are broken down by these enzymes (Bukowska et al., 2022). Third, AMF release organic acids (like oxalic acid) that solubilize PAHs and make them accessible for microbial breakdown via improving plant biomass and root exudation. Due to enhanced root exudates and microbial activity, AMF-inoculated alfalfa (Medicago sativa) decreased phenanthrene concentrations up to 50% (Zhou et al., 2009; Li et al., 2023). AMF-inoculated plants routinely outperform non-inoculated systems in PAH elimination, with degradation rates rising by 15–60% depending on plant-fungi combinations, according to a meta-analysis of 28 research studies.

4.2. Phthalate Esters (PAEs)

Widely employed as plasticizers, PAEs are endocrine disruptors that seep into soils through agricultural films, wastewater, and packaging. Plant detoxification support, microbial consortia activation, and bioavailability modulation are all involved in AMF-mediated PAE breakdown (Zou et al., 2025). AMF secretes the glycoprotein globulin, which binds to PAEs such as di(2-ethylhexyl)phthalate (DEHP) to decrease their toxicity and mobility (González-Chávez et al., 2004). *Glomus versiforme* enriched soil microbial communities that could metabolize DEHP-related genes (pht, hyd) while reducing DEHP accumulation in lettuce (*Lactuca sativa*) shoots by 35% (Ma et al., 2018). Additionally, AMF attracts bacteria that break down PAE, like *Rhodococcus* and *Sphingomonas*. Enhanced *Sphingomonas populations* were associated with a 25% greater DEHP breakdown rate in AMF-colonized maize (*Zea mays*) than in non-mycorrhizal plants. Furthermore, AMF reduces oxidative stress brought on by PAE by upregulating plant antioxidant systems, such as glutathione-S-transferases (Lu et al., 2023). Some authors found that increasing superoxide dismutase activity, AMF inoculation decreased the toxicity of di-n-butyl phthalate (DBP) in cucumbers (*Cucumis sativus*). Scalability for agricultural remediation was established by field applications, such as a contaminated vegetable farm in China, where AMF inoculation in conjunction with pumpkin (*Cucurbita pepo*) decreased soil DBP levels by 45% in a single growing season (Wang et al., 2016a).

4.3. Petroleum Hydrocarbons

There are hazardous alkanes, aromatics, and asphaltenes in crude oil and refined petroleum products. AMF improves remediation by inducing enzymes, rhizodegradation, and phytostabilization. AMF's hyphae carry hydrocarbons to the root zones, where they are broken down by specialist microorganisms like Achromobacter and Alcanivorax. Some authors argue that ryegrass inoculated with *Glomus intraradices* destroyed 60% of the total petroleum hydrocarbons (TPHs) in diesel-contaminated soil, while non-inoculated systems only degraded 40% (Zeng et al., 2025).

AMF also increases plant resistance to hydrocarbon stress by improving phosphorus and other nutrient absorption and decreasing the bioavailability of metal co-contaminants. In a study conducted in the Niger Delta, *Paspalum vaginatum* colonized by AMF sequestered lead and nickel while reducing soil TPHs by 55% (Adesipo et al., 2020). Additionally, cytochrome P450 enzymes, which oxidize hydrocarbons like toluene and benzene, are produced by plants when AMF is present. In a Medicago truncatula-AMF symbiosis, 70% of benzene was broken down in 30 days, exceeding non-symbiotic systems by 20% (Rajtor and Piotrowska-Seget, 2016). Examples of industrial uses include the repair of oil spills in Ecuador, where switchgrass (*Panicum virgatum*) treated with AMF decreased TPHs by 50–70% in highly contaminated areas (Ryszka et al., 2019).

4.4. Pesticides

Pesticides that persist in soils, such as organophosphates and chlorinated chemicals, endanger both human health and ecosystems. Through phytovolatilization, enzymatic hydrolysis, and microbial recruitment, AMF enhances pesticide breakdown. Microbes like Streptomyces and Arthrobacter that break down pesticides are drawn to AMF (Pathak et al., 2022). Enhanced *Pseudomonas* populations were associated with a 30% increase in atrazine mineralization in sunflower (Helianthus annuus) colonized by AMF (Fan et al., 2020).

AMF increases phosphatase and esterase activity, which speeds up enzymatic hydrolysis. Within 60 days, AMF-inoculated tomatoes (*Solanum lycopersicum*) decreased chlorpyrifos residues by 40% (Wołejko et al., 2022; Vargas et al., 2024). Moreover, AMF reduces soil residues by facilitating the transpiration and absorption of volatile pesticides (like lindane). *Cucurbita pepo*, or zucchini, inoculated with AMF volatilized 15% more lindane than controls (Gío-Trujillo et al., 2024). Because AMB decreases pesticide persistence without the use of chemical additives, regulatory agencies such as the European Food Safety Authority (EFSA) now advise its use in integrated pest control (Vincent, 2017).

4.5. Challenges and Innovations

AMB has a number of difficulties, such as contaminant specificity, where complex pesticide combinations or high-molecular-weight PAHs may call for specialized fungal strains, and field scalability constraints brought on by soil heterogeneity and climate variability. Because defined techniques for large-scale AMF inoculation are still lacking, regulatory obstacles still continue to exist. These obstacles are being addressed by developments such as CRISPR-edited AMF strains designed to overexpress enzymes like laccase and biochar-AMF composites, which improve fungal colonization while adsorbing contaminants. By clarifying plant-fungal-microbial interactions, emerging technologies like metagenomics and isotope tracing enhance AMB techniques and open the door to precision remediation catered to particular pollutants and habitats.

5. AM Fungi in Phytoremediation Systems

Phytoremediation using plants to clean up environmental pollutants has gained attention as a cost-effective and environmentally friendly approach to addressing soil contamination. The integration of AMF into phytoremediation systems substantially improves their effectiveness through several mechanisms:

- *Enhanced Plant Establishment:* AMF significantly improves plant growth and survival in contaminated soils through multiple symbiotic mechanisms. This considerably increases plant growth and survival in polluted soils. In nutrient-depleted conditions, these improves nutrient acquisition, especially phosphorus, by increasing the root absorptive surface area via extraradical hyphae, allowing plants to form strong root systems (Yang et al., 2016; Zhao et al., 2024). As an example, *Zea mays* (maize) infected with AMF exhibits enhanced phosphorus absorption via upregulated phosphate transporters (PT1), reversing growth inhibition caused by heavy metals such as cadmium (Cd). Due to better nutrient uptake and less metal transfer to shoots, field studies show that AMF-colonized maize produces 20–30% more biomass in Cd-contaminated soils (Lu et al., 2023; Zhao et al., 2024).

- *Stress Alleviation:* Plant survival and performance in contaminated environments are enhanced by the presence of AM fungi, which assists plants deal with the stress brought on by pollutants. AMF colonization increases a plants' resistance to abiotic stresses (such as drought and heavy metals) via a variety of complex biochemical and molecular processes:
    (i) Antioxidant Defense, to counteract reactive oxygen species (ROS), AMF increases the activity of antioxidant enzymes such as catalase (CAT) and superoxide dismutase (SOD)(Leudo et al., 2020; Yilmaz et al., 2023).
    (ii) Hormonal Regulation: To fine tune stress responses, AMF modulates phytohormones such as abscisic acid (ABA) and jasmonic acid (JA). According to Wahab et al., colonized *Robinia pseudoacacia* exhibits enhanced ABA mediated stomatal control during drought, reducing water loss while preserving photosynthesis (Wahab et al., 2023).
- *Water Relations:* In dry or compacted soils, AMF hyphae increase root access to water, which is essential for plant survival in damaged habitats. In dry conditions, extraradical hyphae provide hydration to roots by directly absorbing water from soil micropores. For instance, hyphal water transfer in AMF-colonized wheat results in 20–30% increased drought survival rates (Wu et al., 2024a; Laanisto et al., 2025)
- *Increased Root Surface Area:* Arbuscular mycorrhizal fungi have vast extraradical hyphal network functions as a biological amplifier, significantly expanding the root surface area that is accessible for pollution uptake and interception. AMF hyphae can reach microsites that are inaccessible to roots by reaching up to 20 cm beyond root zones. This effectively expands the rhizosphere into a hyphosphere that improves contact with pollutants and nutrients (Ma et al., 2022b). In addition to directly adsorbing heavy metals like Cd and Pb onto chitin hyphal walls, this network hosts bacteria that break down pollutants, such as Pseudomonas and Dyella, in its biofilm-rich microenvironments, which promotes the breakdown of hydrocarbons (Reza Boorboori and Zhang, 2022; Ma et al., 2022b; Li et al., 2024).
- *Soil Structure Improvement:* During phytoremediation, arbuscular mycorrhizal fungi greatly improve soil structural integrity by stabilizing the hyphal network and aggregating glomalin. AMF hyphae release the glycoprotein glomalin, which functions as a hydrophobic "biological glue," transforming soil microaggregates (less than 250 μm) into water-stable macroaggregates (greater than 2 mm) that enhance water retention and prevent erosion. This procedure is essential in degraded soils since glomalin resistance guarantees aggregate stability over the long term, even in areas affected by heavy metals or hydrocarbons (Zhao et al., 2024). At the same time, the vast extraradical hyphal network physically enmeshes soil particles, forming a scaffold that improves porosity and decreases

compaction—two essential elements for microbial activity and root growth in remediation zones (Klironomos and Ursic, 1998; Wang et al., 2022).

Field studies demonstrate that AMF inoculation in mine tailings or oil-contaminated soils increases macroaggregate formation by 20–30%, directly correlating with reduced nutrient leaching and improved plant survival rates (Ai et al., 2022). Improved soil structure promotes microbial diversity, which speeds up pollutant breakdown and nutrient cycling. These processes not only stabilize remediation sites but also establish a self-sustaining cycle. When combined with AMF, grasses have shown themselves to be especially successful phytoremediation partners because of their fibrous root systems. When colonized by AMF, their broad root architecture produces an optimal environment for pollutant breakdown by providing a wide surface area for root-soil contact (Saldarriaga et al., 2023).

The integration of vetiver grass (*Chrysopogon zizanioides*) with arbuscular mycorrhizal fungus (AMF), which has been shown to be very successful in Pb phyto-stabilization in polluted soils and mine-tailings. By producing glomalin and sequestering hyphal material, AMF colonization increases vetiver's root biomass and metal retention capacity by immobilizing lead in soil aggregates and root tissues (Wu et al., 2011). Because AMF preferentially retained Pb in roots through cell wall binding and vacuolar compartmentalization, vetiver inoculated with AMF in Pb/Zn mine tailings, for instance, showed 8.1–13.8% increased biomass and decreased Pb transfer to shoots when compared to non-mycorrhizal plants. Field tests further demonstrate that AMF-vetiver systems, which are essential for long-term site rehabilitation, decrease Pb bioavailability and erosion through glomalin-stabilized soil aggregates (Wu et al., 2011).

6. Soil Fertility Enhancement

Beyond their direct role in pollutant removal, AMF are fundamental drivers of soil fertility (physical, chemical and biological fertility), providing multiple ecosystem services that indirectly enhance the success of soil remediation processes.

By the one hand, AMF hyphal networks are essential for maintaining the stability of soil structure. Glomalin, a heat- and water-resistant glycoprotein produced by the extraradical mycelia, functions as a "glue" that keeps soil particles together in stable micro- and macroaggregates (Fall et al., 2022). These aggregates decrease erosion and surface runoff while increasing soil porosity, water infiltration, and gas exchange (Guest et al., 2022). The persistence of dead hyphae and the ongoing renewal of AMF mycelia provide organic matter and physical binding agents that sustain soil aggregation over time (Li et al., 2025). Fertile and resilient soils depend on root growth and microbial habitat quality, which are both strengthened by this improved soil structure and lower the danger of compaction (Fall et al., 2022).

Regarding the chemical fertility of the soils, AFM improve nutrient cycling and bioavailability, particularly P, a key limiting nutrient in many acidic soils. Through symbiotic associations, AMF increase plant access to soluble P by: solubilizing insoluble phosphate compounds, mobilizing phosphate from phosphate fertilizers, and facilitating synergistic interactions with phosphate-solubilizing bacteria (PSB).

Plants cannot directly absorb a large portion of the phosphorus found in soils because it is present in insoluble forms like calcium phosphate, aluminum phosphate, and iron phosphate. By releasing organic acids (such as citric and oxalic acids) and enzymes like phosphatases, which reduce soil pH and chemically mobilize P into plant-accessible soluble orthophosphate forms, AMF helps to solubilize these compounds (Zapata et al., 2024; Pang et al., 2024; García-Berumen et al., 2025).

AMF enhance the effectiveness of rock phosphate fertilizers by mobilizing phosphate ions from these relatively insoluble mineral sources. The fungi's hyphal networks extend far beyond root surfaces, increasing soil exploration and accessing phosphate locked in rock phosphate particles. This mobilization improves fertilizer efficiency and reduces the need for chemical P inputs (López et al., 2020; Hazzoumi et al., 2022; Boutasknit et al., 2024).

In order to further increase P availability, AMF interacts with PSB, which can produce organic acids and siderophores that mineralize organic P compounds and chelate metal ions that bind phosphate (Minaxi et al., 2013; Lu et al., 2023). While AMF gives PSB carbon substrates through fungal exudates, PSB can penetrate the rhizosphere and even employ AMF hyphae as conduits to reach distant soil microsites rich in insoluble phosphorus. Compared to either organism alone, this mutualistic interaction improves the solubilization, mineralization, and uptake of P (Nacoon et al., 2020; Etesami et al., 2021).

AMF helps soils efficiently cycle C and N in addition to P. Through hyphal turnover and carbon inputs, AMF symbiosis increases soil organic matter buildup, decreases nitrogen leaching, and improves nitrogen fixation in legumes (Fall et al., 2022; Luo et al., 2023). For instance, AMF hyphae proliferate extensively in organic-rich soil patches, acquiring nitrogen released from decomposing organic matter, e.g., Glomus hoi and *Glomus mosseae* can take up N from organic sources and transfer a portion to their host plants, enhancing plant N nutrition (Hodge and Fitter, 2010). By increasing the number of nitrogen-fixing genes and enzyme activity, AMF inoculation improves nitrogen uptake and lowers nitrogen loss through gaseous emissions or leaching (Gou et al., 2024). Regarding the contribution to the C cycle, AMF indirectly influence soil carbon turnover by modifying rhizosphere chemistry and providing carbon substrates that stimulate microbial communities involved in organic matter decomposition and nutrient cycling (Wu et al., 2024b).

AMF profoundly influence the composition, diversity, and activity of soil microbial communities primarily through their secretions and by modifying the rhizosphere environment. The hyphosphere is a special microhabitat created by AMF extraradical hyphae, which reach deep into the surrounding soil beyond the root surface. Carbon-rich exudates from AMF hyphae, such as low-molecular-weight sugars, organic acids, and polymeric substances, are abundant in this zone and provide soil bacteria with energy and signaling molecules (Emmett et al., 2021; Saldarriaga et al., 2023). These exudates cause changes in the richness and composition of microbial communities by selectively stimulating or inhibiting particular bacterial and fungal species. For instance, it has been demonstrated that AMF hyphae restrict certain bacterial groups that are less advantageous for soil function while enriching others like Burkholderiales and Actinomycetales, which are linked to nutrient cycle and pollutant breakdown (Qin et al., 2016). Recognizing and harnessing the integral role of AMF in soil microbial ecology is essential for advancing sustainable agriculture, ecosystem restoration, and effective soil remediation strategies.

7. Future Research Directions

Even though our understanding of arbuscular mycorrhizal fungi's (AMF) function in soil pollution removal has advanced significantly, there are still a number of important information gaps that need to be filled in order to fully utilize AMF's potential in environmental remediation.

Screening of High-Efficiency Fungi: Finding and growing AMF species or strains with exceptional remediation capacities for particular contaminants is a top research focus. With over 170 AMF species identified globally, each with distinct physiological and biochemical characteristics, targeted screening can identify fungi that have improved capacities to break down, immobilize, or promote the uptake of specific contaminants, such as organic pollutants or heavy metals. High-throughput, physiological, and advanced molecular screening methods should be used to evaluate the pollutant-remediation capabilities of various AMF isolates in both controlled and field settings. With this strategy, customized AMF inoculants that are optimized for site-specific remediation difficulties can be developed.

Co-existing AMF Efficiency: Since most soils naturally include several AMF species that may interact either antagonistically or synergistically, studying AMF consortia is another exciting line of inquiry. Knowing how various AMF species coexist—or compete—with one another within the same rhizosphere may help identify combinations that optimize the effectiveness of pollution removal. The dynamics of AMF community structure, functional complementarity, and the combined effect on soil health, pollutant uptake, and plant growth should be the main topics of investigation. Multi-strain AMF inoculants that capitalize on the unique characteristics of several fungi may be developed as a result of this research, potentially surpassing single-species uses in challenging, real-world settings.

Soil Aging Effects: Further research is also required to determine how soil aging affects AMF-mediated remediation procedures. The bioavailability and mobility of pollutants, as well as the colonization and operation of AMF, can be impacted by soil aging, which includes alterations in the physical, chemical, and biological characteristics of soil over time. Understanding how aged soils impact AMF's persistence, adaptability, and remedial efficacy will require long-term research. As soils and their contaminants change, these insights will be essential for enhancing long-term remediation tactics and guaranteeing that pollutant removal stays effective.

Plant Uptake Mechanisms: The particular processes by which AMF affects plant absorption of different contaminants also require more investigation. Although the ability of AMF to both promote and inhibit the entry of pollutants into plant tissues is well documented, little is known about the molecular, physiological, and cellular mechanisms underlying these effects. Research ought to concentrate on the control of transporter proteins, the function of fungus exudates (such as glomalin), and the signaling pathways that connect plant roots with AMF and regulate the uptake, sequestration, and detoxification of pollutants. With a better grasp of these processes, more focused and efficient phytoremediation techniques can be developed, which may eventually lead to the stability or selective removal of particular pollutants.

The identification of high-efficiency fungal strains, the optimization of multi-species consortia, and the clarification of long-term soil aging consequences on pollutant stability are all ways that addresses these research goals and will advance AMF-assisted remediation. Plant-fungal systems can be improved for the selective removal of contaminants by comprehending molecular interactions such as stress signaling, root exudates, and transporter proteins. By producing low-cost, scalable bioremediation technologies that are compatible with land restoration and agriculture, these developments will lessen the need for energy-intensive techniques. We can increase carbon sequestration, improve soil biodiversity, and reduce cleanup costs by combining AMF applications with ecological resilience. Rebuilding nutrient cycles and ecosystem health while providing sustainable, nature-based solutions for contaminated places is in line with global sustainability goals. By facilitating the shift from lab-based insights to field-ready methods, it guarantees long-term efficacy and financial sustainability. AMF-driven approaches, in the end, provide a comprehensive framework for pollution reduction that balances environmental restoration with human activity.

8. Conclusion

Arbuscular mycorrhizal fungi (AMF) represent a powerful biological tool for enhancing pollutant removal from contaminated soils. Through their diverse mechanisms of action—direct metabolism, enzymatic degradation, enhanced rhizospheric activity, facilitated plant uptake, metal binding, and compartmentalization—they significantly improve remediation outcomes for both organic pollutants and heavy metals. Through a variety of processes that promote the health of plants and soil, AMF are essential in improving the removal of pollutants from contaminated soils. By immobilizing harmful ions by biosorption, glomalin synthesis, and internal compartmentalization, these fungi enhance nutrient absorption, especially phosphorus, and boost plant resistance to heavy metals. Additionally, they promote the microbial breakdown of organic pollutants such as polycyclic aromatic hydrocarbons, phthalates, pesticides, and petroleum hydrocarbons, as well as the chemical transformation of toxins, such as the reduction of Cr(VI) to the less hazardous Cr(III). Depending on the concentration of the pollutant, AMF exhibits two different remediation strategies: at low concentrations, they improve the selective phytoextraction of beneficial metals like copper and zinc, while at high concentrations, they encourage phytostabilization by enclosing toxic metals in the root zone and preventing their translocation to aerial plant parts. AMF also promotes beneficial microbial communities in the rhizosphere, increases water retention, decreases erosion, and improves soil physical structure through glomalin-induced aggregation. Because of these combined effects, AMF is a powerful agent for phytoremediation and phytostabilization, particularly in areas that have been degraded and contaminated by metals. It also helps to maintain soil fertility over the long term. Additionally, they improve the chemical and biological quality of soil by their participation in the cycling of carbon, nitrogen, and phosphorus.

Identifying high-efficiency AMF strains, comprehending interspecies interactions, and modifying field applications to varying soil conditions are some of the issues that still need to be addressed despite their potential. The molecular and physiological processes underlying pollutant uptake and detoxification should be the main focus of future study in order to fully utilize AMF's promise in scalable, environmentally friendly remediation methods.

References

Adesipo, A.A., Freese, D. & Nwadinigwe, A.O. (2020). Prospects of in-situ remediation of crude oil contaminated lands in Nigeria. Scientific African 8, e00403. https://doi.org/10.1016/J.SCIAF.2020.E00403

Ahammed, G.J., Shamsy, R., Liu, A. & Chen, S. (2023). Arbuscular mycorrhizal fungi-induced tolerance to chromium stress in plants. Environmental Pollution 327, 121597. https://doi.org/10.1016/j.envpol.2023.121597

Ahmed, N., Li, J., Li, Y. et al. (2025). Symbiotic synergy: How Arbuscular Mycorrhizal Fungi enhance nutrient uptake, stress tolerance, and soil health through molecular mechanisms and hormonal regulation. IMA Fungus 16, e144989. https://doi.org/10.3897/IMAFUNGUS.16.144989

Ai, Y.-J., Li, F.-P., Yang, J.-Q. et al. (2022). Research progress and potential functions of AMF and GRSP in the ecological remediation of metal tailings. Sustainability 14, 9611. https://doi.org/10.3390/SU14159611

Akter, S., Mahmud, U., Shoumik, B.A., & Al, Khan, M.Z. (2025). Although invisible, fungi are recognized as the engines of a microbial powerhouse that drives soil ecosystem services. Archives of Microbiology 207, 1–28. https://doi.org/10.1007/S00203-025-04285-4

Banerjee, S., Mandal, J., Sarkar, D. et al. (2024). A review and meta-analysis of the efficacy of arbuscular mycorrhizal fungi in remediating toxic metals in mine-affected soils. Frontiers in Environmental Science 12, 1532169. https://doi.org/10.3389/FENVS.2024.1532169/XML/NLM

Begum, N., Qin, C., Ahanger, M.A. et al. (2019). Role of arbuscular mycorrhizal fungi in plant growth regulation: Implications in abiotic stress tolerance. Frontiers in Plant Science 10. https://doi.org/10.3389/fpls.2019.01068

Boorboori, M.R. & Zhang, H.Y. (2022). Arbuscular mycorrhizal fungi are an influential factor in improving the phytoremediation of arsenic, cadmium, lead, and chromium. Journal of Fungi (Basel) 8, 176–176. https://doi.org/10.3390/JOF8020176

Boorboori, M.R. & Zhang, H.Y. (2024). Correction to: Arbuscular mycorrhizal fungi are an influential factor in improving the phytoremediation of arsenic, cadmium, lead, and chromium Journal of Fungi 8(2), 176, https://doi.org/10.3390/JOF10060376

Boutasknit, A., Ait-El-Mokhtar, M., Fassih, B. et al. (2024). Effect of arbuscular mycorrhizal fungi and rock phosphate on growth, physiology, and biochemistry of carob under water stress and after rehydration in vermicompost-amended soil. Metabolites 14, 202. https://doi.org/10.3390/metabo14040202

Bukowska, B., Mokra, K. & Michałowicz, J. (2022). Benzo[a]pyrene—Environmental occurrence, human exposure, and mechanisms of toxicity. International Journal of Molecular Sciences 23, 6348. https://doi.org/10.3390/IJMS23116348

Cheng, H., Wang, J., Tu, C. et al. (2021). Arbuscular mycorrhizal fungi and biochar influence simazine decomposition and leaching. GCB Bioenergy 13, 708–718. https://doi.org/10.1111/GCBB.12802

Colombo, R.P., Silvani, V.A., Benavidez, M.E. et al. (2024). Different behavior of two strains of the arbuscular mycorrhizal fungus Rhizophagus intraradices on Senecio bonariensis Hook. & Arn. against heavy metal soil pollution: A pilot-scale test. International Journal of Phytoremediation 26, 1741–1748. https://doi.org/10.1080/15226514.2024.2353389;PAGE:STRING:ARTICLE/CHAPTER

Crişan, I., Balestrini, R. & Pagliarani, C. (2024). The current view on heavy metal remediation: The relevance of the plant interaction with arbuscular mycorrhizal fungi. Plant Stress 12, 100439. https://doi.org/10.1016/J.STRESS.2024.100439

Cruz, Y., Camargo, G., Gallego, J.L. & Saldarriaga, J.F. (2019). A kinetic modelling of the growth rate of Lolium perenne for phytotoxicity bioassays. Chemical Engineering Transactions 74, 1441–1446. https://doi.org/10.3303/CET1974241

Cruz, Y., Villar, S., Gutiérrez, K. et al. (2021). Gene expression and morphological responses of Lolium perenne L. exposed to cadmium (Cd^{2+}) and mercury (Hg^{2+}). Scientific Reports 11, 11257. https://doi.org/10.1038/s41598-021-90826-y

Dagher, D.J., de la Providencia, I.E., Pitre, F.E. et al. (2020). Arbuscular mycorrhizal fungal assemblages significantly shifted upon bacterial inoculation in non-contaminated and petroleum-contaminated environments. Microorganisms 8, 602. https://doi.org/10.3390/microorganisms8040602

Dhalaria, R., Kumar, D., Kumar, H. et al. (2020). Arbuscular mycorrhizal fungi as potential agents in ameliorating heavy metal stress in plants. Agronomy 10, https://doi.org/10.3390/AGRONOMY10060815

Dinakarkumar, Y., Ramakrishnan, G., Gujjula, K.R. et al. (2024). Fungal bioremediation: An overview of the mechanisms, applications and future perspectives. Environmental Chemistry and Ecotoxicology 6, 293–302. https://doi.org/10.1016/j.enceco.2024.07.002

Emmett, B.D., Lévesque-Tremblay, V. & Harrison, M.J. (2021). Conserved and reproducible bacterial communities associate with extraradical hyphae of arbuscular mycorrhizal fungi. The ISME Journal 15, 2276–2288. https://doi.org/10.1038/s41396-021-00920-2

Etesami, H. & Glick, B.R. (2023). Exploring the potential: Can mycorrhizal fungi and hyphosphere silicate-solubilizing bacteria synergistically alleviate cadmium stress in plants? Current Research in Biotechnology 6, 100158. https://doi.org/10.1016/J.CRBIOT.2023.100158

Etesami, H., Jeong, B.R. & Glick, B.R. (2021). Contribution of arbuscular mycorrhizal fungi, phosphate – Solubilizing bacteria, and silicon to P uptake by plant. Frontiers in Plant Science 12. https://doi.org/10.3389/fpls.2021.699618

Fall, A.F., Nakabonge, G., Ssekandi, J. et al. (2022). Roles of arbuscular mycorrhizal fungi on soil fertility: Contribution in the improvement of physical, chemical, and biological properties of the soil. Frontiers in Fungal Biology 3. https://doi.org/10.3389/ffunb.2022.723892

Fan, X., Chang, W., Sui, X. et al. (2020). Changes in rhizobacterial community mediating atrazine dissipation by arbuscular mycorrhiza. Chemosphere 256, 127046. https://doi.org/10.1016/J.CHEMOSPHERE.2020.127046

FAO. (2017). Guidelines for the Registrationof Microbial, Botanical and Semiochemical Pestcontrol Agents for Plant Protection and Public Health Uses. Rome

Gao, Y., Li, Q., Ling, W. & Zhu, X. (2011). Arbuscular mycorrhizal phytoremediation of soils contaminated with phenanthrene and pyrene. Journal of Hazardous Materials 185, 703–709. https://doi.org/10.1016/j.jhazmat.2010.09.076

García-Berumen, J.A., Flores de la Torre, J.A., de los Santos-Villalobos, S. et al. (2025). Phosphorus dynamics and sustainable agriculture: The role of microbial solubilization and innovations in nutrient management. Current Research in Microbial Sciences 8, 100326. https://doi.org/10.1016/j.crmicr.2024.100326

George, N.P. & Ray, J.G. (2023). The inevitability of arbuscular mycorrhiza for sustainability in organic agriculture—A critical review. Frontiers in Sustainable Food Systems 7. https://doi.org/10.3389/fsufs.2023.1124688

Ghorui, M., Chowdhury, S., Balu, P. & Burla, S. (2024). Arbuscular mycorrhizal inoculants and its regulatory landscape. Heliyon 10, e30359. https://doi.org/10.1016/J.HELIYON.2024.E30359

Gío-Trujillo, J.A., Aracely Pacheco-López, N., Cristóbal-Alejo, J. et al. (2024). Behavior of Cucurbita pepo L. var. "Grey Zucchini", in native arbuscular mycorrhizal fungi propagation in soils with different management. Biotecnia 26, 128–135. https://doi.org/10.18633/BIOTECNIA.V26.1972

González-Chávez, M.C., Carrillo-González, R., Wright, S.F. & Nichols, K.A. (2004). The role of glomalin, a protein produced by arbuscular mycorrhizal fungi, in sequestering potentially toxic elements. Environmental Pollution 130, 317–323. https://doi.org/10.1016/J.ENVPOL.2004.01.004

Goswami, V., Deepika, S., Chandra, R. et al. (2023). Arbuscular mycorrhizas accelerate the degradation of colour containing organic pollutants present in distillery spent wash leachates. Journal of Hazardous Materials 452, 131291. https://doi.org/10.1016/j.jhazmat.2023.131291

Gou, X., Hu, Y., Ni, H. et al. (2024). Arbuscular mycorrhizal fungi alleviate erosional soil nitrogen loss by regulating nitrogen cycling genes and enzymes in experimental agro-ecosystems. Science of The Total Environment 906, 167425. https://doi.org/10.1016/j.scitotenv.2023.167425

Guest, E.J., Palfreeman, L.J., Holden, J. et al. (2022). Soil macroaggregation drives sequestration of organic carbon and nitrogen with three-year grass-clover leys in arable rotations. Science of The Total Environment 852, 158358. https://doi.org/10.1016/j.scitotenv.2022.158358

Hao, B., Zhang, Z., Bao, Z. et al. (2022). Claroideoglomus etunicatum affects the structural and functional genes of the rhizosphere microbial community to help maize resist Cd and La stresses. Environmental Pollution 307, 119559. https://doi.org/10.1016/J.ENVPOL.2022.119559

Hasanuzzaman, M., Bhuyan, M.H.M.B., Zulfiqar, F. et al. (2020). Reactive oxygen species and antioxidant defense in plants under abiotic stress: Revisiting the crucial role of a universal defense regulator. Antioxidants 9, 681. https://doi.org/10.3390/ANTIOX9080681

Hazzoumi, Z., Azaroual, S.E., El Mernissi, N. et al. (2022). Effect of arbuscular mycorrhizal fungi isolated from rock phosphate mine and agricultural soil on the improvement of wheat plant growth. Frontiers in Microbiology 13. https://doi.org/10.3389/fmicb.2022.881442

Hodge, A. & Fitter, A.H. (2010). Substantial nitrogen acquisition by arbuscular mycorrhizal fungi from organic material has implications for N cycling. Proceedings of the National Academy of Sciences 107, 13754–13759. https://doi.org/10.1073/pnas.1005874107

Iffis, B., St-Arnaud, M. & Hijri, M. (2014). Bacteria associated with arbuscular mycorrhizal fungi within roots of plants growing in a soil highly contaminated with aliphatic and aromatic petroleum hydrocarbons. FEMS Microbiol Letters 358, 44–54. https://doi.org/10.1111/1574-6968.12533

Jiang, Q.Y., Zhuo, F., Long, S.H. et al. (2016). Can arbuscular mycorrhizal fungi reduce Cd uptake and alleviate Cd toxicity of Lonicera japonica grown in Cd-added soils? Scientific Reports 6, 1–9. https://doi.org/10.1038/SREP21805;SUBJMETA=169,172,4081,704;KWRD=ENVIRONMENTAL+CHEMISTRY,ENVIRONMENTAL+IMPACT

Kafle, A., Timilsina, A., Gautam, A. et al. (2022). Phytoremediation: Mechanisms, plant selection and enhancement by natural and synthetic agents. Environmental Advances 8, 100203. https://doi.org/10.1016/J.ENVADV.2022.100203

Kalamulla, R., Karunarathna, S.C., Tibpromma, S. et al. (2022). Arbuscular mycorrhizal fungi in sustainable agriculture. Sustainability 14, 12250. https://doi.org/10.3390/SU141912250

Khaliq, A., Perveen, S., Alamer, K.H. et al. (2022). Arbuscular mycorrhizal fungi symbiosis to enhance plant–soil interaction. Sustainability 14, 7840. https://doi.org/10.3390/SU14137840

Klironomos, J.N. & Ursic, M. (1998). Density-dependent grazing on the extraradical hyphal network of the arbuscular mycorrhizal fungus, Glomus intraradices, by the collembolan, Folsomia candida. Biology and Fertility of Soils 26, 250–253. https://doi.org/10.1007/S003740050375/METRICS

Kuang, Q., Wu, Y., Gao, Y. et al. (2025). Arbuscular mycorrhizal fungi mitigate cadmium stress in maize. Ecotoxicology and Environmental Safety 289, 117600. https://doi.org/10.1016/J.ECOENV.2024.117600

Kumar, A., Yadav, A.N., Mondal, R. et al. (2021). Myco-remediation: A mechanistic understanding of contaminants alleviation from natural environment and future prospect. Chemosphere 284, 131325. https://doi.org/10.1016/j.chemosphere.2021.131325

Laanisto, L., Pavanetto, N., Puglielli, G. et al. (2025). Contrasting mycorrhizal functionality in abiotic stress tolerance of woody species. Scientific Report 15, 1–9. https://doi.org/10.1038/S41598-025-93787-8;SUBJMETA=158,2668,2676,449,631,851,857;KWRD=EVOLUTIONARY+ECOLOGY,MACROECOLOGY,PLANT+ECOLOGY,PLANT+SYMBIOSIS

Laishram, B., Devi, O.R., Dutta, R. et al. (2025). Plant-microbe interactions: PGPM as microbial inoculants/biofertilizers for sustaining crop productivity and soil fertility. Current Research in Microbial Sciences 8, 100333. https://doi.org/10.1016/J.CRMICR.2024.100333

Leudo, A.M., Cruz, Y., Montoya-Ruiz, C. et al. (2020). Mercury phytoremediation with Lolium perenne-Mycorrhizae in contaminated soils. Sustainability 12, 3795. https://doi.org/10.3390/su12093795

Li, M.-Y., Wang, W., Yin, H.-H. et al. (2025). The functional role of arbuscular mycorrhizal fungi in enhancing soil organic carbon stocks and stability in dryland. Soil and Tillage Research 248, 06443. https://doi.org/10.1016/j.still.2024.106443

Li, X., Kang, X., Zou, J. et al. (2023). Allochthonous arbuscular mycorrhizal fungi promote Salix viminalis L.–mediated phytoremediation of polycyclic aromatic hydrocarbons characterized by increasing the release of organic acids and enzymes in soils. Ecotoxicology and Environmental Safety 249, 114461. https://doi.org/10.1016/J.ECOENV.2022.114461

Li, X., Song, C., Kang, X. et al. (2024). Assembly and functional profile of rhizosphere microbial community during the Salix viminalis-AMF remediation of polycyclic aromatic hydrocarbon polluted soils. Journal of Environmental Management 370, 122503. https://doi.org/10.1016/J.JENVMAN.2024.122503

López, J.E., Gallego, J.L., Vargas-Ruiz, A. et al. (2020). Aspergillus tubingensis and Talaromyces islandicus solubilize rock phosphate under saline and fungicide stress and improve Zea mays growth and phosphorus nutrition. Journal of Soil Science and Plant Nutrition. https://doi.org/10.1007/s42729-020-00315-w

Lu, Y., Yan, Y., Qin, J. et al. (2023). Arbuscular mycorrhizal fungi enhance phosphate uptake and alter bacterial communities in maize rhizosphere soil. Frontiers in Plant Science 14, 1206870. https://doi.org/10.3389/FPLS.2023.1206870/BIBTEX

Luo, X., Liu, Y., Li, S. & He, X. (2023). Interplant carbon and nitrogen transfers mediated by common arbuscular mycorrhizal networks: Beneficial pathways for system functionality. Frontiers in Plant Science 14. https://doi.org/10.3389/fpls.2023.1169310

Ma, T., Zhou, W., Chen, L. et al. (2018). Toxicity of phthalate esters to lettuce (Lactuca sativa) and the soil microbial community under different soil conditions. PLoS One 13, e0208111. https://doi.org/10.1371/JOURNAL.PONE.0208111

Ma, Y., Ankit, Tiwari J. & Bauddh, K. (2022a). Plant-mycorrhizal fungi interactions in phytoremediation of geogenic contaminated soils. Frontiers in Microbiology 13, 843415. https://doi.org/10.3389/FMICB.2022.843415/XML/NLM

Malicka, M., Magurno, F., Posta, K. et al. (2021). Differences in the effects of single and mixed species of AMF on the growth and oxidative stress defense in Lolium perenne exposed to hydrocarbons. Ecotoxicology and Environmental Safety 217, 112252. https://doi.org/10.1016/J.ECOENV.2021.112252

Mansoor, S., Ali, A., Kour, N. et al. (2023). Heavy metal induced oxidative stress mitigation and ROS scavenging in plants. Plants 12, 3003. https://doi.org/10.3390/PLANTS12163003

Meglouli, H., Fontaine, J., Lounès-Hadj Sahraoui, A. (2020). Dioxins/furans disturb the life cycle of the arbuscular mycorrhizal fungus, Rhizophagus irregularis and chicory root elongation grown under axenic conditions. International Journal of Phytoremediation 22, 1497–1504. https://doi.org/10.1080/15226514.2020.1784089,

Minaxi, Saxena, J., Chandra, S. & Nain, L. (2013). Synergistic effect of phosphate solubilizing rhizobacteria and arbuscular mycorrhiza on growth and yield of wheat plants. Journal of Soil Science and Plant Nutrition 511–525. https://doi.org/10.4067/S0718-95162013005000040

Montoliu-Nerin, M., Sánchez-García, M., Bergin, C. et al. (2021). In-depth phylogenomic analysis of arbuscular mycorrhizal fungi based on a comprehensive set of de novo genome assemblies. Frontiers in Fungal Biology 2, 716385. https://doi.org/10.3389/FFUNB.2021.716385/BIBTEX

Moreira, H., Pereira, S.I.A., Mench, M. et al. (2021). Phytomanagement of metal(loid)-contaminated soils: Options, efficiency and value. Frontiers in Environmental Science 9, 661423. https://doi.org/10.3389/FENVS.2021.661423/XML/NLM

Muhammad, M., Waheed, A., Wahab, A. et al. (2024). Soil salinity and drought tolerance: An evaluation of plant growth, productivity, microbial diversity, and amelioration strategies. Plant Stress 11, 100319. https://doi.org/10.1016/J.STRESS.2023.100319

Nacoon, S., Jogloy, S., Riddech, N. et al. (2020). Interaction between phosphate solubilizing bacteria and arbuscular mycorrhizal fungi on growth promotion and tuber inulin content of helianthus tuberosus L. Scientific Reports 10, 4916. https://doi.org/10.1038/s41598-020-61846-x

Ndeko, A.B., Géant, Chuma, B. et al. (2024). Soil fertility and crop biofortification: Is there potential improvement through arbuscular mycorrhizal symbiosis in Sub-Saharan Africa? Discover Soil 1, 1 1–33. https://doi.org/10.1007/S44378-024-00019-Y

Pang, F., Li, Q., Solanki, M.K. et al. (2024). Soil phosphorus transformation and plant uptake driven by phosphate-solubilizing microorganisms. Frontiers in Microbiology 15. https://doi.org/10.3389/fmicb.2024.1383813

Patel, A.B., Shaikh, S., Jain, K.R. et al. (2020). Polycyclic aromatic hydrocarbons: Sources, toxicity, and remediation approaches. Frontiers in Microbiology 11, 562813. https://doi.org/10.3389/FMICB.2020.562813/XML/NLM

Pathak, V.M., Verma, V.K., Rawat, B.S. et al. (2022). Current status of pesticide effects on environment, human health and it's eco-friendly management as bioremediation: A comprehensive review. Frontiers in Microbiology 13, 962619. https://doi.org/10.3389/FMICB.2022.962619

Phour, M. & Sindhu, S.S. (2024). Arbuscular mycorrhizal fungi: An eco-friendly technology for alleviation of salinity stress and nutrient acquisition in sustainable agriculture. Arbuscular Mycorrhizal Fungi in Sustainable Agriculture: Nutrient and Crop Management 223–267. https://doi.org/10.1007/978-981-97-0300-5_11

Qin, H., Brookes, P.C. & Xu, J. (2016). Arbuscular mycorrhizal fungal hyphae alter soil bacterial community and enhance polychlorinated biphenyls dissipation. Frontiers in Microbiology 7. https://doi.org/10.3389/fmicb.2016.00939

Rajtor, M. & Piotrowska-Seget, Z. (2016). Prospects for arbuscular mycorrhizal fungi (AMF) to assist in phytoremediation of soil hydrocarbon contaminants. Chemosphere 162, 105–116. https://doi.org/10.1016/J.CHEMOSPHERE.2016.07.071

Reza Boorboori, M. & Zhang, H.-Y. (2022). Arbuscular mycorrhizal fungi are an influential factor in improving the phytoremediation of arsenic, cadmium, lead, and chromium. Journal of Fungi 8, 176–176. https://doi.org/10.3390/JOF8020176

Riaz, M., Kamran, M., Fang, Y. et al. (2021). Arbuscular mycorrhizal fungi-induced mitigation of heavy metal phytotoxicity in metal contaminated soils: A critical review. Journal of Hazardous Materials 402, 123919. https://doi.org/10.1016/j.jhazmat.2020.123919

Ryszka, P., Zarzyka-Ryszka, M., Anielska, T. et al. (2019). Arbuscular mycorrhizal fungi from petroleum-impacted sites in the Polish Carpathians. International Biodeterioration and Biodegradation 138, 50–56. https://doi.org/10.1016/J.IBIOD.2019.01.001

Saboor, A., Ali, M.A., Hussain, S. et al. (2021). Zinc nutrition and arbuscular mycorrhizal symbiosis effects on maize (Zea mays L.) growth and productivity. Saudi Journal of Biological Sciences 28, 6339–6351. https://doi.org/10.1016/J.SJBS.2021.06.096

Saia, S. & Jansa, J. (2022). Editorial: Arbuscular mycorrhizal fungi: The bridge between plants, soils, and humans. Front Plant Sci 13:875958. https://doi.org/10.3389/fpls.2022.875958

Saldarriaga, J.F., López, J.E., Díaz-García, L. & Montoya-Ruiz, C. (2023). Changes in Lolium perenne L. rhizosphere microbiome during phytoremediation of Cd- and Hg-contaminated soils. Environmental Science and Pollution Research 30, 49498–49511. https://doi.org/10.1007/S11356-023-25501-Y/METRICS

Sánchez-Castro, I., Molina, L., Prieto-Fernández, M.Á. & Segura, A. (2023). Past, present and future trends in the remediation of heavy-metal contaminated soil – Remediation techniques applied in real soil-contamination events. Heliyon 9, e16692. https://doi.org/10.1016/J.HELIYON.2023.E16692

Schüssler, A. & Walker, C. (2010). The Glomeromycota: A Species List with New Families and New Genera. Gloucester: Royal Botanic Garden, Edinburgh.

Son, Y., Martínez, C.E. & Kao-Kniffin, J. (2024). Three important roles and chemical properties of glomalin-related soil protein. Frontiers in Soil Science 4, 1418072. https://doi.org/10.3389/FSOIL.2024.1418072/BIBTEX

Sun, D., Shang, X., Cao, H. et al. (2024). Physio-biochemical mechanisms of arbuscular mycorrhizal fungi enhancing plant resistance to abiotic stress. Agriculture 14, 2361. https://doi.org/10.3390/agriculture14122361

Sun, K., Jiang, H.J., Pan, Y.T. et al. (2023). Hyphosphere microorganisms facilitate hyphal spreading and root colonization of plant symbiotic fungus in ammonium-enriched soil. ISME Journal 17, 1626–1638. https://doi.org/10.1038/S41396-023-01476-Z

Sun, K., Liu, J., Gao, Y. et al. (2014). Isolation, plant colonization potential and phenanthrene degradation performance of the endophytic bacterium Pseudomonas sp. Ph6-gfp. Scientific Reports 4, 1–8. https://doi.org/10.1038/SREP05462

Swartz, J.M., Cardenas, B.T., Mohrig, D. & Passalacqua, P. (2022). Tributary channel networks formed by depositional processes. Nature Geoscience 15, 216–221. https://doi.org/10.1038/S41561-022-00900-X;TECHMETA=134;SUBJMETA=215,2151,241,242,3930,704;KWRD=GEOMORPHOLOGY,HYDROGEOLOGY,HYDROLOGY,SEDIMENTOLOGY

Tan, S.Y., Jiang, Q.Y., Zhuo, F. et al. (2015). Effect of inoculation with Glomus versiforme on cadmium Accumulation, antioxidant activities and phytochelatins of Solanum photeinocarpum. PLoS One 10,e0132347. https://doi.org/10.1371/JOURNAL.PONE.0132347

Tchounwou, P.B., Yedjou, C.G., Patlolla, A.K. & Sutton, D.J. (2012). Heavy metals toxicity and the environment. EXS 101, 133. https://doi.org/10.1007/978-3-7643-8340-4_6

Tedersoo, L., Magurno, F., Alkahtani, S. & Mikryukov, V. (2024). Phylogenetic classification of arbuscular mycorrhizal fungi: New species and higher-ranking taxa in Glomeromycota and Mucoromycota (class Endogonomycetes). MycoKeys 107, 273-325 107:273–325. https://doi.org/10.3897/MYCOKEYS.107.125549

Vargas, A., López, J.E., Jaimes, A. & Saldarriaga, J.F. (2024). Phytoremediation of Hg and chlorpyrifos contaminated soils using Phaseolus vulgaris L. with biochar, mycorrhizae, and compost amendments. Environmental Geochemistry and Health 46, 478. https://doi.org/10.1007/s10653-024-02244-4

Wahab, A., Muhammad, M., Munir, A. et al. (2023). Role of arbuscular mycorrhizal fungi in regulating growth, enhancing productivity, and potentially influencing ecosystems under abiotic and biotic stresses. Plants 12, 3102. https://doi.org/10.3390/PLANTS12173102

Wang, F., Zhang, L., Zhou, J. et al. (2022). Exploring the secrets of hyphosphere of arbuscular mycorrhizal fungi: Processes and ecological functions. Plant and Soil 2022, 1–22. https://doi.org/10.1007/S11104-022-05621-Z

Wang, L., Sun, X., Chang, Q. et al. (2016a). Effect of di-n-butyl phthalate (DBP) on the fruit quality of cucumber and the health risk. Environmental Science and Pollution Research 23, 24298–24304. https://doi.org/10.1007/S11356-016-7658-1/METRICS

Wang, Q., Hong, H., Liao, R. et al. (2021). Glomalin-related soil protein: The particle aggregation mechanism and its insight into coastal environment improvement. Ecotoxicology and Environmental Safety 227, 112940. https://doi.org/10.1016/J.ECOENV.2021.112940

Wang, Q., Liu, M., Wang, Z. et al. (2023). The role of arbuscular mycorrhizal symbiosis in plant abiotic stress. Frontiers in Microbiology 14, 1323881. https://doi.org/10.3389/fmicb.2023.1323881

Wang, Z.G., Bi, Y.L., Jiang, B. et al. (2016b). Arbuscular mycorrhizal fungi enhance soil carbon sequestration in the coalfields, northwest China. Scientific Report 6, 1–11. https://doi.org/10.1038/SREP34336;SUBJMETA=158,172,2165,4081,704;KWRD=CLIMATE-CHANGE+ECOLOGY,ENVIRONMENTAL+IMPACT

Wei, Z., Sixi, Z., Xiuqing, Y. et al. (2023). Arbuscular mycorrhizal fungi alter rhizosphere bacterial community characteristics to improve Cr tolerance of Acorus calamus. Ecotoxicology and Environmental Safety 253, 114652. https://doi.org/10.1016/j.ecoenv.2023.114652

Willis, A., Rodrigues, B.F. & Harris, P.J.C. (2013). The ecology of arbuscular mycorrhizal fungi. CRC Critical Reviews in Plant Sciences 32, 1–20. https://doi.org/10.1080/07352689.2012.683375;JOURNAL:JOURNAL:BPTS20;REQUESTEDJOURNAL:JOURNAL:BPTS20;PAGE:STRING:ARTICLE/CHAPTER

Wołejko, E., Łozowicka, B., Jabłońska-Trypuć, A. et al. (2022). Chlorpyrifos occurrence and toxicological risk assessment: A review. International Journal of Environmental Research and Public Health 19, 12209. https://doi.org/10.3390/IJERPH191912209

Wu, C., Bi, Y. & Zhu, W. (2024a). Is the amount of water transported by arbuscular mycorrhizal fungal hyphae negligible? Insights from a compartmentalized experimental study. Plant Soil 499, 537–552. https://doi.org/10.1007/S11104-024-06477-1/METRICS

Wu, S., Fu, W., Rillig, M.C. et al. (2024b). Soil organic matter dynamics mediated by arbuscular mycorrhizal fungi – An updated conceptual framework. New Phytologist 242, 1417–1425. https://doi.org/10.1111/nph.19178

Wu, S., Zhang, X., Sun, Y. et al. (2015). Transformation and immobilization of chromium by arbuscular mycorrhizal fungi as revealed by SEM–EDS, TEM–EDS, and XAFS. Environmental Science & Technology 49, 14036–14047. https://doi.org/10.1021/acs.est.5b03659

Wu, S.C., Wong, C.C., Shu, W.S. et al. (2011). Mycorrhizo-remediation of lead/zinc mine tailings using vetiver: A field study. International Journal of Phytoremediation 13, 61–74. https://doi.org/10.1080/15226511003671353

Yan, A., Wang, Y., Tan, S.N. et al. (2020). Phytoremediation: A promising approach for revegetation of heavy metal-polluted land. Frontiers in Plant Science 11, 513099. https://doi.org/10.3389/FPLS.2020.00359/XML/NLM

Yang, F., Han, J., Lin, R. et al. (2024). Regulation of the rhizosphere microenvironment by arbuscular mycorrhizal fungi to mitigate the effects of cadmium contamination on perennial ryegrass (Lolium perenne L.). Microorganisms 12, 2335. https://doi.org/10.3390/microorganisms12112335

Yang, Y., He, C., Huang, L. et al. (2017). The effects of arbuscular mycorrhizal fungi on glomalin-related soil protein distribution, aggregate stability and their relationships with soil properties at different soil depths in lead-zinc contaminated area. PLoS One 12, e0182264. https://doi.org/10.1371/JOURNAL.PONE.0182264

Yang, Y., Liang, Y., Han, X. et al. (2016). The roles of arbuscular mycorrhizal fungi (AMF) in phytoremediation and tree-herb interactions in Pb contaminated soil. Scientific Report 6, 1–14. https://doi.org/10.1038/SREP20469;SUBJMETA=168,171,2668,326,449,61,631;KWRD=ENVIRONMENTAL+BIOTECHNOLOGY,ENVIRONMENTAL+MICROBIOLOGY,PLANT+ECOLOGY

Yilmaz, A., Yildirim, E., Yilmaz, H. et al. (2023). Use of arbuscular mycorrhizal fungi for boosting antioxidant enzyme metabolism and mitigating saline stress in sweet basil (Ocimum basilicum L.). Sustainability (Switzerland) 15, 5982. https://doi.org/10.3390/SU15075982/S1

Zapata, D., López, J.E. & Saldarriaga, J.F. (2024). Plant growth-promoting and biocontrol potential of Aspergillus tubingensis and Talaromyces islandicus. Journal of Soil Science and Plant Nutrition. https://doi.org/10.1007/s42729-024-01633-z

Zeng, Y., Wang, S., Huang, F. et al. (2025). Fate of polycyclic aromatic hydrocarbons in the phytoremediation of different hydrocarbon contaminated soils with cotton, ryegrass, tall fescue, and wheat. Frontiers in Plant Science 16, 1550234. https://doi.org/10.3389/FPLS.2025.1550234

Zhao, S., Yan, L., Kamran, M. et al. (2024). Arbuscular mycorrhizal fungi-assisted phytoremediation: A promising strategy for cadmium-contaminated soils. Plants 13, 3289. https://doi.org/10.3390/plants13233289

Zhou, X.B., Cébron, A., Béguiristain, T. & Leyval, C. (2009). Water and phosphorus content affect PAH dissipation in spiked soil planted with mycorrhizal alfalfa and tall fescue. Chemosphere 77, 709–713. https://doi.org/10.1016/J.CHEMOSPHERE.2009.08.050

Zou, L., Yao, N., Qiu, S. et al. (2025). From plasticizers to pollutants: The ecological consequences of PAEs in agricultural soils. Environmental Pollution 373. https://doi.org/10.1016/J.ENVPOL.2025.126198

CHAPTER
10

The Role of Mycorrhization in the Production of Flavonoids, Plant Defense-related Molecules

Caio Bezerra Barreto[1,2], Jefferlone Lopes da Silva Filho[1] and Fábio Sérgio Barbosa da Silva[1,2*]

[1] Laboratório de Análises, Pesquisas e Estudos em Micorrizas (LAPEM/UPE) – Centro de Pesquisas do Instituto de Ciências Biológicas, Universidade de Pernambuco, Rua Arnóbio Marques, 310, Santo Amaro, Recife, Brazil, 50100-130.

[2] Programa de Pós-graduação em Biologia Celular e Molecular Aplicada – ICB/UPE, Rua Arnóbio Marques, 310, Santo Amaro, Recife, Brazil, 50100-130.

1. Flavonoids: Antioxidant Molecules of Plant Defense

The diversity of bioactive compounds in plants (Altemimi et al., 2017) has drawn attention due to their pharmacological properties, which contribute to reduced risk of disease (Naczk and Shahidi, 2006), inflammatory processes (Manjeet and Ghosh, 1999) and protection against free radicals. Such effects are mainly associated with secondary metabolites which are specialized for functions that are part of the plant life cycle, such as attracting pollinating agents (Forcone et al., 1997; Silva et al., 2003) and protecting against biotic and abiotic factors (Kaur et al., 2024; Sonbol et al., 2025).

Among the groups of secondary metabolites, phenolic compounds stand out (Aharoni and Galili, 2011); these contain an aromatic ring bonded to hydroxyls in their base structure (Alara et al., 2021). These biomolecules are directly related to the shikimic acid pathway (Macheroux et al., 1999), this acid acts as a common intermediary for the biosynthesis of aromatic amino acids, such as phenylalanine (Dev et al., 2012), the substrate for phenylalanine ammonia-lyase (PAL), a key enzyme in the phenylpropanoid pathway (Vogt, 2010).

Among the classes of synthesized phenolics, flavonoids stand out for being widely distributed in plant tissues. These are structured by two benzene rings united by a central ring that, according to characteristics such as saturation, may be grouped in seven flavonoid subfamilies (Sandoval et al., 2020). Such bioactive compounds

*Corresponding author: fabio.barbosa@upe.br

DOI: 10.1201/9781003669777-10

are directly related to plant antimicrobial protection (Cowan, 1999; Pok et al., 2022), resulting in a mechanism of defense against pathogens. Furthermore, the antioxidant activity associated with flavonoids is widely recognized, as these compounds are effective in neutralizing free radicals, such as singlet oxygen and other reactive species (Shahidi et al., 1992).

Flavonoids have their biosynthesis affected by stressing factors through molecular mechanisms related to genes that cause enzymatic modulation related to the phenylpropanoid pathway (Khakdan et al., 2017). Other important functions of these metabolites are in the response against phytopathogens, by mitigating their effects, and acting as chemical messengers, especially regarding rhizospheric microorganisms (Shen et al., 2022). Even with so many benefits, these biomolecules are produced in low concentrations in plants (Muniz et al., 2023).

In this context, using biostimulants containing beneficial rhizosphere microorganisms to potentiate flavonoid accumulation can be a sustainable and efficient alternative to optimize plant response to adverse conditions. According to Khalid et al. (2022), the application of the mutualistic fungus in *Artemisia annua* L. enhanced enzymatic antioxidant and phenolic compounds production.

Another group of microorganisms of the edaphic environment are the arbuscular mycorrhizal fungi (AMF) (Brundrett and Tedersoo, 2018). The association with these *Glomeromycota* representatives (Wijayawardene et al., 2024) has the potential to enhance tolerance to multiple stressing factors (Güneş et al., 2024), favoring the cultivation of plant species in adverse conditions.

2. Mycorrhizal Symbiosis Modulating the Production of Flavonoids in Stress Conditions

Owning to the potential for modulating host physiology (Wu et al., 2023), the AMF inoculation may be explored as a tool to favor tolerance to drought (Benaffari and Meddich, 2023), waterlogging (Liang et al., 2023), and extreme temperatures (Zhang et al., 2023). Such benefits are obtained, in some cases, by producing secondary molecules through mechanisms such as the activation of metabolic pathways of antioxidant compounds (Liu et al., 2022a) in response to phytohormones and the increase of reactive species (Zhang et al., 2013).

In this context, many studies attest that AMF inoculation increases tolerance to adverse conditions, focusing on the mechanisms for the establishment of mycorrhizal benefits (Table 1). Thereby, mycorrhizal symbiosis improved biosynthesis of flavonoids in about 75% of the studies, and optimized the antioxidant potential in over 50% of the cases (Table 1), confirming the benefits of mycorrhization to increase tolerance to stress conditions. Such findings reinforce utilizing mycorrhizal biostimulants in optimizing the content of antioxidant molecules and the efficiency in favoring the tolerance to some adverse conditions (Table 1).

Observing the experiments conducted in the last five years (Table 1), the effect of AMF inoculation was evaluated against salinity, drought, waterlogging, metal pollution, microplastics, and extreme temperatures. Concerning plant species, over ten families were studied (Figure 1a), mainly in controlled conditions (Figure 1b).

Table 1. Research conducted in the last five years (2021 to March 2025) that evaluated how inoculating arbuscular mycorrhizal fungi (AMF) affects flavonoid production in plants cultivated in stressing conditions

Plant species	Plant parts	Condition	Stresses	AMF tested	Mycorrhizal parameters	Antioxidant assays	Enhanced parameters	Reference
Triticum aestivum L.	Aerial part	Greenhouse	Biotic	*Rhizoglomus irregularis* (Błaszk., Wubet, Renker & Buscot) Sieverd., GA Silva & Oehl	Colonization	FRAP	Proteins Polyphenols Chlorophylls Tocopherol Tolerance Growth Photosynthesis	Sonbol et al., 2025
Lycium barbarum L.	Leaves	Greenhouse	Biotic	*Rhizoglomus intraradices* (NC Schenck & GS Sm.) Sieverd., GA Silva & Oehl	Colonization	None	Tolerance Phenolics	Li et al., 2024
Zea mays L.	Leaves Roots	Field	Metal pollution	*Acaulospora* sp. *Rhizoglomus aggregatum* (N.C. Schenck & G.S. Sm.) Sieverd.	Colonization	DPPH Enzymatic	Flavonoids Tolerance Antioxidant activity Growth	Kaur et al., 2024
Capsicum annuum L.	Fruits	Field	Water stress	*R. irregularis* *Funneliformis mosseae* (T.H. Nicolson & Gerd.) C. Walker & A. Schüßler *Funneliformis geosporum* (T.H. Nicolson & Gerd.) C. Walker & A. Schüßler	Colonization	DPPH	None	Calvo et al., 2024

(Contd.)

(*Contd.*)

Plant species	Plant parts	Condition	Stresses	AMF tested	Mycorrhizal parameters	Antioxidant assays	Enhanced parameters	Reference
Helianthus annuus L.	Leaves	Greenhouse	Salinity	*F. mosseae*	None	DPPH	Flavonoids Growth Photosynthesis Antioxidant activity Tolerance	Borjian et al., 2024
Tamarix gallica L.	Leaves	Greenhouse	Salinity	*Septoglomus constrictum* (Trappe) Sieverd., GA Silva & Oehl *F. mosseae* *Funneliformis caledonium* (T.H. Nicolson & Gerd.) C. Walker & A. Schüßler *Rhizoglomus fasciculatum* (Thaxt.) Sieverd., GA Silva & Oehl *F. geosporum* *Funneliformis coronatus* (Giovann.) C. Walker & A. Schüßler *Oehlia diaphana* (J.B. Morton & C. Walker) Błaszk., Kozłowska & Dalpé	Colonization	DPPH	Flavonoids Proanthocyanidins Phenols	Bencherif et al., 2024

				Gigaspora gigantea (T.H. Nicolson & Gerd.) Gerd. & Trappe				
Glycine max L.	Seeds	Greenhouse	Micro-plastic	*R. irregularis*	None	None	Proteins Lipids Carbohydrates Phenols Flavonoids Tolerance	Asad et al., 2024
T. aestivum	Aerial part	Greenhouse	Metal pollution	*R. irregularis*	Colonization	FRAP	Phenols Flavonoids Tolerance Growth Photosynthesis	Alsherif et al., 2024
Avena sativa L.	Aerial part	Greenhouse	Metal pollution	*R. irregularis*	Colonization	FRAP Enzymatic	Photosynthesis Flavonoids Phenols Tocopherol Antioxidant activity Growth	Albqmi et al., 2024
Prunus persica L.	Roots	Greenhouse	Water stress	*F. mosseae*	Colonization	DPPH Enzymatic •OH O_2•−	Flavonoids Antioxidant activity Growth Tolerance	Liang et al., 2023
Robinia pseudo-acacia L.	Leaves	Greenhouse	Metal pollution	*F. mosseae*	Colonization	None	Phenolics	Huang et al., 2024

(*Contd.*)

(*Contd.*)

Plant species	Plant parts	Condition	Stresses	AMF tested	Mycorrhizal parameters	Antioxidant assays	Enhanced parameters	Reference
Campeiostachys nutans (Griseb.) B.R.Baum, J.L.Yang & C.Ye	Roots	Greenhouse	Cold	*R. irregularis*	Colonization	Enzymatic	None	Zhang et al., 2023
Monarda citriodora Cerv. ex Lag.	Leaves	Greenhouse	Metal pollution	*R. fasciculatum*	Colonization	DPPH	Tolerance	Luxmi et al., 2023
Satureja khuzistanica Jamzad	Leaves	Greenhouse	Salinity	*R. fasciculatum* *R. intraradices*	None	DPPH	Tolerance Proteins Carbohydrates Phenols Flavonoids Antioxidant activity	Saadatfar and Jafari, 2023
Lavandula angustifolia Mill.	Leaves	Greenhouse	Metal pollution	*F. mosseae*	Colonization	FRAP DPPH Enzymatic	Flavonoids Phenols Tolerance Antioxidant activity	Rasouli et al., 2023
Solanum lycopersicum L.	Fruits	Greenhouse	Water stress	*R. irregularis*	Colonization	Enzymatic	Tolerance Yield Antioxidant activity Flavonoids Polyphenols	Soussani et al., 2023

Lactuca sativa L.	Leaves	Greenhouse	Biotic	*F. mosseae*	Colonization	ABTS FRAP	None	Santin et al., 2023
Phragmites australis (Cav.) Trin. ex Steud.	Roots	Greenhouse	Metal pollution	*F. mosseae*	Colonization	Enzymatic	Tolerance Chlorophylls Antioxidant activity Phenols Flavonoids	You et al., 2023
Theobroma cacao L.	Roots	Greenhouse	Metal pollution	*R. intraradices* *Simiglomus hoi* (S.M. Berch & Trappe) G.A. Silva, Oehl & Sieverd. *Gigaspora margarita* W.N. Becker & I.R. Hall *Scutellospora dipurpurescens* J.B. Morton & Koske	Colonization	Enzymatic	Tolerance	Ghomsi et al., 2023
Glycyrrhiza glabra L.	Aerial part	Greenhouse	Water stress	*Entrophospora etunicata* (W.N. Becker & Gerd.) Błaszk., Niezgoda, B.T. Goto & Magurno	None	None	Tolerance Flavonoids Photosynthesis	Haghighi et al., 2023
Chenopodium quinoa Willd.	Seeds	Greenhouse	Water stress	*Acaulospora delicata* C. Walker, C.M. Pfeiff. & Bloss *Acaulospora laevis*	None	DPPH	Tolerance Phenols Flavonoids Antioxidant activity	Benaffari and Meddich, 2023

(*Contd.*)

(*Contd.*)

Plant species	Plant parts	Condition	Stresses	AMF tested	Mycorrhizal parameters	Antioxidant assays	Enhanced parameters	Reference
				Gerd. & Trappe *Acaulospora* sp. *Entrophospora claroidea* (N.C. Schenck & G.S. Sm.) Błaszk., Niezgoda, B.T. Goto & Magurno *R. aggregatum* *Rhizoglomus clarum* (T.H. Nicolson & N.C. Schenck) Sieverd., G.A. Silva & Oehl *Viscospora deserticola* (Trappe, Bloss & J.A. Menge) Tedersoo & Magurno *Glomus heterosporum* G.S. Sm. & N.C. Schenck *Glomus macrocarpum* Tul. & C. Tul., Giorn. *Glomus microcarpum* Tul. & C. Tul. *Glomus* sp. *Diversispora versiformis* (P. Karst.) Oehl, G.A. Silva &				

				Sieverd. *R. intraradices* *Paraglomus bolivianum* (Sieverd. & Oehl) Oehl & G.A. Silva				
T. aestivum	Aerial part	Greenhouse	Metal pollution	*R. irregularis*	Colonization Hyphae length	FRAP Enzymatic	Photosynthesis Tolerance Chlorophylls Growth Antioxidant activity Proteins Phenols Flavonoids Tocopherol	Albqmi et al., 2023
Robinia pseudoacacia L.	Roots	CO_2 Incubator	Metal pollution	*F. mosseae*	Colonization	None	Carbohydrates Phenols	Wang et al., 2023
L. sativa	Aerial part Roots	Greenhouse	Water stress	*A. delicata* *A. laevis* *Acaulospora* sp. *E. claroidea* *R. aggregatum* *R. clarum* *V. deserticola* *G. heterosporum* *F. geosporum*	Colonization	DPPH Enzymatic	Tolerance Flavonoids Antioxidant activity Proteins Carbohydrates	Ouhaddou et al., 2023

(*Contd.*)

(*Contd.*)

Plant species	Plant parts	Condition	Stresses	AMF tested	Mycorrhizal parameters	Antioxidant assays	Enhanced parameters	Reference
				G. microcarpum *D. versiformis* *Glomus* sp. *E. etunicata* *P. bolivianum*				
Beta vulgaris L.	Leaves	Field	Water stress	*R. intraradices*	None	Enzymatic	Chlorophylls Phenols Antioxidant activity	Mir Mahmoudi et al., 2023
Poncirus trifoliata L.	Leaves Roots	Greenhouse	Water stress	*F. mosseae*	Colonization	None	Flavonoids Growth	Liu et al., 2022b
T. aestivum	Leaves	Field	Water stress	Not defined	None	DPPH Enzymatic	Growth Antioxidant activity Phenols Flavonoids	Vafa et al., 2022
T. aestivum	Aerial part	Greenhouse	Salinity	Not defined	None	DPPH Enzymatic	Carbohydrates Chlorophylls Antioxidant activity Phenols Flavonoids Tolerance	Sadak and Dawood, 2023
Salvia abrotanoides (Kar.)	Leaves	Greenhouse	Water stress	*R. irregularis* *F. mosseae*	Colonization	Enzymatic	Phenols Antioxidant	Afshari et al., 2022

Sytsmas; *Salvia yangii* B. T. Drew							activity Flavonoids Chlorophylls	
Coleus forskohlii (Willd.) Briq.	Leaves	Greenhouse	Water stress	*E. claroidea*	Colonization GRSP Phosphatase	DPPH FRAP Enzymatic	Tolerance Phenols Flavonoids Antioxidant activity	Geneva et al., 2023
Vitis vinifera L.	Leaves	Greenhouse	Salinity	*F. mosseae*	None	Enzymatic	Phenols Flavonoids Carbohydrates Proteins Antioxidant activity Tolerance	Karimi and Noori, 2022
G. glabra	Aerial part Roots	Greenhouse	Metal pollution	*F. mosseae*	Colonization GRSP	None	Carbohydrates Phenols Flavonoids Tolerance Chlorophylls	Yeganeh et al., 2023
P. trifoliata	Roots	Greenhouse	Water stress	*F. mosseae*	Colonization	DPPH •OH O_2•−	Antioxidant activity Flavonoids Tolerance	Liu et al., 2022a
Fagopyrum esculentum Moench	Leaves	Greenhouse	Water stress	*R. fasciculatum* *R. irregularis* *F. mosseae*	None	DPPH FRAP	Tolerance Chlorophylls Phenols Flavonoids	Mohammadi et al., 2022

(*Contd.*)

(*Contd.*)

Plant species	Plant parts	Condition	Stresses	AMF tested	Mycorrhizal parameters	Antioxidant assays	Enhanced parameters	Reference
C. quinoa	Leaves Seeds	Field	Water stress	*A. delicata* *A. laevis* *Acaulospora* sp. *R. intraradices* *D. boliviana* *E. claroidea* *R. aggregatum* *R. clarum* *V. deserticola* *G. macrocarpum* *G. heterosporum* *G. microcarpum* *D. versiformis* *Glomus* sp.	Colonization	DPPH Enzymatic	Chlorophylls Phenols Flavonoids Antioxidant activity Tolerance	Toubali et al., 2022
Hordeum vulgare L.	Seeds	Greenhouse	Water stress	*Pacispora franciscana* Sieverd. & Oehl *F. mosseae* *F. geosporum* *R. irregularis* *Glomus tenebrosum* (Thaxt.) S.M. Berch *Glomus* sp.	Colonization	None	Tolerance Phenols Flavonoids	Jerbi et al., 2022
S. lycopersicum	Fruits	Greenhouse	Salinity	*F. mosseae*	None	Enzymatic	Carbohydrates Proteins Antioxidant activity	Zhou et al., 2022

							Flavonoids Proanthocyanidins Tolerance	
T. aestivum	Leaves	Field	Water stress	*F. mosseae*	None	Enzymatic	Carbohydrates Chlorophylls Flavonoids Phenols	Taft et al., 2021
Dracocephalum moldavica L.	Aerial part Roots	Greenhouse	Salinity	*F. mosseae* *E. etunicata*	Colonization	DPPH	Antioxidant activity Tolerance Phenols Flavonoids	Alizadeh et al., 2021
Nicotiana tabacum L.	Leaves	Greenhouse	Water stress	*D. versiformis*	Colonization	Enzymatic	Antioxidant activity Chlorophylls Carbohydrates Phenols Flavonoids Tolerance	Begum et al., 2022
G. max	Seeds	Field	Water stress	*A. laevis* *V. deserticola* *R. irregularis*	Colonization	Enzymatic	Carbohydrates Phenols Flavonoids Proteins Tolerance	Sheteiwy et al., 2021

(Contd.)

(*Contd.*)

Plant species	Plant parts	Condition	Stresses	AMF tested	Mycorrhizal parameters	Antioxidant assays	Enhanced parameters	Reference
N. tabacum	Leaves	Greenhouse	Water stress	*D. versiformis*	Colonization GRSP	Enzymatic	Chlorophylls Phenols Flavonoids Antioxidant activity Tolerance	Begum et al., 2021
Solanum nigrum L.	Leaves	Greenhouse	Metal pollution	*R. intraradices*	None	Enzymatic	Chlorophylls Antioxidant activity Phenols Flavonoids Tolerance	Wang et al., 2021
Sesamum indicum L.	Leaves	Field	Water stress	*F. mosseae* *R. irregularis*	Colonization Glomero-spores	None	Chlorophylls Phenols Tolerance	Gholinezhad and Darvish-Zadeh, 2021
V. vinifera	Fruits	Field	Water stress	*R. intraradices* *F. mosseae* *R. aggregatum* *E. etunicata*	Colonization Dependency	None	Tolerance	Torres et al., 2021
Olea europaea L.	Leaves Roots	Greenhouse	Salinity	*V. deserticola* *G. margarita*	Colonization Dependency	None	Carbohydrates Chlorophylls Phenols Flavonoids Tolerance	Hassena et al., 2021

Eleusine coracana L.	Leaves	Greenhouse	Water stress	*R. intraradices*	Colonization	Enzymatic	Carbohydrates Phenols Flavonoids Antioxidant activity Tolerance	Tyagi et al., 2021

ABTS = 2,2́-azinobis (3-ethylbenzothiazoline-6-sulphonic acid); AMF = Arbuscular mycorrhizal fungi; DPPH = 2,2-diphenyl-1-picrylhydrazyl; FRAP = ferric reducing antioxidant power; GRSP = Glomalin-related soil proteins; •OH = Hydroxyl; O_2•− = Superoxide.

The papers were retrieved after searching the Web of Science, using the search terms "Flavonoid" AND "Arbuscular mycorrhizal fungi" AND "Plant stress", considering the date range from 2021 to March 2025. Of the 104 papers retrieved, reviews and papers that did not inoculate AMF, did not dose flavonoids or did not investigate stress conditions were discarded, resulting in 47 studies used in the creation of this table.

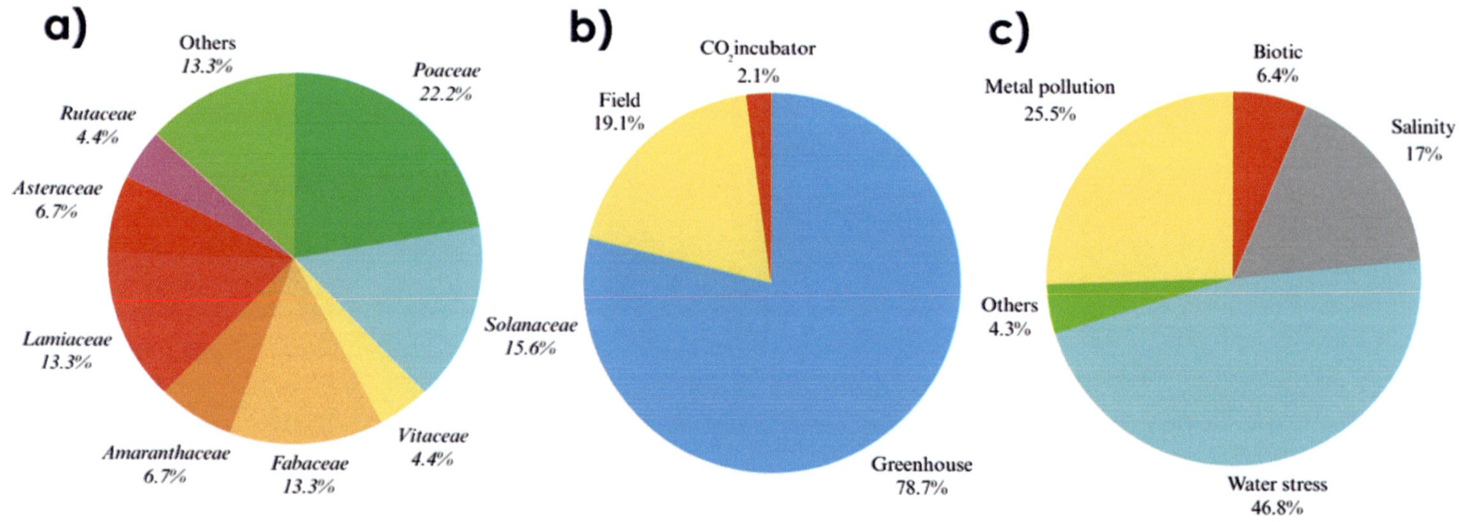

Fig. 1. Main plant families and experimental conditions of studies that evaluated the effect of mycorrhizal inoculation against stressing conditions. (a) Plant families; (b) Experimental conditions; (c) Types of tested stress. Data representing less than 2% of the total were grouped under "Others". Number of studies: 47 (Table 1). Created with Canva.com.

Representatives of *Poaceae, Solanaceae,* and *Fabaceae* were the most studied in stress conditions (Figure 1a; Table 1). That is due to the investigation of optimal cultivation conditions of commercial crops (Taft et al., 2021; Sheteiwy et al., 2021; Zhou et al., 2022; Sadak and Dawood, 2023; Soussani et al., 2023; Asad et al., 2024). In this context, establishing cultivation protocols, using mycorrhizal biostimulants, is an efficient option to alleviate losses in production related to water stress (Soussani et al., 2023).

In this regard, water stress was the most investigated one (Figure 1c), with the majority of studies considering drought conditions, except Liang et al. (2023), who evaluated the mycorrhization role in waterlogged environments. The problems of the drought effects on yield (Toubali et al., 2022) and the concerns about climate change are factors that may have influenced research targeting this theme.

In the considered studies, the leaves were the most evaluated plant tissue, corresponding to over 40% of the evaluated materials. On the other hand, roots, fruits, and seeds were also investigated regarding the production of flavonoids in mycorrhizal plants cultivated under stressing conditions (Figure 2a).

In the analyzed plant tissues, spectrophotometric methods were the most employed to quantify flavonoids and reported in over 85% of the studies (Figure 2b). In parallel, considering the antioxidant activity assays, enzymatic and non-enzymatic methods were used (Figure 2c).

Among the evaluated studies, *Glomeraceae* isolates, especially *Funneliformis mosseae* (T.H. Nicolson & Gerd.) C. Walker & A. Schüßler, were the most common; the three most inoculated isolates belong to *Glomeraceae* (Figure 3). Likely, the offer of commercial inoculants with representatives of this family may have influenced the choice of these isolates, facilitating the reproduction of the cultivation protocols proposed to many stress conditions (Table 1).

3. Final Considerations

To cope with the resulting stresses from abiotic factors, the action of herbivores and pathogens, plants produce bioactive compounds to tolerate environmental conditions. Among these, flavonoids are outstanding antioxidants. Thus, strategies that enhance their production may be employed to minimize the deleterious effects of such conditions. In this context, applying AMF as a biostimulant is an outstanding alternative, as it modulates the host phytochemistry by establishing a mutualistic symbiotic relationship that results in a better fitness to face adverse conditions.

This chapter confirmed the efficiency of cultivation protocols using AMF to enhance the accumulation of flavonoids and increase the antioxidant potential, attenuating stress conditions. Nevertheless, some approaches, such as quantifying proteic glycoconjugates or glomerospores and combinations of AMF isolates, were underexplored. While mycorrhizal parameters may help to understand the tolerance mechanisms, only about 5% of the presented research evaluated the production of GRSP. Furthermore, AMF consortia were efficient in improving tolerance, however, none of the studies compared the effects of using AMF as isolates or in combinations in the same experiment. Such approaches, in addition to the continuation of field experiments, are relevant to develop cultivation protocols applying mycorrhizal

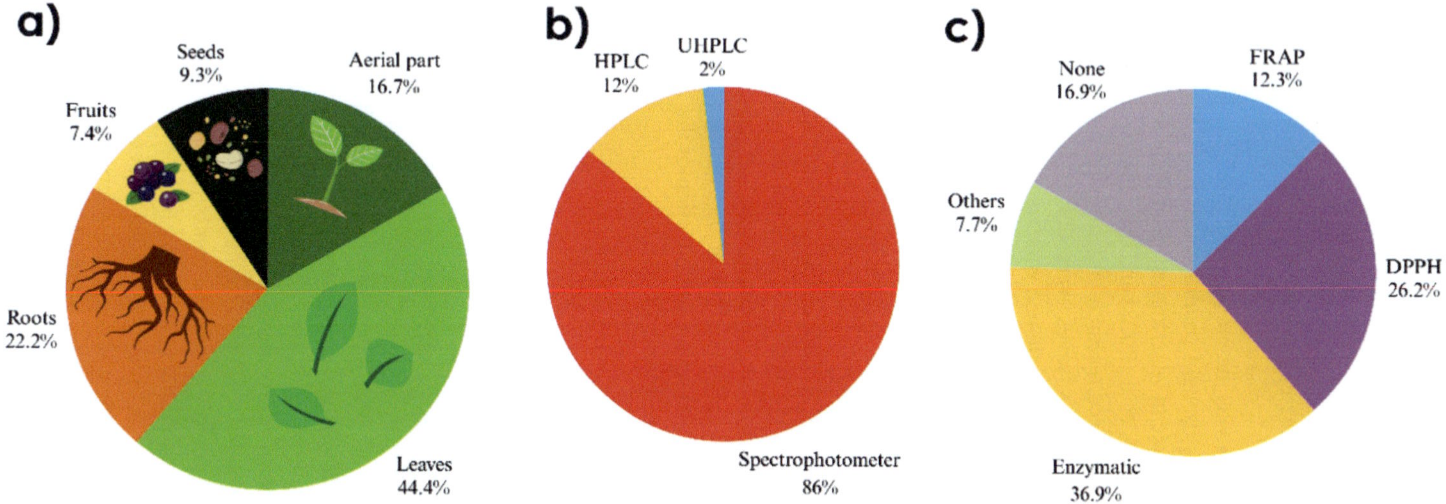

Fig. 2. Parts of the phytomass evaluated, flavonoid dosing methods, and *in vitro* antioxidant assays from studies that evaluated the mycorrhizal inoculation effect in plants under biotic and/or abiotic stress. (a) Most evaluated plant parts; (b) Flavonoid quantification methods; (c) *In vitro* antioxidant assays. Data representing less than 2% of the total were grouped as "Others". DPPH = 2,2-diphenyl-1-picrylhydrazyl; FRAP = Ferric reducing antioxidant power; HPLC = High performance liquid chromatography; UHPLC = Ultra-high performance liquid chromatography. Number of studies: 47 (Table 1). Created with Canva.com.

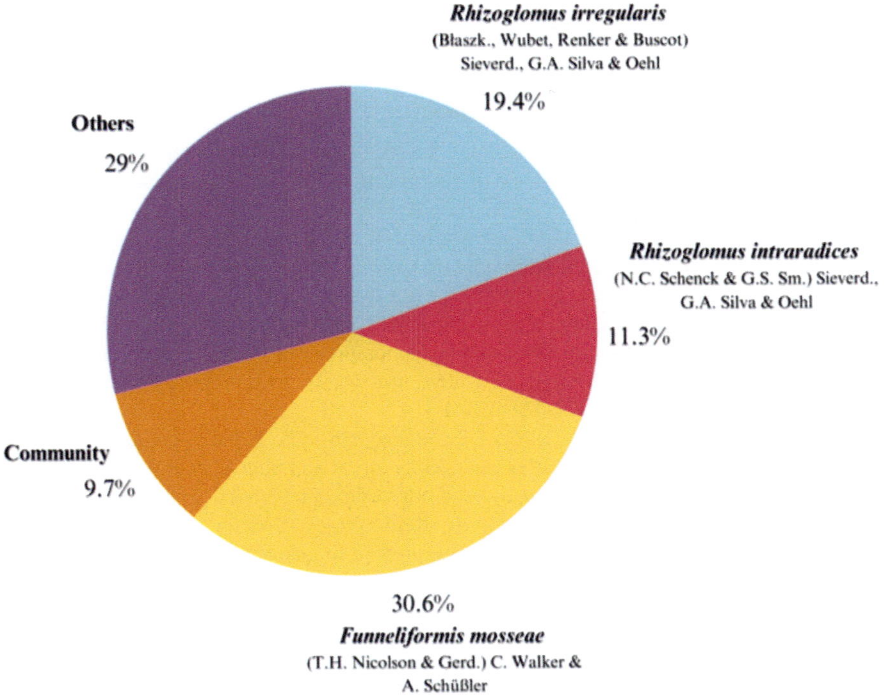

Fig. 3. Main arbuscular mycorrhizal fungi species used in studies that evaluated the effect of mycorrhization on flavonoid production and antioxidant activity of plants under biotic and/or abiotic stress. Data representing up to 4% of the total were grouped as "Others". In turn, the studies using soil-inoculum obtained from cultivation areas or on-farm multiplication were grouped as "Community". Number of studies: 47 (Table 1). Created with Canva.com.

technology for enhancing accumulation of flavonoids, as a mitigator of the losses caused by stressing conditions.

Acknowledgments

The authors thank the Fundação de Amparo à Ciência e Tecnologia do Estado de Pernambuco (FACEPE) (doctoral scholarship to Caio Bezerra Barreto, IBPG-0037-2.02/25) and to the Conselho Nacional de Desenvolvimento Científico e Tecnológico (CNPq) (productivity fellowship to Fábio Sérgio Barbosa da Silva, 306119/2024-2) and a Scientific Initiation scholarship (PIBIC/CNPq/UPE) to Jefferlone Lopes da Silva Filho.

References

Afshari, M., Rahimmalek, M., Sabzalian, M.R., Szumny, A., Matkowski, A. & Jezierska-Domaradzka, A. (2022). Mycorrhizal colonization modulates the essential oil profile and enzymatic and non-enzymatic antioxidants to mitigate the adverse effects of water deficit in *Salvia* subg. *Perovskia*. Biology 11, 1757. https://doi.org/10.3390/biology11121757

Aharoni, A. & Galili, G. (2011). Metabolic engineering of the plant primary–secondary metabolism interface. Current Opinion in Biotechnology 22, 239–244. https://doi.org/10.1016/j.copbio.2010.11.004

Alara, O.R., Abdurahman, N.H. & Ukaegbu, C.I. (2021). Extraction of phenolic compounds: A review. Current Research in Food Science 4, 200–214. https://doi.org/10.1016/j.crfs.2021.03.011

Albqmi, M., Selim, S., Al-Sanea, M.M., Alnusaire, T.S., Almuhayawi, M.S., Jaouni, S.K.A. et al. (2023). Interactive effect of arbuscular mycorrhizal fungi (AMF) and olive solid waste on wheat under arsenite toxicity. Plants 12, 1100. https://doi.org/10.3390/plants12051100

Albqmi, M., Selim, S., Bouqellah, N.A., Alnusaire, T.S., Almuhayawi, M.S. et al. (2024). Improving plant adaptation to soil antimony contamination: The synergistic contribution of arbuscular mycorrhizal fungus and olive mill waste. BMC Plant Biology 24, 364. https://doi.org/10.1186/s12870-024-05044-1

Alizadeh, S., Gharagoz, S.F., Pourakbar, L., Moghaddam, S.S. & Jamalomidi, M. (2021). Arbuscular mycorrhizal fungi alleviate salinity stress and alter phenolic compounds of Moldavian balm. Rhizosphere 19, 100417. https://doi.org/10.1016/j.rhisph.2021.100417

Alsherif, E.A., Sonbol, H., AbdElgawad, H., Ramadan, A., Korany, S.M., Crecchio, C. et al. (2024). Arbuscular mycorrhizal fungi improve tolerance of wheat plants under soil Europium contamination. Plant and Soil 505, 881–895. https://doi.org/10.1007/s11104-024-06936-9

Altemimi, A., Lakhssassi, N., Baharlouei, A., Watson, D.G. & Lightfoot, D.A. (2017). Phytochemicals: Extraction, isolation, and identification of bioactive compounds from plant extracts. Plants 6, 42. https://doi.org/10.3390/plants6040042

Asad, M., Khan, Z., Shah, T., Shah, M.A., Imran, A., Rasool, S. et al. (2024). Mitigation of microplastic toxicity in soybean by synthetic bacterial community and arbuscular mycorrhizal fungi interaction: Altering carbohydrate metabolism, hormonal transduction, and genes associated with lipid and protein metabolism. Plant Stress 14, 100631. https://doi.org/10.1016/j.stress.2024.100631

Begum, N., Akhtar, K., Ahanger, M.A., Iqbal, M., Wang, P., Mustafa, N.S. et al. (2021). Arbuscular mycorrhizal fungi improve growth, essential oil, secondary metabolism, and yield of tobacco (*Nicotiana tabacum* L.) under drought stress conditions. Environmental Science and Pollution Research 28, 45276–45295. https://doi.org/10.1007/s11356-021-13755-3

Begum, N., Wang, L., Ahmad, H., Akhtar, K., Roy, R., Khan, M.I. et al. (2022). Co-inoculation of arbuscular mycorrhizal fungi and the plant growth-promoting rhizobacteria improve growth and photosynthesis in tobacco under drought stress by up-regulating antioxidant and mineral nutrition metabolism. Microbial Ecology 83, 971–988. https://doi.org/10.1007/s00248-021-01815-7

Benaffari, W. & Meddich, A. (2023). Inoculation of arbuscular mycorrhizal fungi and plant growth promoting rhizobacteria mediated water stress tolerance in quinoa seeds by modulating the phenolics compounds, stress markers, osmolytes accumulation and antioxidant defense. Gesunde Pflanzen 75, 2121–2132. https://doi.org/10.1007/s10343-023-00849-w

Bencherif, K., Tifour, M., Dalpè, Y. & Lounès-Hadj Sahraoui, A. (2024). Native arbuscular mycorrhizal inoculation enhances secondary metabolite contents in *Tamarix gallica* grown in saline soil. Symbiosis 94, 71–80. https://doi.org/10.1007/s13199-024-01015-0

Borjian, M., Jafarinia, M. & Ghotbi-Ravandi, A.A. (2024). Biofertilizer-mediated salt tolerance in sunflower: synergistic effects of *Glomus mosseae* and *Pseudomonas fluorescens*. Plant and Soil 1–20. https://doi.org/10.1007/s11104-024-07013-x

Brundrett, M.C. & Tedersoo, L. (2018). Evolutionary history of mycorrhizal symbioses and global host plant diversity. New Phytologist 220, 1108–1115. https://doi.org/10.1111/nph.14976

Calvo, A., Reitz, T., Sillo, F., Montesano, V., Cañizares, E., Zampieri, E. et al. (2024). Interactions between an arbuscular mycorrhizal inoculum and the root-associated microbiome in shaping the response of *Capsicum annuum* "Locale di Senise" to different irrigation levels. Plant and Soil 508, 1–23. https://doi.org/10.1007/s11104-024-06806-4

Chen, J., Zhao, Q., Xie, K., Wang, M., Li, L., Zeng, D. et al. (2025). A mycorrhiza-induced UDP-Glucosyl transferase negatively regulates the arbuscular mycorrhizal symbiosis. Plant, Cell & Environment 48, 1643–1655. https://doi.org/10.1111/pce.15241

Cowan, M.M. (1999). Plant products as antimicrobial agents. Clinical Microbiology Reviews 12, 564–582. https://doi.org/10.1128/cmr.12.4.564

Dev, A., Tapas, S., Pratap, S. & Kumar, P. (2012). Structure and function of enzymes of shikimate pathway. Current Bioinformatics 7, 374–391. https://doi.org/10.2174/157489312803900983

Forcone, A., Galetto, L. & Bernardello, L. (1997). Floral nectar chemical composition of some species from Patagonia. Biochemical Systematics and Ecology 25, 395–402. https://doi.org/10.1016/s0305-1978(97)00030-6

Geneva, M., Kirova, E., Sichanova, M., Stancheva, I. & Zayova, E. (2023). Physiological analysis of drought stress influenced by *Claroideoglomus claroideum* inoculation of *in vitro* or seed-propagated *Coleus forskohlii* Briq plants. Biologia 78, 641–654. https://doi.org/10.1007/s11756-022-01231-3

Gholinezhad, E. & Darvishzadeh, R. (2021). Influence of arbuscular mycorrhiza fungi and drought stress on fatty acids profile of sesame (*Sesamum indicum* L.). Field Crops Research 262, 108035. https://doi.org/108035. 10.1016/j.fcr.2020.108035

Ghomsi, P.G.T., Eke, P., Kepngop, L.R.K., Kamdem, T.H.W., Nguemnang, L.C.M. , Nya, V.D. et al. (2023). Propagation host affects the mycorrhiza-mediated Cd toxicity alleviation in cocoa (*Theobroma cacao* L.) seedlings and restoration of soil aerobic mesophyll microflora. Journal of Soil Science and Plant Nutrition 23, 2694–2707. https://doi.org/10.1007/s42729-023-01226-2

Ghosh, B. (1999). Quercetin inhibits LPS-induced nitric oxide and tumor necrosis factor-α production in murine macrophages. International Journal of Immunopharmacology 21, 435–443. https://doi.org/10.1016/S0192-0561(99)00024-7

Güneş, H., Demir, S., Durak, E.D. & Boyno, G. (2024). The effect of arbuscular mycorrhizal fungal species *Funneliformis mosseae* and biochar against *Verticillium dahliae* in pepper plants under salt stress. European Journal of Plant Pathology 170, 669–686. https://doi.org/10.1007/s10658-024-02926-w

Haghighi, T.M., Saharkhiz, M.J., Ramezanian, A. & Zarei, M. (2023). The use of silicon and mycorrhizal fungi to mitigate changes in licorice leaf micromorphology, chlorophyll fluorescence, and rutin content under water-deficit conditions. Plant Physiology and Biochemistry 197, 107662. https://doi.org/10.1016/j.plaphy.2023.107662

Hassena, A.B., Zouari, M., Trabelsi, L., Decou, R., Amar, F.B., Chaari, A. et al. (2021). Potential effects of arbuscular mycorrhizal fungi in mitigating the salinity of treated wastewater in young olive plants (*Olea europaea* L. cv. Chetoui). Agricultural Water Management 245, 106635. https://doi.org/10.1016/j.agwat.2020.106635

Huang, L., Jia, X., Zhao, Y., Feng, X., Yang, X., Zhang, C. et al. (2024). Effects of elevated CO_2 on phenolics in black locust seedlings inoculated with arbuscular mycorrhizal fungi under cadmium pollution by ^{13}C isotopic tracer technique. Plant and Soil 494, 547–566. https://doi.org/10.1007/s11104-023-06303-0

Jerbi, M., Labidi, S., Laruelle, F., Tisserant, B., Jeddi, F.B. & Sahraoui, A.L.H. (2022). Mycorrhizal biofertilization improves grain yield and quality of hulless Barley (*Hordeum vulgare* ssp. *nudum* L.) under water stress conditions. Journal of Cereal Science 104, 103436. https://doi.org/10.1016/j.jcs.2022.103436

Karimi, R. & Noori, A. (2022). *Streptomyces rimosus* rhizobacteria and *Glomus mosseae* mycorrhizal fungus inoculation alleviate salinity stress in grapevine through morphophysiological changes and nutritional balance. Scientia Horticulturae 305, 111433. https://doi.org/10.1016/j.scienta.2022.111433

Kaur, T., Tejoprakash, N. & Reddy, M.S. (2024). Arbuscular mycorrhizal fungi ameliorate selenium stress and increase antioxidant potential of *Zea mays* in seleniferous soil. Biological Trace Element Research, 1–20. https://doi.org/10.1007/s12011-024-04410-2

Khakdan, F., Nasiri, J., Ranjbar, M. & Alizadeh, H. (2017). Water deficit stress fluctuates expression profiles of 4Cl, C3H, COMT, CVOMT and EOMT genes involved in the biosynthetic pathway of volatile phenylpropanoids alongside accumulation of methylchavicol and methyleugenol in different Iranian cultivars of basil. Journal of Plant Physiology 218, 74–83. https://doi.org/10.1016/j.jplph.2017.07.012

Khalid, M., Saeed, U.R., Haoxin, T.A.N., Lantian, S.U., Pei, Z.H.O.U. & Nan, H.U.I. (2022). Mutualistic fungus *Piriformospora indica* modulates cadmium phytoremediation properties of host plant via concerted action of enzymatic and non-enzymatic biochemicals. Pedosphere 32, 256–267. https://doi.org/10.1016/S1002-0160(21)60014-0

Li, N., Chen, W., Wang, B., Zhang, C., Wang, Y., Li, R. et al. (2024). Arbuscular mycorrhizal fungi improve the disease resistance of *Lycium barbarum* to root rot by activating phenylpropane metabolism. Frontiers in Plant Science 15, 1459651. https://doi.org/10.3389/fpls.2024.1459651

Liang, S.M., Hashem, A., Abd-Allah, E.F. & Wu, Q.S. (2023). Root-associated symbiotic fungi enhance waterlogging tolerance of peach seedlings by increasing flavonoids and activities and gene expression of antioxidant enzymes. Chemical and Biological Technologies in Agriculture 10, 124. https://doi.org/10.1186/s40538-023-00500-w

Liu, X.Q., Cheng, S., Aroca, R., Zou, Y.N. & Wu, Q.S. (2022a). Arbuscular mycorrhizal fungi induce flavonoid synthesis for mitigating oxidative damage of trifoliate orange under water stress. Environmental and Experimental Botany 204, 105089. https://doi.org/10.1016/j.envexpbot.2022.105089

Liu, Z., Cheng, S., Liu, X.Q., Kuča, K., Hashem, A., Al-Arjani, A.B.F. et al. (2022b). Cloning of a CHS gene of *Poncirus trifoliata* and its expression in response to soil water deficit and arbuscular mycorrhizal fungi. Frontiers in Plant Science 13, 1101212. https://doi.org/10.3389/fpls.2022.1101212

Luxmi, S., Singh, R., Ahmed, S., Gandhi, S.G. & Bhanwaria, R. (2023). *Glomus fasciculatum* an arbuscular mycorrhizal fungus alleviate the adverse effects of lead, arsenic, nickel, and improves growth parameters of *Monarda citriodora* Cerv. ex Lag (lemon beebalm). Rhizosphere 27, 100753. https://doi.org/10.1016/j.rhisph.2023.100753

Macheroux, P., Schmid, J., Amrhein, N. & Schaller, A. (1999). A unique reaction in a common pathway: mechanism and function of chorismate synthase in the shikimate pathway. Planta 207, 325–334. https://doi.org/10.1007/s004250050489

Manjeet, R. & Ghosh, B. (1999). Quercetin inhibits LPS-induced nitric oxide and tumor necrosis factor-α production in murine macrophages. International Journal of Immunopharmacology 21, 435–443. https://doi.org/10.1016/S0192-0561(99)00024-7

Mir Mahmoudi, T., Hamze, H. & Golabi Lak, I. (2023). Impact of biofertiliser and zinc nanoparticles on enzymatic, biochemical, and agronomic properties of sugar beet under different irrigation regimes. Zemdirbyste-Agriculture 110, 217–224. https://doi.org/10.13080/z-a.2023.110.025

Mohammadi, E., Fattahi, M., Barin, M. & Ashrafi-Saeidlou, S. (2022). Arbuscular mycorrhiza and vermicompost alleviate drought stress and enhance yield, total flavonoid concentration, rutin content, and antioxidant activity of buckwheat (*Fagopyrum esculentum* Moench). South African Journal of Botany 148, 588–600. https://doi.org/10.1016/j.sajb.2022.05.020

Muniz, B.C., Kapoor, R., Silva Almeida, J.R.G., Bastos Filho, C.J.A. & Silva, F.S.B. (2023). Arbuscular mycorrhizae increase but vermicompost decrease the sun protection factor (SPF) in leaves of *Hymenaea martiana* Hayne seedlings. Rhizosphere 27, 100781. https://doi.org/10.1016/j.rhisph.2023.100781

Naczk, M. & Shahidi, F. (2006). Phenolics in cereals, fruits and vegetables: Occurrence, extraction and analysis. Journal of Pharmaceutical and Biomedical Analysis 41, 1523–1542. https://doi.org/10.1016/j.jpba.2006.04.002

Ouhaddou, R., Ech-Chatir, L., Anli, M., Ben-Laouane, R., Boutasknit, A. & Meddich, A. (2023). Secondary metabolites, osmolytes and antioxidant activity as the main attributes enhanced by biostimulants for growth and resilience of lettuce to drought stress. Gesunde Pflanzen 75, 1737–1753. https://doi.org/10.1007/s10343-022-00827-8

Pok, P.S., García Londoño, V.A., Vicente, S., Pacin, A., Alzamora, S.M. & Resnik, S.L. (2022). Citrus flavonoids against *Fusarium verticillioides* in post-harvest maize: Minimization of fumonisins and alteration of fungal ultrastructure. Journal of Applied Microbiology 132, 2234–2248. https://doi.org/10.1111/jam.15373

Rasouli, F., Hassanpouraghdam, M.B., Pirsarandib, Y., Aazami, M.A., Asadi, M., Ercisli, S. et al. (2023). Improvements in the biochemical responses and Pb and Ni phytoremediation of lavender (*Lavandula angustifolia* L.) plants through *Funneliformis mosseae* inoculation. BMC Plant Biology 23, 252. https://doi.org/10.1186/s12870-023-04265-0

Saadatfar, A. & Jafari, S.H. (2023). Effects of mycorrhizal fungi and methyl jasmonate on salt stress tolerance and phytochemical traits of *Satureja Khuzistanica*. Journal of Plant Growth Regulation 42, 7266–7279. https://doi.org/10.1007/s00344-023-11015-x

Sadak, M.S. & Dawood, M.G. (2023). Biofertilizer role in alleviating the deleterious effects of salinity on wheat growth and productivity. Gesunde Pflanzen, 75, 1207–1219. https://doi.org/10.1007/s10343-022-00783-3

Sandoval, V., Sanz-Lamora, H., Arias, G., Marrero, P.F., Haro, D. & Relat, J. (2020). Metabolic impact of flavonoids consumption in obesity: From central to peripheral. Nutrients 12, 2393. https://doi.org/10.3390/nu12082393

Santin, M., Zeni, V., Grassi, A., Ricciardi, R., Pieracci, Y., Di Giovanni, F. et al. (2023). Do changes in *Lactuca sativa* metabolic performance, induced by mycorrhizal symbionts and leaf UV-B irradiation, play a role towards tolerance to a polyphagous insect pest? Environmental Science and Pollution Research 30, 56207–56223. https://doi.org/10.1007/s11356-023-26218-8

Shahidi, F., Janitha, P.K. & Wanasundara, P.D. (1992). Phenolic antioxidants. Critical Reviews in Food Science & Nutrition 32, 67–103. https://doi.org/10.1080/10408399209527581

Shen, N., Wang, T., Gan, Q., Liu, S., Wang, L. & Jin, B. (2022). Plant flavonoids: Classification, distribution, biosynthesis, and antioxidant activity. Food Chemistry 383, 132531. https://doi.org/10.1016/j.foodchem.2022.132531

Sheteiwy, M.S., Abd Elgawad, H., Xiong, Y.C., Macovei, A., Brestic, M., Skalicky, M. et al. (2021). Inoculation with *Bacillus amyloliquefaciens* and mycorrhiza confers tolerance to drought stress and improve seed yield and quality of soybean plant. Physiologia Plantarum 172, 2153–2169. https://doi.org/10.1111/ppl.13454

Silva, E.M., Dean, B.B. & Hiller, L.K. (2003). Honey bee (*Hymenoptera*: *Apidae*) foraging in response to preconditioning with onion flower scent compounds. Journal of Economic Entomology 96, 1510–1513. https://doi.org/10.1093/jee/96.5.1510

Sonbol, H., Korany, S.M., Nhs, M., Abdi, I., Maridueña-Zavala, M.G., Alsherif, E.A. et al. (2025). Exploring the benefits of AMF colonization for improving wheat growth, physiology and metabolism, and antimicrobial activity under biotic stress from aphid infection. BMC Plant Biology 25, 198. https://doi.org/10.1186/s12870-025-06196-4

Soussani, F.E., Boutasknit, A., Ben-Laouane, R., Benkirane, R., Baslam, M. & Meddich, A. (2023). Arbuscular mycorrhizal fungi and compost-based biostimulants enhance fitness, physiological responses, yield, and quality traits of drought-stressed tomato plants. Plants 12, 1856. https://doi.org/10.3390/plants12091856

Taft, H.Z., Abbaspour, H., Peyvandi, M. & Majd, A. (2021). Arbuscular mycorrhizal symbiosis improves growth, physiological, and biochemical properties of wheat under different irrigation regimes in a semiarid area. Russian Journal of Plant Physiology 68, 1135–1142. https://doi.org/10.1134/S1021443721060200

Torres, N., Yu, R. & Kurtural, S.K. (2021). Arbuscular mycorrhizal fungi inoculation and applied water amounts modulate the response of young grapevines to mild water stress in a hyper-arid season. Frontiers in Plant Science 11, 622209. https://doi.org/10.3389/fpls.2020.622209

Toubali, S., Ait-El-Mokhtar, M., Boutasknit, A., Anli, M., Ait-Rahou, Y., Benaffari, W. et al. (2022). Root reinforcement improved performance, productivity, and grain bioactive quality of field-droughted quinoa (*Chenopodium quinoa*). Frontiers in Plant Science 13, 860484. https://doi.org/10.3389/fpls.2022.860484

Tyagi, J., Shrivastava, N., Sharma, A.K., Varma, A. & Pudake, R. (2021). Effect of Rhizophagus intraradices on growth and physiological performance of Finger Millet (*Eleusine coracana* L.) under drought stress. Plant Science Today 8, 912–923. https://doi.org/10.14719/pst.2021.8.4.1240

Vafa, Z.N., Sohrabi, Y., Mirzaghaderi, G. & Heidari, G. (2022). The effect of rhizobia in improving the protective mechanisms of wheat under drought and supplementary irrigation conditions. Frontiers in Sustainable Food Systems 6, 1073240. https://doi.org/10.3389/fsufs.2022.1073240

Vogt, T. (2010). Phenylpropanoid biosynthesis. Molecular Plant 3, 2–20. https://doi.org/10.1093/mp/ssp106

Wang, G., Wang, L., Ma, F., Yang, D. & You, Y. (2021). Earthworm and arbuscular mycorrhiza interactions: Strategies to motivate antioxidant responses and improve soil functionality. Environmental Pollution 272, 115980. https://doi.org/10.1016/j.envpol.2020.115980

Wang, Y., Zhang, C., Wang, L., Zhao, Y., Gao, Y. & Jia, X. (2023). Influence of arbuscular mycorrhizal fungi on low molecular weight soluble compounds in the rhizosphere soil of black locust seedlings grown in cadmium-contaminated soils under elevated CO_2 scenarios. Plant and Soil 486, 469–486. https://doi.org/10.1007/s11104-023-05885-z

Wijayawardene, N.N., Hyde, K.D., Mikhailov, K.V., Péter, G., Aptroot, A., Pires-Zottarelli, C.L. et al. (2024). Classes and phyla of the kingdom Fungi. Fungal Diversity 128, 1–165. https://doi.org/10.1007/s13225-024-00540-z

Wu, Q.S., Silva, F.S., Hijri, M. & Kapoor, R. (2023). Arbuscular mycorrhiza-mediated augmentation of plant secondary metabolite production. Frontiers in Plant Science 14, 1150900. https://doi.org/10.3389/fpls.2023.1150900

Yeganeh, E., Vatankhah, E., Toghranegar, Z. & Amanifar, S. (2023). Arbuscular mycorrhiza alters metal uptake and the physio-biochemical responses of *Glycyrrhiza glabra* in a lead contaminated soil. Gesunde Pflanzen 75, 921–937. https://doi.org/10.1007/s10343-022-00752-w

You, Y., Wang, L., Ju, C., Wang, X. & Wang, Y. (2023). How does phosphorus influence Cd tolerance strategy in arbuscular mycorrhizal-*Phragmites australis* symbiotic system? Journal of Hazardous Materials 452, 131318. https://doi.org/10.1016/j.jhazmat.2023.131318

Zhang, H., Qi, H., Lu, G., Zhou, X., Wang, J., Li, J., Zheng, K., Fan, Y., Zhou, H., Wang, J. & Wu, C. (2023). Non-targeted metabolomics analysis reveals the mechanism of arbuscular mycorrhizal symbiosis regulating the cold-resistance of *Elymus nutans*. Frontiers in Microbiology 14, 1134585. https://doi.org/10.3389/fmicb.2023.1134585

Zhang, R. Q., Zhu, H. H., Zhao, H. Q. & Yao, Q. (2013). Arbuscular mycorrhizal fungal inoculation increases phenolic synthesis in clover roots via hydrogen peroxide, salicylic acid and nitric oxide signaling pathways. Journal of Plant Physiology 170, 74–79. http://doi.org/10.1016/j.jplph.2012.08.022

Zhou, W., Zhang, M., Tao, K. & Zhu, X. (2022). Effects of arbuscular mycorrhizal fungi and plant growth-promoting rhizobacteria on growth and reactive oxygen metabolism of tomato fruits under low saline conditions. Biocell 46, 2575–2582. https://doi.org/10.32604/biocell.2022.021910

CHAPTER

11

Arbuscular Mycorrhizal Fungi and Plant Antioxidant Systems Under Heat Stress

Gökhan Boyno[1*], Necmettin Teniz[2], Çeknas Erdinç[2], Younes Rezaee Danesh[1] and Semra Demir[1]

[1] Department of Plant Protection, Faculty of Agriculture, Van Yuzuncu Yil University, Van 65090, Türkiye

[2] Department of Agricultural Biotechnology, Faculty of Agriculture, Van Yuzuncu Yil University, Van 65090, Türkiye

1. Introduction

Climate change has become one of the most important environmental stress factors threatening agricultural production worldwide. In particular, increasing temperature values and prolonged heat waves jeopardize agricultural sustainability by negatively affecting plant growth, development and crop yield (Lesk et al., 2016; IPCC, 2021). Heat stress causes a decrease in photosynthesis capacity in plants, disruption of water balance, damage to cellular structures and oxidative stress due to excessive production of reactive oxygen species (ROS) (Bita and Gerats, 2013; Hasanuzzaman et al., 2013). This situation directly affects not only plant physiology but also food security and economic development (FAO, 2021). In this context, the development of plant systems resistant to heat stress is of great importance.

Plants try to adapt to these adverse conditions by developing a number of physiological, biochemical and molecular mechanisms against heat stress. The most remarkable among these responses are the synthesis of heat shock proteins, stabilization of cell membranes, production of osmotic regulatory compounds and activation of antioxidant defence systems (Khan and Shahwar 2020; Kolupaev et al., 2023). Antioxidant defense systems protect the plant from cellular damage by detoxifying ROS through various molecules, especially enzymes such as superoxide dismutase (SOD), catalase (CAT), ascorbate peroxidase (APX) (Berwal et al., 2021; Rajput et al., 2021). However, these defense systems may be insufficient in many cases and therefore external supportive factors are needed.

*Corresponding author: gokhanboyno@yyu.edu.tr

DOI: 10.1201/9781003669777-11

Arbuscular mycorrhizal fungi (AMF), through their symbiotic relationship with plant roots, increase the nutrient and water uptake of plants and enable them to develop resistance against various environmental stress factors (Demir et al., 2022; Wahab et al., 2023). The favorable effects of AMF on plants under heat stress have been increasingly clearly demonstrated in recent years (Jian et al., 2024; Sousa et al., 2024). AMF increases root surface area, allowing more efficient water uptake, facilitates the transport of vital nutrients such as phosphorus, and supports plant metabolism in general. Moreover, an increase in the activity of antioxidant enzymes was observed with AMF colonization, which helped to control ROS levels (Boyno et al., 2022). However, some AMF species are reported to modulate gene expression of heat shock proteins specific to heat stress, affect hormonal balance (e.g. ABA, JA, SA) and direct the plant towards a more resistant phenotype (Ma et al., 2022).

In this chapter, the effects of heat stress on plants and how AMF develop a defense mechanism against this stress will be examined in detail. Firstly, the natural responses of plants to heat stress will be discussed and then the effects of AMF on plant antioxidant systems will be explained at the molecular level. The regulatory role of AMF in plant-cell metabolism will be evaluated within the framework of its reducing effect on heat-induced ROS accumulation and oxidative damage. In the light of recent studies in the literature, variations observed in different plant species and different AMF strains will be supported by examples and the potential of AMF in agricultural applications will be discussed. Thus, the critical role of AMF-based biological solutions in terms of sustainable agricultural systems in the face of the pressures caused by climate change in agriculture will be revealed.

2. Heat Stress and Its Impact on Plant Physiology

2.1. Cellular and Molecular Effects of Heat Stress

High temperature conditions cause serious disruptions in the basic structure and function of plant cells. This situation deeply affects not only metabolic processes but also cellular integrity and genetic arrangements.

Impairment of Membrane Structure and Integrity: Under heat stress, the lipid composition of cell membranes is physically altered. Under normal conditions, biological membranes have an ordered structure in the liquid-crystalline phase of phospholipids (Angelova et al., 2021). However, high temperatures disrupt this equilibrium, causing the membrane to exceed the phase transition temperature (Raza et al., 2023). This leads to an increase in membrane permeability and a decrease in membrane potential. As a result, ion leakage (especially ions such as K^+ and Ca^{2+}), metabolite loss and disturbances in intracellular water balance occur (Neligan, 2021). In addition, the membrane components sterols and unsaturated fatty acids are determinants of heat tolerance (He and Ding, 2020). Some plants try to increase membrane stability by reducing the unsaturation of membrane lipids in response to heat stress (He and Ding, 2020).

Effects on Protein Structure and Functionality: Proteins are extremely sensitive to temperature. Especially in structural and enzymatic proteins, protein denaturation, aggregate formation and loss of function occur with increasing temperature (Rajan et al., 2021). In response to this situation, plants produce various molecular chaperones, especially Heat Shock Proteins (HSPs) (Waters and Vierling, 2020; Pareek et al., 2021). HSPs maintain proteostasis by participating in the refolding of damaged proteins or directing them to protein degradation systems (Lang et al., 2021). HSPs also stabilize cytoskeletal elements and maintain the integrity of membrane structures of organelles such as mitochondria and chloroplasts (Table 1).

Table 1. Common HSP families in plants and their major roles

HSP family	Functional description	Key features/Localization
HSP100	Resolves aggregated proteins and assists in their refolding.	Crucial for re-establishing proteostasis after severe heat stress.
HSP90	Facilitates proper protein folding and is involved in signal transduction.	Interacts with kinases and transcription factors.
HSP70	Assists in folding newly synthesized or denatured proteins.	Versatile function; found in the cytoplasm, nucleus, and organelles.
sHSP (Small HSP)	Protects proteins from denaturation, especially under heat stress.	Localized in organelles like chloroplasts, mitochondria, and peroxisomes.

Transcriptional Regulation and Gene Expression: Under heat stress, changes in the expression of many genes are observed in plants. This process is mostly regulated by heat shock factors (HSFs) (Haider et al., 2022). HSFs bind to heat shock element (HSE) sequences in the promoter regions of HSP genes and initiate transcription of these genes (Haider et al., 2022). Some HSFs also regulate processes such as ROS detoxification, antioxidant production, hormone responses (e.g. ABA) (Andrási et al., 2021). Modern transcriptome analyses show that heat stress-induced genes are not limited to HSPs. Cell wall reorganizers, secondary metabolite synthesizers, heat-sensitive kinases and signaling pathways are also important parts of these responses.

Effects on Cell Growth, Division and Development Processes: One of the long-term effects of heat stress is severe reductions in plant growth rate. The main reason for this is the slowing down or complete cessation of the cell cycle due to temperature. Especially S phase (DNA replication) and M phase (mitosis) are heat sensitive processes (Ovejero et al., 2020). Under high temperature, DNA synthesis decreases, mitotic division rate decreases and cell growth is arrested (Saini et al., 2022). This is manifested by a decrease in root and shoot elongation, a reduction in leaf size and ultimately a decrease in plant biomass. In addition, while the levels of some growth hormones (e.g. IAA, GA) may change depending on temperature, the production of ABA (abscisic acid) and ethylene, which are stress hormones, increases (EL Sabagh et al., 2022). This hormonal imbalance causes a shift in the balance of plant growth and development in favor of stress tolerance.

2.2. Disruption of Photosynthesis and Water Balance

Heat stress has direct and indirect negative effects on the main components of the photosynthesis process in plants. The efficiency of photosynthesis is critical for plant growth and development; therefore, disturbances in this process lead to significant adverse effects on the overall physiological status of the plant.

Thermal Sensitivity of Photosynthetic Complexes: Photosystem II (PSII), which is involved in the early stages of the photosynthesis process, is highly sensitive to high temperatures. Heat stress causes damage to structural components such as the D1 protein in the reaction centre of PSII (Wang et al., 2022). This damage leads to disruptions in the electron transfer chain, resulting in a decrease in ATP and NADPH production (Suzuki, 2023). Functional defects in PSII lead to a condition called photoinhibition and a decrease in photosynthetic capacity. In addition, photosystem I (PSI), electron transport chain and lipids in chloroplast membranes are also adversely affected by high temperature (Zhang et al., 2023). With lipid peroxidation, membrane structures are damaged, membrane permeability is altered and this leads to a general impairment of chloroplast function (Zhang et al., 2023).

Decreased Rubisco Activity and Carboxylation: Rubisco enzyme, which functions in the carboxylation stage of photosynthesis, loses its activity under heat stress (Wijewardene et al., 2021). Especially since the thermal stability of Rubisco activase enzyme, which activates Rubisco, is low, this enzyme loses its function at high temperature (Wijewardene et al., 2021; Qu et al., 2023). As a result, the CO_2 binding capacity of Rubisco decreases and the carboxylation rate decreases (Wijewardene et al., 2021). This leads to significant reductions in the carbon capture capacity of photosynthesis. Furthermore, the photorespiration rate increases under heat stress conditions. This is because Rubisco is directed towards the oxygenation reaction, further reducing the efficiency of photosynthesis (Wijewardene et al., 2021). Increased photorespiration leads to energy and carbon loss.

Effects on Stomatal Behaviour and Gas Exchange: One of the important effects of heat stress on plants is on stomatal behaviour (Bjerring Jensen et al., 2024). High temperature can cause stomatal cells to close. Stomatal closure is a short-term defense mechanism to reduce water loss from the leaves (transpiration) (Gahir et al., 2021). However, stomatal closure directly reduces the rate of photosynthesis as it limits CO_2 uptake. Reduced CO_2 uptake negatively affects photosynthetic carbon metabolism and results in a reduced net photosynthetic rate (Moreno et al., 2024). Prolonged stomatal closure severely restricts plant growth performance.

Water Loss, Turgor Pressure and Cellular Effects: Under heat stress, atmospheric vapor pressure increases and this accelerates water loss through evaporation from leaves. If the water in the soil cannot be delivered to the plant sufficiently, the turgor pressure in the cells decreases. Turgour pressure keeps the cell plump by stretching the cell wall; its decrease causes slow cell expansion and wilting of tissues (Bhat, 2025). This water stress leads to the triggering of stomatal closure and disrupts the regulation of cellular metabolism (Gahir et al., 2021). Intracellular water loss leads to ion imbalances, impaired protein stability and decreased enzyme activity (Lorenzo et al., 2019).

Plant Water Balance and Adaptation Mechanisms: Increased water loss with heat stress requires plants to use water efficiently. Plants try to increase water uptake by shortening the stomatal opening time or deepening the root system. They also synthesize osmoprotectants (such as proline, betaine) to increase intracellular water holding capacity. These metabolites help maintain turgor by balancing the intracellular water potential (Singh et al., 2015). Furthermore, an increase in stress hormones such as abscisic acid (ABA) promotes stomatal closure and reduces water loss (Bharath et al., 2021). However, these adaptation mechanisms may be limited depending on the severity and duration of stress.

2.3. Oxidative Stress and Reactive Oxygen Species (ROS) Production

One of the most devastating effects of high temperature stress on plants is the triggering of oxidative stress. Heat stress negatively affects oxygen utilization in organelles such as mitochondria, where energy metabolism takes place intensively in plant cells, chloroplast, the center of photosynthesis, and peroxisomes, where detoxification processes are carried out (Suzuki, 2023). This leads to increased production of reactive oxygen species (ROS) as a result of the binding of stray electrons to oxygen in electron transport chains (Suzuki, 2023).

Major ROS Types and Mechanisms of Formation:

- **Superoxide anion (O_2·-):** Formed as a result of single electron reduction in electron transfer chains. In chloroplasts and mitochondria, it originates from complexes such as photosystem II and NADH dehydrogenase (Kozuleva, 2022).
- **Hydrogen peroxide (H_2O_2):** Generated by the dismutation of superoxide anion or by oxidative enzyme activities in peroxisomes. H_2O_2 is more stable than other ROS and can also act as an intracellular signalling molecule (Andrés et al., 2022).
- **Hydroxyl radical (–OH):** It is the most reactive type of ROS and is usually formed by the reaction of H_2O_2 with metal ions (Fe^{2+}, Cu^{+}) (Fenton reaction). It causes very rapid and widespread damage to cellular components (Nakai and Tsuruta, 2021).

Mechanisms of Cellular Damage by ROS: Because ROS are highly reactive, they attack cellular components, leading to oxidative damage. These damages include:

- **Lipid Peroxidation:** Fatty acids that provide the structural integrity of cell membranes are oxidized by ROS. Lipid peroxidation leads to increased membrane permeability, disruption of ion balance and loss of cell membrane function (Valgimigli, 2023).
- **Protein Oxidation:** Oxidation of the amino acid side chains in proteins impairs the function of enzymes and structural proteins and inhibits cellular metabolism.
- **DNA Damage:** ROS create base modifications, single or double chain breaks in DNA. This compromises the integrity of the genetic material and can cause cell death or mutations (Domínguez et al., 2021).
- **Carbohydrate Oxidation:** Leads to structural disruptions in the cytoskeleton and cell wall components, reducing cellular durability (Li et al., 2025).

The Severity of Oxidative Stress and its Effects on Plant Health: If heat stress is too severe or prolonged, the antioxidant systems of plants may become inadequate. In this case, cellular damage and death are inevitable. In particular, membrane damage caused by lipid peroxidation can disrupt the integrity of the cell membrane, leading to ion imbalance, loss of metabolic function and eventually programmed cell death (apoptosis) or necrosis (Sahu et al., 2022). Oxidative stress also affects cell signalling pathways and alters gene expression, sometimes triggering stress responses and sometimes leading to impaired cellular function (Sadiq, 2023).

3. Plant Antioxidant Defence Systems

Oxidative stress caused by environmental stresses such as high temperatures in plants leads to the accumulation of reactive oxygen species (ROS) that damage cells. Plants have developed a complex, multi-layered antioxidant defense system to reduce the effects of these harmful molecules. This system consists of both enzymatic and non-enzymatic components. Additionally, there is dynamic regulation and cross-talk between these defense mechanisms.

3.1. Enzymatic Antioxidants

Enzymatic antioxidants are protein-based molecules that minimize the harmful effects of ROS in plants (Mandal et al., 2022). These enzymes protect cell structures by converting ROS such as superoxide and hydrogen peroxide, which are formed in the oxidative stress environment within cells, into less toxic compounds. Localized in various organelles of the plant, they prevent the accumulation of ROS and form a critical line of defense against oxidative damage (Mittler et al., 2022). These enzymes typically work in complementary functions, with the following key enzymes standing out (Table 2).

Superoxide Dismutase (SOD): SOD is the first step in ROS detoxification and directly converts superoxide anions ($O_2{\bullet}^-$) to hydrogen peroxide (H_2O_2) (Batool et al., 2022). Superoxide is one of the most reactive and harmful types of ROS in cells and is produced in high concentrations, particularly in mitochondria, chloroplasts and the cytoplasm (Napolitano et al., 2021). SOD prevents the spread of harmful effects within the cell by rapidly converting these species. Plants have different isoforms: Cu/Zn-SOD (cytoplasm and chloroplasts), Mn-SOD (mitochondria), Fe-SOD (chloroplasts), etc. (Batool et al., 2022).

Catalase (CAT): Hydrogen peroxide produced by SOD can still be toxic to cells. At this stage, catalase comes into play and rapidly breaks down hydrogen peroxide into water (H_2O) and oxygen (O_2), protecting the cell from the harmful effects of H_2O_2 (Takio et al., 2021). Catalase is particularly concentrated in the peroxisome organelle and is effective in areas with high hydrogen peroxide concentrations. Catalase's high reaction rate is critical for maintaining cellular ROS balance (Medica et al., 2025).

Ascorbate Peroxidase (APX): APX takes electrons from ascorbate (vitamin C) to reduce hydrogen peroxide (Noctor, 2025). This converts H_2O_2 into water. APX is found primarily in chloroplasts, mitochondria, and the cytoplasm. In photosynthetic

Table 2. Major enzymatic antioxidants involved in plant defence against heat-induced oxidative stress

Enzyme	Primary function	Cellular location	Role and characteristics
Superoxide Dismutase (SOD)	Converts superoxide anion ($O_2^{\bullet -}$) into hydrogen peroxide (H_2O_2)	Cytoplasm, Mitochondria, Chloroplast	First line of defense against ROS; exists in multiple isoforms (Cu/Zn-SOD, Mn-SOD, Fe-SOD)
Catalase (CAT)	Breaks down hydrogen peroxide (H_2O_2) into water and oxygen	Peroxisome	Rapidly detoxifies high levels of H_2O_2; protects cells from toxic effects of hydrogen peroxide
Ascorbate Peroxidase (APX)	Reduces H_2O_2 to water using ascorbate as electron donor	Chloroplast, Cytoplasm, Mitochondria	Controls ROS accumulation in photosynthetic organelles; plays a key role in the ascorbate-glutathione cycle
Glutathione Reductase (GR)	Converts oxidized glutathione (GSSG) back to reduced form (GSH)	Cytoplasm, Chloroplast, Mitochondria	Maintains cellular redox balance; sustains glutathione cycle using NADPH

organelles, APX actively clears ROS when ROS production increases due to light stress (Zandi and Schnug 2025). APX plays an important role in the ascorbate-glutathione cycle, which helps the plant cope with oxidative stress (Mishra et al., 2023).

Glutathione Reductase (GR): GR ensures the continuous renewal of GSH, the reduced form of glutathione (Lapenna 2023). Glutathione is one of the most important redox buffers in the cell and can directly neutralise ROS. However, when the oxidized form of glutathione (GSSG) is formed, the GR enzyme uses NADPH to reduce GSSG back to GSH. This process is essential for maintaining redox balance in the cell and the continuity of the ascorbate-glutathione cycle. GR activity determines long-term resilience in managing oxidative stress in the cell (Couto et al., 2016).

Coordination and Importance of Enzymatic Antioxidants: The activities of these enzymes balance intracellular ROS levels and prevent oxidative damage to cellular macromolecules (DNA, protein, lipid). While each enzyme targets specific types of ROS, the combined action of these enzymes enables the plant to survive under oxidative stress conditions. For example, SOD converts superoxide to H_2O_2, and then CAT and APX detoxify this H_2O_2 (Ighodaro and Akinloye, 2018). GR, on the other hand, replenishes the antioxidants used in this process and completes the cycle (Ighodaro and Akinloye 2018). Thus, redox homeostasis is maintained in the intracellular environment. Under heat stress and other environmental challenges, the gene expression and activity of these enzymes increase, thereby enhancing the plant's stress tolerance (Haider et al., 2021). For this reason, enzymatic antioxidants are considered a vital defense mechanism in plants.

3.2. Non-Enzymatic Antioxidants

Non-enzymatic antioxidants are low molecular weight compounds that can directly neutralise ROS (Irato and Santovito, 2021). These molecules are among the fundamental components of the plant antioxidant defense system and play an important role in maintaining cellular redox balance, limiting oxidative damage, and enhancing the plant's resistance to environmental stress factors by working in conjunction with enzymatic antioxidant systems.

One of the key compounds in this group is ascorbate (vitamin C). Ascorbate is water-soluble, has six carbon atoms, and is a powerful electron donor (Mittu et al., 2022). It has the capacity to directly neutralize reactive molecules such as $^1O_2$, $O_2{\bullet}^-$ and •OH, which are types of ROS. Additionally, ascorbate acts as an electron donor for the ascorbate peroxidase (APX) enzyme and facilitates the reactivation of other antioxidants such as oxidised tocopherol (Sharma et al., 2022). It is found in various regions within the cell, including the cytoplasm, chloroplasts, mitochondria, vacuoles, and apoplast. Ascorbate levels and redox status are key factors in the plant's defense capacity against environmental conditions such as heat stress (Xiao et al., 2021).

Another important molecule is glutathione. Glutathione is a tripeptide compound composed of the amino acids glutamate, cysteine, and glycine (Vašková et al., 2023). Both its free form (GSH) and oxidised form (GSSG) perform critical functions in the cell. It detoxifies ROS species such as H_2O_2 with the help of glutathione peroxidase (GPX) and sustains the cellular antioxidant cycle by facilitating the conversion of ascorbate to its reduced form in the ascorbate-glutathione cycle (Singh and Satheeshkumar, 2023). It also regulates intracellular redox balance and affects stress-related signaling pathways (Singh and Satheeshkumar 2023). Glutathione is found in various cellular compartments such as the cytoplasm, chloroplasts, mitochondria, nucleus, and peroxisomes (Zechmann 2020). The GSH:GSSG ratio is directly related to the degree of oxidative stress in the plant and is considered an indicator of antioxidant capacity (Zechmann 2020).

Carotenoids are pigment compounds with lipophilic properties. This group includes molecules such as β-carotene, lutein and zeaxanthin (Riaz et al., 2021). In addition to neutralizing reactive species such as singlet oxygen ($^1O_2$), they protect the photosynthetic apparatus by converting excess light energy into heat (non-photochemical quenching mechanism). They also contribute to maintaining structural integrity by protecting cell membranes from lipid peroxidation (Riaz et al., 2021). They are particularly abundant in thylakoid membranes and support the plant's heat tolerance by reducing oxidative damage associated with heat stress (Zahra et al., 2023).

Flavonoids and other phenolic compounds are secondary metabolites with a polyphenolic structure. Components such as quercetin, kaempferol and anthocyanins belong to this group. These molecules directly neutralize ROS and reactive nitrogen species (RNS); they form chelates with transition metals, thereby preventing harmful ROS production processes such as the Fenton reaction (Hajam et al., 2023). They also protect membrane integrity and exert a regulatory effect on hormonal signaling pathways under stress conditions. They accumulate in regions such as the cell wall, vacuole, epidermal cells, and apoplast. Flavonoid accumulation limits the harmful effects of oxidative stress by increasing the stability of cell membranes, especially under high temperatures, and also acts as a protective barrier against UV radiation (Ferreyra et al., 2021).

3.3. Regulation and Cross-talk of Antioxidant Pathways under Heat Stress

In plant cells, antioxidant systems function as a multi-layered and dynamic defense network consisting of both enzymatic and non-enzymatic components. This system is rapidly activated by complex regulatory mechanisms when exposed to environmental stress conditions, particularly heat stress, and aims to maintain cellular homeostasis through intermolecular cooperation.

Under heat stress, increased metabolic rate in cells and imbalances in the electron transport chain in energy centres such as mitochondria and chloroplasts increase the production of reactive oxygen species (ROS) (Suzuki, 2023). This situation leads to an intensification of oxidative stress. The plant cell regulates its antioxidant system both at the transcriptional level (through gene expression) and through post-translational modifications (e.g., phosphorylation, redox changes) in order to limit this damage.

During heat stress, superoxide dismutase (SOD) enzymes form the first line of defense by converting superoxide radicals ($O_2\bullet^-$) into hydrogen peroxide (H_2O_2) (Mylostyva et al., 2021). Enzymes such as catalase (CAT) and ascorbate peroxidase (APX) then convert H_2O_2 into water and oxygen to eliminate its toxic effects (Mylostyva et al., 2021). The cooperation between these enzymes ensures the effective elimination of ROS. Non-enzymatic antioxidants such as ascorbate and glutathione complement these processes, contributing to both the direct detoxification of ROS and the regeneration of enzymes (Rudenko et al., 2023). In particular, the ascorbate-glutathione cycle (AsA-GSH cycle) plays a critical role in organelles such as chloroplasts, where ROS production is intense. This cycle involves the conversion of ascorbate used by APX into dehydroascorbate and the regeneration of glutathione (from GSSG to GSH) by glutathione reductase (GR). This ensures that antioxidant molecules are reusable and maintains the continuity of the system.

In addition to directly neutralizing ROS, carotenoids and flavonoids act as photoprotective molecules that suppress ROS production and protect membrane structure (Alhaithloul et al., 2021). Carotenoids, especially in chloroplast membranes, limit ROS production due to light damage, while phenolic compounds contribute to the reduction of oxidative stress both inside and outside the cell (Riaz et al., 2021). When faced with heat stress, there is coordination between these systems at the level of gene expression. For example, high temperature signals trigger the activation of antioxidant genes through transcription factors (e.g., HSFA1, DREB, WRKY) (Cheng et al., 2021; Wang and Zhu, 2023). In this way, genes such as SOD, APX, GR and CAT are rapidly activated and protective defense systems are activated. Hormonal signaling pathways (e.g., salicylic acid, jasmonate, ethylene) also play an important role in regulating these responses (Zhao et al., 2021).

4. Arbuscular Mycorrhizal Fungi: Biology and Function

4.1. Taxonomy and Symbiotic Mechanisms

Arbuscular mycorrhizal fungi (AMF) are obligate biotrophic microorganisms and belong to the phylum Glomeromycota (Boyno and Demir, 2022). They form a mutualistic relationship with about 80% of land plants. Taxonomically, AMFs were initially classified according to spore morphology and wall structure, but this classification has now been further clarified thanks to molecular phylogenetic studies using genetic analyses such as SSU rRNA. Thanks to these analyses, important genera such as *Rhizophagus, Funneliformis, Claroideoglomus, Acaulospora* and *Gigaspora* have been identified.

The symbiotic interaction between AMF and host plant begins when strigolactones released from plant roots stimulate fungal hyphae branching and metabolic activity (Boyno et al., 2023). In response, AMFs release signaling molecules called "Myc factors", which are sensed by plant root cells (Boyno and Demir, 2022). Subsequently, appressorium structures form on the root surface and hyphae penetrate the root cortex cells, where they form highly branched structures called arbuscules. These arbuscules are symbiotic exchange points where phosphorus, nitrogen and trace elements are transported to the plant, in exchange for carbon (sugars and fatty acids) from the plant. Arbuscule development and function are tightly regulated at the genetic and cellular level by both the plant and the fungus (Bhantana et al., 2021).

4.2. AMF-Mediated Nutrient Uptake and Signaling

One of the main benefits of AMF colonization is a more efficient uptake of phosphorus (P) by the plant, especially low mobility phosphorus (P) (Bhantana et al., 2021). The external hyphae network of AMF transports nutrients such as phosphorus, nitrogen, zinc and copper to the plant by reaching soil regions inaccessible to roots. The transfer of these nutrients to the plant is mediated by mycorrhiza-specific phosphate transporters (e.g., MtPT4, OsPT11) located on the periarbuscular membrane (Boyno et al., 2023).

Besides nutrient uptake, AMF colonization leads to systemic changes in signaling pathways within the plant. In particular, phytohormone pathways such as jasmonate (JA), salicylic acid (SA) and abscisic acid (ABA) are modulated by this symbiosis (Kandalgaonkar and Barvkar, 2025). This results in a state of "preparedness" in the plant's immune system, which allows the defense against pathogens to be strengthened without generating a full hypersensitive response. This effect occurs through fine-tuned changes in gene expression and hormonal cross-talk, so that the plant can effectively balance growth and defense (Fig. 1).

4.3. AMF Colonization under Heat Stress Conditions

High temperature is one of the main abiotic stress factors that directly affect plant physiology, metabolism and cellular structure, causing serious losses in growth and yield. Decreased photosynthesis rate, impaired fluidity of cell membranes, protein denaturation, decreased enzyme activity and accumulation of reactive oxygen species

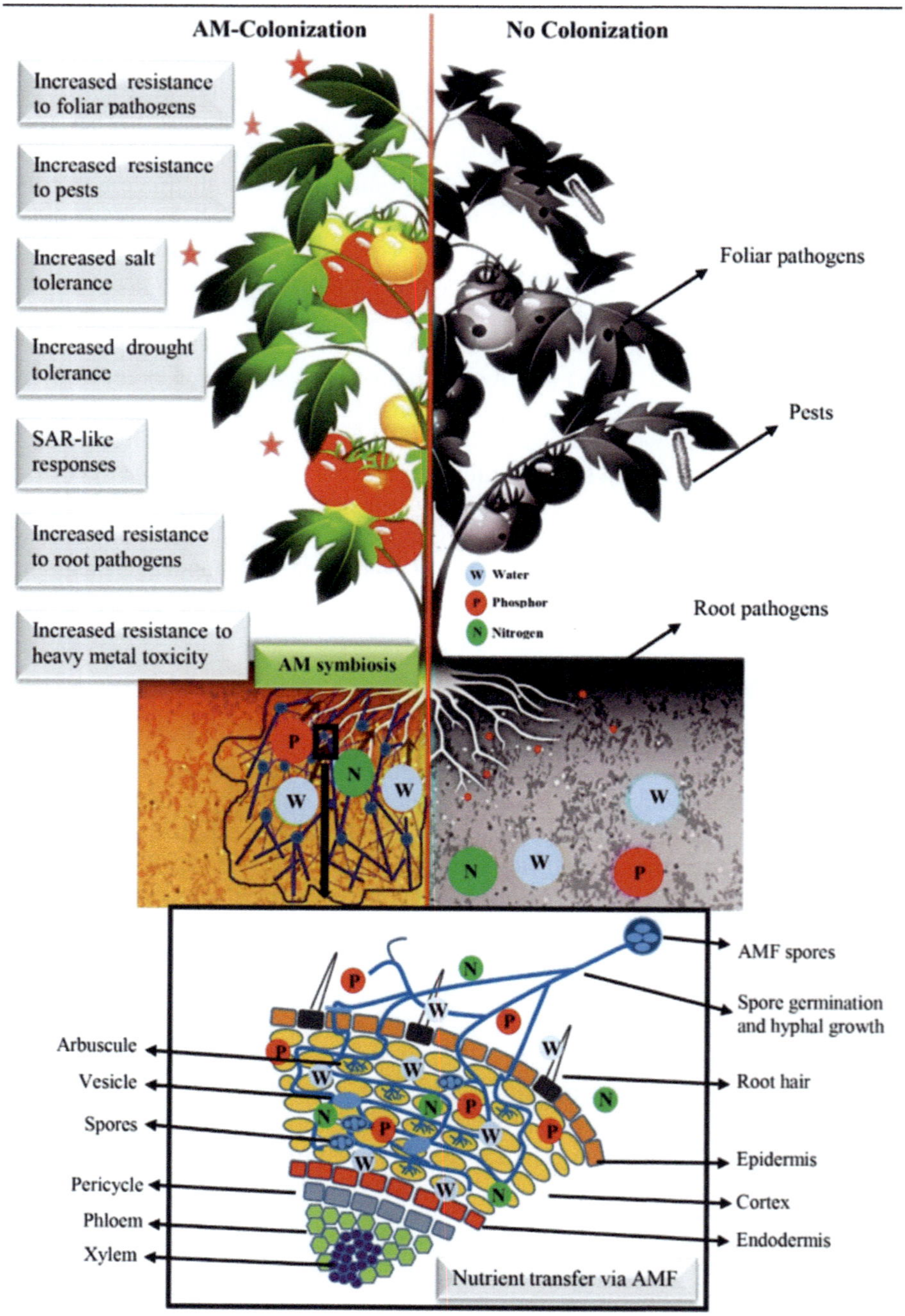

Fig. 1. Graphical overview of symbiotic mechanisms and physiological impacts of AMF on host plants (Figure by Boyno, G.) (Manteghi et al., 2022).

(ROS) are among the main consequences of this stress. These adverse effects disrupt the intracellular redox balance, leading to the emergence of oxidative stress. AMFs play a critical role in enhancing plant tolerance to such stress conditions through symbiotic relationships with the plant (Demir et al., 2022). AMFs not only increase the water and nutrient uptake capacity of the plant, but also activate protective mechanisms against heat stress at the cellular level. However, temperature stress can affect not only the plant but also the development cycle of the AMF itself (Jian et al., 2024). In particular, at temperatures above 30 °C, the sporulation rate may decrease, spore germination may be delayed, external hyphae elongation may slow down and arbuscle formation may be impaired in some AMF species (Mathur et al., 2021). Therefore, the success of the symbiotic interaction depends on the temperature tolerance of both the AMF species and the host plant.

Some AMF species tolerant to high temperatures are able to maintain symbiotic relationships and thus protect the plant at various levels. This protection is primarily achieved by strengthening the antioxidant defense system of the plant. AMF colonization increases gene expression and activity of antioxidant enzymes such as superoxide dismutase (SOD), catalase (CAT), ascorbate peroxidase (APX) and glutathione reductase (GR) under heat stress (Haddidi et al., 2020; Ma et al., 2022). These enzymes prevent oxidative damage to cell structures by detoxifying heat-induced ROS. In addition, the amount of osmolytes such as proline and soluble sugars that accumulate in the cell under heat stress is also increased through AMF (Liu et al., 2023). These compounds maintain intracellular osmotic balance, stabilize proteins and contribute to the maintenance of membrane integrity.

Photosynthesis is one of the most sensitive physiological processes to temperature stress. High temperature reduces photosynthetic efficiency, especially by disrupting the structural and functional integrity of photosystem II (PSII) (Agrawal and Jajoo, 2021). AMF colonization supports the maintenance of photosynthesis by contributing to the maintenance of chlorophyll a and b levels (Khan et al., 2022). Furthermore, through regulation of stomatal conductance, water use efficiency is increased and transpiration losses are reduced (Bhardwaj et al., 2023). In this way, the plant can realize a more balanced gas exchange against thermal stress.

At the soil level, the benefits of AMF are also remarkable. The glycoprotein substance glomalin secreted by AMF improves soil structure by increasing the stability of soil aggregates (Agnihotri et al., 2022). This increases the water holding capacity of the soil, reduces evaporation and keeps the root zone cooler and more moist. AMF also increases the branching of the root system and root surface area, facilitating both nutrient uptake and water absorption.

Studies at the gene expression level show that AMF up-regulates genes associated with heat shock proteins (HSP), aquaporins, LEA proteins and antioxidant enzymes in plants under heat stress (Abdelrahman et al., 2017; González-Morales et al., 2021; Jian et al., 2024). Thus, cellular protection systems become more efficient. In conclusion, AMF colonization enables the plant to mount a comprehensive defense response to heat stress at the physiological, biochemical and molecular levels. This not only protects plant health, but also supports sustainable agricultural production, making AMF a strategic biological tool to combat climate change.

5. AMF-Induced Modulation of Antioxidant Systems

Arbuscular mycorrhizal fungi (AMF) not only increase nutrient and water uptake through their symbiotic relationships with plants, but also play an important role in developing resistance to stress conditions at the molecular and cellular levels (Boyno and Demir, 2022). Among these contributions, the regulation of antioxidant defense systems is paramount. AMF colonization reduces the harmful effects of oxidative stress by modulating the levels and efficacy of both enzymatic and non-enzymatic antioxidant defense components (Afshari et al., 2022). This regulation helps protect cell structures by limiting the accumulation of reactive oxygen species (ROS).

5.1. Enhancement of Enzymatic Antioxidants by AMF

AMF plays a very important role in helping plants gain resistance to environmental stress conditions through mutually beneficial symbiotic relationships with plants. This symbiosis is particularly effective in activating antioxidant defense systems that control molecules that can damage cells, such as ROS (Boyno et al., 2022). Enzymatic antioxidant systems are one of the most critical defense components involved in the detoxification of ROS (Rajput et al., 2021). This system is of vital importance in preventing oxidative stress and maintaining redox balance at the cellular level.

As a result of AMF colonization, the expression of enzymes such as superoxide dismutase (SOD), catalase (CAT), ascorbate peroxidase (APX), glutathione reductase (GR) and monodehydroascorbate reductase (MDHAR), which are key components of this system, increases at the gene level, and their enzymatic activities also increase significantly in parallel (Haddidi et al., 2020; Ma et al., 2022). This increase enables the plant to become more resistant to a wide variety of stress factors such as temperature, drought, salinity, heavy metal toxicity, or pathogen attacks. For example, under environmental stress factors, ROS production increases intensively in intracellular organelles such as mitochondria, chloroplasts, and peroxisomes (Corpas et al., 2015). Among these ROS, superoxide anion (O_2^-), hydrogen peroxide (H_2O_2), hydroxyl radical (•OH), and singlet oxygen ($^1O_2$) are particularly reactive and can cause serious damage to cellular macromolecules such as proteins, nucleic acids, and membrane lipids (Edge and Truscott, 2021). Enzymatic antioxidants form the first line of defense in eliminating these harmful species: SOD converts superoxide radicals into hydrogen peroxide; then enzymes such as CAT and APX reduce this hydrogen peroxide to water, a less harmful molecule. GR and MDHAR, on the other hand, play central roles in maintaining redox homeostasis by participating in the ascorbate-glutathione cycle.

Research has shown that AMF colonization not only increases the activity of these enzymes, but also regulates gene expression through epigenetic mechanisms (Denzer et al., 2020; He et al., 2020; McCaw et al., 2020; Bind et al., 2022). In particular, DNA methylation changes in the promoter regions of antioxidant enzyme genes under stress conditions indicate that AMF also plays a regulatory role at the epigenetic level (He et al., 2020). This suggests that symbiotic interactions may cause permanent changes in plant stress memory. Additionally, it has been shown that some AMF species can specifically induce the expression of certain antioxidant genes more strongly. For example, while SOD and APX expression increased more significantly in plants colonized by *Glomus intraradices* (Gong et al., 2015), higher levels of CAT and GR were reported in plants in symbiosis with *Rhizophagus irregularis* (Fuentes, et al., 2016). This indicates the presence of AMF-specific effects and that the symbiotic relationship is supported by specialized adaptation mechanisms.

In conclusion, AMF's enhancement of enzymatic antioxidant systems is of strategic importance not only as a cellular defense mechanism but also in terms of maintaining the plant's overall physiological stability, ensuring the sustainability of growth parameters and safeguarding agricultural productivity. In this context, AMF can be considered a biological 'natural supplement system' for the cultivation of plants with high stress tolerance.

5.2. Modulation of Non-Enzymatic Antioxidant Levels

AMF play a powerful role in combating oxidative stress not only through enzymatic antioxidant mechanisms, but also by regulating the levels and effectiveness of non-enzymatic low-molecular-weight antioxidant compounds. Non-enzymatic antioxidants form a critical line of defense for plant cell survival and biochemical integrity by directly neutralizing ROS in the cellular environment. The main compounds in this group include ascorbate (AsA), glutathione (GSH), tocopherols (vitamin E), carotenoids, and flavonoids. AsA and GSH are central in maintaining intracellular redox balance (Sachdev et al., 2023). AsA can directly detoxify ROS such as superoxide anion, hydrogen peroxide, and singlet oxygen (Xiao et al., 2021). GSH, on the other hand, participates in ROS detoxification with the help of enzymes such as glutathione reductase and ascorbate peroxidase, and also facilitates the regeneration of other antioxidant compounds (Averill-Bates, 2023). AMF colonization increases the biosynthesis of these molecules and ensures that intracellular redox cycles function more efficiently, maintaining high AsA/dehydroascorbate (DHA) and GSH/oxidised glutathione (GSSG) ratios (Bisht and Garg, 2024). These ratios are considered indicators of antioxidant capacity and redox homeostasis at the cellular level. AMF's ability to maintain these redox ratios at high levels, thereby keeping plant metabolism active, balanced, and stress-resistant, is one of the molecular advantages of the symbiotic relationship.

AMF also promotes the biosynthesis of phenolic compounds and flavonoids. These compounds act not only as antioxidants but also as physical and chemical shields against environmental stresses such as UV radiation, pathogens, heavy metals and salinity. The synthesis of flavonoids is initiated by the activation of various phenylpropanoid pathway enzymes, primarily phenylalanine ammonia lyase (PAL) (Yang et al., 2022). AMF induces these pathways by stimulating PAL enzyme activity and increases flavonoid production (Fotovvat et al., 2024). These molecules provide direct protective effects by capturing free radicals in the cytoplasm and also strengthen the cell structurally by participating in the synthesis of lignin and suberin-like compounds in the cell wall. This dual effect increases the plant's resistance to both biotic and abiotic stress conditions.

AMF colonization has also been reported to positively affect carotenoid and tocopherol accumulation. Carotenoids protect photosystem II, one of the primary sources of ROS in chloroplasts; they also limit photooxidative damage by preventing excessive energy accumulation in conditions of high light energy (Izuhara et al., 2020). Tocopherols, on the other hand, protect membrane integrity by reducing lipid peroxidation (Mesa and Munné-Bosch, 2023). AMF's promotion of the biosynthesis of these lipophilic antioxidants contributes to the stability of membrane structures even under stress (Wahab et al., 2023). Finally, some studies have shown that AMF not only increases the levels of these non-enzymatic compounds quantitatively but also regulates their intracellular distribution, helping to provide localized defense in specific stress regions (e.g. chloroplasts, mitochondria, nuclear periphery) (Chandrasekaran 2022; Sun et al., 2024). This highlights that the regulatory effect of AMF on the antioxidant system is highly sophisticated, targeted, and multi-layered.

5.3. AMF and ROS Scavenging Efficiency

Plant cells increase the production of ROS under environmental stress conditions, such as drought, extreme heat, salinity, or pathogen attacks. While ROS play important roles in cellular signalling at low levels, their accumulation can lead to serious cellular damage, such as lipid peroxidation, protein denaturation and DNA damage, resulting in harmful effects that can lead to cell death (Ozougwu, 2016). Therefore, it is vital for plants not only to control ROS production but also to effectively eliminate the ROS that are produced.

AMF not only increases the amount of antioxidant enzymes or metabolites, but also significantly improves the ROS scavenging capacity of plants by ensuring the effectiveness, localization and dynamic regulation of these components. This effect is most clearly observed in organelles such as chloroplasts, mitochondria and peroxisomes, where ROS production is naturally high. Numerous studies have demonstrated that under AMF colonization, ROS accumulation is limited in these organelles, peroxidation products such as malondialdehyde (MDA), which are indicators of oxidative stress, remain at lower levels, and membrane stability is thus maintained (Wu et al., 2014; Zou et al., 2021; Huang et al., 2023). In photosynthetic tissues in particular, the presence of AMF preserves the efficiency of light harvesting in photosystem II (PSII) (Mathur et al., 2019). This is directly related to both the maintenance of the structural integrity of chlorophyll pigments and the reduction of photooxidative stress. ROS production in chloroplasts increases particularly under high light intensity or heat stress (Pospíšil, 2016). AMF facilitates the rapid detoxification of ROS by increasing the local activity of SOD, APX, and glutathione cycle enzymes in these regions, thereby maintaining the energy balance necessary for photosynthesis (Yang et al., 2015). Similarly, in mitochondria and peroxisomes, AMF has been shown to reduce superoxide and hydrogen peroxide levels, thereby maintaining mitochondrial membrane potential and metabolic activity (Begum et al., 2019).

It has been reported that AMF modulates the gene expression and activity of ROS-producing enzymes such as NADPH oxidase in plant cells (Srivastava et al., 2025). NADPH oxidases rapidly trigger ROS production in plants in response to pathogen perception or stress signals (Srivastava et al., 2025). However, a continuous or excessive response is more harmful than beneficial. AMF helps establish a balance by regulating the activity of these enzymes, promoting ROS production only when necessary, so that ROS can continue to play its role as a defense signal without harming the cell (Wahab et al., 2023). In this context, AMF offers a dual-action mechanism that both limits ROS production and enhances detoxification.

ROS scavenging efficiency is also related to the integration of local signalling networks. Under the influence of AMF, a more coordinated antioxidant response develops between cells. For example, the equalization of ROS concentrations in different tissues, such as roots and leaves, allows for the development of not only a local but also a systemic stress response (Choudhary et al., 2020). AMF's effect in this regard helps plants develop faster, more balanced, and energy-efficient defense responses to environmental stress factors (Wahab et al., 2023). Finally, thanks to this symbiotic relationship, the plant can redirect its energy resources to fundamental physiological processes such as growth and development rather than overactivating its defense system. This situation provides a significant advantage, especially in terms of maintaining agricultural productivity.

6. Molecular Mechanisms of AMF-Heat Stress Interaction

Arbuscular mycorrhizal fungi (AMF) not only increase nutrient uptake through their symbiotic relationship with plant roots, but also contribute to the development of resistance in plants against environmental stress conditions. In particular, how AMF regulate the plant's molecular responses under abiotic stress factors such as high temperatures has been an intensively studied topic in recent years (Jian et al., 2024). This interaction involves the interplay of multiple biochemical and molecular pathways, including antioxidant systems, heat shock proteins (HSPs), and plant hormone signals.

6.1. Gene Expression Changes in Antioxidant Pathways

High temperature stress causes metabolic imbalances in plant cells, dramatically increasing the production of reactive oxygen species (ROS) (Devireddy et al., 2021). ROS such as superoxide anion (O_2^-), hydrogen peroxide (H_2O_2) and hydroxyl radical (•OH) cause lipid peroxidation in cell membranes, protein denaturation and DNA damage when accumulated at high levels, leading to the destruction of cellular structure (Devireddy et al., 2021). To counteract these harmful effects, plants activate their antioxidant defense systems. These systems include both enzymatic (e.g., superoxide dismutase, catalase) and non-enzymatic (e.g., ascorbate, glutathione) components.

Arbuscular mycorrhizal fungi (AMF) colonization is emerging as an important biological factor that increases plants' tolerance to high temperature stress (Mathur et al., 2018). AMF not only improves nutrient uptake and water balance by integrating into the plant's root system but also activates the plant's molecular defense pathways (Boyno et al., 2023). In particular, significant increases in the expression levels of genes associated with antioxidant defence systems have been observed. Genes associated with key antioxidant enzymes such as superoxide dismutase (SOD), catalase (CAT), ascorbate peroxidase (APX), and glutathione reductase (GR) are upregulated following AMF colonisation, significantly increasing ROS detoxification capacity (Afshar and Abbaspour, 2023). The activation of these enzymes at the gene level contributes to the protection of the cellular structure by effectively clearing ROS. These genetic regulations are not limited to temporary transcriptional changes. Recent studies have revealed changes in methylation patterns in the promoter regions of antioxidant genes in AMF-colonized plants (Tang et al., 2021; Afshar and Abbaspour, 2023). DNA methylation is a mechanism that plays a critical role in the epigenetic control of gene expression. A decrease in methylation in gene promoters allows the gene to become more transcriptionally active. In this context, AMF's restructuring of antioxidant gene methylation levels enables these genes to be activated in a more permanent and sustainable manner (Cicatelli et al., 2014). This allows the plant to prepare in advance not only for current stress conditions but also for similar environmental challenges it may encounter in the future. The resulting 'epigenetic stress memory' enhances the plant's adaptive capacity.

6.2. AMF-Mediated Regulation of Heat Shock Proteins (HSPs)

Heat shock proteins (HSPs) are molecular chaperones that maintain protein stability at the cellular level, are synthesized under stress conditions, and play a critical role in the adaptation of living organisms to environmental changes (Jeyachandran et al., 2023). In plants, HSPs maintain proteostasis by preventing protein misfolding, aggregation, and denaturation, especially under high temperature stress (Khan and Shahwar, 2025). These proteins also play a role in the refolding of damaged proteins or their degradation via the proteasome pathway.

AMF colonization enhances the plant's response to heat stress not only through antioxidant systems but also through molecular chaperone systems. Studies have shown that AMF-colonized plants exhibit significant increases in the expression levels of many isoforms belonging to the HSP gene family, particularly HSP60, HSP70, HSP90, and HSP101 (Ceylan et al., 2023; Tian et al., 2023). These gene products maintain organelle-specific proteome stability by localizing both in the cytosol and in organelles such as chloroplasts, mitochondria, and the endoplasmic reticulum. In particular, HSP70 and HSP90 are key chaperones involved in protein folding, translocation, and complex formation (Luengo et al., 2019). AMF colonization increases transcriptional activation in the promoter regions of these HSPs, thereby preventing protein damage under stress conditions (Afshar and Abbaspour, 2023). HSP60, on the other hand, plays a role in the recovery of thermotolerance following thermal stress; the activation of this protein by AMF enables the plant to recover quickly after temporary stress (Gadkar et al., 2006).

At the molecular level, this regulation is usually carried out through transcription factors belonging to the HSF (Heat Shock Transcription Factors) family. In particular, it has been shown that members such as HSFA1 and HSFA2 are activated at higher levels in plants associated with AMF (Wang et al., 2015; Lin et al., 2018; Afshar and Abbaspour 2023). These transcription factors trigger gene transcription by binding to the 'heat shock element' (HSE) regions of HSP gene promoters, thereby triggering gene transcription. AMF's enhancement of these factors' activity systematically regulates HSP production, providing a proactive defense during stress. However, AMF's effect on regulating HSPs is not merely a transient stress response but also provides a 'pre-conditioning' effect. In plants colonized by AMF, basal expression levels of HSPs are higher, indicating that the plant has a more resilient molecular infrastructure even before stress begins (Branco et al., 2022). This highlights the importance of AMF applications in terms of long-term adaptation strategies, particularly in agricultural areas experiencing chronic temperature increases.

6.3. Hormonal Signaling and Cross-Talk

Plants have developed a highly sophisticated hormonal signaling network to survive environmental stress factors. This network detects environmental stimuli and converts them into internal physiological responses, ensuring the coordinated execution of growth, development and stress defense mechanisms. AMF modulate both the sensitivity and efficacy of these phytohormonal signaling pathways, enabling the plant to respond to stress conditions in a more balanced and adaptive manner. In particular, hormones such as abscisic acid (ABA), jasmonic acid (JA), salicylic acid (SA), and strigolactones (SLs) play a central role in the protective effect of AMF-based symbiosis against heat stress.

ABA (Abscisic Acid): ABA is one of the most important hormones that enables plants to survive abiotic stresses, particularly heat, drought and salt stress (Dar et al., 2017). Under high temperature stress, stomata must close to prevent water loss in plant tissues (Malini et al., 2023). AMF colonization promotes stomatal closure by increasing ABA biosynthesis and reduces water loss through transpiration. This process is achieved through the upregulation of genes involved in ABA biosynthesis, such as 9-cis-epoxycarotenoid dioxygenase (NCED), by AMF (Chen et al., 2024). Thus, AMF not only increases the water uptake capacity of roots but also minimizes water loss in leaves, thereby enhancing the plant's water use efficiency.

JA (Jasmonic Acid) and SA (Salicylic Acid): JA and SA typically play a role in the defense responses developed by plants against biotic stresses (e.g. pathogens); however, a growing number of studies have revealed that these two hormones also perform important functions under abiotic stress conditions (Li and Ahammed, 2023). Under heat stress, these hormones are involved in numerous processes, including ROS production, cell death signaling, and the regulation of antioxidant defense genes. AMF modulates JA and SA signaling pathways both individually and through cross-talk between them. For example, JA is involved in the activation of certain transcription factors (e.g., MYC2) in response to heat stress, while SA is more closely associated with heat-induced programmed cell death and the production of heat shock proteins (Cha et al., 2020; Bhatla and Lal, 2023). AMF regulates the balance of these hormones, enabling the plant to continue growing while effectively managing its defense responses.

SLs (Strigolactones): Strigolactones are terpenoid-based hormones that are important for both plant development (particularly lateral root development and shoot growth) and symbiotic interactions (Singh et al., 2023). SL levels typically increase under stress conditions, and this increase promotes the establishment of symbiotic relationships with AMF. Increased SL production under high temperature conditions leads to changes in root system architecture, enabling more effective colonization of roots by AMF (Luo et al., 2020). Additionally, SLs play a supportive role in plant physiology by aiding adaptation to heat stress. For example, it has been reported that SLs synergistically work with ABA in regulating stomatal aperture, while also providing heat protection by maintaining photosynthetic efficiency and chlorophyll stability (Li et al., 2022).

The effect of AMF on these hormones may arise by directly regulating gene expression in biosynthetic pathways or by affecting the receptors and transporters of these hormones. Modulation of these hormonal pathways allows the plant to simultaneously regulate not only its short-term defense responses under stress but also its developmental processes. For example, while growth typically ceases under high temperatures, the maintenance of hormonal balance through AMF prevents growth from completely stopping (Zhu et al., 2021).

7. Case Studies and Experimental Evidence

7.1. Plant-Specific Responses: Examples from Wheat, Maize, and Tomatoes

AMF has the potential to improve plant performance under various abiotic stress conditions thanks to the symbiotic relationships they establish with plants. However, this symbiotic interaction varies depending on plant species, and not all plant-AMF combinations yield equally effective results. Plants' responses to AMF colonization are determined by numerous factors, including genetic structure, root morphology, physiological adaptation capacity, defense system dynamics, and molecular compatibility of symbiotic interactions (Wahab et al., 2023). In the context of heat stress, responses to AMF applications are being studied using important crop species such as wheat (*Triticum aestivum* L.), maize (*Zea mays* L.), and tomato (*Solanum lycopersicum* L.) as examples.

Wheat is a model cereal that has a high capacity for symbiotic interaction with AMF and is frequently used in abiotic stress research. Studies conducted under high temperature conditions have shown that AMF preserves the levels of photosynthetic pigments such as chlorophyll and carotenoids in wheat, increases membrane stability, and reduces the accumulation of oxidative stress markers such as H_2O_2 and MDA (Huang et al., 2023). Additionally, significant increases in the activity of antioxidant enzymes such as superoxide dismutase (SOD), catalase (CAT), and peroxidase (POD) have been recorded (Savicka, M., & Škute 2010)). Furthermore, it has been reported that agronomic traits such as spike number, thousand-grain weight, and grain yield are significantly maintained or improved in plants inoculated with AMF despite temperature stress. In particular, species such as *Funneliformis mosseae* and *Rhizophagus intraradices* stand out in enhancing wheat's resistance to heat stress (Ganugi et al., 2019).

Corn is a plant that is highly suitable for AMF colonization due to its extensive and developed root system. In corn plants exposed to heat stress, it has been shown that AMF application improves root and leaf water potential, reduces leaf temperature and increases water use efficiency (Mathur et al., 2018). Additionally, by promoting proline accumulation, osmotic regulation mechanisms were supported, carbon metabolism balance was maintained, and plant growth was preserved. Furthermore, the expression levels of genes associated with heat shock proteins such as CsHsp70s and CsPIPs were increased through AMF, thereby maintaining cellular protein homeostasis despite temperature stress (Liu et al., 2023). It has also been reported that AMF contributes to photosynthetic performance by maintaining photosystem II activity (Fv/Fm) in maize (Mathur and Jajoo, 2020).

Tomato is one of the model vegetable plants commonly used in abiotic stress research and has a strong symbiotic relationship with AMF. In tomato plants, physiological processes such as stomatal closure and transpiration regulation under high temperature conditions are effectively modulated by AMF colonization (Abdel Latef and Chaoxing, 2011). AMF optimizes the plant's hormonal balance by affecting the levels and signaling pathways of phytohormones such as ABA, JA and SA, which play an important role in the abiotic stress response (Haddidi et al., 2020; Ahmed et al., 2025). This enables stomatal control, reduction of water loss and activation of cellular defense mechanisms. Additionally, significant increases in the levels of redox buffers such as glutathione (GSH) and ascorbate (AsA) in tomato roots were observed under heat stress, contributing to the prevention of oxidative damage (Liu et al., 2018). It has also been reported that AMF application has positive effects on quality parameters such as fruit number, fruit size, and shelf life (Chafai et al., 2023).

As can be seen from these plant examples, the protective effects of AMF against heat stress may vary depending on the plant species; these effects vary at both the physiological and molecular levels. Therefore, in order for AMF applications to be effectively utilized in agricultural production, product-based strategies must be developed and specific plant-AMF pairings optimized. This approach will play a critical role in the construction of sustainable agricultural systems in the fight against climate change.

7.2. AMF Strain-Specific Differences in Heat Tolerance

AMF are important symbiotic organisms with the potential to increase the resistance of plants to various abiotic stress conditions. However, the protective effect of AMF against heat stress can vary greatly not only between species but also between strains within species. The varying levels of effectiveness of AMF strains against heat stress are based on the interaction of factors such as their genetic diversity, thermal tolerance range, root colonization capacity, symbiotic efficiency, and molecular compatibility with the host plant.

Funneliformis mosseae is one of the species frequently highlighted in the literature due to its superior adaptation capacity under high temperature conditions in its symbiotic interactions with different plants (Gavito and Azcón–Aguilar, 2012). This AMF species can successfully maintain root colonization even at high soil temperatures, while also activating antioxidant defense systems in plants (e.g., superoxide dismutase, peroxidase, glutathione reductase) (Li, 2024). Additionally, it has been shown that *F. mosseae* promotes proline accumulation under heat stress, thereby supporting osmotic balance and maintaining photosynthetic efficiency (Tuo et al., 2015). With these characteristics, it stands out as an ideal AMF biofertilizer candidate, particularly for plants grown under arid and hot climatic conditions (Jha and Songachan, 2023).

On the other hand, some commonly used species, such as **Rhizophagus irregularis,** may be effective against moderate heat stress, but may lose their effectiveness under prolonged or extremely high temperature conditions (Romero-Munar et al., 2023). It has been reported that some strains of *R. irregularis* experience a decrease in root colonization efficiency at high temperatures, and that heat-related disruptions occur in symbiotic carbon-nutrient exchange processes (Romero-Munar et al., 2023). However, it has been noted that some genetic variants may be more tolerant, and that this may be related to the level of adaptation to local environmental conditions (Peña et al., 2020).

Septoglomus constrictum is an effective species against combined stress conditions such as high temperature and drought (Duc and Posta, 2018). This fungus protects cell membranes by reducing lipid peroxidation, lowers H_2O_2 levels, and increases water use efficiency through stomatal regulation (Szentpéteri et al., 2024). In particular, it provides advantages such as maintaining leaf water potential and sustaining photosystem II efficiency in tomatoes (Szentpéteri et al., 2024).

Diversispora versiformis is another important AMF species that exhibits resistance to heat stress. It has been shown that this fungus increases the expression of HSP70 (heat shock protein) and PIP (plasma membrane intrinsic protein) genes under heat stress, thereby maintaining photosynthetic efficiency (Tian et al., 2023). It also supports biomass increase (Tian et al., 2023).

In studies conducted with plants such as strawberries and soybeans, **Gigaspora margarita** has been found to maintain cell membrane stability against heat stress and increase the activity of enzymes such as CuZn-SOD (Zhu et al., 2017). Similarly, **Acaulospora spp.** supports cell structure and antioxidant defense through its symbiosis with thermophilic plants (da Silva et al., 2022).

These differences highlight the importance of using mycorrhizal consortia composed of combinations of different species and strains rather than a single AMF species. Such consortia can maintain symbiotic activity across a wide temperature range and more effectively regulate antioxidant defense, hormonal balance, and the carbon-nutrient cycle in plants (Meena et al., 2017). Additionally, the use of regionally adapted endemic AMF species may yield more successful results under temperature stress specific to local soil and climate conditions. Therefore, when selecting AMF strains to be used against temperature stress, not only species-level but also strain-level characteristics should be considered; molecular identification techniques (e.g., phylogenetic analyses based on the ITS region, abundance measurements using qPCR) and functional stress tests must be used. This approach will enable the development of more sustainable and effective solutions for integrating AMF into agricultural systems.

8. Practical Applications, Future Prospects, and Conclusion

In order to ensure the sustainability of plant production in the face of rising global temperatures and climate change, the integration of arbuscular mycorrhizal fungi (AMF) into agricultural practices has become a strategy that has attracted attention in recent years. The physiological, molecular, and hormonal advantages that AMF provide to plants against heat stress have placed these symbiotic organisms at the centre of climate change-resilient agricultural systems. In this context, AMF-based biotechnological applications are being developed, with a focus on sustainable solutions for the future.

AMF-based biological formulations have begun to occupy an important place in the category of microbial fertilizers. Developed specifically for use in agricultural areas under heat stress, these formulations are generally in the form of live inoculums containing spores, hyphae and root fragments. The type and strains of AMF used in the formulations are selected based on the target plant species, local ecological conditions, and application purpose. For example, species such as *Funneliformis mosseae* or *Claroideoglomus etunicatum* are known to perform well in arid and hot regions (Alam and Roy, 2024). These biological products are typically applied through seed coating, soil application, or drip irrigation systems, facilitating the establishment of symbiosis between the plant and AMF in the early stages and contributing to the development of stress tolerance. Such applications also minimize negative environmental impacts by reducing the use of traditional chemical fertilizers. Additionally, some commercial products utilize synergistic microbial consortia that include beneficial bacteria alongside AMF, further enhancing bio-effectiveness.

Climate-Smart Agriculture (CSA) is a framework that aims to increase agricultural production while adapting to climate change and reducing greenhouse gas emissions (Kabato et al., 2025). AMF has become an important component of this approach. AMF practices play an effective role, particularly in improving water use efficiency, enhancing soil structure, promoting carbon sequestration, and implementing low-input sustainable production models.

Under the CSA, the use of AMF is accompanied by support for reduced tillage (minimum tillage), organic matter applications and integrated plant nutrition strategies. This holistic approach not only increases crop yields but also preserves long-term soil health by protecting biodiversity. Furthermore, considering that AMF is effective in coping with stresses such as drought, salinity, and heavy metals, it is possible to produce integrated solutions against multiple stress factors (Demir et al., 2022).

There are some limiting factors in the widespread use of AMF-based applications (Boyno et al., 2024). First, the wide genetic diversity of AMF in nature makes it difficult to select and propagate effective strains. In addition, the production of AMF spores in vitro is still technically limited, and special infrastructure is required to produce high quantities of high-quality inoculum. Factors such as AMF failing to colonize at the expected level under field conditions or interacting with the soil microbiome can also affect application success.

In the future, characterising endemic AMF species with local adaptation at the molecular and functional levels will be of great importance. Additionally, metagenomic and metatranscriptomic approaches will enable us to better understand the dynamic interactions between AMF and plants. Enhancing the symbiotic capacity of AMF through genetic engineering techniques such as CRISPR/Cas9 offers potential for more effective and environmentally friendly agricultural practices (Boyno et al., 2023). Furthermore, the integrated use of AMF with other beneficial microorganisms (e.g., *Trichoderma, Bacillus* species) could enhance the resilience of biological formulations against multiple stress conditions. Future studies should also address issues such as economic feasibility, product-based impact analyses, and the long-term effects on soil health in the integration of AMF into agricultural systems.

The temperature increases caused by global climate change have become one of the most important abiotic stress factors threatening the sustainability of agricultural production. In particular, negative effects such as a decrease in photosynthesis capacity, disruption of water balance, accumulation of reactive oxygen species (ROS), and imbalance in cellular metabolism seriously limit plant growth and product quality (Fahad et al., 2017). In this context, there is an increasing need for strategies that go beyond traditional agronomic approaches and incorporate biologically based solutions.

As discussed in detail in this chapter, AMF stand out as versatile biological agents that increase plants' resistance to heat stress. AMF not only facilitate nutrient uptake through the mutualistic symbiosis they establish with plant roots, but also shape the plant's molecular and cellular stress adaptation. The activation of antioxidant defense systems, regulation of heat shock proteins (HSPs), and modulation of hormonal signaling pathways form the basis of this symbiotic relationship's protective effects against heat stress. Additionally, the variations exhibited by AMF at the species and strain levels highlight the uniqueness of their interactions with different plant species and emphasize the importance of species-specific applications. For example, while some AMF species such as *F. mosseae* and *C. etunicatum* exhibit high temperature tolerance, species such as *Rhizophagus irregularis* show optimal effects under more moderate conditions (Nie et al., 2024). However, the selection of AMF strains adapted to local climate and soil conditions is considered the key to success in field applications.

In light of this information, it has been demonstrated that AMF should be considered an active stress regulator not only in natural ecosystems but also in agricultural production systems. Findings from studies conducted in both controlled environments and field conditions indicate that AMF applications should be prioritised in climate change adaptation strategies. In particular, the development of AMF-based biological products (bioformulations) within the framework of climate-smart agriculture is of great importance in terms of both supporting plant health and ensuring agricultural sustainability.

In conclusion, AMF applications have the potential to offer environmentally friendly, low-cost, and effective biological solutions for the agriculture of the future. However, to fully realize this potential, interdisciplinary research, field validation, and more in-depth analyses at the genetic and epigenetic levels are required. A better understanding of the molecular mechanisms of AMF will contribute to the development of plant-microorganism systems resistant to climate change and ensure food security in the long term.

References

Abdel Latef, A.A.H. & Chaoxing, H. (2011). Arbuscular mycorrhizal influence on growth, photosynthetic pigments, osmotic adjustment and oxidative stress in tomato plants subjected to low temperature stress. Acta Physiologiae Plantarum 33, 1217–1225.

Abdelrahman, M., El-Sayed, M., Jogaiah, S., Burritt, D.J. & Tran, L.S.P. (2017). The "STAY-GREEN" trait and phytohormone signaling networks in plants under heat stress. Plant Cell Reports 36, 1009–1025.

Afshar, A.S. & Abbaspour, H. (2023). Mycorrhizal symbiosis alleviate salinity stress in pistachio plants by altering gene expression and antioxidant pathways. Physiology and Molecular Biology of Plants 29(2), 263–276.

Afshari, M., Rahimmalek, M., Sabzalian, M.R., Szumny, A., Matkowski, A. & Jezierska-Domaradzka, A. (2022). Mycorrhizal colonization modulates the essential oil profile and enzymatic and non-enzymatic antioxidants to mitigate the adverse effects of water deficit in Salvia subg. Perovskia. Biology 11(12), 1757.

Agnihotri, R., Sharma, M.P., Prakash, A., Ramesh, A., Bhattacharjya, S., Patra, A.K., ... & Kuzyakov, Y. (2022). Glycoproteins of arbuscular mycorrhiza for soil carbon sequestration: Review of mechanisms and controls. Science of the Total Environment 806, 150571.

Agrawal, D. & Jajoo, A. (2021). Study of high temperature stress induced damage and recovery in photosystem II (PSII) and photosystem I (PSI) in Spinach leaves (Spinacia oleracia). Journal of Plant Biochemistry and Biotechnology 30(3), 532–544.

Ahmed, N., Li, J., Li, Y., Deng, L., Deng, L., Chachar, M., ... & Tu, P. (2025). Symbiotic synergy: How arbuscular mycorrhizal fungi enhance nutrient uptake, stress tolerance, and soil health through molecular mechanisms and hormonal regulation. IMA Fungus 16, e144989.

Alam, M.Z. & Roy, M.D. (2024). The reduction of abiotic stress in food crops through climate-smart mycorrhiza-enriched biofertilizer. AIMS Microbiology, 10(3), 674.

Alhaithloul, H.A., Galal, F.H. & Seufi, A.M. (2021). Effect of extreme temperature changes on phenolic, flavonoid contents and antioxidant activity of tomato seedlings (Solanum lycopersicum L.). PeerJ, 9, e11193.

Andrási, N., Pettkó-Szandtner, A. & Szabados, L. (2021). Diversity of plant heat shock factors: regulation, interactions, and functions. Journal of Experimental Botany 72(5), 1558–1575.

Andrés, C.M.C., Pérez de la Lastra, J.M., Juan, C.A., Plou, F.J. & Pérez-Lebeña, E. (2022). Chemistry of hydrogen peroxide formation and elimination in mammalian cells, and its role in various pathologies. Stresses 2(3), 256–274.

Angelova, A., Angelov, B., Drechsler, M., Bizien, T., Gorshkova, Y.E. & Deng, Y. (2021). Plasmalogen-based liquid crystalline multiphase structures involving docosapentaenoyl derivatives inspired by biological cubic membranes. Frontiers in Cell and Developmental Biology 9, 617984.

Averill-Bates, D.A. (2023). The antioxidant glutathione. *In*: Vitamins and Hormones (Vol. 121, pp. 109–141). Academic Press.

Batool, R., Umer, M.J., Hussain, B., Anees, M. & Wang, Z. (2022). Molecular mechanisms of superoxide dismutase (SODs)-mediated defense in controlling oxidative stress in plants. In: Antioxidant Defense in Plants: Molecular Basis of Regulation (pp. 157–179). Singapore: Springer Nature Singapore.

Begum, N., Ahanger, M.A., Su, Y., Lei, Y., Mustafa, N.S.A., Ahmad, P. & Zhang, L. (2019). Improved drought tolerance by AMF inoculation in maize (Zea mays) involves physiological and biochemical implications. Plants 8(12), 579.

Berwal, M.K., Kumar, R., Prakash, K., Rai, G.K. & Hebbar, K.B. (2021). Antioxidant defense system in plants against abiotic stress. *In*: Abiotic Stress Tolerance Mechanisms in Plants (pp. 175–202). CRC Press.

Bhantana, P., Rana, M.S., Sun, X.C., Moussa, M.G., Saleem, M.H., Syaifudin, M., ... & Hu, C.X. (2021). Arbuscular mycorrhizal fungi and its major role in plant growth, zinc nutrition, phosphorous regulation and phytoremediation. Symbiosis 84, 19–37.

Bharath, P., Gahir, S. & Raghavendra, A.S. (2021). Abscisic acid-induced stomatal closure: An important component of plant defense against abiotic and biotic stress. Frontiers in Plant Science 12, 615114.

Bhardwaj, A.K., Chandra, K.K. & Kumar, R. (2023). Water stress changes on AMF colonization, stomatal conductance and photosynthesis of Dalbergia sissoo seedlings grown in entisol soil under nursery condition. Forest Science and Technology 19(1), 47–58.

Bhat, G. (2025). Dry Land Horticulture: Thriving in Tough Conditions. Educohack Press.

Bhatla, S.C. & Lal, M.A. (2023). Jasmonic acid. *In*: Plant Physiology, Development and Metabolism (pp. 473–478). Singapore: Springer Nature Singapore.

Bind, S., Bind, S., Sharma, A.K. & Chaturvedi, P. (2022). Epigenetic modification: A key tool for secondary metabolite production in microorganisms. Frontiers in Microbiology 13, 784109.

Bisht, A. & Garg, N. (2024). Harnessing the role of arbuscular mycorrhizae in arresting nodular senescence by modulating osmolyte synthesis and ascorbate-glutathione pool in cadmium stressed pigeon pea. Plant Growth Regulation 102(2), 409–427.

Bita, C.E. & Gerats, T. (2013). Plant tolerance to high temperature in a changing environment: Scientific fundamentals and production of heat stress-tolerant crops. Frontiers in Plant Science 4, 273. https://doi.org/10.3389/fpls.2013.00273

Bjerring Jensen, N., Vrobel, O., Akula Nageshbabu, N., De Diego, N., Tarkowski, P., Ottosen, C.O. & Zhou, R. (2024). Stomatal effects and ABA metabolism mediate differential regulation of leaf and flower cooling in tomato cultivars exposed to heat and drought stress. Journal of Experimental Botany 75(7), 2156–2175.

Boyno, G. & Demir, S. (2022). Plant-mycorrhiza communication and mycorrhizae in inter-plant communication. Symbiosis 86(2), 155–168.

Boyno, G., Ansari, R.A., Durak, E.D., Güneş, H., Çevik, R. & Demir, S. (2024). Arbuscular mycorrhizal technology in sustainable agriculture: Current knowledge and challenges in agroforestry. Mycorrhizal Symbiosis and Agroecosystem Restoration 173–195.

Boyno, G., Demir, S. & Danesh, Y.R. (2022). Effects of some biological agents on the growth and biochemical parameters of tomato plants infected with Alternaria solani (Ellis & Martin) Sorauer. European Journal of Plant Pathology 162(1), 19–29.

Boyno, G., Rezaee Danesh, Y., Demir, S., Teniz, N., Mulet, J.M. & Porcel, R. (2023). The complex interplay between arbuscular mycorrhizal fungi and strigolactone: Mechanisms, sinergies, applications and future directions. International Journal of Molecular Sciences 24(23), 16774.

Branco, S., Schauster, A., Liao, H.L. & Ruytinx, J. (2022). Mechanisms of stress tolerance and their effects on the ecology and evolution of mycorrhizal fungi. New Phytologist 235(6), 2158–2175.

Ceylan, Y., Altunoglu, Y.C. & Horuz, E. (2023). HSF and HSP gene families in sunflower: A comprehensive genome-wide determination survey and expression patterns under abiotic stress conditions. Protoplasma 260(6), 1473–1491.

Cha, J.Y., Kang, S.H., Ali, I., Lee, S.C., Ji, M.G., Jeong, S.Y., ... & Kim, W.Y. (2020). Humic acid enhances heat stress tolerance via transcriptional activation of Heat-Shock Proteins in Arabidopsis. Scientific Reports, 10(1), 15042.

Chafai, W., Haddioui, K., Serghini-Caid, H., Labazi, H., AlZain, M.N., Noman, O., ... & Khalid, A. (2023). Impact of arbuscular mycorrhizal fungal strains isolated from soil on the growth, yield, and fruit quality of tomato plants under different fertilization regimens. Horticulturae 9(9), 973.

Chandrasekaran, M. (2022). Arbuscular mycorrhizal fungi mediated alleviation of drought stress via non-enzymatic antioxidants: A meta-analysis. Plants 11(19), 2448.

Chen, W., Niu, T., Lian, W., Ye, T., Sun, Q. & Zhang, J. (2024). Involvement of endogenous IAA and ABA in the regulation of arbuscular mycorrhizal fungus on rooting of tea plant (*Camellia sinensis* L.) cuttings. BMC Plant Biology 24(1), 1266.

Cheng, Z., Luan, Y., Meng, J., Sun, J., Tao, J. & Zhao, D. (2021). WRKY transcription factor response to high-temperature stress. Plants 10(10), 2211.

Choudhary, A., Kumar, A. & Kaur, N. (2020). ROS and oxidative burst: Roots in plant development. Plant Diversity 42(1), 33–43.

Cicatelli, A., Todeschini, V., Lingua, G., Biondi, S., Torrigiani, P. & Castiglione, S. (2014). Epigenetic control of heavy metal stress response in mycorrhizal versus non-mycorrhizal poplar plants. Environmental Science and Pollution Research 21, 1723–1737.

Corpas, F.J., Gupta, D.K. & Palma, J.M. (2015). Production sites of reactive oxygen species (ROS) in organelles from plant cells. Reactive Oxygen Species and Oxidative Damage in Plants Under Stress, 1–22.

Couto, N., Wood, J. & Barber, J. (2016). The role of glutathione reductase and related enzymes on cellular redox homoeostasis network. Free Radical Biology and Medicine, 95, 27–42.

da Silva, K.J., Fernandes, J.A., Magurno, F., Leandro, L., Goto, B.T. & Theodoro, R.C. (2022). Phylogenetic review of Acaulospora (Diversisporales, Glomeromycota) and the homoplasic nature of its ornamentations. Journal of Fungi 8(9), 892.

Dar, N.A., Amin, I., Wani, W., Wani, S.A., Shikari, A.B., Wani, S.H. & Masoodi, K.Z. (2017). Abscisic acid: A key regulator of abiotic stress tolerance in plants. Plant Gene, 11, 106–111.

Demir, S., Danesh, Y.R., Boyno, G. & Najafi, S. (2022). Arbuscular mycorrhizal fungi in biotic and abiotic stress conditions: Function and management in horticulture. Sustainable Horticulture, 157–183.

Denzer, L., Schroten, H. & Schwerk, C. (2020). From gene to protein—How bacterial virulence factors manipulate host gene expression during infection. International Journal of Molecular Sciences 21(10), 3730.

Devireddy, A.R., Tschaplinski, T.J., Tuskan, G.A., Muchero, W. & Chen, J.G. (2021). Role of reactive oxygen species and hormones in plant responses to temperature changes. International Journal of Molecular Sciences 22(16), 8843.

Domínguez, R., Pateiro, M., Munekata, P.E., Zhang, W., Garcia-Oliveira, P., Carpena, M., ... & Lorenzo, J.M. (2021). Protein oxidation in muscle foods: A comprehensive review. Antioxidants 11(1), 60.

Duc, N.H. & Posta, K. (2018). Inoculation with Septoglomus constrictum improves tolerance to heat shock in tomato plants. COLUMELLA – Journal of Agricultural and Environmental Sciences 5(2), 7–14.

Edge, R. & Truscott, T.G. (2021). The reactive oxygen species singlet oxygen, hydroxy radicals, and the superoxide radical anion—Examples of their roles in biology and medicine. Oxygen 1(2), 77–95.

EL Sabagh, A., Islam, M.S., Hossain, A., Iqbal, M.A., Mubeen, M., Waleed, M., ... & Abdelhamid, M.T. (2022). Phytohormones as growth regulators during abiotic stress tolerance in plants. Frontiers in Agronomy 4, 765068.

Fahad, S., Bajwa, A.A., Nazir, U., Anjum, S.A., Farooq, A., Zohaib, A., ... & Huang, J. (2017). Crop production under drought and heat stress: Plant responses and management options. Frontiers in Plant Science 8, 1147.

Ferreyra, M.L.F., Serra, P. & Casati, P. (2021). Recent advances on the roles of flavonoids as plant protective molecules after UV and high light exposure. Physiologia Plantarum 173(3), 736–749.

Food and Agriculture Organization of the United Nations (FAO). (2021). The state of food security and nutrition in the world 2021. https://www.fao.org/publications/sofi/2021/en/

Fotovvat, M., Najafi, F., Khavari-Nejad, R.A., Talei, D. & Rejali, F. (2024). Investigating the simultaneous effect of chitosan and arbuscular mycorrhizal fungi on growth, phenolic compounds, PAL enzyme activity and lipid peroxidation in Salvia nemorosa L. Plant Physiology and Biochemistry 210, 108617.

Fuentes, A., Almonacid, L., Ocampo, J.A. & Arriagada, C. (2016). Synergistic interactions between a saprophytic fungal consortium and Rhizophagus irregularis alleviate oxidative stress in plants grown in heavy metal contaminated soil. Plant and Soil 407, 355–366.

Gadkar, V. & Rillig, M.C. (2006). The arbuscular mycorrhizal fungal protein glomalin is a putative homolog of heat shock protein 60. FEMS Microbiology Letters 263(1), 93–101.

Gahir, S., Bharath, P. & Raghavendra, A.S. (2021). Stomatal closure sets in motion long-term strategies of plant defense against microbial pathogens. Frontiers in Plant Science 12, 761952.

Ganugi, P., Masoni, A., Pietramellara, G. & Benedettelli, S. (2019). A review of studies from the last twenty years on plant–arbuscular mycorrhizal fungi associations and their uses for wheat crops. Agronomy 9(12), 840.

Gavito, M.E. & Azcón–Aguilar, C. (2012). Temperature stress in arbuscular mycorrhizal fungi: a test for adaptation to soil temperature in three isolates of Funneliformis mosseae from different climates. Agricultural and Food Science 21(1), 2–11.

Gong, M., You, X. & Zhang, Q. (2015). Effects of Glomus intraradices on the growth and reactive oxygen metabolism of foxtail millet under drought. Annals of Microbiology 65, 595–602.

González-Morales, S., Solís-Gaona, S., Valdés-Caballero, M.V., Juárez-Maldonado, A., Loredo-Treviño, A. & Benavides-Mendoza, A. (2021). Transcriptomics of biostimulation of plants under abiotic stress. Frontiers in Genetics 12, 583888.

Haddidi, I., Duc, N.H., Tonk, S., Rápó, E. & Posta, K. (2020). Defense enzymes in mycorrhizal tomato plants exposed to combined drought and heat stresses. Agronomy 10(11), 1657.

Haider, S., Iqbal, J., Naseer, S., Yaseen, T., Shaukat, M., Bibi, H., ... & Mahmood, T. (2021). Molecular mechanisms of plant tolerance to heat stress: Current landscape and future perspectives. Plant Cell Reports, 1–25.

Haider, S., Raza, A., Iqbal, J., Shaukat, M. & Mahmood, T. (2022). Analyzing the regulatory role of heat shock transcription factors in plant heat stress tolerance: A brief appraisal. Molecular Biology Reports 49(6), 5771–5785.

Hajam, Y.A., Lone, R. & Kumar, R. (2023). Role of plant phenolics against reactive oxygen species (ROS) induced oxidative stress and biochemical alterations. In: Plant Phenolics in Abiotic Stress Management (pp. 125–147). Singapore: Springer Nature Singapore.

Hasanuzzaman, M., Nahar, K., Alam, M.M., Roychowdhury, R. & Fujita, M. (2013). Physiological, biochemical, and molecular mechanisms of heat stress tolerance in plants. International Journal of Molecular Sciences 14(5), 9643–9684. https://doi.org/10.3390/ijms14059643

He, J.D., Zou, Y.N., Wu, Q.S. & Kuča, K. (2020). Mycorrhizas enhance drought tolerance of trifoliate orange by enhancing activities and gene expression of antioxidant enzymes. Scientia Horticulturae 262, 108745.

He, M. & Ding, N.Z. (2020). Plant unsaturated fatty acids: Multiple roles in stress response. Frontiers in Plant Science 11, 562785.

Huang, S., Gill, S., Ramzan, M., Ahmad, M.Z., Danish, S., Huang, P., ... & Alharbi, S.A. (2023). Uncovering the impact of AM fungi on wheat nutrient uptake, ion homeostasis, oxidative stress, and antioxidant defense under salinity stress. Scientific Reports 13(1), 8249.

Ighodaro, O.M. & Akinloye, O.A. (2018). First line defence antioxidants-superoxide dismutase (SOD), catalase (CAT) and glutathione peroxidase (GPX): Their fundamental role in the entire antioxidant defence grid. Alexandria Journal of Medicine 54(4), 287–293.

Intergovernmental Panel on Climate Change (IPCC). (2021). Climate Change 2021: The Physical Science Basis. Contribution of Working Group I to the Sixth Assessment Report of the Intergovernmental Panel on Climate Change. Cambridge University Press. https://www.ipcc.ch/report/ar6/wg1/

Irato, P. & Santovito, G. (2021). Enzymatic and non-enzymatic molecules with antioxidant function. Antioxidants 10(4), 579.

Izuhara, T., Kaihatsu, I., Jimbo, H., Takaichi, S. & Nishiyama, Y. (2020). Elevated levels of specific carotenoids during acclimation to strong light protect the repair of photosystem II in Synechocystis sp. PCC 6803. Frontiers in Plant Science 11, 1030.

Jeyachandran, S., Chellapandian, H., Park, K. & Kwak, I.S. (2023). A review on the involvement of heat shock proteins (extrinsic chaperones) in response to stress conditions in aquatic organisms. Antioxidants 12(7), 1444.

Jha, S.S. & Songachan, L.S. (2023). Exploring the potential of arbuscular mycorrhizal fungi as a biofertilizer. Studies in Fungi 8(1).

Jian, P., Zha, Q., Hui, X., Tong, C. & Zhang, D. (2024). Research progress of arbuscular mycorrhizal fungi ımproving plant resistance to temperature stress. Horticulturae 10(8), 855.

Kabato, W., Getnet, G.T., Sinore, T., Nemeth, A. & Molnár, Z. (2025). Towards climate-smart agriculture: Strategies for sustainable agricultural production, food security, and greenhouse gas reduction. Agronomy 15(3), 565.

Kandalgaonkar, K.N. & Barvkar, V.T. (2025). Intricate phytohormonal orchestration mediates mycorrhizal symbiosis and stress tolerance. Mycorrhiza 35(2), 1–20.

Khan, Y., Shah, S. & Tian, H. (2022). The roles of arbuscular mycorrhizal fungi in influencing plant nutrients, photosynthesis, and metabolites of cereal crops—A review. Agronomy 12(9), 2191.

Khan, Z. & Shahwar, D. (2020). Role of heat shock proteins (HSPs) and heat stress tolerance in crop plants. Sustainable Agriculture in the Era of Climate Change 211–234.

Kolupaev, Y.E., Yastreb, T.O., Ryabchun, N.I., Yemets, A.I., Dmitriev, O.P. & Blume, Y.B. (2023). Cellular mechanisms of the formation of plant adaptive responses to high temperatures. Cytology and Genetics 57(1), 55–75.

Kozuleva, M. (2022). Recent advances in the understanding of superoxide anion radical formation in the photosynthetic electron transport chain. Acta Physiologiae Plantarum 44(9), 92.

Lang, B.J., Guerrero, M.E., Prince, T.L., Okusha, Y., Bonorino, C. & Calderwood, S.K. (2021). The functions and regulation of heat shock proteins; Key orchestrators of proteostasis and the heat shock response. Archives of Toxicology 95(6), 1943–1970.

Lapenna, D. (2023). Glutathione and glutathione-dependent enzymes: From biochemistry to gerontology and successful aging. Ageing Research Reviews 92, 102066.

Lesk, C., Rowhani, P. & Ramankutty, N. (2016). Influence of extreme weather disasters on global crop production. Nature 529(7584), 84–87. https://doi.org/10.1038/nature16467

Li, J., Liang, W., He, X.Q. & Qian, W. (2025). DNA damage triggers heritable alterations in DNA methylation patterns in Arabidopsis. Molecular Plant 18(3), 501–512.

Li, X. (2024). The role of Funneliformis mosseae symbiosis on cotton plants under lead toxicity: Molecular and physiological aspects. Russian Journal of Plant Physiology 71(3), 83.

Li, Y., Gao, Z., Lu, J., Wei, X., Qi, M., Yin, Z. & Li, T. (2022). SlSnRK2. 3 interacts with SlSUI1 to modulate high temperature tolerance via Abscisic acid (ABA) controlling stomatal movement in tomato. Plant Science 321, 111305.

Li, Z. & Ahammed, G.J. (2023). Salicylic acid and jasmonic acid in elevated CO_2-induced plant defense response to pathogens. Journal of Plant Physiology 286, 154019.

Lin, K.F., Tsai, M.Y., Lu, C.A., Wu, S.J. & Yeh, C.H. (2018). The roles of Arabidopsis HSFA2, HSFA4a, and HSFA7a in the heat shock response and cytosolic protein response. Botanical Studies 59, 1–9.

Liu, T., Hu, X., Zhang, J., Zhang, J., Du, Q. & Li, J. (2018). H_2O_2 mediates ALA-induced glutathione and ascorbate accumulation in the perception and resistance to oxidative stress in Solanum lycopersicum at low temperatures. BMC Plant Biology 18, 1–10.

Liu, X.R., Rong, Z.Y., Tian, X., Hashem, A., Abd_Allah, E.F., Zou, Y.N. & Wu, Q.S. (2023). Mycorrhizal fungal effects on plant growth, osmolytes, and CsHsp70s and CsPIPs expression in leaves of cucumber under a short-term heat stress. Plants 12(16), 2917.

Lorenzo, I., Serra-Prat, M. & Yébenes, J.C. (2019). The role of water homeostasis in muscle function and frailty: a review. Nutrients 11(8), 1857.

Luengo, T.M., Mayer, M.P. & Rüdiger, S.G. (2019). The Hsp70–Hsp90 chaperone cascade in protein folding. Trends in Cell Biology 29(2), 164–177.

Luo, H., Xu, H., Chu, C., He, F. & Fang, S. (2020). High temperature can change root system architecture and intensify root interactions of plant seedlings. Frontiers in Plant Science 11, 160.

Ma, W.Y., Qin, Q.Y., Zou, Y.N., Kuča, K., Giri, B., Wu, Q.S., ... & Xu, Y.J. (2022). Arbuscular mycorrhiza induces low oxidative burst in drought-stressed walnut through activating antioxidant defense systems and heat shock transcription factor expression. Frontiers in Plant Science 13, 1089420.

Malini, M.K., Karwa, S., Priyadarsini, P., Kumar, P., Nagar, S., Kumar, M., ... & Pal, M. (2023). Abscisic-acid-modulated stomatal conductance governs high-temperature stress tolerance in rice accessions. Agriculture 13(3), 545.

Mandal, M., Sarkar, M., Khan, A., Biswas, M., Masi, A., Rakwal, R., ... & Sarkar, A. (2022). Reactive Oxygen Species (ROS) and Reactive Nitrogen Species (RNS) in plants – Maintenance of structural individuality and functional blend. Advances in Redox Research 5, 100039.

Manteghi, A., Danesh, Y.R., Hesar, A.M., Demir, S., Boyno, G., Catani, L. & Semprucci, F. (2022). Is the arbuscular mycorrhizal fungus Funneliformis mosseae a suitable agent to control criconematid populations? Diversity 14(11), 898.

Mathur, S. & Jajoo, A. (2020). Arbuscular mycorrhizal fungi protects maize plants from high temperature stress by regulating photosystem II heterogeneity. Industrial Crops and Products, 143, 111934.

Mathur, S., Agnihotri, R., Sharma, M.P., Reddy, V.R. & Jajoo, A. (2021). Effect of high-temperature stress on plant physiological traits and mycorrhizal symbiosis in maize plants. Journal of Fungi 7(10), 867.

Mathur, S., Sharma, M.P. & Jajoo, A. (2018). Improved photosynthetic efficacy of maize (Zea mays) plants with arbuscular mycorrhizal fungi (AMF) under high temperature stress. Journal of Photochemistry and Photobiology B: Biology 180, 149–154.

Mathur, S., Tomar, R.S. & Jajoo, A. (2019). Arbuscular mycorrhizal fungi (AMF) protects photosynthetic apparatus of wheat under drought stress. Photosynthesis Research 139, 227–238.

McCaw, B.A., Stevenson, T.J. & Lancaster, L.T. (2020). Epigenetic responses to temperature and climate. Integrative and Comparative Biology 60(6), 1469–1480.

Medica, A.J., Swegen, A., Seifi-Jamadi, A., McIntosh, K. & Gibb, Z. (2025). Catalase in unexpected places: Revisiting H_2O_2 detoxification pathways in Stallion spermatozoa. Antioxidants 14(6), 718.

Meena, K.K., Sorty, A.M., Bitla, U.M., Choudhary, K., Gupta, P., Pareek, A., ... & Minhas, P.S. (2017). Abiotic stress responses and microbe-mediated mitigation in plants: The omics strategies. Frontiers in Plant Science 8, 172.

Mesa, T. & Munné-Bosch, S. (2023). α-Tocopherol in chloroplasts: nothing more than an antioxidant? Current Opinion in Plant Biology 74, 102400.

Mishra, N., Jiang, C., Chen, L., Paul, A., Chatterjee, A. & Shen, G. (2023). Achieving abiotic stress tolerance in plants through antioxidative defense mechanisms. Frontiers in Plant Science 14, 1110622.

Mittler, R., Zandalinas, S.I., Fichman, Y. & Van Breusegem, F. (2022). Reactive oxygen species signalling in plant stress responses. Nature Reviews: Molecular Cell Biology, 23(10), 663–679.

Mittu, B., Bhat, Z.R., Chauhan, A., Kour, J., Behera, A. & Kaur, M. (2022). Ascorbic acid. *In*: Nutraceuticals and Health Care (pp. 289–302). Academic Press.

Moreno, H.D., Rokitta, S., Tremblay, N., Boersma, M., Groß, E., Klip, H.C., ... & Meunier, C.L. (2024). Higher temperature, increased CO_2, and changing nutrient ratios alter the carbon metabolism and induce oxidative stress in a cosmopolitan diatom. Limnology and Oceanography 69(1), 121–139.

Mylostyva, D., Prudnikov, V., Kolisnyk, O., Lykhach, A., Begma, N., Kalinichenko, O., ... & Mylostyvyi, R. (2021). Biochemical changes during heat stress in productive animals with an emphasis on the antioxidant defense system. Journal of Animal Behaviour and Biometeorology 10(1).

Nakai, K. & Tsuruta, D. (2021). What are reactive oxygen species, free radicals, and oxidative stress in skin diseases? International Journal of Molecular Sciences 22(19), 10799.

Napolitano, G., Fasciolo, G. & Venditti, P. (2021). Mitochondrial management of reactive oxygen species. Antioxidants 10(11), 1824.

Neligan, P.J. (2021). Fluid and electrolyte balance. Anaesthesia & Intensive Care Medicine 22(3), 169–173.

Nie, W., He, Q., Guo, H., Zhang, W., Ma, L., Li, J. & Wen, D. (2024). Arbuscular mycorrhizal fungi: Boosting crop resilience to environmental stresses. Microorganisms 12(12), 2448.

Noctor, G. (2025). Hydrogen peroxide, ascorbate, and glutathione: Building the Foyer–Halliwell–Asada pathway. Planta 261(6), 1–11.

Ovejero, S., Bueno, A. & Sacristán, M.P. (2020). Working on genomic stability: From the S-phase to mitosis. Genes, 11(2), 225.

Ozougwu, J.C. (2016). The role of reactive oxygen species and antioxidants in oxidative stress. International Journal of Research 1(8), 1–8.

Pareek, A., Mishra, D., Rathi, D., Verma, J.K., Chakraborty, S. & Chakraborty, N. (2021). The small heat shock proteins, chaperonin 10, in plants: An evolutionary view and emerging functional diversity. Environmental and Experimental Botany 182, 104323.

Peña, R., Robbins, C., Corella, J.C., Thuita, M., Masso, C., Vanlauwe, B., ... & Sanders, I.R. (2020). Genetically different isolates of the arbuscular mycorrhizal fungus Rhizophagus irregularis induce differential responses to stress in cassava. Frontiers in Plant Science 11, 596929.

Pospíšil, P. (2016). Production of reactive oxygen species by photosystem II as a response to light and temperature stress. Frontiers in Plant Science 7, 1950.

Qu, Y., Mueller-Cajar, O. & Yamori, W. (2023). Improving plant heat tolerance through modification of Rubisco activase in C3 plants to secure crop yield and food security in a future warming world. Journal of Experimental Botany 74(2), 591–599.

Rajan, R., Ahmed, S., Sharma, N., Kumar, N., Debas, A. & Matsumura, K. (2021). Review of the current state of protein aggregation inhibition from a materials chemistry perspective: Special focus on polymeric materials. Materials Advances 2(4), 1139–1176.

Rajput, V.D., Harish, Singh, R.K., Verma, K.K., Sharma, L., Quiroz-Figueroa, F.R., ... & Mandzhieva, S. (2021). Recent developments in enzymatic antioxidant defence mechanism in plants with special reference to abiotic stress. Biology 10(4), 267.

Raza, A., Charagh, S., Abbas, S., Hassan, M.U., Saeed, F., Haider, S., ... & Varshney, R.K. (2023). Assessment of proline function in higher plants under extreme temperatures. Plant Biology 25(3), 379–395.

Riaz, M., Zia-Ul-Haq, M. & Dou, D. (2021). Chemistry of carotenoids. Carotenoids: Structure and Function in the Human Body, 43–76.

Romero-Munar, A., Aroca, R., Zamarreño, A.M., García-Mina, J.M., Perez-Hernández, N. & Ruiz-Lozano, J.M. (2023). Dual inoculation with Rhizophagus irregularis and Bacillus megaterium improves maize tolerance to combined drought and high temperature stress by enhancing root hydraulics, photosynthesis and hormonal responses. International Journal of Molecular Sciences 24(6), 5193.

Rudenko, N.N., Vetoshkina, D.V., Marenkova, T.V. & Borisova-Mubarakshina, M.M. (2023). Antioxidants of non-enzymatic nature: their function in higher plant cells and the ways of boosting their biosynthesis. Antioxidants 12(11), 2014.

Sachdev, S., Ansari, S.A. & Ansari, M.I. (2023). Antioxidant defensive mechanisms to regulate cellular redox homeostatic balance. *In*: Reactive Oxygen Species in Plants: The Right Balance (pp. 143–172). Singapore: Springer Nature Singapore.

Sadiq, I.Z. (2023). Free radicals and oxidative stress: Signaling mechanisms, redox basis for human diseases, and cell cycle regulation. Current Molecular Medicine 23(1), 13–35.

Sahu, P.K., Jayalakshmi, K., Tilgam, J., Gupta, A., Nagaraju, Y., Kumar, A., ... & Rajawat, M.V.S. (2022). ROS generated from biotic stress: Effects on plants and alleviation by endophytic microbes. Frontiers in Plant Science 13, 1042936.

Saini, K., Dwivedi, A. & Ranjan, A. (2022). High temperature restricts cell division and leaf size by coordination of PIF4 and TCP4 transcription factors. Plant Physiology 190(4), 2380–2397.

Savicka, M. & Škute, N. (2010). Effects of high temperature on malondialdehyde content, superoxide production and growth changes in wheat seedlings (Triticum aestivum L.). Ekologija 56(1), 26–33.

Sharma, S.K., Singh, D., Pandey, H., Jatav, R.B., Singh, V. & Pandey, D. (2022). An overview of roles of enzymatic and nonenzymatic antioxidants in plant. Antioxidant Defense in Plants: Molecular Basis of Regulation 1–13.

Singh, A. & Satheeshkumar, P.K. (2024). Reactive oxygen species (ROS) and ROS scavengers in plant abiotic stress response. *In*: Stress Biology in Photosynthetic Organisms: Molecular Insights and Cellular Responses (pp. 41–63). Singapore: Springer Nature Singapore.

Singh, D., Mathur, S. & Ranjan, R. (2023). Roles of Strigolactone in plant growth and development. *In*: Strigolactones, Alkamides and Karrikins in Plants (pp. 19–33). CRC Press.

Singh, M., Kumar, J., Singh, S., Singh, V.P. & Prasad, S.M. (2015). Roles of osmoprotectants in improving salinity and drought tolerance in plants: A review. Reviews in Environmental Science and Bio/Technology 14, 407–426.

Sousa, B., Soares, C., Sousa, F., Martins, M., Mateus, P., Rodrigues, F., ... & Fidalgo, F. (2024). Enhancing tomato plants' tolerance to combined heat and salt stress – The role of arbuscular mycorrhizae and biochar. Science of The Total Environment 948, 174860.

Srivastava, A.K., Singh, R.D., Pandey, G.K., Mukherjee, P.K. & Foyer, C.H. (2025). Unravelling the molecular dialogue of beneficial microbe-plant interactions. Plant, Cell & Environment 48(4), 2534–2548.

Sun, D., Shang, X., Cao, H., Lee, S.J., Wang, L., Gan, Y. & Feng, S. (2024). Physio-biochemical mechanisms of arbuscular mycorrhizal fungi enhancing plant resistance to abiotic stress. Agriculture 14(12), 2361.

Suzuki, N. (2023). Fine tuning of ROS, redox and energy regulatory systems associated with the functions of chloroplasts and mitochondria in plants under heat stress. International Journal of Molecular Sciences 24(2), 1356.

Szentpéteri, V., Virág, E., Mayer, Z., Duc, N.H., Hegedűs, G. & Posta, K. (2024). First peek into the transcriptomic response in heat-stressed tomato ınoculated with Septoglomus constrictum. Plants 13(16), 2266.

Takio, N., Yadav, M. & Yadav, H.S. (2021). Catalase-mediated remediation of environmental pollutants and potential application: A review. Biocatalysis and Biotransformation 39(6), 389–407.

Tang, M., Xu, L., Wang, Y., Dong, J., Zhang, X., Wang, K., ... & Liu, L. (2021). Melatonin-induced DNA demethylation of metal transporters and antioxidant genes alleviates lead stress in radish plants. Horticulture Research, 8.

Tian, X., Liu, X.Q., Liu, X.R., Li, Q.S., Abd_Allah, E.F. & Wu, Q. S. (2023). Mycorrhizal cucumber with Diversispora versiformis has active heat stress tolerance by up-regulating expression of both CsHsp70s and CsPIPs genes. Scientia Horticulturae 319, 112194.

Tuo, X.Q., Li, S., Wu, Q.S. & Zou, Y.N. (2015). Alleviation of waterlogged stress in peach seedlings inoculated with Funneliformis mosseae: Changes in chlorophyll and proline metabolism. Scientia Horticulturae 197, 130–134.

Valgimigli, L. (2023). Lipid peroxidation and antioxidant protection. Biomolecules 13(9), 1291.

Vašková, J., Kočan, L., Vaško, L. & Perjési, P. (2023). Glutathione-related enzymes and proteins: A review. Molecules 28(3), 1447.

Wahab, A., Muhammad, M., Munir, A., Abdi, G., Zaman, W., Ayaz, A., ... & Reddy, S.P.P. (2023). Role of arbuscular mycorrhizal fungi in regulating growth, enhancing productivity, and potentially influencing ecosystems under abiotic and biotic stresses. Plants 12(17), 3102.

Wang, G., Xing, M., Hu, T., Ji, M., Li, X., Amombo, E., ... & Fu, J. (2022). Photosystem II photochemical adjustment of tall fescue against heat stress after melatonin priming. Journal of Plant Physiology 275, 153758.

Wang, Q. & Zhu, Z. (2023). Transcription factors in the regulation of plant heat responses. Critical Reviews in Plant Sciences 42(6), 385–398.

Wang, Y., Cai, S., Yin, L., Shi, K., Xia, X., Zhou, Y., ... & Zhou, J. (2015). Tomato HsfA1a plays a critical role in plant drought tolerance by activating ATG genes and inducing autophagy. Autophagy 11(11), 2033–2047.

Waters, E.R. & Vierling, E. (2020). Plant small heat shock proteins – Evolutionary and functional diversity. New Phytologist 227(1), 24–37.

Wijewardene, I., Shen, G. & Zhang, H. (2021). Enhancing crop yield by using Rubisco activase to improve photosynthesis under elevated temperatures. Stress Biology 1(1), 2.

Wu, Q.S., Zou, Y.N. & Abd-Allah, E.F. (2014). Mycorrhizal association and ROS in plants. In: Oxidative Damage to Plants (pp. 453–475). Academic Press.

Xiao, M., Li, Z., Zhu, L., Wang, J., Zhang, B., Zheng, F., ... & Zhang, Z. (2021). The multiple roles of ascorbate in the abiotic stress response of plants: Antioxidant, cofactor, and regulator. Frontiers in Plant Science 12, 598173.

Yang, L., Li, W.C., Fu, F.L., Qu, J., Sun, F., Yu, H. & Zhang, J. (2022). Characterization of phenylalanine ammonia-lyase genes facilitating flavonoid biosynthesis from two species of medicinal plant Anoectochilus. PeerJ 10, e13614.

Yang, Y., Han, X., Liang, Y., Ghosh, A., Chen, J. & Tang, M. (2015). The combined effects of arbuscular mycorrhizal fungi (AMF) and lead (Pb) stress on Pb accumulation, plant growth parameters, photosynthesis, and antioxidant enzymes in Robinia pseudoacacia L. PloS ONE 10(12), e0145726.

Zahra, N., Hafeez, M.B., Ghaffar, A., Kausar, A., Al Zeidi, M., Siddique, K.H. & Farooq, M. (2023). Plant photosynthesis under heat stress: Effects and management. Environmental and Experimental Botany 206, 105178.

Zandi, P. & Schnug, E. (2022). Reactive oxygen species, antioxidant responses and implications from a microbial modulation perspective. Biology 11(2), 155.

Zechmann, B. (2020). Subcellular roles of glutathione in mediating plant defense during biotic stress. Plants 9(9), 1067.

Zhang, L., Chang, Q., Hou, X., Wang, J., Chen, S., Zhang, Q., ... & Liu, J. (2023). The effect of high-temperature stress on the physiological indexes, chloroplast ultrastructure, and photosystems of two herbaceous peony cultivars. Journal of Plant Growth Regulation 42(3), 1631–1646.

Zhao, B., Liu, Q., Wang, B. & Yuan, F. (2021). Roles of phytohormones and their signaling pathways in leaf development and stress responses. Journal of Agricultural and Food Chemistry 69(12), 3566–3584.

Zhu, T., Fonseca De Lima, C.F. & De Smet, I. (2021). The heat is on: How crop growth, development, and yield respond to high temperature. Journal of Experimental Botany 72(21), 7359–7373.

Zhu, X., Song, F. & Liu, F. (2017). Arbuscular mycorrhizal fungi and tolerance of temperature stress in plants. *In*: Wu, Q.S. [ed.). Arbuscular Mycorrhizas and Stress Tolerance of Plants. Springer, Singapore.

Zou, Y.N., Wu, Q.S. & Kuča, K. (2021). Unravelling the role of arbuscular mycorrhizal fungi in mitigating the oxidative burst of plants under drought stress. Plant Biology 23, 50–57.

CHAPTER
12

Vermicompost and Arbuscular Mycorrhizal Fungi: Can they Act in Synergism to Improve Phenolic Biosynthesis?

João Gabriel Lira de Carvalho[1,2], Eduarda Lins Falcão[1,2] and Fábio Sérgio Barbosa da Silva[1,2*]

[1] Laboratório de Análises, Pesquisas e Estudos em Micorrizas (LAPEM/UPE) – Centro de Pesquisas do Instituto de Ciências Biológicas, Universidade de Pernambuco, Rua Arnóbio Marques, 310, Santo Amaro – Recife, PE – Brazil, 50100-130

[2] Programa de Pós-graduação em Biologia Celular e Molecular Aplicada – ICB/UPE, Rua Arnóbio Marques, 310, Santo Amaro – Recife, PE – Brazil, 50100-130

1. Introduction

Current agricultural practices increasingly aim for methods that are less harmful to the environment (Jin et al., 2022), avoiding the use of excessive inorganic inputs into the soil (He et al., 2024). Nevertheless, chemical fertilizers are used every year (Baghel, 2024). Despite the essential need to reduce these practices, the global demand for agriculture must be met, especially due to the population increase (OECD/FAO, 2023).

Therefore, it is important to consider sustainable technologies that assist the producer of plant raw materials, which is one of the initial steps in the implementation of agriculturally friendly practices (Alonso-Martínez et al., 2024). Thus, this chapter discusses two alternatives: vermicompost and arbuscular mycorrhizal fungi (AMF) inoculation. Despite distinct methods of obtaining, both cultivation tools are relevant in enhancing the edaphic quality (Castellini et al., 2024; Falcão et al., 2023), the phytomass accumulation (Paymaneh et al., 2023), and the production of plant bioactive compounds, such as phenolic compounds (Carvalho et al., 2024).

In this context, the metabolites of phenolic origin are indispensable, considering that the market for these antioxidant compounds alone was worth around 4.6 billion dollars in 2024 (Pulidindi and Ahuja, 2024). Moreover, these metabolites can be present in food composition as antioxidant additives, preservatives, and flavoring agents (Zeb, 2020; Albuquerque et al., 2021). Additionally, the phenolic compounds

*Corresponding author: fabio.barbosa@upe.br

DOI: 10.1201/9781003669777-12

are also relevant in the pharmaceutical (Scott et al., 2022) and cosmetics industries (Bharadjava et al., 2023), segments that use extracts from plants that accumulate phenolic compounds. This need to use plant bioactive molecules reaffirms the search for biostimulants that guarantee the supply of sustainable plant raw material to the industrial sectors.

Therefore, the following topics will discuss the physiological importance of phenolic to plants, how the use of vermicompost and mycorrhizal biotechnology can improve the production of phenolic metabolites and whether the combined application of these supplies is a viable option for acquiring phytomass with an outstanding accumulation of total plant phenolic compounds, based on experimental results available in the literature.

2. Phenolic Compounds: Biosynthesis and Contribution to Plant Physiology

2.1 General Classification and Biosynthesis

Phenolic compounds are among the most diverse molecules in plant tissues and are found in all plant organs (e.g., roots, stems, flowers, leaves, and fruits) (Chen et al., 2018; Goufo et al., 2020; Feng et al., 2021; Malarz et al., 2023). These molecules have a benzene ring coupled with hydroxyls (Heleno et al., 2015). Their classification depends mainly on the number of hydroxyls and the conformation of aromatic rings (Saltveit, 2018). Thus, phenols can be sub-grouped as: phenolic acids (Rashmi and Negi, 2020), flavonoids (Panche et al., 2016), anthocyanins (Sunil and Shetty, 2022), proanthocyanidins (Dixon and Sarnala, 2020), tannins (Grundhöfer et al., 2001), coumarins (Stefanachi et al., 2018), lignins (Yoon et al., 2015), and stilbenes (Teka et al., 2022).

To be synthesized, the shikimate pathway has to be activated. For this, an aldolic-type condensation between phosphoenolpyruvate and erythrose-4-phosphate (Vogt, 2010), that, after hydroxylation, decarboxylation reactions, and molecular reorganization, results in the shikimic acid (Marchiosi et al., 2020). From this point, gallic acid can be biosynthesized, which further can give origin to tannins, polyphenols subclassified as gallotannins and ellagitannins (Gross, 2008). However, aromatic amino acids are necessary to form other phenolic-type metabolites, such as phenylalanine (Zhu et al., 2024). For this, the shikimic acid needs to undergo phosphorylation and incorporation of phosphoenolpyruvate to synthesize 5-enolpyruvylshikimate-3-phosphate, which is subsequently converted into chorismate by enzymatic action (Santos-Sánchez et al., 2019). Then, this latter molecule is rearranged to prephenate, which is decarboxylated and receives an amino group, producing the phenylamine (Hu et al., 2022).

After this step, a key enzyme responsible for initiating the pathway related to flavonoid and other phenolic compounds, phenylalanine ammonia lyase, deaminates the phenylalanine and cinnamic acid is formed (Sharma et al., 2024) and is hydroxylated to *p*-coumaric acid and converted into *p*-coumaroyl-CoA due to 4-coumarate CoA ligase action (Deng and Lu, 2017). This molecule is then condensed with three malonyl-CoA molecules, given chalcone synthase activity, producing a flavonoid-type structure: naringenin chalcone (Zhuang et al., 2023). The further steps in this pathway can produce subtypes of flavonoids, such as flavonols, flavones, dihydroflavonols, and isoflavonones (Zagoskina et al., 2023), depending on the aromatic ring saturation and hydroxyl groups (Lemos et al., 2019). Proanthocyanidins are derived from the condensation of flavan-3-ol and flavan-3,4-diol units (Yu et al., 2020; 2023).

It is worth mentioning that lignin, coumarins, and stilbenes are also derived from aromatic amino acids (Yoon et al., 2015; Stefanachi et al., 2017; Teka et al., 2022), and some examples of these phenolic subgroups are the monolignols (Peracchi et al., 2024), umbelliferone (Mazimba, 2017), and resveratrol (Giovinazzo et al., 2012), respectively. Reviews that describe the biosynthesis or relevance of these metabolites can be found in the papers written by Dixon and Barros (2019), Garg et al. (2020), and Teka et al. (2022).

Phenolic metabolites are accumulated in plant tissues for several reasons, playing a role not only in plant defense, and oxidative balance, which will be briefly discussed in the next sections.

2.2 Phenolics in Biotic and Abiotic Stress

Since plants can thrive in several environmental conditions, it is understandable that these organisms possess mechanisms, involving secondary metabolites, to overcome challenges related to herbivores (Divekar et al., 2022), viruses (Jiang et al., 2025), and pathogenic bacteria and fungi (Ramaroson et al., 2022).

In this perspective, when insects damage leaves, for example, some mechanisms related to phenol production come into play (Barbehenn and Constabel, 2011). For instance, high phenolic concentration can lead to toxicity and increase the mortality rate of herbivores, as seen in the research of Dixit et al. (2017), who observed phenol-induced production in *Gossypium* sp. after infestation with *Helicoverpa armigera* and *Spodoptera litura* larvae.

Notwithstanding, phenolics not only play a role against biotic stresses, but also help the plant to face harsh conditions caused by abiotic stresses such as high levels of ultraviolet rays (Jan et al., 2022), salinity (Linić et al., 2019), and drought (Kubra et al., 2021). They also improve the antioxidant status of the plant, since they are considered non-enzymatic antioxidants (Rudenko et al., 2023). Therefore, phenolics are essential to plant survival and possess industrial interest. The last two sections will discuss how vermicompost and mycorrhizal biostimulants may increase the biosynthesis of such relevant metabolites.

3. Why Does Vermicompost Increase the Phenolic Compounds Biosynthesis?

The application of organic fertilizers is well established in seedling production, with vermicompost standing out as a viable option considering its low cost and high sustainability (Mendes et al., 2012). This fertilizer is produced through vermicomposting, an alternative way of managing organic waste (Lirikum et al., 2022), whose decomposition is accelerated by earthworms (Enebe and Erasmus,

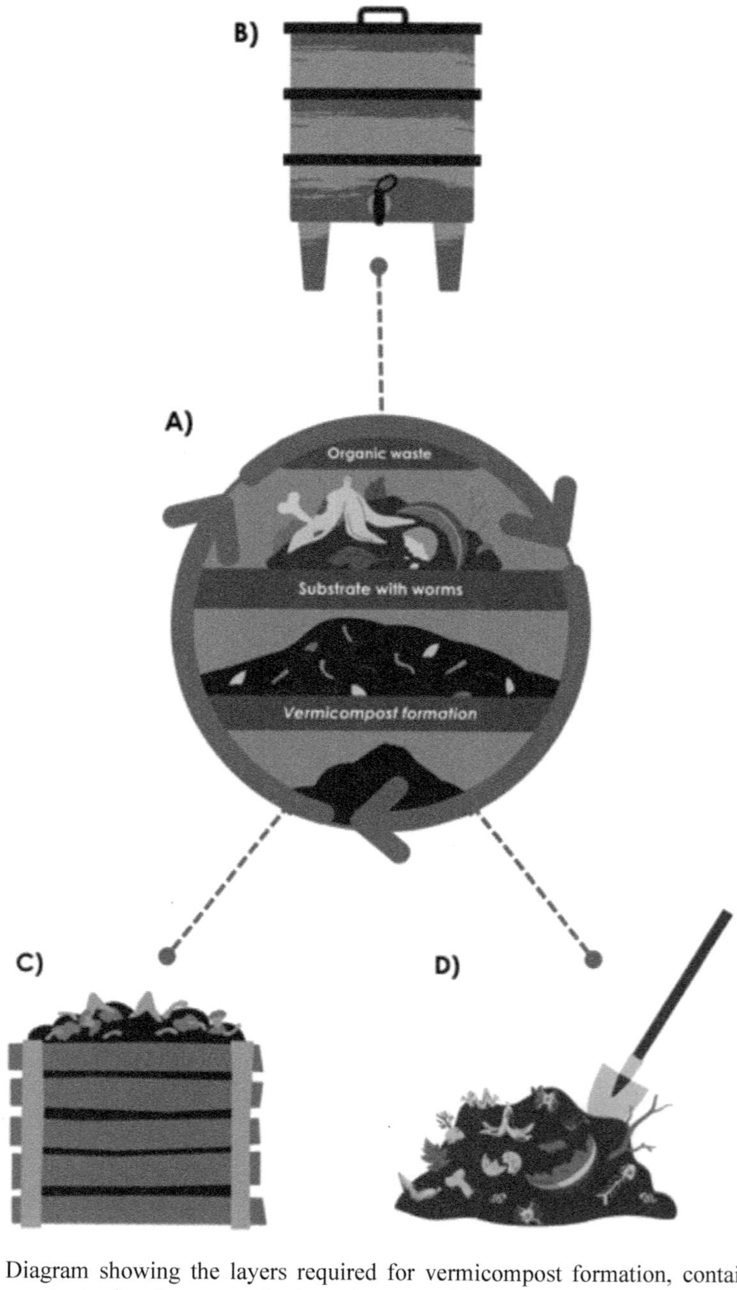

Fig. 1. Diagram showing the layers required for vermicompost formation, containing the organic waste to be decomposed, the substrate with earthworms and microorganisms, and the vermicompost formed (A), the types of vermicomposting systems, such as in compartmentalized boxes (B), in vertical wooden boxes (C), and on the ground (D) (Aquino et al., 2005; Mupambwa and Mnkeni, 2018; Lirikum et al., 2022; Enebe and Erasmus, 2023; Ratnasari et al., 2023). Figure made in Canva.com.

2023). In vermicompost production, it is important to consider some steps, such as the separation of waste materials (Chanu et al., 2018), e.g., domestic wastes, agricultural wastes, and sewage sludge (Maharjan et al., 2022; Vuković et al., 2021; Oyege and Balaji Bhaskar, 2023). Then, a layered bioreactor system is established, which can be done in a variety of ways, from compartmentalized boxes on vertical screens, or even on the ground in a bed-like shape (Aquino et al., 2005; Ratnasari et al., 2023). Regardless of the form or storage location, the system consists of one or more layers of organic waste to be decomposed, a substrate layer containing earthworms and microorganisms, which will digest the organic matter and mineralize the nutrients (Mupambwa and Mnkeni, 2018), and also a chamber to collect the vermicompost (Figure 1). At this stage, it is also possible to collect the leachate from the system, known as vermiwash, which can also be used as a liquid fertilizer (Awadhpersad et al., 2021; Gudeta et al., 2021). It is worth noting that for the technology to work properly, the layers must be fed, turned, and humidified periodically, promoting the aeration and humidity required by the system (Zhou et al., 2022).

In this context, vermicompost is obtained as the final product (Fig. 1), an organic substrate that can be used as a fertilizer (Bhat et al., 2018). This substrate can then be applied to the soil, promoting several benefits (Table 1), such as improving the porosity, aeration, water retention (Singh et al., 2020), the microbial community (Zhang et al., 2023), and nutrient availability (Zhao et al., 2020). The utilization of vermicompost in plant growth can also induce high rates of growth (Deepthi et al., 2021), photosynthetic activity (Toor et al., 2024), and metabolite content (Lahbouki et al., 2022a,b), producing a phytomass of quality (Table 1). These benefits are mostly due to the substrate composition, which has phytohormones (Zhang et al., 2024), besides providing organic matter and nutrients (Wako and Muleta, 2022), and improving soil attributes (Rehman et al., 2023). It is worth noting that it is possible to obtain vermicompost tea, which consists of a solution of the substrate and water, which can be applied directly to the plant (Zhou et al., 2022; Campana et al., 2025; Yin et al., 2025). However, the quality and effects of vermicompost by-products may depend on the vermicompost source (Zarei et al., 2018).

Table 1. Summary of the main benefits of vermicompost and by-products application in plant production

Benefit	References
Growth	Joshi et al., 2015; Raza et al., 2023
Nutrient mineralization and availability	Nurhidayati et al., 2018; Hoque et al., 2022
Improvement of soil attributes	Das et al., 2022
Abiotic stress tolerance	Toor et al., 2023; Vambe et al., 2023
Disease and pest control	Yatoo et al., 2021; Tikoria et al., 2023
Remediation of degraded and contaminated soil	Lin et al., 2016; Liu et al., 2019
Metabolite production	Souffront et al., 2022
Enhancement of soil biodiversity	Iqbal et al., 2024

4. How do Arbuscular Mycorrhizal Fungi Improve the Accumulation of Phenolic Compounds?

The mycorrhizal biostimulants are comprised of *Glomeromycota* representatives, obligate symbiotic fungi (Wijayawardene et al., 2024) capable of associating with more than 80% of land plants (Brundrett and Tedersoo, 2018). For symbiosis to occur properly, the environmental conditions of the soil must be suitable for germination of the glomerospore (Giovannetti, 2000; Camargo-Ricalde, 2002), starting the asymbiotic/pre-symbiotic phase (Koltai et al., 2009). After germination, the mycelium may find the root surface (Bonfante and Genre, 2010), where an appressorium for fixation and penetration by the fungus in the root will be formed (Harrison, 2005). This is followed by the symbiotic phase (Koltai et al., 2009), in which the fungi colonize the root cortex, and inter- and intracellular structures are formed (Piliarová et al., 2019). These intracellular structures are: the arbuscules, branched modified hyphae where the nutrient exchange between the partners occurs (Luginbuehl and Oldroyd, 2017), and the vesicles, storage structures on the intraradical mycelium present in some AMF families (Ai et al., 2022). After that, the production of extraradical mycelium begins (Delaeter et al., 2024), where new spores will be produced and dispersed in the soil (Bücking et al., 2012), maintaining the AMF life cycle (Fig. 2).

From this association, they may provide various benefits to plants (Table 2), such as improved nutrient uptake (Moreira et al., 2016), which can be expressed in plant growth parameters (Droh et al., 2016; Qiu et al., 2020), higher production and quality of fruit and flowers (Cao et al., 2021; Copetta et al., 2021), greater tolerance to pathogens (Wang and Chen, 2024) and stresses, such as soil salinity (Guo et al., 2022), and enhanced production of primary and secondary metabolites (Carvalho et al., 2024; Metwally and Abdelhameed, 2024), as well as providing a better soil environment (Passos et al., 2021).

Table 2. Summary of some benefits due to arbuscular mycorrhizal fungi (AMF) application in plant cultivation

Benefit	References
Improvement of photosynthetic rate	Begum et al., 2022; Ye et al., 2022
Plant growth	Khaliq et al., 2022; Saha et al., 2022
Nutrient uptake	Huang et al., 2023; Akter et al., 2024
Optimization of antioxidant activity	Nahuelcura et al., 2022; Abdelhahim et al., 2022
Modulation of edaphic attributes	Bi et al., 2021; Falcão et al., 2023; Xu et al., 2024
Quality of fruit and flowers	Zhou et al., 2023a; Andrzejak and Janowska, 2021
Accumulation of metabolites	Machiani et al., 2022; Johny et al., 2021
Abiotic stress tolerance	Chandrasekaran et al., 2021; Hamzehzadeh et al., 2025
Pathogen defence	Weng et al., 2022; Pu et al., 2022
Phytomass accumulation	Falcão et al., 2022

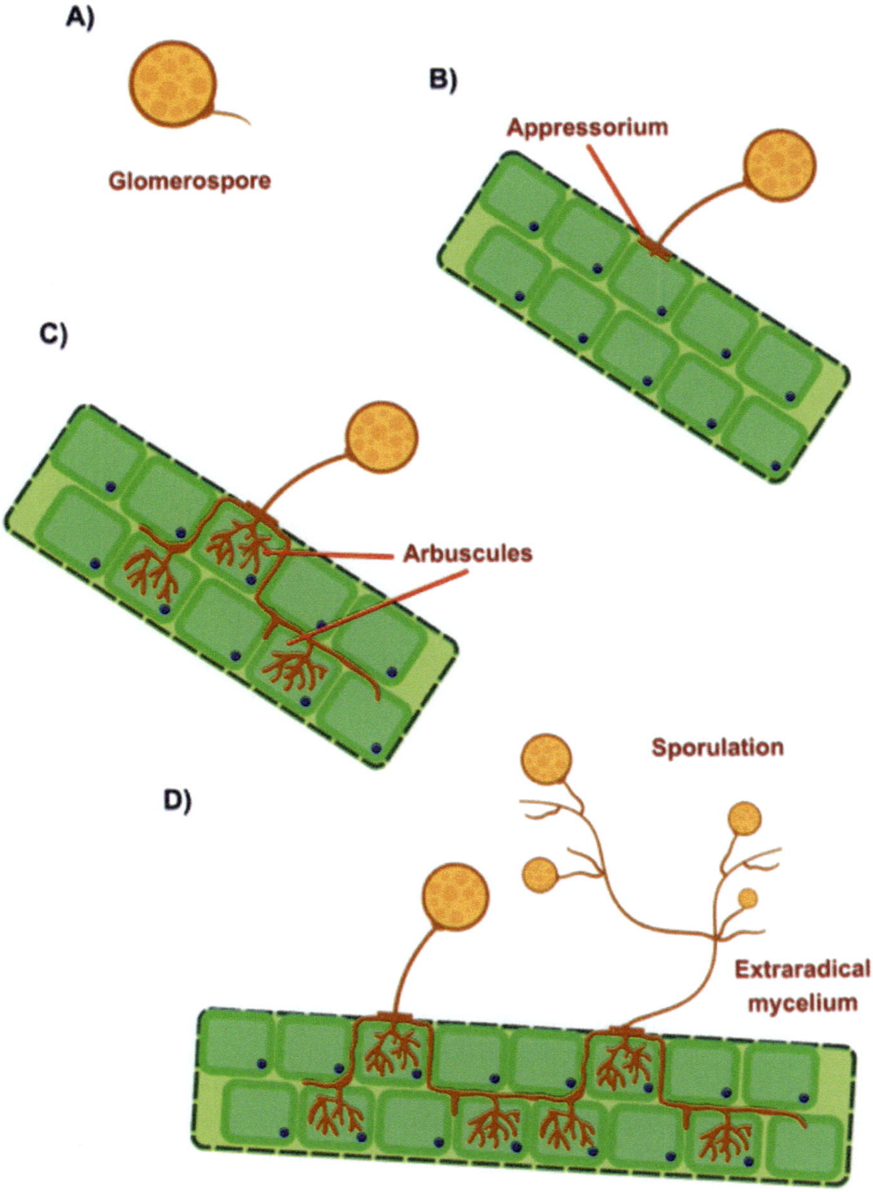

Fig. 2. Stages of the AMF life cycle. Asymbiotic phase: glomerospore germination (A), near the root, under favorable conditions of temperature and humidity, and formation of the appressorium after contact with root surface (B), for attachment and penetration of the symbiont. Symbiotic phase: formation of intercellular and then intracellular mycelium, forming structures called arbuscules (C), a site of nutrient exchange between the partners; production of extraradicular mycelium, generating new spores (D), which will propagate, maintaining the AMF life cycle (Bonfante and Genre, 2010; Bücking et al., 2012; Montoliu-Nerin et al., 2020). Figure made in Canva.com.

However, before being used for plant cultivation, these biostimulants need to go through a multiplication process to obtain appreciable quantities of infective units, such as spores and hyphae (Agnihotri et al., 2023). To do this, the most commonly used method is in pots containing an inert substrate such as sand, perlite, and/or vermiculite (Corazon-Guivin et al., 2022), using grass representatives as photobionts (INVAM, 2025). After flowering begins, watering is ceased, and AMF sporulation is induced, with the subsequent drying of the substrate, the roots are sectioned and mixed to obtain the soil-inoculum (Nadieline et al., 2024). In addition to this method, other practices can be used to produce mycorrhizal inocula, such as aeroponics and hydroponics (Sudheer et al., 2023), the use of transformed roots (Maitra et al., 2021) and plant hormones to stimulate the sporulation of mycorrhizal isolates (Tanaka et al., 2022).

In a similar way to organic substrates, AMF act by improving the production of plant compounds through various mechanisms, such as greater absorption of nutrients, especially phosphorus, which partakes in the metabolic pathways of energy production (Khan et al., 2021) and the formation of intermediates from primary and secondary metabolism (Santos et al., 2021). Other mechanisms are the increase in photosynthetic capacity (Mathur et al., 2018) and plant sugar levels (Ran et al., 2021), effects provided by the fungus, which subsidize the production of compounds. In addition to nutritional mechanisms, AMF are also capable of activating metabolic pathways related to the production of phenolic compounds (Mandal et al., 2013) by modulating the expression of signaling molecules, key enzymes (Li et al., 2024), and plant hormones (Cui et al., 2019). In addition, epigenetic modulation and gene overexpression are other factors stimulated by AMF, enabling the distinguished metabolite production (Fasani et al., 2023). It is also worthy to highlight the beneficial effects of the fungus on the edaphic environment, positively modulating rhizospheric attributes (Falcão et al., 2023), which govern the production of bioactive molecules in plants symbiotically associated with AMF. These and other mycorrhizal mechanisms for increasing the production of plant bioactives were summarized by Muniz et al. (2024) (Fig. 3).

Despite the extensive list of mycorrhizal benefits (Table 2), some variables can interfere with the AMF activity in the rhizosphere, one of which is the source and proportion of organic matter in the soil (Silva-Flores et al., 2019; Moura et al., 2024). Thus, it is known that the formation of aggregates, due to the presence of organic matter, favors glomerospore production (Soti et al., 2023). This benefit to the biology of AMF can also have an indirect positive influence on other soil microorganisms, which benefit from the deposition of glomalin-related soil proteins (GRSP) (Zhou et al., 2023b). However, the high availability of organic matter in the soil, coupled with the Phosphorus level, can mitigate the positive effect provided by the AMF (Carvalho et al., 2024). It is therefore essential to understand how vermicompost can modulate the synthesis of phenolics, a group of metabolites widely applied in several industrial sectors (Chiocchio et al., 2021).

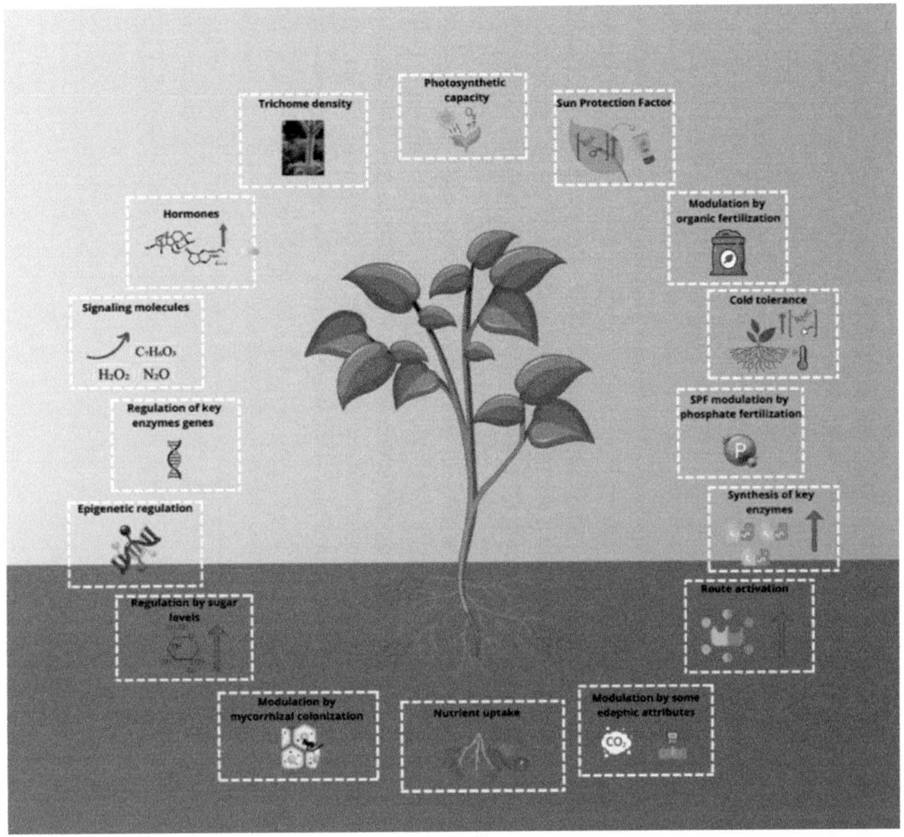

Fig. 3. Mechanisms that explain the improvement in the biosynthesis of secondary compounds in response to mycorrhization. Figure from Muniz et al., 2024. Reproduced with permission of the authors.

5. Effects of Arbuscular Mycorrhizal Fungi and Vermicompost Combined Apply to the Accumulation of Phenolic Compounds

Bearing in mind that the use of AMF combined with vermicompost can result in different effects on the physiology of the photobiont (Fig. 4; Table 1), a search was carried out on the Web of Science and Google Scholar using the search terms "AMF", "vermicompost" and "phenolic compounds", without restricting the publication date. Fifteen papers were retrieved and nine studies documented synergistic behavior between the vermicompost and the AMF tested, reporting increased production of phenolic compounds (Table 3). These results were related to the greater absorption by the AMF of the nutrients made available by the vermicompost, as well as the increased photosynthetic rate. Other factors that may have influenced this are the dose of vermicompost used in the substrate (Sarathambal et al., 2024). On the other hand, antagonistic behaviors were also recorded in four studies; however, these effects are not yet well elucidated. This pattern may be related to the AMF isolate used, as well as to the plant's mycorrhizal dependence (Habte and Manjunath, 1991; Carvalho et al., 2022). To a lesser extent, a probable neutral effect was also verified, in which the vermicompost and the AMF acted separately (Muniz et al., 2023). In addition, in one study, the mitigating effect of these technologies on phenolic production was found, with the benefits reported only when they are applied separately (Carvalho et al., 2024).

In this regard, it is worth testing different plant species, AMF isolates, and vermicompost dosages to elucidate better the existing relationships and the factors governing these effects. Based on these studies, it is possible to identify the best alternative for supplying the industry with the plant material containing the greatest accumulation of phenolic compounds. It is worth noting that these are low-cost (Silva and Silva, 2020; Carvalho et al., 2024) and sustainable technologies: the vermicompost as a form of organic waste management (Kamar Zaman and Yaacob, 2022) and the AMF use has the potential to reduce soil chemical fertilizers supply (Trejo et al., 2021; Ishaq et al., 2025). Thus, they may be viable biotechnologies in crop production, guaranteeing quality phytomass in a way that does not harm the soil.

6. Final Considerations

Both mycorrhizal agrobiotechnology and the application of vermicompost in plant growth substrates have the potential to integrate cultivation strategies, aiming for the augmented production of phenolic compounds. Based on this, the application of these technologies is a promising alternative to industries that use plant material to obtain phenolic compounds since they are sustainable options, have low costs, and are widely tested in isolated ways or combined. However, as there are reports of antagonistic and mitigating effects of these agrobiotechnological tools, the choice of fungi, the vermicompost dose, and the plant host should be considered when desiring an optimized production of plant phenolics.

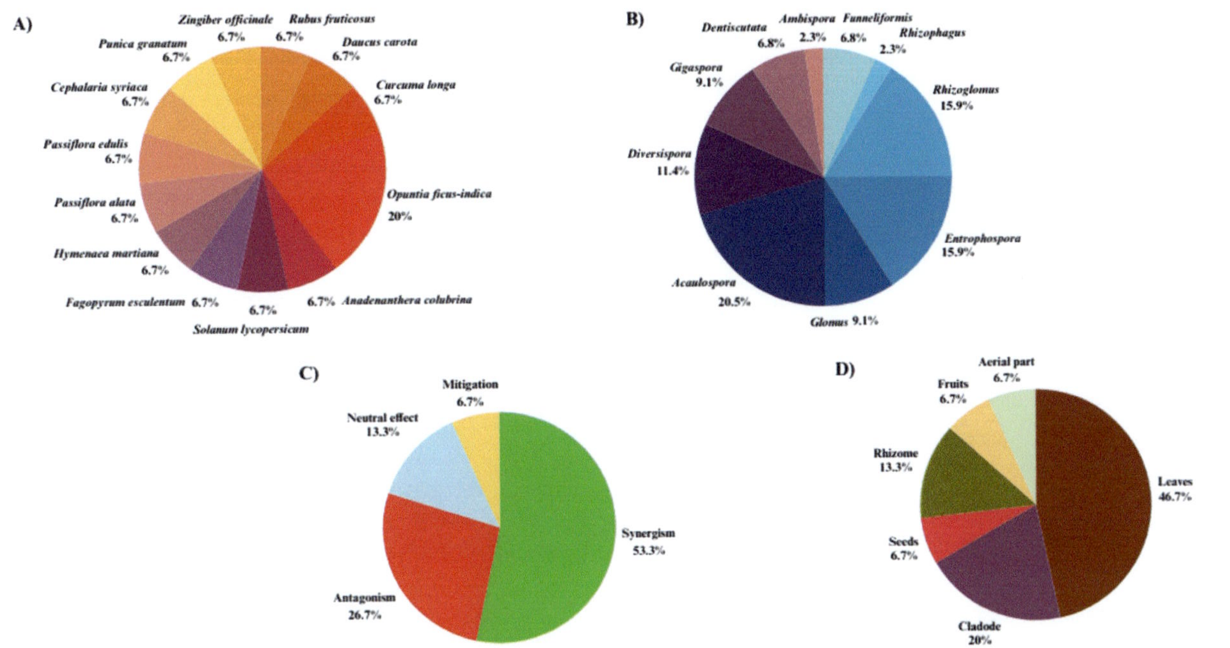

Fig. 4. Percentage of species (A), AMF genera (B), effects between AMF and vermicompost (C), and plant parts (D) present in scientific papers that evaluated the production of total phenolics in mycorrhizal plants grown in substrate with vermicompost. The papers considered were Oliveira et al., 2015, Abud-Archila et al., 2018, Rahimi et al., 2019, Ferrari et al., 2020, Oliveira et al., 2020, Silva and Silva, 2020, Lahbouki et al., 2022a,b; 2023; 2024, Mohammadi et al., 2022, Ahamad et al., 2023, Muniz et al., 2023, Sarathambal et al., 2024, Carvalho et al., 2024. *Rubus Fruticosus* Lour., *Daucus carota* L., *Anadenanthera colubrina* (Vell.) Brenan, *Curcuma longa* L., *Opuntia ficus-indica* (L.) Mill., *Solanum lycopersicum* L., *Fagopyrum esculentum* Moench, *Hymenaea martiana* Hayne, *Passiflora alata* Curtis, *Passiflora edulis* f. *flavicarpa* Deg. *Cephalaria syriaca* L., *Zingiber officinale* Rosc., *Punica granatum* L.

Table 3. Studies evaluating the production of total phenolics in mycorrhized plants grown in vermicompost substrate

Plant species	AMF genera	Plant part	Effect	References
Passiflora alata Curtis	*Acaulospora; Ambispora; Entrophospora; Gigaspora*	Aerial parts	Synergism	Oliveira et al., 2015
Rubus Fruticosus Lour.	*Funneliformis*	Leaves	Synergism	Abud-Archila et al., 2018
Cephalaria syriaca L.	*Rhizoglomus*	Seeds	Synergism	Rahimi et al., 2019
Punica granatum L.	*Acaulospora; Gigaspora*	Leaves	Synergism	Silva and Silva, 2020
Opuntia ficus-indica (L.) Mill.	*Acaulospora; Glomus; Rhizoglomus; Diversispora; Entrophospora; Gigaspora*	Cladodes	Synergism	Lahbouki et al., 2022a
O. ficus-indica	*Acaulospora; Glomus; Rhizoglomus, Dentiscutata; Entrophospora; Diversispora*	Cladodes	Synergism	Lahbouki et al., 2023
Daucus carota L.	*Funneliformis*	Leaves	Synergism	Ahamad et al., 2023
Zingiber officinale Rosc.	*Rhizoglomus*	Rhizome	Synergism	Sarathambal et al., 2024
Curcuma longa L.	*Rhizophagus; Entrophospora*	Rhizome	Antagonism	Ferrari et al., 2020
Passiflora edulis f. *flavicarpa* Deg.	*Acaulospora*	Leaves	Antagonism	Oliveira et al., 2020
Fagopyrum esculentum Moench	*Rhizoglomus; Funneliformis*	Leaves	Antagonism	Mohammadi et al., 2022
O. ficus-indica	*Acaulospora; Glomus; Rhizoglomus, Dentiscutata; Entrophospora; Diversispora*	Cladodes	Antagonism	Lahbouki et al., 2022b
Hymenaea martiana Hayne	*Acaulospora; Entrophospora; Gigaspora*	Leaves	Neutral	Muniz et al., 2023
Solanum lycopersicum L.	*Acaulospora; Glomus; Rhizoglomus; Dentiscutata; Entrophospora; Diversispora*	Fruits	Neutral	Lahbouki et al., 2024
Anadenanthera colubrina (Vell.) Brenan	*Acaulospora; Gigaspora*	Leaves	Mitigation	Carvalho et al., 2024

Acknowledgments

The authors thank the *Conselho Nacional de Desenvolvimento Científico e Tecnológico* for the productivity fellowship given to Fábio Sérgio Barbosa da Silva (313859/2021-3) and the doctoral scholarship to Eduarda Lins Falcão (140405/2024-0) and the *Fundação de Amparo à Ciência e Tecnologia do Estado de Pernambuco* (IBPG-0039-2.02/25) for granting the master's scholarship to João Gabriel Lira de Carvalho.

References

Abdelhalim, T.S., Tia, N.A., Elkhatim, K.A.S., Othman, M.H., Joergensen, R.G., Almaiman, S.A. et al. (2022). Exploring the potential of arbuscular mycorrhizal fungi (AMF) for improving health-promoting phytochemicals in sorghum. Rhizosphere 24, 100596. doi: 10.1016/j.rhisph.2022.100596

Abud-Archila, M., Luján-Hidalgo, M.C., López-Pérez, J.M., Ordaz-Rivera, J., Ruiz-Valdiviezo, V.M., Oliva-Llaven, M.Á. et al. (2018). Growth and fruit chemical characteristics of blackberry (*Rubus fruticosus*) cultivated with vermicompost, *Glomus mosseae* and Phosphate rock. Compost Sci Util 26, 225–231. doi: 10.1080/1065657X.2018.1471367

Agnihotri, R., Sharma, M., Bucking, H., Dames, J. & Bagyaraj, D. (2023). Methods for assessing the quality of AM fungal bio-fertilizer: Retrospect and future directions. WJMB 38, 97. doi: 10.1007/s11274-022-03288-3

Ahamad, L., Bhat, A.H., Kumar, H., Rana, A., Hasan, M.N., Ahmed, I. et al. (2023). From soil to plant: Strengthening carrot defenses against *Meloidogyne incognita* with vermicompost and arbuscular mycorrhizal fungi biofertilizers. Front Microbiol 14, 1206217. doi: 10.3389/fmicb.2023.1206217

Ai, Y.-J., Li, F.-P., Yang, J.-Q., Lu, S. & Gu, H.-H. (2022). Research progress and potential functions of AMF and GRSP in the ecological remediation of metal tailings. Sustainability 14, 9611. doi: 10.3390/su14159611

Akter, S., Kamruzzaman, M., Sarder, M.P., Amin, M.S., Joardar, J.C., Islam, M.S. et al. (2024). Mycorrhizal fungi increase plant nutrient uptake, aggregate stability and microbial biomass in the clay soil. Symbiosis 93, 163–176. doi: 10.1007/s13199-024-00994-4

Albuquerque, B.R., Heleno, S.A., Oliveira, M.B.P.P., Barros, L. & Ferreira, I.C.F.R. (2021). Phenolic compounds: Current industrial applications, limitations and future challenges. Food & Function 12, 14–29. doi: 10.1039/d0fo02324h

Alonso-Martínez, D., Jiménez-Parra, B. & Cabeza-García, L. (2024). Theoretical framework to foster and assess sustainable agriculture practices: Drivers and key performance indicators. Environ Sustain Ind 23, 100434. doi: 10.1016/j.indic.2024.100434

Andrzejak, R. & Janowska, B. (2021). Yield and quality of inflorescences in the *Zantedeschia albomaculata* (Hook.) Baill. 'Albomaculata' after the treatment with AMF and GA_3. Agronomy 11, 644. doi: 10.3390/agronomy11040644

Aquino, A.M., Oliveira, A.M.G. & Loureiro, D.C. (2005). Integrando compostagem e vermicompostagem na reciclagem de resíduos orgânicos domésticos. Circular Técnica (EMBRAPA).

Awadhpersad, V.R.R., Ori, L. & Adil Ansari, A. (2021). Production and effect of vermiwash and vermicompost on plant growth parameters of tomato (*Lycopersicon esculentum* Mill.) in Suriname. Int J Recycl Org Waste Agric 10, 397–413. doi: 10.30486/IJROWA.2021.1911898.1148

Baghel, R. (2024). Chemical fertilizers market size, global forecast report 2037. *In*: Research Nester. https://www.researchnester.com/reports/chemical-fertilizers-market/329. Accessed 15 Jan 2025

Barbehenn, R.V. & Constabel, P.C. (2011). Tannins in plant–herbivore interactions. Phytochemistry 72, 1551–1565. doi: 10.1016/j.phytochem.2011.01.040

Begum, N., Wang, L., Ahmad, H. et al. (2022). Co-inoculation of arbuscular mycorrhizal fungi and the plant growth-promoting rhizobacteria improve growth and photosynthesis in tobacco under drought stress by up-regulating antioxidant and mineral nutrition metabolism. Microb Ecol 83, 971–988. doi: 10.1007/s00248-021-01815-7

Bharadvaja, N., Gautam, S. & Singh, H. (2023). Natural polyphenols: A promising bioactive compounds for skin care and cosmetics. Mol Biol Rep 50, 1817–1828. doi: 10.1007/s11033-022-08156-9

Bhat, S.A., Singh, J. & Vig, A.P. (2018). Earthworms as organic waste managers and biofertilizer producers. Waste Bio. Valori 9, 1073–1086. doi: 10.3389/fmicb.2023.1206217

Bi, Y., Wang, K., Du, S., Ma, S., Zhang, J. & Xie, L. (2021). Shifts in arbuscular mycorrhizal fungal community composition and edaphic variables during reclamation chronosequence of an open-cast coal mining dump. Catena 203, 105301. doi: 10.1016/j.catena.2021.105301

Bonfante, P. & Genre, A. (2010). Mechanisms underlying beneficial plant–fungus interactions in mycorrhizal symbiosis. Nat Commun 1, 48. doi: 10.1038/ncomms1046

Brundrett, M.C. & Tedersoo, L. (2018). Evolutionary history of mycorrhizal symbioses and global host plant diversity. New Phytol 220, 1108–1115. doi: 10.1111/nph.14976

Bücking, H., Liepold, E. & Ambilwade, P. (2012). The role of the mycorrhizal symbiosis in nutrient uptake of plants and the regulatory mechanisms underlying these transport processes. Plant Sci 4, 108–132. doi: 10.5772/52570

Camargo-Ricalde, S.L. (2002). Dispersal, distribution and establishment of arbuscular mycorrhizal fungi: A review. Bot Sci 71, 33–34. doi: 10.17129/botsci.1661

Campana, E., Ciriello, M., Lentini, M., Rouphael, Y. & De Pascale, S. (2025). Sustainable agriculture through compost tea: Production, application, and impact on horticultural crops. Horticulturae 11, 433. doi: 10.3390/horticulturae11040433

Cao, M.-A., Wang, P., Hashem, A., Wirth, S., Abd_Allah, E.F. & Wu, Q.-S. (2021). Field inoculation of arbuscular mycorrhizal fungi improves fruit quality and root physiological activity of citrus. Agriculture 11, 1297. doi: 10.3390/agriculture11121297

Carvalho, J.G.L., Falcão, E.L., Bastos Filho, C.J.A., Wu, Q.-S. & Silva, F.S.B. (2024). Separate application of vermicompost and mycorrhization enhances antioxidant production and SPF in *Anadenanthera colubrina* (Vell.) Brenan seedlings. Microbiol Res 15, 2713–2725. doi: 10.3390/microbiolres15040180

Carvalho, L.G.V., Santos, S.C., Lourente, E.R.P., Trovato, V.W., Santos, C.C. & Rui, R.F. (2022). Jatobazeiro seedlings associated with arbuscular mycorrhizal fungi. Rev Bras Frutic 44, e-006. doi: 10.1590/0100-29452022006

Castellini, M., Bondì, C., Giglio, L. & Iovino, M. (2024). Impact of vermicompost addition on water availability of differently textured soils. Heliyon 10, e35699. doi: 10.1016/j.heliyon.2024.e35699

Chandrasekaran, M., Boopathi, T. & Manivannan, P. (2021). Comprehensive assessment of ameliorative effects of AMF in alleviating abiotic stress in tomato plants. J Fungi 7, 303. doi: 10.3390/jof7040303

Chanu, L.J., Hazarika, S., Choudhury, B.U., Ramesh, T., Balusamy, A., Moirangthem, P. et al. (2018). A guide to vermicomposting-production process and socioeconomic aspects. Extension Bull 81, 30

Chen, G.-L., Chen, S.-G., Xiao, Y. & Fu, N.-L. (2018). Antioxidant capacities and total phenolic contents of 30 flowers. Ind Crop Prod 111, 430–445. doi: 10.1016/j.indcrop.2017.10.051

Chiocchio, I., Mandrone, M., Tomasi, P., Marincich, L. & Poli, F. (2021). Plant secondary metabolites: An opportunity for circular economy. Molecules 26, 495. doi: 10.3390/molecules26020495

Copetta, A., Todeschini, V., Massa, N., Bona, E., Berta, G. & Lingua, G. (2021). Inoculation with arbuscular mycorrhizal fungi improves melon (*Cucumis melo*) fruit quality under field conditions and plant performance in both field and greenhouse. Plant Biosyst 155, 1063-1074. doi: 10.1080/11263504.2020.1813831

Corazon-Guivin, M.A., Vallejos-Tapullima, A., Aguila, S.R.-D., Rondinel-Mendoza, N.V., Hernández-Cuevas, L.V., Carvajal-Vallejos, F.M. et al. (2022). Influence of substrate properties on communities of arbuscular mycorrhizal fungi isolated from agroecosystems in Peru. J Soil Sci Plant Nutr 22, 4784–4797. doi: 10.1007/s42729-022-00960-3

Cui, L., Guo, F., Zhang, J., Yang, S., Meng, J., Geng, Y. et al. (2019). Synergy of arbuscular mycorrhizal symbiosis and exogenous Ca^{2+} benefits peanut (*Arachis hypogaea* L.) growth through the shared hormone and flavonoid pathway. Sci Rep 9, 1628. doi: 10.1038/s41598-019-52630-7

Das, D., Abbhishek, K., Banik, P. & Swain, D.K. (2022). Comparative evaluation of changes in soil bio-chemical properties after application of traditional and enriched vermicompost. Environ Technol Innov 28, 102956. doi: 10.1016/j.eti.2022.102956

Deepthi, M.P., Nivethitha, S., Saminathan, K., Narendhirakannan, R.T., Karmegam, N. & Kathireswari, P. (2021). Effect of vermiwash prepared from livestock biowaste as vermiponics medium on the growth and biochemical indices of *Amaranthus viridis* L. Environ Technol Innov 21, 101300. doi: 10.1016/j.eti.2020.101300

Delaeter, M., Magnin-Robert, M., Randoux, B. & Lounès-Hadj Sahraoui, A. (2024). Arbuscular mycorrhizal fungi as biostimulant and biocontrol agents: A review. Microorganisms 12, 1281. doi: 10.3390/microorganisms12071281

Deng, Y. & Lu, S. (2017). Biosynthesis and regulation of phenylpropanoids in plants. Crit Rev Plant Sci 36, 257–290. doi: 10.1080/07352689.2017.1402852

Divekar, P.A., Narayana, S., Divekar, B.A., Kumar, R., Gadratagi, B.G., Ray, A. et al. (2022). Plant secondary metabolites as defense tools against herbivores for sustainable crop protection. Int J Mol Sci 23, 2690. doi: 10.3390/ijms23052690

Dixit, G., Praveen, A., Tripathi, T., Yadav, V.K. & Verma, P.C. (2017). Herbivore-responsive cotton phenolics and their impact on insect performance and biochemistry. J Asia-Pac Entomol 20, 341–351. doi: 10.1016/j.aspen.2017.02.002

Dixon, R.A. & Barros, J. (2019). Lignin biosynthesis: Old roads revisited and new roads explored. Open Biol 9, 190215. doi: 10.1098/rsob.190215

Dixon, R.A. & Sarnala, S. (2020). Proanthocyanidin Biosynthesis – A matter of protection. Plant Physiol 184, 579–591. doi: 10.1104/pp.20.00973

Droh, G., Kouassi, A.B., Kouadjo, Z.G.C., Zeze, A., Nguetta, A.S.P. & Sanders, I.R. (2016). Effects of two types of AMF on growth of cocoa seedlings (*Theobroma cacao* L.) in greenhouses. GJAR 3, 157–164.

Enebe, M.C. & Erasmus, M. (2023). Vermicomposting technology – A perspective on vermicompost production technologies, limitations and prospects. J Environ Manage 345, 118585. doi: 10.1016/j.jenvman.2023.118585

Falcão, E.L., Bastos Filho, C.J.A. & Silva, F.S.B. (2022). Arbuscular mycorrhizal fungi enhance the Sun Protection Factor (SPF) biosynthesis in *Anadenanthera colubrina* (Vell.) Brenan leaves. Rhizosphere 24, 100595. doi: 10.1016/j.rhisph.2022.100595

Falcão, E.L., Muniz, B.C., Bastos Filho, C.J.A., Kapoor, R. & Silva, F.S.B. (2023). Soil microbial respiration and pH modulated by arbuscular mycorrhizal fungi influence the biosynthesis of health-promoting compounds in *Anadenanthera colubrina* (Vell.) Brenan. Rhizosphere 26, 100685. doi: 10.1016/j.rhisph.2023.100685

Fasani, E., Giannelli, G., Varotto, S., Visioli, G., Bellin, D., Furini, A. et al. (2023). Epigenetic control of plant response to heavy metals. Plants 12, 3195. doi: 10.3390/plants12183195

Feng, S., Yi, J., Li, X., Wu, X., Zhao, Y., Ma, Y. et al. (2021). Systematic review of phenolic compounds in apple fruits: Compositions, distribution, absorption, metabolism, and processing stability. J Agric Food Chem 69, 7–27. doi: 10.1021/acs.jafc.0c05481

Ferrari, M.P.S., Queiroz, M.S., Andrade, M.M., Alberton, O., Gonçalves, J.E., Gazim, Z.C. et al. (2020). Substrate-associated mycorrhizal fungi promote changes in terpene composition, antioxidant activity, and enzymes in *Curcuma longa* L. acclimatized plants. Rhizosphere 13, 100191. doi: 10.1016/j.rhisph.2020.100191

Garg, S.S., Gupta, J., Sharma, S. & Sahu, D. (2020). An insight into the therapeutic applications of coumarin compounds and their mechanisms of action. Eur J Pharm Sci 152, 105424. doi: 10.1016/j.ejps.2020.105424.

Giovannetti, M. (2000). Spore germination and pre-symbiotic mycelial growth. *In*: Kapulnik, Y., Douds, D.D. [eds.]. Arbuscular Mycorrhizas: Physiology and Function. Springer, Dordrecht. doi: 10.1007/978-94-017-0776-3_3

Giovinazzo, G., Ingrosso, I., Paradiso, A. et al. (2012). Resveratrol biosynthesis: Plant metabolic engineering for nutritional improvement of food. Plant Foods Hum Nutr 67, 191–199. doi: 10.1007/s11130-012-0299-8

Goufo, P., Singh, R.K. & Cortez, I. (2020). A reference list of phenolic compounds (including stilbenes) in grapevine (*Vitis vinifera* L.) roots, woods, canes, stems, and leaves. Antioxidants 9, 398. doi: 10.3390/antiox9050398

Gross, G.G. (2008). From lignins to tannins: Forty years of enzyme studies on the biosynthesis of phenolic compounds. Phytochem 69, 3018–3031. doi: 10.1016/j.phytochem.2007.04.031.

Grundhöfer, P., Niemetz, R., Schilling, G. & Gross, G.G. (2001). Biosynthesis and subcellular distribution of hydrolyzable tannins. Phytochem 57, 915–-927. doi: 10.1016/S0031-9422(01)00099-1.

Gudeta, K., Julka, J.M., Kumar, A., Bhagat, A. & Kumari, A. (2021). Vermiwash: An agent of disease and pest control in soil, a review. Heliyon 7, e06434. doi: 10.1016/j.heliyon.2021.e06434

Guo, S., Wang, Q., Tang, L., Zhang, T., Li, J., Xiao, Y. et al. (2022). Inoculation with arbuscular mycorrhizal fungi reinforces tea plant's tolerance to salinity. J Plant Growth Regul 41, 3498–3517. doi: 10.1007/s00344-021-10529-6

Habte, M. & Manjunath, A. (1991). Categories of vesicular-arbuscular mycorrhizal dependency of host species. Mycorrhiza 1, 3–12. doi: 10.1007/BF00205896

Hamzehzadeh, H., Abbaspour, H., Safipour Afshar, A. et al. (2025). AMF-mediated salinity adaptation in pistachio plants: Photosynthetic efficiency and ionic balance. Biologia 80, 1–17. doi: 10.1007/s11756-024-01810-6

Harrison, M.J. (2005). Signaling in the arbuscular mycorrhizal symbiosis. Annu Rev Microbiol 59, 19–42. doi: 10.1146/annurev.micro.58.030603.123749

He, H., Peng, M., Hou, Z. & Li, J. (2024). Unlike chemical fertilizer reduction, organic fertilizer substitution increases soil organic carbon stock and soil fertility in wheat fields. J Sci Food Agric 104, 2798–2808. doi: 10.1002/jsfa.13167

Heleno, S.A., Martins, A., Queiroz, M.J.R.P. & Ferreira, I.C.F.R. (2015). Bioactivity of phenolic acids: Metabolites versus parent compounds: A review. Food Chem 173, 501–513. doi: 10.1016/j.foodchem.2014.10.057

Hoque, T.S., Hasan, A.K., Hasan, M.A., Nahar, N., Dey, D.K., Mia, S. et al. (2022). Nutrient release from vermicompost under anaerobic conditions in two contrasting soils of Bangladesh and its effect on wetland rice crop. Agriculture 12, 376. doi: 10.3390/agriculture12030376

Hu, W., Sarengaowa, Guan Y. & Feng, K. (2022). Biosynthesis of phenolic compounds and antioxidant activity in fresh-cut fruits and vegetables. Front Microbiol 13, 906069. doi: 10.3389/fmicb.2022.906069

Huang, S., Gill, S., Ramzan, M. et al. (2023). Uncovering the impact of AM fungi on wheat nutrient uptake, ion homeostasis, oxidative stress, and antioxidant defense under salinity stress. Sci Rep 13, 8249. doi: 10.1038/s41598-023-35148-x

INVAM. (2025). Host plant choices. *In*: INVAM. https://invam.ku.edu/host-plant-choices. Accessed 11 Jan 2025

Iqbal, A., Hussain, Q., Mo, Z., Hua, T., Mustafa, A.E.-Z.M.A. & Tang, X. (2024). Vermicompost supply enhances fragrant-rice yield by improving soil fertility and eukaryotic microbial community composition under environmental stress conditions. Microorganisms 12, 1252 doi: 10.3390/microorganisms12061252

Ishaq, L.F., Amalia, F.C., Benggu, Y.I., Tae, A.S.J.A. & Airthur, M.M. (2025). The potential of arbuscular mycorrhizal fungi as biofertilizer to reduce chemical fertilizer use in calcareous soil. IOP Conference Series: Earth and Environmental Science 1482, 012015. doi: 10.1088/1755-1315/1482/1/012015

Jan, R., Khan, M.-A., Asaf, S., Lubna, Waqas M., Park, J.-R., Asif, S. et al. (2022). Drought and UV radiation stress tolerance in rice is improved by overaccumulation of non-enzymatic antioxidant flavonoids. Antioxidants 11, 917. doi: 10.3390/antiox11050917

Jiang, T., Hao, T., Chen, W., Li, C., Pang, S., Fu, C. et al. (2025). Reprogrammed plant metabolism during viral infections: Mechanisms, pathways and implications. Mol Plant Pathol 26, e70066. doi: 10.1111/mpp.70066

Jin, L., Jin, N., Wang, S., Li, J., Meng, X., Xie, Y. et al. (2022). Changes in the microbial structure of the root soil and the yield of Chinese baby cabbage by chemical fertilizer reduction with bio-organic fertilizer application. Microbiol Spectr 10, e01215-22. doi: 10.1128/spectrum.01215-22

Johny, L., Cahill, D.M. & Adholeya, A. (2021). AMF enhance secondary metabolite production in ashwagandha, licorice, and marigold in a fungi-host specific manner. Rhizosphere 17, 100314. doi: 10.1016/j.rhisph.2021.100314

Joshi, R., Singh, J. & Vig, A.P. (2015). Vermicompost as an effective organic fertilizer and biocontrol agent: Effect on growth, yield and quality of plants. Rev Environ Sci Biotechno 14, 137–159. doi: 10.1007/s11157-014-9347-1

Kamar Zaman, A.M. & Yaacob, J.S. (2022). Exploring the potential of vermicompost as a sustainable strategy in circular economy: Improving plants' bioactive properties and boosting agricultural yield and quality. Environ Sci Pollut Res 29, 12948–12964. doi: 10.1007/s11356-021-18006-z

Khaliq, A., Perveen, S., Alamer, K.H., Zia Ul Haq, M., Rafique, Z., Alsudays, I.M. et al. (2022). Arbuscular mycorrhizal fungi symbiosis to enhance plant–soil interaction. Sustainability 14, 7840. doi: 10.3390/su14137840

Khan, N., Ray, R.L., Sargani, G.R., Ihtisham, M., Khayyam, M. & Ismail, S. (2021). Current progress and future prospects of agriculture technology: Gateway to sustainable agriculture. Sustainability 13, 4883. doi: 10.3390/su13094883

Koltai, H., Gadkar, V. & Kapulnik, Y. (2009). Biochemical and practical views of arbuscular mycorrhizal fungus-host association in horticultural crops. *In*: Horticultural Reviews (p. 257–287). John Wiley & Sons, Ltd. doi: 10.1002/9780470527238.ch5

Kubra, G., Khan, M., Munir, F., Gul, A., Shah, T., Hussain, A. et al. (2021). Expression characterization of flavonoid biosynthetic pathway genes and transcription factors in peanut under water deficit conditions. Front Plant Sci 12, 680368. doi: 10.3389/fpls.2021.680368

Lahbouki, S., Anli, M., El Gabardi, S., Ait-El-Mokhtar, M.M., Ben-Laouane, R., Boutasknit, A. et al. (2022a). Evaluation of arbuscular mycorrhizal fungi and vermicompost supplementation on growth, phenolic content and antioxidant activity of prickly pear cactus (*Opuntia ficus-indica*). Plant Biosyst 156, 882–892. doi: 10.1080/11263504.2021.1947408

Lahbouki, S., Ben-Laouane, R., Anli, M., Boutasknit, A., Ait-Rahou, Y., Ait-El-Mokhtar, M. et al. (2022b). Arbuscular mycorrhizal fungi and/or organic amendment enhance the tolerance of prickly pear (*Opuntia ficus-indica*) under drought stress. J Arid Environ 199, 104703. doi: 10.1016/j.jaridenv.2021.104703

Lahbouki, S., Ben-Laouane, R., Outzourhit, A. & Meddich, A. (2023). The combination of vermicompost and arbuscular mycorrhizal fungi improves the physiological properties and chemical composition of *Opuntia ficus-indica* under semi-arid conditions in the field. Arid Land Res Manag. doi: 0.1080/15324982.2022.2115952

Lahbouki, S., Hashem, A., Kumar, A., Abd_Allah, E.F. & Meddich, A. (2024). Integration of horse manure vermicompost doses and arbuscular mycorrhizal fungi to improve fruit quality, and soil fertility in tomato field facing drought stress. Plants 13, 1449. doi: 10.3390/plants13111449

Lemos, V.C., Reimer, J.J. & Wormit, A. (2019). Color for life: Biosynthesis and distribution of phenolic compounds in pepper (*Capsicum annuum*). Agriculture 9, 81. doi: 10.3390/agriculture9040081

Li, N., Chen, W., Wang, B., Zhang, C., Wang, Y., Li, R. et al. (2024). Arbuscular mycorrhizal fungi improve the disease resistance of *Lycium barbarum* to root rot by activating phenylpropane metabolism. Front Plant Sci 15, 1459651. doi: 10.3389/fpls.2024.1459651

Lin, Z., Bai, J., Zhen, Z., Lao, S., Li, W., Wu, Z. et al. (2016). Enhancing pentachlorophenol degradation by vermicomposting associated bioremediation. Ecol Eng 87, 288–294. doi: 10.1016/j.ecoleng.2015.12.004

Linić, I., Šamec, D., Grúz, J., Vujčić Bok, V., Strnad, M. & Salopek-Sondi, B. (2019). Involvement of Phenolic Acids in short-term adaptation to salinity stress is species-specific among *Brassicaceae*. Plants 8, 155. doi: 10.3390/plants8060155

Lirikum, Kakati L.N., Thyug, L. & Mozhui, L. (2022). Vermicomposting: An eco-friendly approach for waste management and nutrient enhancement. Trop Ecol 63, 325–337. doi: 10.1007/s42965-021-00212-y

Liu, B., Wu, C., Pan, P., Fu, Y., He, Z., Wu, L. & Li, Q. (2019). Remediation effectiveness of vermicompost for a potentially toxic metal-contaminated tropical acidic soil in China. Ecotoxicol Environ Saf 182, 109394. doi: 10.1016/j.ecoenv.2019.109394

Luginbuehl, L.H. & Oldroyd, G.E.D. (2017). Understanding the arbuscule at the heart of endomycorrhizal symbioses in plants. Curr Biol 27, R952–R963. doi: 10.1016/j.cub.2017.06.042

Machiani, A.M., Javanmard, A., Machiani, H.R. & Sadeghpour, A. (2022). Arbuscular mycorrhizal fungi and changes in primary and secondary metabolites. Plants 11, 2183. doi: 10.3390/plants11172183

Maharjan, K.K., Noppradit, P. & Techato, K. (2022). Suitability of vermicomposting for different varieties of organic waste: A systematic literature review (2012–2021). Org Agric 12, 581–602. doi: 10.1007/s13165-022-00413-2

Maitra, P., Al-Rashid, J., Barman, N.C., Khan, M.N.M., Mandal, D., Rasul, N.M. et al. (2021). Sand particle size and phosphorus amount affect *Rhizophagus irregularis* spore production using *in vitro* propagated spore as a starter Inoculum in Rhizosphere of maize (*Zea mays*) Plantlets. J Fungi 7, 846. doi: 10.3390/jof7100846

Malarz, J., Yudina, Y.V. & Stojakowska, A. (2023). Hairy root cultures as a source of phenolic antioxidants: Simple phenolics, phenolic acids, phenylethanoids, and hydroxycinnamates. Int J Mol Sci 24, 6920. doi: 10.3390/ijms24086920

Mandal, S., Evelin, H., Giri, B., Singh, V.P. & Kapoor, R. (2013). Arbuscular mycorrhiza enhances the production of stevioside and rebaudioside-A in *Stevia rebaudiana* via nutritional and non-nutritional mechanisms. Appl Soil Ecol 72, 187–194. doi: 10.1016/j.apsoil.2013.07.003

Marchiosi, R., Santos, W.D., Constantin, R.P. et al. (2020). Biosynthesis and metabolic actions of simple phenolic acids in plants. Phytochem Rev 19, 865–906. doi: 10.1007/s11101-020-09689-2

Mathur, S., Sharma, M.P. & Jajoo, A. (2018). Improved photosynthetic efficacy of maize (*Zea mays*) plants with arbuscular mycorrhizal fungi (AMF) under high temperature stress. J Photochem Photobiol B 180, 149–154. doi: 10.1016/j.jphotobiol.2018.02.002

Mazimba, O. (2017). Umbelliferone: Sources, chemistry and bioactivities review. Bull Fac Pharm Cairo Univ 55, 223-232. doi: 10.1016/j.bfopcu.2017.05.001.

Mendes, C.B., Lima, G.F., Alves, V.N., Coelho, N.M.M., Dragunski, D.C. & Tarley, C.R.T. (2012). Evaluation of vermicompost as a raw natural adsorbent for adsorption of pesticide methylparathion. Environ Technol Rev 33, 167–172. doi: 10.1080/09593330.2011.554890

Metwally, R.A. & Abdelhameed, R.E. (2024). Co-application of arbuscular mycorrhizal fungi and nano-$ZnFe_2O_4$ improves primary metabolites, enzymes and NPK status of pea (*Pisum sativum* L.) plants. J Plant Nutr 47, 468–486. doi: 10.1080/01904167.2023.2280121

Mohammadi, E., Fattahi, M., Barin, M. & Ashrafi-Saeidlou, S. (2022). Arbuscular mycorrhiza and vermicompost alleviate drought stress and enhance yield, total flavonoid concentration, rutin content, and antioxidant activity of buckwheat (*Fagopyrum esculentum* Moench). S Afr J Bot 148, 588–600. doi: 10.1016/j.sajb.2022.05.020

Montoliu-Nerin, M., Sánchez-García, M., Bergin, C., Grabherr, M., Ellis, B., Kutschera, V.E. et al. (2020). Building *de novo* reference genome assemblies of complex eukaryotic microorganisms from single nuclei. Sci Rep, 10, 1303. doi: 10.1038/s41598-020-58025-3

Moreira, S.L.S., Prates Júnior, P., Fernandes, R.B.A., Cunha, A.C.M.M. & Campos. A.N.R. (2016). Growth and nutrients uptake in *Euterpe edulis* Martius inoculated with arbuscular mycorrhizal fungi. PAT 46, 169–176. doi: 10.1590/1983-40632016v4639547

Moura, J.B., Ramos, M.L.G., Konrad, M.L.F., Saggin Júnior, O.J. & Silva, S.D. (2024). Mycorrhizal fungi arbuscular in organic and conventional sugarcane systems. Sci Rep 14, 14322. doi: 10.1038/s41598-024-65358-w

Muniz, B.C., Falcão, E.L., Luz, R.C.R., Barreto, C.B. & Silva, F.S.B. (2024). Overview of brazilian studies on phytochemistry of mycorrhizal species. *In*: Falcão, E.L. and Silva, F.S.B. [eds.]. The Potential of Mycorrhizal Technology in Improving the Production of Plant Bioactive Compounds: What is the Overview of Studies Conducted in Brazil? 1st ed Atena Editora, Brazil.

Muniz, B.C., Kapoor, R., Almeida, J.R.G.S., Bastos Filho, C.J.A. & Silva, F.S.B. (2023). Arbuscular mycorrhizae increase but vermicompost decrease the sun protection factor (SPF) in leaves of *Hymenaea martiana* Hayne seedlings. Rhizosphere 27, 100781. doi: 10.1016/j.rhisph.2023.100781

Mupambwa, H.A. & Mnkeni, P.N.S. (2018). Optimizing the vermicomposting of organic wastes amended with inorganic materials for production of nutrient-rich organic fertilizers: A review. Environ Sci Pollut Res 25, 10577-10595. doi: 10.1007/s11356-018-1328-4

Nadieline, C.V., Quéré, A., Ndiaye, C., Diène, A.A., Rego, F.D., Sadio, O. et al. (2024). Development of on-farm AMF inoculum production for sustainable agriculture in Senegal. PloS ONE, 19, e0310065. doi: 10.1371/journal.pone.0310065

Nahuelcura, J., Ruiz, A., Gomez, F. & Cornejo, P. (2022). The effect of arbuscular mycorrhizal fungi on the phenolic compounds profile, antioxidant activity and grain yields in wheat cultivars growing under hydric stress. J Sci Food Agric 102, 407–416. doi: 10.1002/jsfa.11370

Nurhidayati, N., Machfudz, M. & Murwani, I. (2018). Direct and residual effect of various vermicompost on soil nutrient and nutrient uptake dynamics and productivity of four mustard Pak-Coi (*Brassica rapa* L.) sequences in organic farming system. Int J Recycl Org Waste Agric 7, 173–181. doi: 10.1007/s40093-018-0203-0

OECD/FAO (2023). OECD-FAO Agricultural Outlook 2023-2032, OECD Publishing, Paris, doi: 10.1787/08801ab7-en

Oliveira, M.S., Campos, M.A. & Silva, F.S. (2015). Arbuscular mycorrhizal fungi and vermicompost to maximize the production of foliar biomolecules in *Passiflora alata* Curtis seedlings. J Sci Food Agri 95, 522–528. doi: 10.1002/jsfa.6767

Oliveira, P.T.F., Santos, E.L., Silva, W.A.V., Ferreira, M.R.A., Soares, L.A.L., Silva, F.A. et al. (2020). Use of mycorrhizal fungi releases the application of organic fertilizers to increase the production of leaf vitexin in yellow passion fruit. J Sci Food Agri 100, 1816–1821. doi: 10.1002/jsfa.10197

Oyege, I. & Balaji Bhaskar, M.S. (2023). Effects of vermicompost on soil and plant health and promoting sustainable agriculture. Soil Systems 7, 101. doi: 10.3390/soilsystems7040101

Panche, A.N., Diwan, A.D. & Chandra, S.R. (2016). Flavonoids: An overview. J Nutr Sci 5, e47. doi:10.1017/jns.2016.41

Passos, J.H., Maia, L.C., Assis, D.M.A., Silva, J.A., Oehl, F. & Silva, I.R. (2021). Arbuscular mycorrhizal fungal community structure in the rhizosphere of three plant species of crystalline and sedimentary areas in the Brazilian Dry Forest. Microb Ecol. 82, 104–121. doi: 10.1007/s00248-020-01557-y

Paymaneh, Z., Sarcheshmehpour, M., Mohammadi, H. & Hesni, M.A. (2023). Vermicompost and/or compost and arbuscular mycorrhizal fungi are conducive to improving the growth of pistachio seedlings to drought stress. Appl Soil Ecol 182, 104717. doi: 10.1016/j.apsoil.2022.104717

Peracchi, L.M., Panahabadi, R., Barros-Rios, J., Bartley, L.E. & Sanguinet, K.A. (2024). Grass lignin: Biosynthesis, biological roles, and industrial applications. Front Plant Sci 15, 1343097. doi: 10.3389/fpls.2024.1343097

Piliarová, M., Ondreičková, K., Hudcovicová, M., Mihálik, D. & Kraic, J. (2019). Arbuscular mycorrhizal fungi – Their life and function in ecosystem. Agriculture, 65, 3–15. doi: 10.2478/agri-2019-0001

Pu, C., Ge, Y., Yang, G., Zheng, H., Guan, W., Chao, Z. et al. (2022). Arbuscular mycorrhizal fungi enhance disease resistance of *Salvia miltiorrhiza* to Fusarium wilt. Front Plant Sci. 13, 975558. doi: 10.3389/fpls.2022.975558

Pulidindi, K. & Ahuja, K. (2024). Phenolic antioxidants market – By source, by type, by form, by application and forecast, 2025-2034. *In*: Global Market Insights. https://www.gminsights.com/industry-analysis/phenolic-antioxidants-market. Accessed 15 Jan 2025

Qiu, Y.-J., Zhang, N.-L., Zhang, L.-L., Zhang, X.-L., Wu, A.-P., Huang, J.-Y. et al. (2020). Mediation of arbuscular mycorrhizal fungi on growth and biochemical parameters of *Ligustrum vicaryi* in response to salinity. Physiol Mol Plant Pathol 112, 101522. doi: 10.1016/j.pmpp.2020.101522

Rahimi, A., Moghaddam, S.S., Ghiyasi, M., Heydarzadeh, S., Ghazizadeh, K. & Popović-Djordjević, J. (2019). The influence of chemical, organic and biological fertilizers on agrobiological and antioxidant properties of syrian cephalaria (*Cephalaria Syriaca* L.). Agriculture 9, 122. doi: 10.3390/agriculture9060122

Ramaroson, M.-L., Koutouan, C., Helesbeux, J.-J., Le Clerc, V., Hamama, L., Geoffriau, E. et al. (2022). Role of phenylpropanoids and flavonoids in plant resistance to pests and diseases. Molecules 27, 8371. doi: 10.3390/molecules27238371

Ran, Z., Yang, X., Zhang, Y., Zhou, J. & Guo, L. (2021). Transcriptional responses for biosynthesis of ginsenoside in arbuscular mycorrhizal fungi-treated *Panax quinquefolius* L. seedlings using RNA-seq. Plant Growth Regul 95, 83-96. doi: 10.1007/s10725-021-00727-3

Rashmi, H.B. & Negi, P.S. (2020). Phenolic acids from vegetables: A review on processing stability and health benefits. Food Res Int 136, 109298. doi: 10.1016/j.foodres.2020.109298.

Ratnasari, A., Syafiuddin, A., Mehmood, M.A. & Boopathy, R. (2023). A review of the vermicomposting process of organic and inorganic waste in soils: Additives effects, bioconversion process, and recommendations. Bioresour Technol 21, 101332. doi: 10.1016/j.biteb.2023.101332

Raza, S.T., Zhu, B., Yao, Z., Wu, J., Chen, Z., Ali, Z. et al. (2023). Impacts of vermicompost application on crop yield, ammonia volatilization and greenhouse gases emission on upland in Southwest China. Sci Total Environ 860, 160479. doi: 10.1016/j.scitotenv.2022.160479

Rehman, S.C.F., Aprile, A., Benedetti, M. & Fanizzi, F.P. (2023). Vermicompost: Enhancing plant growth and combating abiotic and biotic stress. Agronomy 13, 1134. doi: 10.3390/agronomy13041134

Rudenko, N.N., Vetoshkina, D.V., Marenkova, T.V. & Borisova-Mubarakshina, M.M. (2023). Antioxidants of non-enzymatic nature: Their function in higher plant cells and the ways of boosting their biosynthesis. Antioxidants 12, 2014. doi: 10.3390/antiox12112014

Saha, H., Kaloterakis, N., Harvey, J.A., Van der Putten, W.H. & Biere, A. (2022). Effects of light quality on colonization of tomato roots by AMF and implications for growth and defense. Plants 11, 861. doi: https://doi.org/10.3390/plants11070861

Saltveit, M.E. (2017). Synthesis and metabolism of phenolic compounds. *In*: Yahia, E.M. [ed.]. Fruit and Vegetable Phytochemicals, 1st ed. (pp 115–124). Wiley.

Santos, E.L., Falcão, E.L. & Silva, F.S.B. (2021). Mycorrhizal technology as a bioinsumption to produce phenolic compounds of importance to the herbal medicine industry. Res Soc Dev 10, e54810212856. doi: 10.33448/rsd-v10i2.12856.

Santos-Sánchez, F.N., Salas-Coronado, R., Hernández-Carlos, B. et al. (2019). Shikimic acid pathway in biosynthesis of phenolic compounds. plant physiological aspects of phenolic compounds. IntechOpen. doi: 10.5772/intechopen.83815.

Sarathambal, C., Srinivasan, V., Jeevalatha, A., Sivaranjani, R., Alagupalamuthirsolai, M., Peeran, M.F. et al. (2024). Unravelling the synergistic effects of arbuscular mycorrhizal fungi and vermicompost on improving plant growth, nutrient absorption, and secondary metabolite production in ginger (*Zingiber officinale* Rosc.). Front Sustain Food Syst 8, 1412610. doi: 10.3389/fsufs.2024.1412610

Scott, K.A., Cox, P.B. & Njardarson, J.T. (2022). Phenols in pharmaceuticals: Analysis of a recurring motif. J Med Chem 65, 7044–7072. doi: 10.1021/acs.jmedchem.2c00223

Sharma, H., Sharma, P., Kumar, A. et al. (2024). Multifaceted regulation of anthocyanin biosynthesis in plants: A comprehensive review. J Plant Growth Regul 43, 3048–3062. doi: 10.1007/s00344-024-11306-x

Silva, F.S.B. & Silva, F.A. (2020). A low cost alternative, using mycorrhiza and organic fertilizer, to optimize the production of foliar bioactive compounds in pomegranates. J Appl Microbiol 128, 513–517. doi: 10.1111/jam.14477

Silva-Flores, P., Bueno, C.G., Neira, J. & Palfner, G. (2019). Factors affecting arbuscular mycorrhizal fungi spore density in the Chilean Mediterranean-type ecosystem. J Soil Sci Plant Nut 19, 42–50. doi: 10.1007/s42729-018-0004-6

Singh, A., Karmegam, N., Singh, G.S., Bhadauria, T., Chang, S.W., Awasthi, M.K. et al. (2020). Earthworms and vermicompost: An eco-friendly approach for repaying nature's debt. Environ Geochem Hlth 42, 1617–1642. doi: 10.1007/s10653-019-00510-4

Soti, P., Kariyat, R. & Racelis, A. (2023). Effective farm management promotes native AMF and benefit organic farming systems. Agric Ecosyst Environ 342, 108240. doi: 10.1016/j.agee.2022.108240

Souffront, D.K.S., Salazar-Amoretti, D. & Jayachandran, K. (2022). Influence of vermicompost tea on secondary metabolite production in tomato crop. Sci Hortic 301, 111135. doi: 10.1016/j.scienta.2022.111135

Stefanachi, A., Leonetti, F., Pisani, L., Catto, M. & Carotti, A. (2018). Coumarin: A natural, privileged and versatile scaffold for bioactive compounds. Molecules 23, 250. doi: 10.3390/molecules23020250

Sudheer, S., Johny, L., Srivastava, S. & Adholeya, A. (2023). The trade-in-trade: Multifunctionalities, current market and challenges for arbuscular mycorrhizal fungal inoculants. Symbiosis 89, 259–272. doi: 10.1007/s13199-023-00905-z

Sunil, L. & Shetty, N.P. (2022). Biosynthesis and regulation of anthocyanin pathway genes. Appl Microbiol Biotechnol 106, 1783–1798. doi: 10.1007/s00253-022-11835-z

Tanaka, S., Hashimoto, K., Kobayashi, Y., Yano, K., Maeda, T., Kameoka, H. et al. (2022). Asymbiotic mass production of the arbuscular mycorrhizal fungus *Rhizophagus clarus*. Commun Biol 5, 43. doi: 10.1038/s42003-021-02967-5

Teka, T., Zhang, L., Ge, X., Li, Y., Han, L. & Yan, X. (2022). Stilbenes: Source plants, chemistry, biosynthesis, pharmacology, application and problems related to their clinical application – A comprehensive review. Phytochem 197, 113128. doi: 10.1016/j.phytochem.2022.113128.

Tikoria, R., Kaur, A. & Ohri, P. (2023). Amelioration of oxidative stress and growth enhancement by application of vermicompost via modulating phyto-constituents in tomato plants during nematode stress. J Soil Sci Plant Nutr 23, 3944–3960. doi: 10.1007/s42729-023-01313-4

Toor, M.D., Kizilkaya, R., Ullah, I., Koleva, L., Basit, A. & Mohamed, H.I. (2023). Potential role of vermicompost in abiotic stress tolerance of crop plants: A review. J Soil Sci Plant Nutr 23, 4765–4787. doi: 10.1007/s42729-023-01476-0

Toor, M.D., Ridvan Kizilkaya, R., Anwar, A., Koleva, L. & Eldesoky, G.E. (2024). Effects of vermicompost on soil microbiological properties in lettuce rhizosphere: An environmentally friendly approach for sustainable green future. Environ Res 243, 117737. doi: 10.1016/j.envres.2023.117737

Trejo, D., Sangabriel-Conde, W., Gavito-Pardo, M.E. & Banuelos, J. (2021). Mycorrhizal inoculation and chemical fertilizer interactions in pineapple under field conditions. Agriculture 11, 934. doi: https://doi.org/10.3390/agriculture11100934

Vambe, M., Coopoosamy, R.M., Arthur, G. & Naidoo, K. (2023). Potential role of vermicompost and its extracts in alleviating climatic impacts on crop production. J Agric Food Res 12, 100585. doi: 10.1016/j.jafr.2023.100585

Vogt, T. (2010). Phenylpropanoid biosynthesis. Molecular Plant 3, 2–20. doi: 10.1093/mp/ssp106

Vuković, A., Velki, M., Ečimović, S., Vuković, R., Štolfa Čamagajevac, I. & Lončarić, Z. (2021). Vermicomposting – Facts, benefits and knowledge gaps. Agronomy 11, 1952. doi: 10.3390/agronomy11101952

Wako, F.L. & Muleta, H.D. (2022). The role of vermicompost application for tomato production: A review. J Plant Nutr 46, 129–144. doi: 10.1080/01904167.2022.2035751

Wang, H. & Chen, Y. (2024). Protecting plants from pathogens through arbuscular mycorrhiza: Role of fungal diversity. Microbiol Res 289, 127919. doi: 10.1016/j.micres.2024.127919

Weng, W., Yan, J., Zhou, M., Yao, X., Gao, A., Ma, C. et al. (2022). Roles of arbuscular mycorrhizal fungi as a biocontrol agent in the control of plant diseases. Microorganisms 10, 1266. doi: 10.3390/microorganisms10071266

Wijayawardene, N.N., Hyde, K.D., Mikhailov, K.V., Péter, G., Aptroot, A., Pires-Zottarelli, C.L.A. et al. (2024). Classes and phyla of the kingdom Fungi. Fungal Diversity 128, 1–165. doi: 10.1007/s13225-024-00540-z

Xu, W., Pan, Q., Zhang, Q. et al. (2024). Synergistic changes in AM fungi and soil abiotic properties in rhizosphere soils of invasive *Solidago canadensis* may confer its stronger dominance in communities. Plant Soil 498, 561–577. doi: 10.1007/s11104-023-06454-0

Yatoo, A.M., Ali, M.N., Baba, Z.A. & Hassan, B. (2021). Sustainable management of diseases and pests in crops by vermicompost and vermicompost tea: A review. Agron Sustain Dev 41, 7. doi: 10.1007/s13593-020-00657-w

Ye, Q., Wang, H. & Li, H. (2022). Arbuscular mycorrhizal fungi improve growth, photosynthetic activity, and chlorophyll fluorescence of *Vitis vinifera* L. cv. Ecolly under drought stress. Agronomy 12, 1563. doi: 10.3390/agronomy12071563

Yin, J., Wang, J., Zhao, L., Cui, Z., Yao, S., Li, G. & Yuan, J. (2025). Compost tea: Preparation, utilization mechanisms, and agricultural applications potential – A comprehensive review. Environ Technol Innov 38, 104137. doi: 10.1016/j.eti.2025.104137

Yoon, J., Choi, H. & An, G. (2015). Roles of lignin biosynthesis and regulatory genes in plant development. J Integr Plant Biol 57, 902–912. doi: 10.1111/jipb.12422

Yu, D., Huang, T., Tian, B. & Zhan, J. (2020). Advances in biosynthesis and biological functions of proanthocyanidins in horticultural plants. Foods 9, 1774. doi: 10.3390/foods9121774

Yu, K., Song, Y., Lin, J. & Dixon, R.A. (2023). The complexities of proanthocyanidin biosynthesis and its regulation in plants. Plant Comm 4, 100498. doi: 10.1016/j.xplc.2022.100498

Zagoskina, N.V., Zubova, M.Y., Nechaeva, T.L., Kazantseva, V.V., Goncharuk, E.A., Katanskaya, V.M. et al. (2023). Polyphenols in plants: Structure, biosynthesis, abiotic stress regulation, and practical applications. Int J Mol Sci 24, 13874. doi: 10.3390/ijms241813874

Zarei, M., Jahandideh Mahjen, A.V.A. & Moridi, A. (2018). Comparison of vermiwash and vermicompost tea properties produced from different organic beds under greenhouse conditions. Int J Recycl Org Waste Agric 7, 25–32. doi: 10.1007/s40093-017-0186-2

Zeb, A. (2020). Concept, mechanism, and applications of phenolic antioxidants in foods. J Food Biochem 44, e13394. doi: 10.1111/jfbc.13394

Zhang, J., Fang, H., Zhao, Y., Zheng, Y., Jiang, J. & Gu, X. (2023). Responses of soil nutrients and rhizosphere microbial communities of a medicinal plant *Pinellia ternata* to vermicompost. 3 Biotech 13, 353. doi: 10.1007/s13205-023-03780-z

Zhang, Y., Yuan, H., Peng, S., Wang, Z., Cai, S., Chen, Z. et al. (2024). Vermicomposting preferably alters fungal communities in wasted activated sludge and promotes the production of plant growth-promoting biostimulants in the vermicompost. Chem Eng J 495, 153232. doi: 10.1016/j.cej.2024.153232

Zhao, F., Zhang, Y., Li, Z., Shi, J., Zhang, G., Zhang, H. & Yang, L. (2020). Vermicompost improves microbial functions of soil with continuous tomato cropping in a greenhouse. J Soils Sediments 20, 380–391. doi: 10.1007/s11368-019-02362-y

Zhou, L.-J., Wang, Y., Alqahtani, M.D. & Wu, Q.-S. (2023a). Positive changes in fruit quality, leaf antioxidant defense system, and soil fertility of beni-madonna tangor citrus (*Citrus nanko* × *C. amakusa*) after field AMF inoculation. Horticulturae 9, 1324. doi: 10.3390/horticulturae9121324

Zhou, X., Wang, T., Wang, J., Chen, S. & Ling, W. (2023b). Research progress and prospect of glomalin-related soil protein in the remediation of slightly contaminated soil. Chemosphere 344, 140394. doi: 0.1016/j.chemosphere.2023.140394

Zhou, Y., Xiao, R., Klammsteiner, T., Kong, X., Yan, B., Mihai, F.-C. et al. (2022). Recent trends and advances in composting and vermicomposting technologies: A review. Bioresour Technol 360, 127591. doi: /10.1016/j.biortech.2022.127591

Zhu, Z., Chen, R. & Zhang, L. (2024). Simple phenylpropanoids: Recent advances in biological activities, biosynthetic pathways, and microbial production. Nat Prod Rep 41, 6–24. doi: 10.1039/D3NP00012E

Zhuang, W.-B., Li, Y.-H., Shu, X.-C., Pu, Y.-T., Wang, X.-J. , Wang, T. & Wang, Z. (2023). The classification, molecular structure and biological biosynthesis of flavonoids, and their roles in biotic and abiotic stresses. Molecules 28, 3599. doi: 10.3390/molecules28083599

Index

P

R

S

T